Free Radical and Antioxidant Protocols

METHODS IN MOLECULAR BIOLOGY™

John M. Walker, Series Editor

108. **Free Radical and Antioxidant Protocols,** edited by *Donald Armstrong, 1998*
107. **Cytochrome P450 Protocols,** edited by *Ian R. Phillips and Elizabeth A. Shephard, 1998*
106. **Receptor Binding Techniques,** edited by *Mary Keen, 1998*
105. **Phospholipid Signaling Protocols,** edited by *Ian Bird, 1998*
104. **Mycoplasma Protocols,** edited by *Roger J. Miles and Robin A. J. Nicholas, 1998*
103. ***Pichia* Protocols,** edited by *David R. Higgins and James Cregg, 1998*
102. **Bioluminescence Methods and Protocols,** edited by *Robert A. LaRossa, 1998*
101. **Myobacteria Protocols,** edited by *Tanya Parish and Neil G. Stoker, 1998*
100. **Nitric Oxide Protocols,** edited by *M. A. Titheradge, 1997*
99. **Human Cytokines and Cytokine Receptors,** edited by *Reno Debets, 1998*
98. **DNA Profiling Protocols,** edited by *James M. Thomson, 1997*
97. **Molecular Embryology:** *Methods and Protocols,* edited by *Paul T. Sharpe and Ivor Mason, 1998*
96. **Adhesion Proteins Protocols,** edited by *Elisabetta Dejana, 1997*
95. **DNA Topology and DNA Topoisomerases:** *II. Enzymology and Topoisomerase Targetted Drugs,* edited by *Mary-Ann Bjornsti, 1998*
94. **DNA Topology and DNA Topoisomerases:** *I. DNA Topology and Enzyme Purification,* edited by *Mary-Ann Bjornsti, 1998*
93. **Protein Phosphatase Protocols,** edited by *John W. Ludlow, 1997*
92. **PCR in Bioanalysis,** edited by *Stephen Meltzer, 1997*
91. **Flow Cytometry Protocols,** edited by *Mark J. Jaroszeski, 1998*
90. **Drug–DNA Interactions:** *Methods, Case Studies, and Protocols,* edited by *Keith R. Fox, 1997*
89. **Retinoid Protocols,** edited by *Christopher Redfern, 1997*
88. **Protein Targeting Protocols,** edited by *Roger A. Clegg, 1997*
87. **Combinatorial Peptide Library Protocols,** edited by *Shmuel Cabilly, 1997*
86. **RNA Isolation and Characterization Protocols,** edited by *Ralph Rapley, 1997*
85. **Differential Display Methods and Protocols,** edited by *Peng Liang and Arthur B. Pardee, 1997*
84. **Transmembrane Signaling Protocols,** edited by *Dafna Bar-Sagi, 1997*
83. **Receptor Signal Transduction Protocols,** edited by *R. A. J. Challiss, 1997*
82. ***Arabidopsis* Protocols,** edited by *José M Martinez-Zapater and Julio Salinas, 1998*
81. **Plant Virology Protocols,** edited by *Gary D. Foster, 1998*
80. **Immunochemical Protocols,** *second edition,* edited by *John Pound, 1998*
79. **Polyamine Protocols,** edited by *David M. L. Morgan, 1998*
78. **Antibacterial Peptide Protocols,** edited by *William M. Shafer, 1997*
77. **Protein Synthesis:** *Methods and Protocols,* edited by *Robin Martin, 1998*
76. **Glycoanalysis Protocols,** edited by *Elizabeth F. Hounsel, 1998*
75. **Basic Cell Culture Protocols,** edited by *Jeffrey W. Pollard and John M. Walker, 1997*
74. **Ribozyme Protocols,** edited by *Philip C. Turner, 1997*
73. **Neuropeptide Protocols,** edited by *G. Brent Irvine and Carvell H. Williams, 1997*
72. **Neurotransmitter Methods,** edited by *Richard C. Rayne, 1997*
71. **PRINS and *In Situ* PCR Protocols,** edited by *John R. Gosden, 1997*
70. **Sequence Data Analysis Guidebook,** edited by *Simon R. Swindell, 1997*
69. **cDNA Library Protocols,** edited by *Ian G. Cowell and Caroline A. Austin, 1997*
68. **Gene Isolation and Mapping Protocols,** edited by *Jacqueline Boultwood, 1997*
67. **PCR Cloning Protocols:** *From Molecular Cloning to Genetic Engineering,* edited by *Bruce A. White, 1996*
66. **Epitope Mapping Protocols,** edited by *Glenn E. Morris, 1996*
65. **PCR Sequencing Protocols,** edited by *Ralph Rapley, 1996*
64. **Protein Sequencing Protocols,** edited by *Bryan J. Smith, 1996*
63. **Recombinant Proteins:** *Detection and Isolation Protocols,* edited by *Rocky S. Tuan, 1996*
62. **Recombinant Gene Expression Protocols,** edited by *Rocky S. Tuan, 1996*
61. **Protein and Peptide Analysis by Mass Spectrometry,** edited by *John R. Chapman, 1996*
60. **Protein NMR Protocols,** edited by *David G. Reid, 1996*
59. **Protein Purification Protocols,** edited by *Shawn Doonan, 1996*
58. **Basic DNA and RNA Protocols,** edited by *Adrian J. Harwood, 1996*
57. **In Vitro Mutagenesis Protocols,** edited by *Michael K. Trower, 1996*
56. **Crystallographic Methods and Protocols,** edited by *Christopher Jones, Barbara Mulloy, and Mark Sanderson, 1996*
55. **Plant Cell Electroporation and Electrofusion Protocols,** edited by *Jac A. Nickoloff, 1995*
54. **YAC Protocols,** edited by *David Markie, 1995*
53. **Yeast Protocols:** *Methods in Cell and Molecular Biology,* edited by *Ivor H. Evans, 1996*
52. **Capillary Electrophoresis:** *Principles, Instrumentation, and Applications,* edited by *Kevin D. Altria, 1996*
51. **Antibody Engineering Protocols,** edited by *Sudhir Paul, 1995*
50. **Species Diagnostics Protocols:** *PCR and Other Nucleic Acid Methods,* edited by *Justin P. Clapp, 1996*
49. **Plant Gene Transfer and Expression Protocols,** edited by *Heddwyn Jones, 1995*
48. **Animal Cell Electroporation and Electrofusion Protocols,** edited by *Jac A. Nickoloff, 1995*
47. **Electroporation Protocols for Microorganisms,** edited by *Jac A. Nickoloff, 1995*

METHODS IN MOLECULAR BIOLOGY™

Free Radical and Antioxidant Protocols

Edited by

Donald Armstrong

State University of New York, Buffalo, NY

Humana Press **Totowa, New Jersey**

999 Riverview Drive, Suite 208
Totowa, New Jersey 07512

This publication is printed on acid-free paper. ∞
ANSI Z39.48-1984 (American Standards Institute)
Permanence of Paper for Printed Library Materials.

Cover illustration: Courtesy of Dr. Donald Armstrong.

Cover design by Patricia F. Cleary.

For additional copies, pricing for bulk purchases, and/or information about other Humana titles, contact Humana at the above address or at any of the following numbers: Tel.: 973-256-1699; Fax: 973-256-8341; E-mail: humana@humanapr.com; or visit our Website: http://humanapress.com

Printed in the United States of America. 10 9 8 7 6 5 4 3 2 1

Library of Congress Cataloging in Publication Data

Main entry under title:

Methods in molecular biology™.

Free radical and antioxidant protocols / edited by Donald Armstrong.
p. cm. -- (Methods in molecular biology™ ; v. 108)
Includes bibliographic references and index.
ISBN 0-89603-472-0 (alk. paper)
1. Free radicals (Chemistry)--Pathophysiology--Laboratory manuals. 2. Antioxidants--Physiological effect--Laboratory manuals. 3. Oxidation, Physiological--Laboratory manuals. 4. Stress (Physiology)--Research--Laboratory manuals. I. Armstrong, Donald 1933–. II. Series: Methods in Molecular Biology (Totowa, NJ) ; 108.
RB170.F68 1998
616.07--dc21

98-16172
CIP

Preface

From a historical perspective, it is interesting and appropriate that 50 years ago, Michaelis *(1)* first proposed that hydroxyl, peroxy radicals, and hydrogen peroxide could be expected as a result of normal oxidative metabolism. Two years later, the role of these reactive oxygen species (ROS) was demonstrated during lipid auto-oxidation, which was attributed to lipid peroxidation detected by the conjugated diene method *(2)* and by histochemistry *(3)*. In 1945, Dam and Granados *(4)* showed increased peroxidation of fats in vitamin E deficiency and in 1953, Tappel *(5)* demonstrated the protective effects of antioxidants (AOX) against oxygen toxicity. In 1954, Gerschman *(6)* described how the combination of hyperbaric oxygen and X-irradiation caused formation of oxygen radicals and showed a correlation with decreased survival time. Her observation that females were significantly less sensitive to this oxidation is the first indication of estrogen protection. In 1956, Harman *(7)* proposed his free radical theory of aging, and subsequently Yagi *(8)* devised a method to measure lipid hydroperoxides (LHP) in human serum, which inaugurated in vivo testing procedures.

From that era until the present, numerous investigators have made important contributions to our theoretical–conceptual understanding of these processes, as well as methodological development, the acute and chronic pathophysiological mechanisms, and therapeutic or nutritional intervention in these processes. Although carbon- and nitrogen-centered radicals play a role in the etiology and pathogenesis of disease, it is primarily ROS that initiate these events and subsequently, the level of aqueous- and lipid-phase antioxidants that influence the level of protection.

The rapid proliferation of biological publications on free radicals, LHP, and AOX in the literature since 1980 is shown in **Fig. 1.**

The idea for a book of techniques describing current state-of-the-art methodology and technology emerged from a 1993 international conference covering the use of free radical assays in diagnostic medicine. Virtually every organ and body fluid have now been studied with respect to the normal concentration of ROS and AOX. The development of experimental models has allowed their utilization in designing studies that examine basic mechanisms of human disease, the effect of AOX, and the assessment of risk factors using in vivo, ex vivo, in vitro, and *in situ* conditions. Summarized in **Table 1**, are the numbers

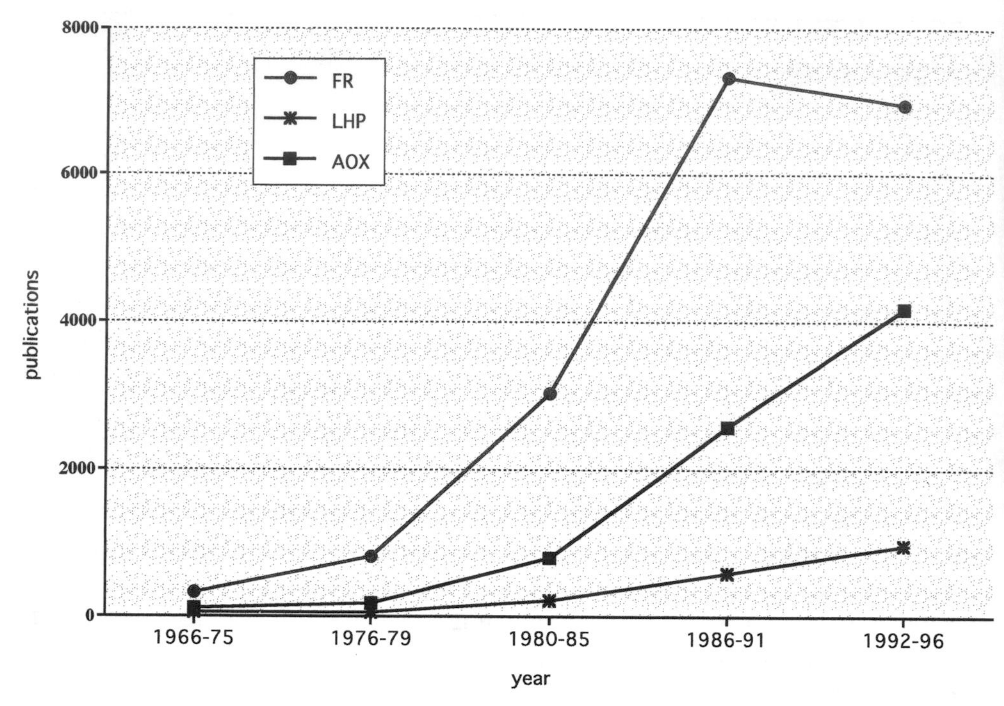

Fig. 1.

of papers published on the various disorders in which oxidative stress has been implicated (from **refs. *9*** and ***10***; *also see* **ref. *11*** for other publications covering general and organ specific diseases, as well as the VERIS Research and Information Service, La Grange, IL 60525 which abstracts antioxidant-related papers).

The preparation of *Free Radical and Antioxidant Protocols* is an undertaking by 70 authors who present step-by-step information on a wide range of assays that can be utilized in the study of primary or secondary oxidative stress. Protocols in routine use are not included; new instrumentation is a key feature of this book. Because the oxidative stress process involves a cascade of complex, interacting events—for example, (1) generation of the primordial radical, (2) formation of lipid and steroid intermediates that propagate the reaction, (3) modulation of signal transduction pathways by upregulation of regulatory enzymes through nuclear transcription factors and gene expression, and (4) termination by radical inactivation—it is clear that more than one method of analysis is often required.

Table 1
Disorders Associated with Oxidative Stress

Neurological
Alzheimers Disease
Down's Syndrome
Amyotrophic Lateral Sclerosis
Schizophrenia
Tardine Dyskinesis
Parkinson's Disease
Huntington's Chorea
Ataxia Telangectasia
Stroke

Ocular
Cataract
Age Related Macular Degeneration
Retinopathy of Prematurity
Light Damage

Endocrine
Diabetes
Infertility
Menopause
Thyroiditis
Neoplasia
Hyperthyroidism
Musculocutaneous
UV Injury
Xeroderma Pigmentosum
Duchenne Muscular Dystrophy

Vascular
Atherosclerosis
Coronary Artery Bypass Grafting
Circulatory Shock
Cardiac and Peripheral Disease

Hemolytic
RBC Fragility
Nephrotic Syndrome
Chronic Renal Failure
Membraneous Glomerulonephritis

Renal
Hemochromatosis
Anemia
Sepsis

Hepatic
Hepatitis
Fatty Necrosis
Fetal Alcohol Syndrome

Autoimmune
Rheumatoid Arthritis
HIV
Systemic Lupus Erythematosis

Pulmonary
Respiratory Distress Syndrome
Cystic Fibrosis
Emphysema
Sarcoid Alveolitis
Chronic Idiopathic Fibrosis

Gastrointestinal
Colitis
Acute Pancreatitis
Gastric Mucosal Erosion
Acute Cholecystitis

Other Conditions
Idiosyncratic Drug Reactions
Neoplasia
Trauma
Ischemic Reoxygenation Injury
Multiple Organ Dysfunction
Aging
Bloom's Syndrome
Transplantation Rejection
Air Pollution
Toxicity
Inflammation
Thermal Injury
Excessive Exercise
Apoptosis
Intermittant Claudication
Obesity
Pre-eclampsia

The reader will find this book a user friendly and helpful resource that is arranged into separate sections related to ROS, LHP, and AOX methodology. Part III describes six procedures that are applicable to both of these functions.

Instructors teaching courses on oxidative stress should find this book equally valuable as an adjunct for the laboratory component. A companion volume on clinical applications is planned for the near future.

I thank the following organizations for financial support and/or the use of institutional facilities during the preparation of this book: Department of Clinical Laboratory Science, University at Buffalo; Cook Institute for Research and Development; Photon Medical Research Center, Hamamatsu University, Japan; Showa University, Japan and the Japanese Ministry of Higher Education. My special appreciation is extented to Professor John Walker for his assistance in the review process, M. Ohbayashi for technical assistance, and C. Armstrong for secretarial assistance.

Donald Armstrong

References

1. Michaelis, L. (1946) *Am. Scientist* **34,** 573.
2. Mead, J. (1952) *Science* **115,** 470.
3. Glavind, J. et al. (1949) *Experientia* **5,** 84,85.
4. Dam, H. and Granados, H. (1945) *Acyta Physiol. Scand.* **10,** 162.
5. Tappel, A. (1953) *Arch. Biochem. Biophys* **42,** 293.
6. Grishman, R. (1954) *Science* **119,** 623.
7. Harman, D. (1956) *J. Gerontol.* **11,** 298.
8. Yagi, K. (1968) *Vitamins* **39,** 105.
9. Armstrong, D. (1994) *Free Radicals in Diagnostic Medicine,* Plenum, New York.
10. Armstrong, D. et al. (1984) *Free Radicals in Molecular Biology, Aging and Disease,* Plenum, New York.
11. Books in Print, vol. 6. (1995) R.R. Bowker's Database Publishing Group, New Providence, NJ, p. 2850.

Contents

Contributors

EHAD ABDEL-RAHMAN • *Molecular Biology Laboratory, Endocrine-Diabetes Center of Western New York, Millard Fillmore Health System, Buffalo, NY*

NADER G. ABRAHAM • *Department of Gene Therapy, New York Medical College, Valhalla, NY*

AHMAD ALJADA • *Endocrinology-Diabetes Center of Western New York, Millard Fillmore Health System, Buffalo, NY*

KAZUMASA AOYAGI • *Division of Nephrology, Department of Internal Medicine, Institute of Clinical Medicine, University of Tsukuba, Japan*

DONALD ARMSTRONG • *Department of Clinical Laboratory Science and Pathology, State University of New York at Buffalo, NY*

SOFIA BARDIN • *Department of Biophysics, Roswell Park Cancer Institute, New York State Department of Health, Buffalo, NY*

HENRYK BOROWY-BOROWSKI • *Gloucester, Ontario, Canada*

DANIEL J. BRACKETT • *Department of Surgery, College of Medicine, The University of Oklahoma Health Sciences Center, Oklahoma, City, OK*

RICHARD W. BROWNE • *Department of Clinical Laboratory Science, State University of New York at Buffalo, NY*

ULF BRUNK • *Department of Pathology, Faculty of Health Sciences, University of Linkoping, Sweden*

TOT BUI • *Department of Pharmaceutics, University of Washington, Seattle, WA*

GERMAN CAMEJO • *Department of Biochemistry, Astra Hassle AB, Molndal, Sweden*

YVES CHANCERELLE • *CRSSA, Unité de Radiobiochimie, La Tronche, France*

JOSIANE CILLARD • *Laboratorie de Biologie Cellulaire et Vegetale, UFR des Sciences Pharmaceutiques, Université de Rennes, France*

PARESH DANDONA • *Department of Medicine, University of Buffalo School of Medicine and Biomedical Sciences, Endocrinology-Diabetes Center of Western New York, Millard Fillmore Hospital, Buffalo, NY*

FRANÇOIS C. DELORI • *The Schepens Eye Research Institute, Harvard University School of Medicine, Boston, MA*

C. KATHLEEN DOREY • *The Schepens Eye Research Institute, Harvard University School of Medicine, Boston, MA*

Mervi Enojärvi • *Department of Biochemistry, Astra Hassle AB, Molndal, Sweden*
Eiji Fujimori • *Department of Applied Chemistry, School of Engineering, Nagoya University, Japan*
Ken Fujimori • *Department of Chemistry, University of Tsukuba, Japan*
Jan Galle • *Abt. Nephrologie, Universitatsklinic, Julius Maximilians Universitat Wurzburg, Germany*
Andrea Ghiselli • *Instituto Nazionale della Nutrizione, Roma, Italy*
Nikolai V. Gorbunov • *Department of Respiratory Research, Walter Reed Army Institute of Research, Washington, DC*
Scott Graves • *NeoRx Corporation, Seattle, WA*
Howard J. Halpern • *Department of Pharmacology and Toxicology, School of Pharmacology, University of Maryland at Baltimore, MD*
Hiroki Haraguchi • *Department of Applied Chemistry, School of Engineering, Nagoya University, Nagoya, Japan*
Tadahisa Hiramitsu • *Photon Medical Research Center, Hamamatsu University School of Medicine, Hamamatsu, Japan*
Kazumi Inagaki • *Department of Applied Chemistry, School of Engineering, Nagoya University, Nagoya, Japan*
Takashi Ito • *Japan Immunoresearch Laboratories Co., Ltd., Takasaki-city, Japan*
Rodney Ho • *Department of Pharmaceutics, University of Washington, Seattle, WA*
Young Soo Hong • *Department of Biochemistry, School of Medicine and Biomedical Sciences, State University of New York at Buffalo, NY*
Valerian E. Kagan • *Department of Environmental and Occupational Health, University of Pittsburgh Graduate School of Public Health, Pittsburgh, PA*
Yasuhiro Kambayashi • *Department of Photon and Free Radical Research, Japan Immunoresearch Laboratories Co., Ltd., Takasaki, Japan*
J. F. Kergonou • *CRSSA, Unité de Radiobiochimie, La Tronche, France*
Akio Koyama • *Division of Nephrology, Department of Internal Medicine, Institute of Clinical Medicine, University of Tsukuba, Japan*
Joachim G. Liehr • *Department of Pharmacology and Toxicology, The University of Texas Medical Branch at Galveston, Galveston, TX*
Mauro Magnani • *Instituto di Chimica Biologica "Giorgio Fornaini" Universita degli Studi di Urbino, Urbino, Italy*
J. Mathieu • *CRSSA, Unité de Radiobiochimie, La Tronche, France*

Seiichi Matsugo • *Department of Chemical and Biochemical Engineering, Faculty of Engineering, Toyama University, Japan*

Paul B. McCay • *National Biomedical Center for Spin Trapping and Free Radicals, Free Radical Biology and Aging Program, Oklahoma Medical Research Foundation, Department of Veterans Affairs Medical Center, Oklahoma City, OK*

Nicholas J. Miller • *FRCPath, Free Radical Research Group, Division of Biochemistry and Molecular Biology, UMDS-Guy's Hospital Medical School, London, UK*

Yoshinori Mizuguchi • *Systems Division, Department 17 Biomedical Applications, Hamamatsu Photonics K.K.d., Hamamatsu, Japan*

Sohji Nagase • *Division of Nephrology, Department of Internal Medicine, Institute of Clinical Medicine, University of Tsukuba, Japan*

Minoru Nakano • *Japan Immunoresearch Laboratories Co., Ltd., Takasaki-city, Japan*

Mitsuharu Narita • *Division of Nephrology, Department of Internal Medicine, Institute of Clinical Medicine, University of Tsukuba, Japan*

Thomas M. Nicotera • *Department of Biophysics, Roswell Park Cancer Institute and University at Buffalo School of Medicine and Biomedical Sciences, New York State Department of Health, Buffalo, NY*

J. Mark Ordy • *Department of Neuroscience, John Carroll University, Cleveland, OH*

George Paganga • *Free Radical Research Group, Division of Biochemistry and Molecular Biology, UMDS-Guy's Hospital Medical School, London, UK*

Mulchand S. Patel • *Department of Biochemistry, School of Medicine and Biomedical Sciences, University at Buffalo, NY*

Charles E. Pippenger • *Cook Institute for Research and Education, Butterworth Hospital and Michigan State University School of Medicine, Grand Rapids, MI*

Sovitj Pon • *Department of Pharmacology and Toxicology, School of Pharmacology, University of Maryland at Baltimore, MD*

Vladimir B. Ritov • *Department of Environmental and Occupational Health, University of Pittsburgh, PA*

Gerald M. Rosen • *Pharmaceutical Sciences Department, Pharmacology and Toxicology Program, School of Pharmacy, University of Maryland at Baltimore, MD*

Luigia Rossi • *Instituto di Chimica Biologica "Giorgio Fornaini," Universita degli Studi di Urbino, Urbino, Italy*

DEODUTTA ROY • *Department of Pharmacology and Toxicology, The University of Texas Medical Branch at Galveston, TX*

DON A. SAMUELSON • *Graduate Studies and Research, Department of Small Animal Clinical Sciences, College of Veterinary Medicine, University of Florida, Gainesville, FL*

LUCIAN SAUCAN • *Vascular Disease Prevention Laboratory, Carl T. Hayden VA Medical Center, Phoenix, AZ*

MARIANNA SIKORSKA • *Institute for Biological Sciences, National Research Council of Canada, Ottawa, Ontario, Canada*

ODILE SERGENT • *Laboratorie de Biologie Cellulaire et Vegetale, UFR des Sciences Pharmaceutiques, Université de Rennes, France*

J. RANDALL SLEMMON • *Department of Biochemistry, University of Rochester, NY*

GLENN SPEHAR • *Department of Biophysics, Roswell Park Cancer Institute, New York State Department of Health, Buffalo, NY*

ISTVAN STADLER • *HEMEX Laboratories, Department of Medicine, State University of New York at Buffalo, Buffalo General Hospital, Buffalo, NY*

SHIZUO TOJO • *Division of Nephrology, Department of Internal Medicine, Institute of Clinical Medicine, University of Tsukuba, Japan*

KULDIP THUSU • *Vascular Lab, Endocrinology-Diabetes Center of Western New York, Millard Fillmore Health System, Buffalo, NY*

VLADIMIR A. TYURIN • *Department of Environmental and Occupational Health, University of Pittsburgh, Pittsburgh, PA*

YULIA Y. TYURINA • *Department of Environmental and Occupational Health, University of Pittsburgh, PA*

TAKAKO UEDA • *Department of Pharmacology, Showa University School of Medicine, Tokyo, Japan*

BOEL WALLIN • *Department of Biochemistry, Astra Hassle AB, Molndal, Sweden*

GEMMA WALLIS • *Department of Surgery, College of Medicine, The University of Oklahoma Health Sciences Center, Oklahoma City, OK*

CHRISTOPH WANNER • *Abt. Nephrologie Universitatsklinic, Julius Maximilians-Universitat Wurzburg, Germany*

THOMAS M. WENGENACK • *Department of Neurology, Mayo Foundation and Clinic, Rochester, MN*

MICHAEL F. WILSON • *Department of Medicine-Cardiology, Millard Fillmore Health Systems, University at Buffalo School of Medicine and Biomedical Sciences, Buffalo, NY*

CLIVE WOODHOUSE • *Receptagen, Inc., Edmonds, WA*
KUNIO YAGI • *Institute for Applied Biochemistry and Nagoya University School of Medicine, Yagi Memorial Park, Mitake, Gifu, Japan*
YORIHIRO YAMAMOTO • *Research Center for Advanced Science and Technology, University of Tokyo, Tokyo, Japan*
DAZHONG YIN • *Faculty of Health Sciences, University of Linkoping, Linkoping, Sweden*

I

Techniques for the Measurement of Oxidative Stress

1

Oxygen Consumption Methods

Xanthine Oxidase and Lipoxygenase

Istvan Stadler

1. Introduction

A biochemical paradox that has been observed for many years is now becoming understood: O_2, essential for the aerobic life forms, can be inappropriately metabolized, becoming toxic to an organism. Mammals derive most of their cellular adenosine triphosphate (ATP) from the controlled four-electron reduction of O_2, to form H_2O by the mitochondrial electron-transport system. Approximately 98% of all O_2 consumed by cells enters the mitochondria, where it is reduced by a terminal oxidase, such as cytochrome oxidase.

However, O_2 can accept less than four electrons to form a reactive O_2^- metabolite, which may be toxic to cells. The reactions producing various reduced species are illustrated on the following system:

$$O_2 + e > O_2^- \text{ (superoxide)}$$

$$O_2 + 2e^- + 2H^+ > H_2O_2 \text{ (hydrogen peroxide)}$$

$$O_2^- + H_2O_2 + H^+ > O_2 + H_2O + {}^*OH \text{ (hydroxyl radical)}$$

$$O_2 + 4e^- + 4H^+ > 2\ H_2O \text{ (water)}$$

The spin restriction on O_2, limits the speed of molecular O_2 reactions. Organic compounds exist in a single state, meaning the outer orbital is closed, with one electron. O_2 however, exists in a triplet diradical state, with three electrons in the outer orbital capable of activity. Spin restriction can be overcome if enough energy of activation is available to excite O_2. If this occurs, a spin inversion or pairing of outer electrons is possible. A singlet O_2 (O_2^*), is

From: *Methods in Molecular Biology, vol. 108: Free Radical and Antioxidant Protocols*
Edited by: D. Armstrong © Humana Press Inc., Totowa, NJ

Table 1
Forms of Injury-Associated Free Radical Production

Injury	Damage
Ischemia/reperfusion	Lipidperoxidation (enzymatic and non)
Thermal injuries	Breaking of DNA, proteins
Trauma, toxins, radiation exposure	Metal ion released, stimulating free-radical formation
Infection, excess exercise	Interference with natural antioxidants

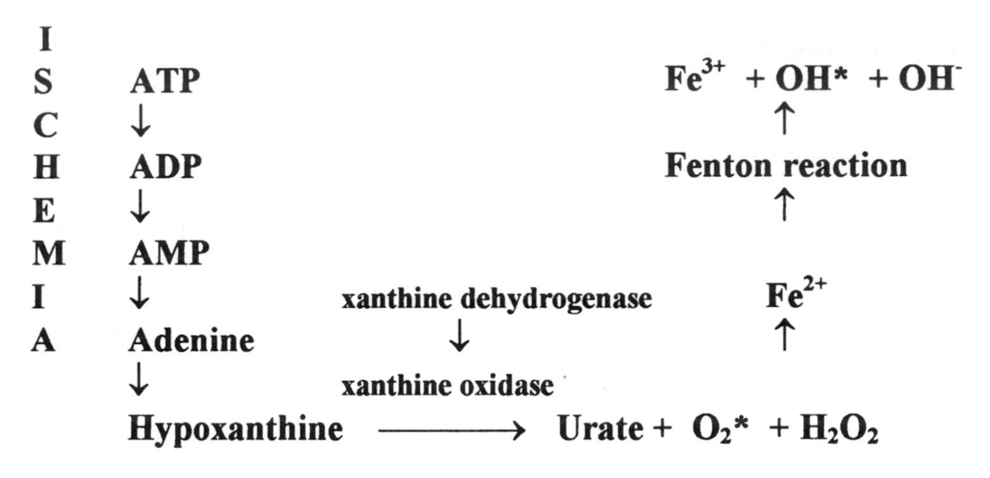

Fig. 1. Sources for oxygen-derived free radicals.

produced, which is a potent oxidant with an extremely short lifetime. Once this oxidative agent is formed, it can cause severe tissue damage *(1,2)*.

The production of toxic oxygen radicals is associated with various forms of injury and/or trauma to an organisms system. **Table 1** highlights some of these injuries and their potential consequences. Few possibilities exist to serve as sources for oxygen derived free radicals *(2,3)*. The pathways illustrated in **Fig. 1**, depict the precise reactions that are necessary.

The reactive oxygen metabolites that form within an organism must contend with the system's natural defense measures. For tracing this quite complicated mechanism, we should examine an element of that system, such as oxygen consumption. Monitoring the oxygen consumption provides a good indication of activity, and the progression of events in this chain reaction. Oxygen consumption is the disappearance of dissolved oxygen from the system, and can be determined by the following three methods:

1. Electrochemical measurements of dissolved oxygen by the Clark electrode.
2. Spectrophotometric detection of oxygen consumption in the presence of indicator compounds like cytochrome C.
3. Spectrophotometric detection of oxygen consumption by the determination of xanthine-oxidase activity.

Various enzymatic processes, that utilize oxygen may be measured by applying the methods that monitor oxygen consumption:

1. Enzymes that produce oxygenated compounds, including toxic ones (e.g., lipidperoxidase and cyclooxyganase).
2. Oxygen-derived free radical-producing enzyme xanthine oxidase (XO).
3. The free-radical scavenger enzymes, i.e., superoxide-dismutases (SODs), catalases, and glutathione peroxidases.
4. The mitochondrial respiration enzymes, such as the cytochrome oxidases.

2. Materials

2.1. Electrochemical Measurements

2.1.1. Oxygen (pO_2) Electrode (Clark Electrode)

1. The application of Clark electrode is a good example of amperometry. The measuring system contains a thermostable cuvet of approx 2–3 mL, in which one wall is replaced by an oxygen-permeable polypropylene or teflon membrane. The membrane separates the sample from the supporting electrolyte (*see* **Note 1**). Behind this membrane, a platinum microelectrode serves as a cathode and a silver-silver chloride electrode serves as an anode. The supporting electrolyte is 0.1 *M* phosphate buffer, pH 7.5, saturated with silver chloride. The electrode pair is charged with +0.65 V. The amount of oxygen entering from the sample through the membrane to the electrolyte will be proportional to the oxygen concentration of the original sample. The entered oxygen will be reduced in the following manner: $O_2 + H_2O + 4\ e^- > 4\ OH^-$ The current flow that results from this reaction is directly proportional to the partial pressure of oxygen in the sample. The sensitivity of this reaction is: $\Delta I = pO_2 = 10^{-4}$ A/Pa.
2. Before using the electrode, the system is calibrated with a known concentration of dissolved oxygen in an applied buffer system. The buffer system is saturated with a mixture of gases containing known oxygen concentrations (for example: 15% oxygen, 5% carbon dioxide, and 80% nitrogen). Applying Dalton's law, the partial pressure in the system is:

$$p(amb) = pO_2 + pCO_2 + pN_2 + pH_2O$$

$$pH_2O \text{ at } 37°C = 47 \text{ mm Hg}$$

3. Taking a barometric reading for actual atmospheric pressure, the partial pressure of oxygen is calculated. Once the partial pressure for oxygen is known, Henry's Law may be applied:

$$c\ dO = \alpha \times pO \text{ where } c = \text{mol/L of dissolved } O_2 \text{ (dO)}$$

$$\alpha = \text{Henry constant}$$

4. Once the concentration of dissolved gas (oxygen) is calculated, the instrument is set. Zero oxygen concentration is produced by adding sodium thiosulfate (a few crystals) to the cuvets, and the zero line is adjusted on the output device. Oxygen consumption (μ*M*) is calculated based on a signal change during the applied reaction.
5. Output device: The Clark electrode should be attached to an output device, such as a potentiometric recorder or PC via interface.

2.2. Spectrophotometric Methods

2.2.1. Indirect Methods

1. An ultraviolet-visible spectrophotometer, with time-course registration (kinetic) capacity. (The SHIMADZU UV-VIS Spectrophotometer-1601 with cuvets is suitable.) (*See* **Note 2**.)
2. 50 m*M* Carbonate buffer with 0.1 m*M* ethylenediamine tetraacetic acid (EDTA), pH 10: 275 mL of 0.1 *M* Na_2CO_3 and 225 mL of 0.1 *M* $NaHCO_3$, 23.4 mg EDTA; add 400 mL type I water, adjust pH to 10.0 with 2.0 *N* HCl or NaOH, adjust the volume to 1000 mL with type I water.
3. Xanthine (2,6 dihydroxypurine) stock solution: Dissolve 15.7 mg xanthine in 30 mL type I water, add 2 drops 1 *M* NaOH if xanthine shows difficulty going into solution (mix 10 min).
4. XO solution (grade III from buttermilk, Sigma, St. Louis, MO, cat. no. X-4500): 0.12–0.16 U/mL XO in 50 m*M* carbonate buffer (approx 10–20 U/mL of enzyme stock solution needed for 50 mL).
5. Cytochrome C solution (from horse heart; Sigma, cat. no. C-2506): 10.5 μ*M* in 50 m*M* carbonate buffer, 3000 U/mg standard SOD-lyophilized powder from bovine erythrocytes (Sigma, cat. no. S-2515).
6. Working solution: 28 mL of 50 m*M* carbonate buffer; 1 mL of xanthine solution, 1 mL of cytochrome C solution. Absorbance of this working solution at 550 nm should be approx 0.200.

2.2.2. Direct Methods

1. Preparation of the substrate (*see* **Note 3**):
   a. Dissolve 20 Mg polyunsaturated fatty acid (Linolenic acid; 9,12,15-octadecatrienoic acid Sigma, cat. no. L-2376) in 0.5 mL 96% ethanol.
   b. Convert the polyunsaturated fatty acid to sodium-linolenate, by adding an equivalent amount of sodium hydroxide (1 *M*).

c. Adjust the volume to 20 mL yielding a solution of 1.0 mg/mL sodium linolenate in 0.2 *M* borate buffer, pH 9.0.

2. Borate buffer preparation: Add 6.18 g boric acid to 300 mL deionized H_2O, adjust pH to 9.0 using 50% NaOH, and adjust volume to 500 mL using deionized H_2O.

2.2.3. Direct Method Using XO

1. Xanthine solution: Add 17.4 mg of xanthine (Sigma, cat. no. X 2502) to 10 mL of type I H_2O. Boil 5 min to dissolve components completely.
2. XO solution: Use 250 U/mL of XO (Sigma, cat. no. X 4500) in 0.1 *M* phosphate buffer, pH 7.4. This XO solution will serve as the standard.

3. Methods

3.1. Electrochemical Measurements of Dissolved Oxygen Using the Clark Electrode

3.1.1. Oxygenase activity with Clark Electrode

1. Add 1.9 mL of 10 mg/mL of lipoxygenase enzyme solution (soybean lipoxygenase Sigma or potato lipoxygenase *[4]*), in .1 *M* borate buffer, pH 9.0: Transfer to a measuring cuvet. One wall of the cuvet consists of a Clark electrode-specific membrane. Thereafter, cuvet is hermetically sealed and equilibrated for 10 min at 37°C.
2. 0.100 mL of 1.0 mg/mL sodium arachidonate (*see* **Note 4**): Inject into the cuvet through the rubber stopcock; start output device (recorder) at same time. Record oxygen consumption for 15 min. The typical curve is depicted in **Fig. 2**.

3.1.2. Potato Lipoxygenase Activity

1. Potato lipoxygenase (a model for 5-lipoxygenase): Prepared according to O'Flaherty *(4)*.
2. Enzyme activity is measured with same instrument as previously described, by applying 0.1 *M* phosphate buffer at pH 6.5. Enzyme concentration is 5.0 mg/mL based on a protein assay.

3.1.3. Cyclooxygenase Activity

Partially purified cyclooxygenase (from sheep vesicula seminalis) ***(4)***, is prepared for this measurement. 1.9 mL of 10 mg/mL protein extract is applied as described in **Subheading 3.1.1.**

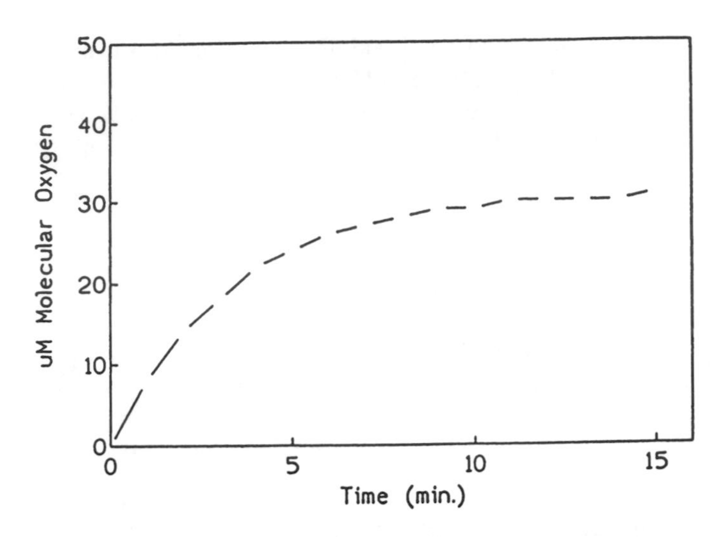

Fig. 2. Oxygen consumed in the presence of lipoxygenase activity.

*3.1.4. Inhibition of Cyclooxygenase Activity and Lipoxygenase Activity by a Nonsteroid Anti-Inflammatory Agent **(5)***

1. Sodium meclofenamate (n-2,6 dichloro-m-tolyl-anthranilate sodium, Meclomen [Warner-Lambert]) inhibits oxygenase-enzyme activity. Enzyme-activity measurements are performed according to **Subheadings 3.2.1., 3.2.2.,** and **3.2.3.** (*see* **Note 5**).
2. Oxygen consumption is recorded in the absence of inhibitor and in presence of different concentrations of inhibitors. A typical inhibition curve is depicted in **Fig. 3** and the kinetics constants are summarized in **Table 2**.

3.2. Spectrophotometric Methods for Measurement of Oxygen Consumption

Three spectrophotometric methods for measuring oxygen consumption catalyzed by oxygenases can be applied:

1. Indirect method: An indicator compound is applied to detect the recently formed oxidative species. Reduced cytochrome C appears to be an excellent indicator because it has a well-defined change in its absorbance at 560 nm (increase) and 405 nm (decrease), on its oxidized or reduced stage. The change in the absorbance indicates the presence of the oxygen-derived oxidative species. These compounds have a short lifetime. This method can be applied to the detection of free-radical scavenger enzymes, such as superoxide dismutase ***(6,7)***.
2. Direct UV method for studying the oxygenase enzymes (mostly lipoxygenases) ***(8–10)***: Enzymes convert the polyunsaturated fatty acids to hydroperoxy-fatty

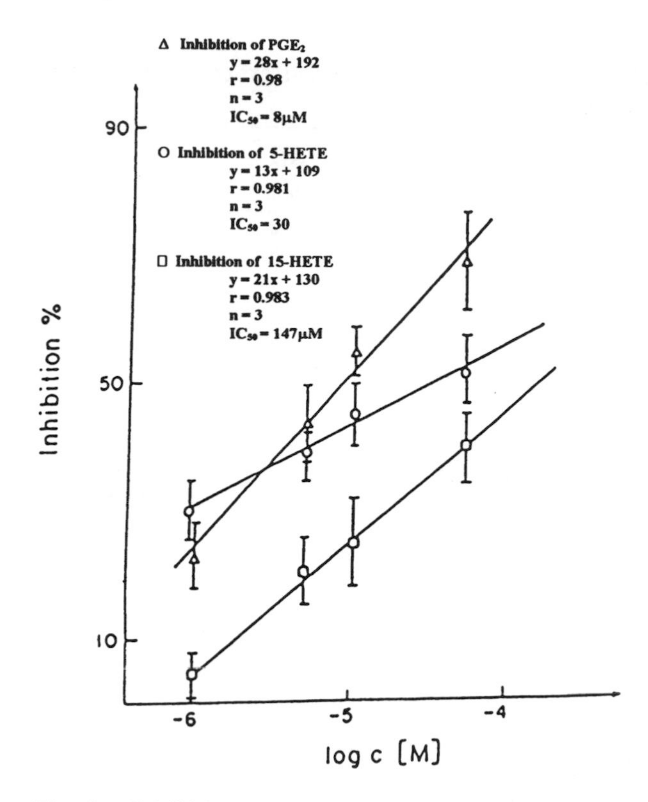

Fig. 3. Inhibition of oxygenase activity by Meclomen.

acids. This oxygenated product (of fatty acids) has an absorption maximum at 234 nm (ε = 25,000/cm) owing to the hydroperoxy group.

3. Direct UV method for XO determination: The XO-catalyzed reaction is one of the free radical-producing processes ***(11)***. Examination of this process yields information concerning oxygen consumption as it relates to the production of free radicals.

3.2.1. Indirect Method of Oxygen Consumption Determination Using Cytochrome C as an Indicator Compound

The presence of the superoxide anion (O_2^-) is a threat to living cells. SOD, a family of metalloenzymes, catalyzes the conversion of the superoxide anion to hydrogen peroxide and oxygen.

In this assay, the oxidation of xanthine by XO, is used to produce the superoxide anions. The formed superoxide reduces the cytochrome C indicator. In the presence of SOD, the reduction rate of cytochrome C is decreased. The decreased reduction rate of cytochrome C is proportional to the existing SOD

Table 2
Kinetic Constants of Meclomen for Inhibition of Cyclooxygenase and Lipoxygenases

Enzymes	IC_{50},[a] μM	Ki, μM	Km, μM	n
Cyclooxygenase (from ram seminal vesicles)	2.5 ± 3.2	15 ± 3.2	10 ± 2.1	3
5-Lipoxygenase (from potato tuber)	50 ± 6.7	25 ± 3.1	40 ± 5.1	4
15-Lipoxygenase (from soybean)	110 ± 1.0	27 ± 3.6	10 ± 1	4

[a]IC_{50} was calculated from the formula: $(IC_{50})/(s) = (Ki)/(Km)$, where Ki is inhibitory constant and Km is the Michaelis Menten Constant.

activity; 1 U SOD activity is equivalent with 50% inhibition in reduction of cytochrome C in a given condition.

3.2.1.1. Sample Collection and Preparation of SOD Activity from Erythrocytes

1. Whole blood is collected (7.0 mL) into sodium EDTA, using a purple-top Becton-Dickinson vacutainer (*see* **Note 6**). Blood can be stored at 4°C for 48 h after collection but, within 48 h the plasma and red blood cells (RBC) should be separated. Packed RBC should be washed three times with 0.9 sodium chloride.
2. Approximately 2.0 mL of packed RBC are suspended in 4.0 mL cold 0.9% sodium chloride and centrifuged for 10 min at 2000*g* (discard supernatant). After the last washing cycle, the packed cells are resuspended in 4.0 mL cold type I water for lysis. The hemolysate can be stored at –70°C until analysis.
3. From the hemolysate, the hemoglobin (Hgb) can be precipitated by adding 1 volume of cold chloroform: abs ethanol = 15:1 cold mixture. The precipitated Hgb and the organic layer are separated by centrifugation and the supernatant is the source of SOD activity. SOD activity is expressed in U/mg protein or U/gr Hgb of the hemolysate.

3.2.1.2. Measurement

1. Set the spectrophotometer to 550 nm in kinetic mode (five 30-s cycles) (*see* **Subheading 2.2.1.**).
2. Pipet 0.9 mL working solution and 0.05 mL buffer into a quartz cuvet with a 1.0-cm light pass.
3. Measure the absorbance at 550 nm and start the reaction with 0.05 mL (0.12–0.16 U/mL) of XO.

4. Record the absorbance in five 30-s cycles (total time = 150 s).
5. Calculate the change in absorbance/min (A/min). This value represents the non-inhibited superoxide production.
6. To test the actual SOD activity, repeat procedure as previously noted, replacing 0.05 mL buffer with 0.05 mL SOD-containing substance.

3.2.1.3. Analysis

1. A calibration curve is set up using the different concentrations of standard SOD powder (from 0.1–2 U/mL).
2. Calculate the A/min in presence of SOD. The presence of SOD decreases the reaction rate (A/min). Inhibition percentage is calculated using the following formula:

$$(A_{NI}–A_{IN}) / A_{NI}) \times 100 = \text{inhibition } \%$$

   where A_{NI} = reaction rate in absence of SOD and A_{IN} = reaction rate in presence of SOD.
3. One unit of SOD activity represents a 50% inhibition of the production of free radicals. Actual activity is expressed as Unit/mg protein, or Unit/g Hgb for RBC hemolyzate.

3.2.2. Direct Method of Lipoxygenase Activity Determination Using Spectrophotometry

Enzyme activity is measured by applying the use of a spectrophotometer and unsaturated fatty acid. The following reaction should be considered:

$$\text{Unsaturated Fatty Acid} + O_2 \rightarrow \text{Hydroperoxide-Fatty Acid.}$$

The product of this lipoxygenase enzyme activity, a hydroperoxide-fatty acid has a UV absorbance owing to the hydroperoxide group. One unit of enzyme has 0.001 Ab. unit/min..

3.2.2.1. Measurement

1. Set the spectrophotometer at 234 nm in the Kinetics mode, if possible.
2. Use 2 quartz cuvets, with a 1.0 cm light path.
3. Prepare the blank cuvet: 1.0 mL buffer.
4. Prepare the test solution: Add 0.95 mL of substrate (1.0 mg/mL sodium linolenate solution) to the test cuvet (*see* **Note 6**).
5. Start the reaction in the test cuvet with 0.05 mL enzyme solution.
6. Record the absorbance increase at 234 nm, against the blank, for 3 min.

3.2.2.2. Analysis

Determine the delta A_{234}/min. This is used to calculate the specific activity. Applying the ε = 25,000/cm absorbance coefficient for formed product, the amount of hydroperoxy-fatty acid can be calculated.

3.2.3. Direct UV Method for XO Determination

Prolonged periods of ischemia cause ATP to be lost, which increases the Ca^{2+} levels in the cytosol. The increased Ca^{2+} levels can activate a Ca^{2+}-dependent protease, which will convert the xanthine-dehydrogenase (XD) to XO by proteolysis. XO can catalyze the following reaction:

$$\text{Hypoxanthine} + 2O_2 + H_2O \rightarrow \text{Xanthine} + O_2^-$$

This proteolytic conversion of XD to XO is found to occur during ischemic periods. The ischemia primes the tissue for reoxygenation damage, because a new enzyme (XO) is formed in addition to one of its required substrates (hypoxanthine). During the reoxygenation—or reperfusion—period, the other substrate O_2 is introduced, and there is a burst of XO generated O_2^- production in reperfused tissues. Experimental models have demonstrated that SOD or allopurinol (an XO inhibitor), protect the reperfused tissues from damage (*see* **Note 7**).

3.2.3.1. Measurement

1. Set the spectrophotometer at 295 nm, using the Kinetics mode if possible.
2. Set up 2 quartz cuvets; a reference cell and a sample cell.
3. Pipet 1.0 mL of 0.1 *M* phosphate buffer pH 7.4 into each of the 2 cuvets.
4. Pipet 15 µL of xanthine into both the sample and reference cuvets.
5. Start the reaction by adding 10 µL of XO to the sample cuvet.
6. Record the absorbance at 295 nm for 3 min, at room temperature (25°C).

3.2.3.2. Analysis

1. The delta A/min can be calculated from the readings.
   Applying ε = 12,300 for uric acid, the uric acid formed is calculated using the following formula:

   $$(\text{DA/min}) / 12{,}300 \times 10^6 = \text{m}M \text{ uric acid}$$

2. One unit of enzyme represents the amount of protein that catalyzes the formation of 1 µ*M* /min uric acid at pH 7.5 and 25°C.

4. Notes

1. The Clark electrode must be calibrated before each period of use to avoid any drift in results. The cuvet system must be airtight; the membrane must be changed daily, to avoid protein accumulation, and replaced intact.
2. Follow the standard guidelines for the spectrophotometer that is available to you, because there may be some variance in running the kinetics programs from dif-

ferent software manufacturers. Select a series of quartz cuvets, and complete all measurements using the same cuvets in the same sequence.

3. To prevent the spontaneous autoxidation of linolenic acid, store under nitrogen or CO_2.
4. Sodium arachidonate may undergo spontaneous autoxidation; to avoid this, store the arachidonate under nitrogen or CO_2.
5. It is important to remember that the enzyme inhibitor must be introduced into the system before adding the substrate, then incubate for 3 min before beginning the test.
6. The whole blood obtained cannot be coagulated. EDTA vacutainer tubes are the collection method recommended. Blood that has been treated with heparin or citrate will not work well.
7. The volume of XO to the sample cuvet must be determined experimentally, prior to beginning the measurements. Start with fresh XO solution and add volumes ranging within 7 µL above and below the suggested 10 µL (3–17 uL). The volume that produces an optimum ΔAbs/min. at 295 nm is the volume that is best for your system.

Acknowledgments

Support for this work has come from the Department of Clinical Laboratory Science, School of Health Related Professions, State University of New York at Buffalo, and The Department of Medicine at Buffalo General Hospital. I thank Agnes Stadler, PhD and my colleagues for their assistance and support. I extend my gratitude to Andrea Barbarossa, a graduate student at the University of Buffalo, for her technical support, and for her assistance in the preparation of this chapter.

References

1. Sies, H. (1991) Oxidative stress: introduction, in *Oxidative Stress: Oxidant and Antioxidant* (Sies, H., ed.), Academic Press, San Diego, CA, pp. 15–22.
2. Halliwell, B. (1994) Free radicals, antioxidants and human disease: curiosity, cause, or consequence? *Lancet* **344,** 721.
3. Kehrer, J. P. (1993) Free radicals as mediators of tissue injury and disease. *Crit. Rev. Toxicol.* **23,** 21–48.
4. O'Flaherty, J. T. (1982) Lipid mediators of inflammation and allergy. *Lab. Invest.* **47,** 314–317.
5. Stadler, I., Kapui, Z., and Ambrus, J. L. (1994) Study on the mechanisms of action of sodium meclofenamic acid (meclomen) a "Double Inhibitor" of the arachidonic acid cascade. *J. Med.* **25(6)**, 371–382.
6. Fluber, J. C., Succari, M., and Cals, M. S. (1992) Semi-automated assay of erythrocyte Cu-Zn superoxide dismutase activity. *Clin. Biochem.* **25,** 115–119.

7. Wheeler, C., Salzman, J. A., Elsayed, N. M., Omaye, S. T., and Korte, D. W. (1990) Automated assay for superoxide dismutase, catalase, glutathione peroxidase and glutathione reductase activity. *Anal. Biochem.* **184,** 193–199.
8. Dormandy, T. L. (1988) In praise of lipid peroxidation. *Lancet* **2,** 1126–1128.
9. Cheeseman, K. H. (1993) Mechanism and effects of lipid peroxidation. *Mol. Aspects Med.* **14(3),** 191–197.
10. Masotti, L., Casali, E., and Galeotie, T. (1988) Lipid peroxidation and tumors. *Free Rad. Biol. Med.* **4,** 377–386.
11. Saugstad, O. D. (1996) Role of xanthine oxidase and its inhibitor in hypoxia: reoxygenation injury. *Pediatrics* **98,** 1.

2

Spin Trapping and Electron Paramagnetic Resonance Spectroscopy

Daniel J. Brackett, Gemma Wallis, Michael F. Wilson, and Paul B. McCay

1. Introduction

Electron paramagnetic resonance (EPR) spectroscopy methodology is a highly selective and sensitive assay for detecting paramagnetic species. Owing to the unpaired electron in the outer orbit, free radicals are paramagnetic species and, when in sufficient quantity, are directly detectable and measurable using EPR spectroscopy. However, many free-radicals species are highly reactive, with relatively short half-lives, and the concentrations found in biochemical systems are usually inadequate for direct detection by EPR spectroscopy. Spin-trapping is a chemical reaction that provides an approach to help overcome this problem. Spin traps are compounds that react covalently with highly transient free radicals to form relatively stable, persistent spin adducts that also possess paramagnetic resonance spectra detectable by EPR spectroscopy. When a spin trap is added to a free radical-generating biochemical reaction, a growing pool of relatively long-lived spin adducts is created as the free radicals react with the spin trap. Detectable EPR spectra are generated by the reaction when the signal strength of the accumulation of adducts reaches the lower limit of sensitivity of the particular spectrometer being utilized.

EPR spectroscopy, combined with spin trap techniques, furnishes a sensitive probe for the assessment and characterization of free-radical production and provides direct, unequivocal evidence of free-radical production in in vitro and complex biologic systems ***(1–5)***. The relative intensity of free-radical formation can be determined because the EPR spectroscopy signal is directly related to the concentration of spin adducts ***(6)***; the height of the peaks of the

From: *Methods in Molecular Biology, vol. 108: Free Radical and Antioxidant Protocols*
Edited by: D. Armstrong © Humana Press Inc., Totowa, NJ

spectra are proportional to the number of radical adduct molecules in the accumulating pool. The capacity to detect and record reliably these signals in a controlled experimental environment permits establishment of the onset and duration of free-radical production by adding the spin trap and sampling the system at selected, sequential time points. Importantly, the capacity to document these parameters permits the expression of the intensity of free-radical generation as a function of time and the establishment of a temporal profile of free-radical production in response to in vitro and in vivo experimental conditions of interest. For example, to determine the approximate onset and duration of radical formation in specific tissue of experimental animals following the administration of an agent that induces free radicals, the spin trap would be administered to groups of animals at selected times following the administration of the agent whose metabolism is suspected of involving free-radical intermediates. The tissues of interest would be collected and prepared for spectral analysis. This experimental strategy provides the opportunity to identify the target sites (i.e., lung, liver, intestine) of the induction agent and to differentiate the intensity, onset, and duration of free-radical generation for each tissue ***(7,8)***. The same experimental procedure can be applied to evaluate onset and duration of free-radical activity in in vitro studies using cellular and subcellular preparations ***(9)***. Additionally, EPR spectroscopy combined with spin-trapping techniques can be employed not only to generate information regarding free-radical formation in a biological system, as previously described, but also to evaluate the effect of an intervention or manipulation of the system on the magnitude, onset, and duration of radical generation in specific sites.

The interpretation of the EPR spectra and the subsequent identification of the free radical associated with the spin adduct involved in the generation of the spectra is the most sophisticated and difficult component of the spin-trapping process. Spectra interpretation requires a working knowledge of how the configuration of the EPR spectra is affected by the structural features of the trapped free radical and the dynamics of its interaction with the spin trap ***(10)***. A definitive discussion of these relationships is not within the scope and space limitations of this chapter. However, it is important to recognize that the EPR spectra of a particular spin adduct has unique characteristics that are dependent on the specific spin trap used and the free radical trapped, and serve as sensitive and specific markers of the presence of a particular free-radical species. It is even possible to detect several different radical adducts that may be present in a system even though their spectra overlap each other.

It is apparent, even from this brief description, that combining spin-trap techniques with EPR spectroscopy provides a powerful tool for elucidating the regulation and mechanisms of the generation of specific free-radical species, for investigating the biochemical interactions of free radicals at the cellular and subcellular level, and for the evaluation of the role of free radicals as significant mediators in a myriad of disease processes. However, we must emphasize that the successful application of this methodology to a specific experimental hypothesis or scientific problem is dependent on access to an established EPR spectroscopy facility staffed by experienced technical personnel, and the support of a collaborator with training and expertise in the interpretation of spectra derived from spin adducts. Nevertheless, the wealth of definitive information regarding free-radical reactions in the investigation of diverse disciplines ranging from molecular biology to integrative physiology to pathological processes justifies the effort required to establish the necessary collaborative relationships.

The objective of this chapter is to describe the procedures leading up to the generation and interpretation of EPR spectra by detailing the materials and experimental methods of three different approaches to the investigation of free-radical biology and its potential effects in biological systems that have been employed in our laboratories. These approaches include:

1. An in vivo experimental protocol for the evaluation of the site, onset, duration, and intensity of the generation of the free radical nitric oxide in response to endotoxin administration ***(8)***;
2. A xanthine oxidase (XO) system to study the effect of catechol adrenergic agents on the generation of hydroxyl radicals ***(11)***; and
3. An in vitro biologic system using hepatic microsomes to elucidate the role of ethanol in radical production in the liver ***(12)***. These procedures are adaptable and applicable to the investigation of any free radical of interest, which would then dictate the choice of the spin trap and the free-radical induction substance to be used in the assay.

2. Materials

2.1. In Vivo Free-Radical Induction

2.1.1. Animal and Spin-Trap Preparation

1. Male Sprague-Dawley rats: 250–350 grams body weight (Sasco, Omaha, NE).
2. Infant laryngoscope with Modified Millar 0 blade.
3. Tracheal tube: PE-240, 9.5 cm total length, 5.5-cm insertion length.
4. Anesthesia machine; isoflurane vaporizer.
5. IsoFlo®; isoflurane, USP (Abbott Laboratories, North Chicago, IL).
6. Rat ventilator (Harvard Instruments Inc, South Natick, MA).
7. Surgical instruments.
8. Silastic, 0.025 in. i.d., 0.047 in. o.d.Venous Catheter.

9. 10 mg/mL Lidocaine hydrochloride (Abbott Laboratories).
10. Lipopolysaccharide B, *E. coli* 0127:BP (Difco Labs, Detroit, MI) or the substance of choice for induction of the free radical to be studied.
11. Diethyldithiocarbamate (DETC): Nitric oxide spin trap (Sigma, St. Louis, MO) or the appropriate spin trap for the free radical of interest.
12. Sterile, nonpyrogenic saline (Baxter Healthcare Corp., Muskegon, MI).
13. Tuberculin syringes (Becton-Dickinson, Franklin Lakes, NJ).

2.1.2. Spin-Adduct Extraction and Concentration

1. Sartorius Top Loading Balance: Model # 1204 MP (Sartorius Division of Brinkmann Inst., Westbury, NY).
2. 50-mL Pyrex beaker.
3. Commercial blender (Waring, Fisher Scientific, Plano,TX).
4. Blender cup: Semi micro s/s Press Fitcover (Eberbach, Ann Arbor, MI).
5. Chloroform-methanol: (2:1). Omnisolv (EM Science, Gibbstown, NJ).
6. 100- and 250-mL graduated cylinders (Kimble Scientific Products, Vineland, NJ).
7. 0.5% sodium chloride solution (Baker Chemical, Phillipsburg, NJ).
8. Direct Torr Vacuum Pump (Sargent-Welch, Skokie, IL) equipped with a 1000-mL Pyrex large cold finger (California Laboratory Equipment, Oakland, CA) and a small Pyrex cold finger.
9. Three evaporators, Model C (California Laboratory Equipment) attached to the vacuum pump.
10. 15-mL conical graduated test tubes (Kimble Scientific).
11. 5 3/4 and 9 in. disposable Pasteur pipets, (Fisher Scientific).
12. Whatman No.1 filter paper (Fisher Scientific).
13. Chromatographic grade nitrogen (Air Liquide America, Houston, TX).

2.1.3. Spectral Analysis

1. EPR spectrometer; Bruker SRC 300E (Billerica, MS).
2. 2.79 ± 0.013 mm, i.d. Quartz tubes (Wilmad Glass, Buona, NJ).

2.2. In Vitro Free-Radical Generating System

2.2.1. XO System and Spin-Trap Preparation

1. Glassware.
2. Phosphate sodium chloride buffer (0.1 *M*, pH 7.4).
3. Type I ferritin from horse spleen (Sigma).
4. Hypoxanthine (Sigma).
5. XO (Sigma).
6. 5,5-dimethyl-1-pyrroline-N-oxide (DMPO) or spin trap of choice depending on free radical of interest (Sigma).
7. Epinephrine, norepinephrine, isoproterenol (Sigma) or substance of interest to be tested in the system to determine its capacity to enhance or inhibit free-radical formation.

8. Shaking incubator: Dubnoff Metabolic (Precision Scientific Group, Chicago, IL).

2.2.2. EPR Spectrometry

1. Electron paramagnetic resonance spectrometer; Bruker SRC 300E.
2. Quartz EPR flat cells (Wilmad).

2.3. In Vitro Cellular/Subcellular System for Free-Radical Induction

2.3.1. Liver Microsome System and Spin-Trapping Preparation

1. Female Sprague-Dawley rats: 140–150 grams (Sasco).
2. Potassium phosphate buffer: 150 and 50 m*M*, pH 7.4.
3. Phenyl-N-t-butylnitrone (PBN) (Sigma) or appropriate spin trap for trapping the radical of interest.
4. Glucose-6-phosphate (Sigma).
5. Nicotinamide adenine dinucleotide phosphate (NADP) (Sigma).
6. Glucose-6-phosphate dehydrogenase (Sigma).

2.3.2. EPR Spectroscopy

1. EPR Spectrometer, Bruker SRC 300E (Billerica, MS).
2. 2.8 mm i.d. Quartz tubes (Wilmad).

3. Methods

3.1. In Vivo Procedures for Free-Radical Induction and Spin Trapping

3.1.1. Animal Instrumentation

1. Induce anesthesia using a Bell jar containing the rapidly metabolized, inhalational anesthetic, isoflurane (5.0%).
2. Intubate the trachea.
3. Connect the tracheal tube to a rodent respirator delivering 2.5% isoflurane combined with 100% O_2 delivered at 600 mL/min.
4. Make a 2.0-cm midline incision through the skin on the ventral side of the neck.
5. Isolate the right jugular vein.
6. Insert a venous catheter for the administration of lipopolysaccharide or saline.
7. Guide the catheter under the skin to exit through an incision in the back of the neck just below the base of the skull.
9. Inject skin edges of the incision with lidocaine.
10. Suture wounds and secure catheter at the exit incision using 4-0 surgical silk.
11. Remove the animal from the respirator and delivery of isoflurane.
12. Upon resumption of normal respiration and demonstration of the righting reflex remove the intratracheal tube.
13. Return animal to home cage.

3.1.2. Experimental Protocol for the Comprehensive Evaluation of In Vivo Free-Radical Activity

1. Allow a 60-min recovery period beginning at cession of the anesthetic.
2. Administer lipopolysaccharide, 20 mg/kg in saline (or the substance selected to evoke the free radical of interest) or the delivery vehicle (control) via the jugular vein catheter for direct introduction into the circulatory system. Other routes of delivery could also be used (e.g., subcutaneous, intraperitoneal, even orally in some cases) depending on the free-radical species to be studied.
3. Permit the reaction to the substance chosen to initiate free-radical formation (lipopolysaccharide in our example study) to develop in the animals for predetermined time periods (hourly for 6 h in our example) to establish a temporal profile of the onset, duration, and intensity of free radical activity (*see* **Notes 1** and **2**).
4. Fifteen min (this may vary depending the rate of up-take and distribution of the particular spin adduct formed) before the end of each designated time period, administer the spin trap appropriate for the free radical of interest (to trap nitric oxide; DETC sodium salt, 500 mg/kg, intraperitoneally and a mixture of $FeSO_4 \cdot 7H_2O$, 7.5 mg/kg, and 37.5 mg/kg of sodium citrate, subcutaneously).
5. At the end of each designated time period, anesthetize (5% isoflurane) the animals and rapidly harvest the tissues hypothesized to have free-radical activity (liver, kidneys, intestines, and plasma in the model used in this discussion) permitting identification of the site of free-radical activity and construction of the temporal profile of onset, duration, and intensity for each organ (*see* **Notes 3–5**).
6. Immediately transfer the tissues to 50 mL cold saline-filled beakers placed in ice.

3.1.3. Preparation of Tissue for Spectral Analysis: Procedures for the Extraction and Concentration of Lipid Soluble Spin Adducts

1. Remove tissue of interest from the ice-cold saline-filled beakers.
2. Blot tissue on a paper towel.
3. Weigh tissue on a top loading Sartorius balance.
4. Mince tissue with scissors.
5. Transfer tissue to a cold blender cup (kept on ice).
6. Add of cold chloroform–methanol (2:1), measured in a graduated cylinder at a volume of 20:1, to the tissue in the ice cold blender cup.
7. Homogenize tissue in a Waring blender for 30 s (longer when necessary, depending on type of tissue).
8. Filter homogenized tissue with Whatman No. 1 filter paper and collect in the appropriate size side arm flask. Assist filtration with "in house" vacuum.
9. Transfer the filtrate to either 125 or 250 mL separator funnels.
10. After 15–30 min add 1/5 vol of 0.5% NaCl to the filtrate. Shake the sample, release the gas, and allow separation to occur for approx 1 h.
11. Drain the bottom layer to 100 or 250 mL round bottom flasks.
12. Remove the chloroform with a rotary evaporator.

13. As the sample approaches dryness add 2–5 mL of absolute ethanol and continue to evaporate the sample to dryness.
14. Using a Pasteur pipet, transfer the sample in a few milliliters of chloroform to a 15 mL graduated conical tube.
15. Concentrate the sample under a stream of nitrogen gas to a final volume of 0.2 mL (volume may be more depending on tissue).
16. Transfer samples to a quartz EPR tube for scanning.

3.1.4. EPR Spectrometry

1. Position the quartz tube containing the filtrate in the scanning cavity of the EPR spectrometer.
2. Spectrometer settings for the nitric oxide study used as the in vivo example in this chapter were determined by experienced EPR facility personnel (*see* **Notes 6** and **7**). Representative values are as follows:
   a. Microwave frequency: 9.7 GHz.
   b. Microwave power: 20 mW.
   c. Modulation frequency: 100 KHz.
   d. Modulation amplitude: 2.018 G.
   e. Time constant: 81.92 ms.
   f. Sweep time: 167.772 s.
   g. Sweep width: 100 G.
   h. Center field: 3420 G.
3. Gains for the individual spectra were adjusted at the discretion of the operator and ranged from 2.5e + 04 to 1.0e + 05. Control samples were scanned at 1.0e + 05.

3.2. In Vitro Free Radical Generating System Procedure

3.2.1. Protocol for Evaluation of the Manipulation of Free Radicals Generated by the XO System Using Spin Trapping

1. Acid-wash all glassware to remove traces of contaminating iron.
2. Charcoal-filter DMPO or the spin trap of choice to remove possible trace contaminants that may produce an EPR signal.
3. Dilute type I ferritin in a phosphate/NaCl buffer.
4. Dissolve hypoxanthine in double-distilled deionized water in a 10 mL volumetric flask after initial addition of 0.1 mL of 10*N* NaOH to solubilize the compound.
5. Dissolve catechol adrenergic agents in double-distilled deionized water in a 10 mL volumetric flask after first adding 0.1 mL of 2 NHCl to solubilize and prevent oxidation. Other agents of interest may also require special treatment.
6. Prepare the basic XO radical system by combining the following solutions: 0.1 *M* phosphate buffer, and 0.09 *M* NaCl at pH 7.4 in a volume of 0.5 mL, with 0.1 mL of 0.4 m*M* hypoxanthine. DMPO at a concentration of 40 m*M* is used to trap superoxide and the hydroxyl radical or the spin trap appropriate for the free

radical to be studied in the system (0.1 mL) and ferritin equivalent to 1 m*M* total iron content at a volume of 0.1 mL.

7. Add the catechol adrenergic agent, 0.4 mm of choice (or the substance hypothesized to modify the formation of free radicals) or the vehicle for delivering the agent (0.1 mL).
8. Initiate the reaction for radical formation by adding 0.1 mL of XO prepared at a concentration of 0.1 U/mL.
9. The final volume of all incubation systems is 1.0 mL.
10. Place in a shaking incubator for 10 min at 37°C in room air.
11. Transfer to a quartz EPR flat cell for scanning.

3.2.2. EPR Spectrometry

1. Position the quartz EPR flat cell containing the solution containing the spin adduct in the scanning cavity of the EPR spectrometer.
2. Spectrometer settings for the catechol adrenergic agent-free radical study used as an example in this chapter were determined by experienced EPR facility personnel. Representative values are as follows:
   a. Microwave frequency: 9.78 GHz.
   b. Microwave power: 20 mW.
   c. Modulation frequency: 100 KHz.
   d. Modulation amplitude: 1.053 G.
   e. Time constant: 164.34 ms.
   f. Sweep time: 83.89 s.
   g. Sweep width: 100 G.
   h. Center field: 3480 G.
3. Gains for the individual spectra were adjusted at the discretion of the operator and ranged from 2.5e + 05 to 1.0e + 06 and each sample was scanned four times with accumulation to reduce noise.

3.3. In Vitro Cellular/Subcellular System for Free Radical Induction

3.3.1. Hepatic Microsome Reaction System and Spin-Trapping Preparation

1. Decapitate rats and rapidly remove livers for preparation of microsomes ***(13)***.
2. Homogenize the livers in 0.15 *M* potassium phosphate buffer at pH of 7.4.
3. Centrifuge the homogenized tissue at 10,000*g* for 15 min.
4. Centrifuge the supernatant fraction at 105,000*g* for 90 min.
5. Wash the microsomal pellet twice by resuspending in phosphate buffer and centrifuge at 105,000*g* for 1 h.
6. The resultant microsomal pellets can be either frozen until needed or used immediately.

7. Resuspend the microsomal pellets in 0.15 mm potassium phosphate buffer, pH of 7.4, at a ratio of 1.0 mg of microsomal protein ***(14)*** to 0.1 mL of buffer solution.
8. Prepare the reaction system by mixing: 1.0–2.0 mg of microsomal protein; ethanol (50 mm) or the substance being evaluated for its capacity to induce radical formation; PBN to trap 1-hydroxyethyl radicals or the appropriate spin trap for the anticipated radical; and the NADPH-generating system consisting of 55 mm glucose-6-phosphate, 0.3 mm NADP, and glucose-6-phosphate dehydrogenase at 0.5 Kornberg U/mL ***(14)***.
9. Add the phosphate buffer to bring the volume to 1.0 mL.
10. Incubate the solution at 37°C for 30 min.
11. Extract the reaction systems by adding 750 µL of toluene.
12. Vortex for 15 s followed by centrifugation at 3000*g* for 10 min.
13. Pipet the top toluene layer.
14. Transfer samples into EPR quartz tubes.
15. Bubble the extract with N_2 for at least 5 min to remove oxygen.

3.3.2. EPR Spectrometry

1. Position the EPR quartz tube in the scanning cavity of the EPR spectrometer.
2. Spectrometer settings for the study involving ethanol-induced free-radical production in hepatic microsomes that was used as an example in this chapter were determined by experienced EPR facility personnel. Settings were similar to those stated in **Subheading 3.2.2.**
3. Gains for the individual spectra were adjusted at the discretion of the operator and ranged from 4.0e + 04 to 5.0e + 04 $\times 10^4$.

4. Notes

1. The duration of time for exposure of spin trap to the free radical studied must be exactly the same, both in vitro and in vivo, when comparing signals between samples.
2. The duration time and the concentration for each spin trap must be adjusted to achieve optimal trapping depending on the specific radical species under study and the distribution of the spin trap in the in vivo studies.
3. Tissue must be removed rapidly and immediately placed in chilled containers in the in vivo studies.
4. Control samples are imperative.
5. Conscious animals should be used in the in vivo studies to avoid the confounding effects created by the metabolism of anesthetics.
6. Access to an EPR spectrometry facility with experienced personnel is critical.
7. Collaboration with an investigator proficient in the evaluation and interpretation of signals generated by free radicals is required for the success of any study involving spin trapping and EPR spectrometry.

Acknowledgments

We gratefully recognize Megan R. Lerner for her invaluable contribution to this work. We also acknowledge the support of the Department of Surgery Research Fund and the Presbyterian Health Foundation, Oklahoma City, OK, the Department of Veterans Affairs Medical Research Service, and the National Institutes of Health, Grant no. HL43151.

References

1. McCay, P. B. (1987) Application of ESR spectroscopy in toxicology. *Arch. Toxicol.* **60,** 133–137.
2. Bolli, R., Patel, B. S., Jeroudi, M. O., Lai, E. K., and McCay, P. B. (1988) Demonstration of free radical generation in "stunned" myocardium of intact dogs with the use of the spin trap α-phenyl N-tert-butyl nitrone. *J. Clin. Invest.* **82,** 476–485.
3. Zweier, J. L. (1988) Measurement of superoxide-derived free radicals in the reperfused heart. *J. Biol. Chem.* **263,** 1353–1357.
4. McCay, P. B. and Poyer, J. L. (1989) General mechanisms of spin trapping in vitro and in vivo, in *CRC Handbook of Free Radicals and Antioxidants in Biomedicine* (Miquel, J., Quintanilha, A. T., and Weber, H., eds.), CRC Press, Boca Raton, FL, pp. 187–191.
5. Brackett, D. J., Lerner, M. R., Wilson, M. F., and McCay, P. B. (1994) Evaluation of in vivo free radical activity during endotoxic shock using scavengers, electron microscopy, spin traps, and electron paramagnetic resonance spectroscopy, in *Free Radicals in Diagnostic Medicine* (Armstrong, D., ed.), Plenum Press, NY, pp. 407–409.
6. Ayscough, P. B. (1967) Numerical double integration of the first derivative curve, in *Electron Spin Resonance in Chemistry,* Methuen Press, London, pp. 442, 443.
7. Brackett, D. J., Lai, E. K., Lerner, M. R., Wilson, M. F., and McCay, P. B. (1989) Spin trapping of free radicals produced in vivo in heart and liver during endotoxemia. *Free Rad. Res. Comm.* **7,** 315–324.
8. Wallis, G., Brackett, D., Lerner, M. R., Kotake, Y., Bolli, R., and McCay, P. B. (1996) In vivo spin trapping of nitric oxide generated in the small intestine, liver, and kidney during the development of endotoxemia: a time-course study. *Shock* **6 (4),** 274–278.
9. Poyer, J. L., McCay, P. B., Lai, E. K., Janzen, E. G., and Davis, E. R. (1980) Confirmation of assignment of the trichloromethyl radical spin adduct detected by spin trapping during ^{13}C-carbon tetrachloride metabolism in vitro and in vivo. *Biochem. Biophys. Res. Commun.* **94,** 1154–1160.
10. Janzen, E. G., Stronks, H. J., DuBose, C. M., Poyer, J. L., and McCay, P. B. (1985) Chemistry and biology of spin-trapping radicals associated with halocarbon metabolism in vitro and in vivo. *Environ. Health Perspect.* **64,** 151–170.
11. Allen, D. R., Wallis, G. L., and McCay, P. B. (1994) Catechol adrenergic agents enhance hydroxyl radical generation in xanthine oxidase systems containing

ferritin: Implications for ischemia/reperfusion. *Arch. Biochem. Biophys.* **315,** 235–243.

12. Reinke, L. A., Rau, J. M., and McCay, P. B. Possible roles of free radicals in alcoholic tissue damage. *Free Rad. Res. Comms.* **9,** 205–211.
13. McCay, P. B., Lai, E. K., Poyer, J. L., DuBose, C. M., and Jensen, E. G. (1984) Oxygen- and carbon-centered free radical formation during carbon tetrachloride metabolism. *J. Biol. Chem.* **259,** 2135–2143.
14. Lowry, O. H., Posebrough, M. J., Farr, A. L., and Randall, R. J. (1951) Protein measurement with the Folin phenol reagent. *J. Biol. Chem.* **193,** 265–275.

3

In Vivo Detection of Free Radicals in Real Time by Low-Frequency Electron Paramagnetic Resonance Spectroscopy

Gerald M. Rosen, Sovitj Pou, and Howard J. Halpern

1. Introduction

During his studies on the properties of oxygen, Priestley *(1)* noted that this gas, an essential ingredient for life processes, appears to "burn out the candle of life too quickly." More than two centuries would elapse, however, before this observation would be associated with Grubbé's *(2)* accounts of redness and irritation on the hands of his workers testing X-ray tubes. By 1954, Gerschman et al. *(3)* suggested that free radicals were the common element linking the observed toxicity of oxygen to the harmful effects of ionizing radiation. The implication of this hypothesis seemed remote at that time. However, within a decade, the search for biologically generated free radicals would lead to the discovery of superoxide and an enzyme that attenuated cellular levels of this free radical *(4,5)*. In the intervening years, free radicals have been recognized as common intermediates in cellular metabolism *(6,7)*, found to play an essential role in host immune response *(8)* and demonstrated to regulate many essential physiologic functions *(9)*.

As the biological significance of free radicals, such as superoxide and hydroxyl radical, became apparent, methodologies were developed to identify these reactive species. One such approach, known as spin trapping *(10)*, involved the addition of free radicals to either nitrosoalkanes or nitrones *(11,12)*. In so doing, the initial unstable free radical is "trapped" as a long-lived nitroxide, which can be observed by electron paramagnetic resonance (EPR) spectroscopy at ambient temperature *(13)*. In the intervening years, spin trapping has been found to be able to simultaneously measure and distinguish among a variety of important biologically generated free radicals *(14,15)*. With

From: *Methods in Molecular Biology, vol. 108: Free Radical and Antioxidant Protocols*
Edited by: D. Armstrong © Humana Press Inc., Totowa, NJ

the advent of lower frequency EPR spectrometers ***(16)***, attention has focused on the in vivo *in situ* spin trapping of free radicals in real time ***(17–19)***. In this review, we will describe our recent studies measuring hydroxyl radical production at the site where it evolved using a low-frequency EPR spectrometer in combination with in vivo spin-trapping techniques ***(18)***.

2. Materials

1. Chelex 100 (Bio-Rad, Hercules, CA).
2. Conalhumin.
3. Diethylenetriaminepentaacetic acid (DTPA).
4. Ethylenediaminetetraacetic acid (EDTA).
5. 4-Pyridyl-1-oxide-N-tert-butylnitrone (4-POBN).
6. 5,5-Dimethyl-1-pyrroline N-oxide (DMPO).
7. Phenyl N-tert-butyl nitrone (PEN).
8. EPR spectrometer.
9. Stripline plastic sample holder, ABS, resin GPX3700 (General Electric, Pittsfield, MA).
10. Nitrones and nitrosoalkanes can be obtained from several commercial sources. To obtain accurate data, these spin traps should be of the highest purity possible. As iron and copper salts are common contaminants of most laboratory buffers, removal of these ions can be accomplished by passing the buffer through an ion-exchange resin, such as Chelex 100 (Bio-Rad) ***(20)***. In some experiments, conalbumin has proven to be a more reliable method at removing iron salts ***(21)***. Verification that redox active metal ions are no longer in the buffer can be accomplished by using a simple ascorbate assay ***(22)***. Inclusion of the metal ion chelator, diethylenetriaminepentaacetic acid (DTPA), but not EDTA renders Fe^{+2}, Cu^{+2} and Mn^{+2} inefficient as hydroxyl radical catalysts ***(23)***. All other reagents should be of the highest quality obtainable.

3. Methods

3.1. In Vivo *In Situ* Detection of Free Radicals

For the better part of a decade, we have refined techniques to allow the detection of free radicals in living animals, culminating in the recent in vivo *in situ* identification of hydroxyl radical in an irradiated tumor of a living mouse ***(18)***. This requires an EPR spectrometer, which can detect free radicals deep in living tissue. Loss of signal with depth is obviated by operating at much lower frequencies, 100–300 MHz ***(24)***. To use the low frequencies necessary for in vivo measurement of free radicals in living tissue deep in animals, we chose to develop *ab initio* a low frequency EPR spectrometer ***(16)***. The frequency used, 250 MHz, is capable of measurements 7 cm deep in tissues ***(16,25,26)***.

Because this is a magnetic resonance technique, the spectrometer is capable of imaging ***(16)***. We have found it possible to detect spin trapped adducts at these frequencies ***(27)***. Even though in vivo spin traps compete for free radicals with other chemical species, the resultant effects may often result in disparate biologic endpoints. Because of this, spin traps should be considered reporters of specific free-radical events, as well as pharmaceutical agents, altering physiologic outcomes.

3.2. Enhanced Stability of Spin-Trapped Adducts

Poor cellular stability of many spin-trapped adducts have made it difficult to verify the presence of specific free radicals ***(28)***, even with the enhanced sensitivity afford by using deuterium-labeled spin traps ***(27,29)***. The search for more stable spin trapped adducts, especially for superoxide and hydroxyl radicals, directly led to the discovery of the spin trapping system 4-POBN plus ethanol (EtOH) ***(30)***. In this reaction, hydroxyl radical abstracts a hydrogen atom from ethanol, resulting in formation of α-hydroxyethyl radical. Spin trapping of this free radical with 4-POBN gives 4-POBN-$CH(CH_3)OH$, exhibiting remarkable stability in homogenous solutions and cell preparations ***(30,31)***.

$$HO^{\bullet} + CH_3CH_2OH \rightarrow CH_3{\cdot}CHOH + H_2O$$

$$CH_3{\cdot}CHOH + \text{4-POBN} \rightarrow \text{4-POBN-}CH(CH_3)OH$$

Formation of 4-POBN-$CH(CH_3)OH$ is specific for hydroxyl radical, because superoxide does not catalyze formation of α-hydroxyethyl radical; rapid with a second-order rate constant of 3.1 x 10^7/M/s; more sensitive towards hydroxyl radical than either DMPO plus EtOH or PBN plus EtOH ***(30)***.

3.3. Design Characteristics of the Low-Frequency EPR Spectrometer

We have developed *ab initio* a low frequency EPR spectrometer, which has the capability to detect free radicals within animal tissues and organs ***(16)***. The schematic of the low-frequency EPR spectrometer is shown in **Fig. 1**. Five design criteria were followed:

1. Very low frequency. 50–500 MHz, presently at 260 ± 0 MHz. Skin depth is 7 cm ***(16)***.
2. Continuous-wave operation. This is desirable owing to the short relaxation time of nitroxides. It means that, unlike pulsed systems, the spectrometer is a relatively highly tuned system. As a result, we have implemented electronic feedback stabilization of both the frequency and the coupling. There is point by point monitoring of and correction for frequency change.

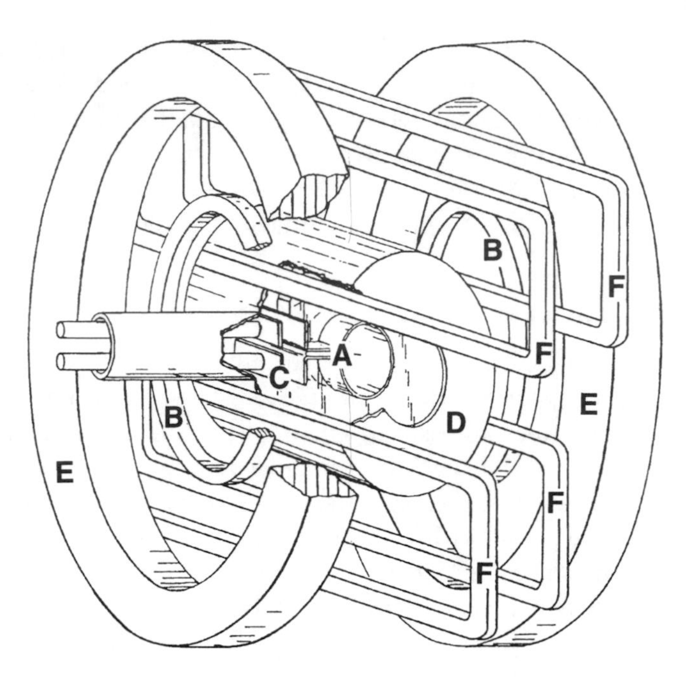

Fig. 1. Design of a low-frequency EPR spectrometer. (A) Stripline sample holder, (B) modulation coils, (C) capactive coupling, (D) radiofrequency shield, (E) main magnet, and (F) horizontal plane defining coils.

3. Open design to accommodate varied samples including animals.
4. A wide variety of resonators of stripline design. Each designed to accommodate a specific sample. This allows filling of the resonator with the sample, giving a high filling factor (filling fraction) to which the EPR signal is directly proportional. We use machineable, platable plastic, ABS.
5. Low main magnetic field. 90 Gauss (0.009T), substantial gradients and the open design argue that main magnetic-field production and gradient production be combined in a single multi-purpose magnet of nonferrous design. Operation with these low magnetic fields allows standard laboratory equipment to give extremely high absolute stability—≈0.5 milligauss—with relatively moderate fractional stability (5 parts in 10^6). The Helmholtz coils provide similar uniformity over the tumor. With these low magnetic fields, eventual designs could include currents installed in walls of clinical rooms without danger to personnel (the maximum magnetic field is less than 1/100 that of an magnetic resonance imaging [MRI] spectrometer) and to other equipment and with minimal interference if any. To generate gradients along the axis of the resonator, we have chosen to display the magnetic coils. This provides uniquely linear gradients. Gradients in the perpendicular horizontal direction are generated by current imbalances.

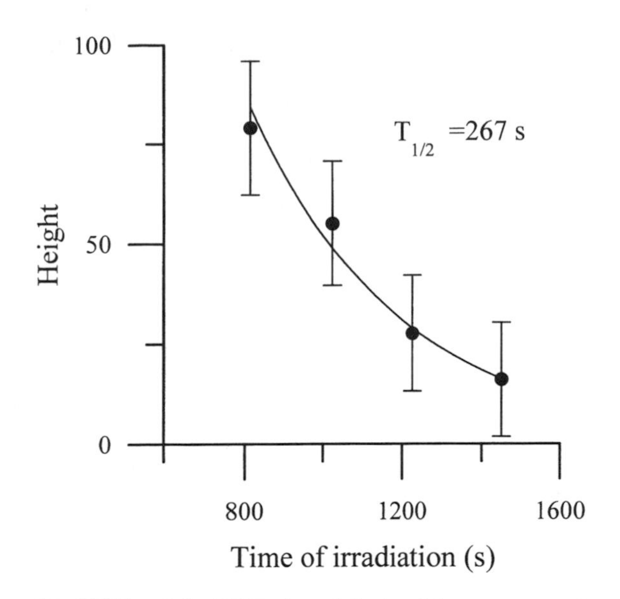

Fig. 2. Spectral half-life of the EPR signal derived from repeated measurements of 4-POBN-CH(CH_3)OH peak height while the tumor on the leg of a mouse resided in the spectrometer (adduct signal vs time). Half-life was calculated to be 267 s.

3.4. Results

3.4.1. *In Vivo* In Situ *Spin Trapping of Radiation Generated Free Radicals with Very Low-Frequency EPR Spectroscopy*

Several years ago, we reported the use of spin trapping to image and identify hydroxyl radicals produced by ionizing radiation in buffered solutions *(27)*. These solutions mimic the difficult spectroscopic penetrability of living tissue. The same low frequency—250 MHz—spectrometer described herein and deuterated spin traps DMPO-d_{11} and DMPO-d_3 were used. Partial deuteration simplified the spectral signature (DMPO-d_3) and full deuteration (DMPO-d_{11}) increased the signal intensity. These results established the feasibility of in vivo spin trapping *(27)*, even though in vivo studies would require the more stable spin-trapped adduct, 4-POBN-CH(CH_3)OH. For these initial experiments, we chose an extremity tumor to deliver high, toxic doses of radiation to a substantial bulk of the tissue with minimal effect on the physiology of the rest of the animal.

The identification of radiolytic-generated hydroxyl radical, as 4-POBN-CH(CH_3)OH, is dependent on the pharmacodynamics properties of the tumor. As 4-POBN is a zwitterion, with an octanol/water partition coefficient of only

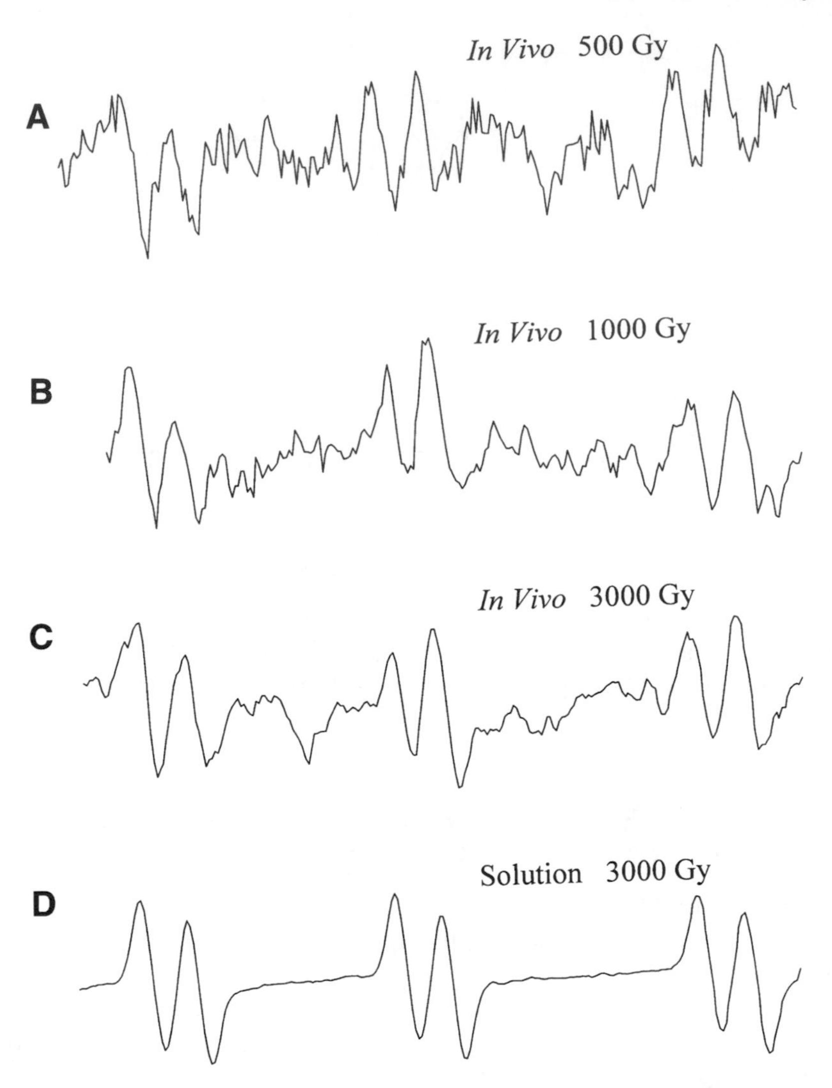

Fig. 3. EPR spectrum of 4-POBN-CH(CH_3)OH obtained by irradiating a tumor in the leg of a mouse, as described in *ref. **18***. (**A**), 500 Gy. (**B**), 1000 Gy. (**C**), 3000 Gy. (**D**), EPR spectrum of 4-POBN-CH(CH_3)OH, generated by irradiating a solution of 4-POBN and ethanol with 3000 Gy.

0.09 ***(30)***, the detection of 4-POBN-CH(CH_3)OH will depend upon the concentration of hydroxyl radical and EtOH in vascular compartments and the rate of spin-trapped adduct diffusion from the site of irradiation. The dynamics of this process is illustrated by following the pharmacokinetics of the positively charged nitroxide, 3-trimethylaminomethyl-2,2,5,5-tetramethyl-1-pyrrolidinyloxyl iodide, with an octanol/water partition coefficient <0.02 ***(18)***. Its $T_{1/2}$ of elimination from a leg tumor increased from 8 to 35 min after placing a tourniquet around the tumor ***(18)***. In a similar experimental design, 4-POBN-$CH(CH_3)OH$, whose octanol/water partition coefficient is 0.22 ***(18)***, was eliminated from the tumor more rapidly, with $T_{1/2}$ approaching 4.5 min (**Fig. 2**).

Taking into account these limitations, has allowed us to directly detect radiolytic-generated hydroxyl radical in a leg tumor in a living mouse with the very low-frequency EPR spectrometer and the 4-POBN plus EtOH spin trapping system ***(18)***. **Fig. 3A** depicts the EPR spectrum of 4-POBN-$CH(CH_3)OH$, generated by irradiating a leg tumor of a mouse with 500 Gy. **Fig. 3B,C** was derived under similar experimental conditions, except the radiation was increased to 1000 and 3000 Gy, respectively. Verification of the reaction was obtained by irradiating an aqueous solution of 4-POBN plus EtOH. These data demonstrate the feasibility of spin trapping hydroxyl radical *in vivo* and the ability to detect this spin trapped adduct in a living animal in real time. This is the first example of the detection of hydroxyl radical in a living animal in real time. Nevertheless, the in vivo *in situ* identification of other free radicals has begun ***(17,19)*** and the future is even more promising. With advances in the design of lower frequency EPR spectrometers and new spin traps, in vivo *in situ* identification of free radicals in real time is within reach.

Acknowledgment

This research was supported in part from a grant from the National Institutes of Health, CA-69538.

References

1. Gilbert, D. L. (1981) Perspective on the History of Oxygen and Life, in *Oxygen and Living Processes. An Interdisciplinary Approach* (Gilbert, D. L., ed.), Springer-Verlag, NY, pp 1–43.
2. Grubbé, E. H. (1933) Priority in the therapeutic use of X-rays. *Radiology* **21**, 156–162.
3. Gerschman, R., Gilbert, D. L., Nye, S. W., Dwyer, P., and Fenn, W. O. (1954) Oxygen poisoning and x-irradiation: a mechanism in common. *Science* **119**, 623–626.
4. McCord, J. M. and Fridovich, I. (1968) The reduction of cytochrome c by milk xanthine oxidase. *J. Biol. Chem.* **243**, 5753–5760.
5. McCord, J. M. and Fridovich, I. (1969) Superoxide dismutase: an enzymic function for erythrocuprein (hemocuprein). *J. Biol. Chem.* **244**, 6049–6055.
6. Fridovich, I. (1978) The biology of oxygen radicals. *Science* **201**, 875–880.
7. Guengerich, F. P. and Macdonald, T. L. (1984) Chemical mechanisms of catalysis by cytochrome P-450: a unified view. *Acc. Chem. Res.* **17**, 9–16.

8. Babior, B. M., Kipnes, R. S., and Curnutte, J. T. (1973) Biological defense mechanisms: the production by leukocytes of superoxide, a potential bactericidal agent. *J. Clin. Invest.* **52**, 741–744.
9. Moncada, S. and Higgs, A. (1993) The L-arginine-nitric oxide pathway. *N. Engl. J. Med.* **329**, 2002–2012.
10. Janzen, E. G. and Blackburn, B. J. (1968) Detection and identification of short-lived free radicals by an electron spin resonance trapping technique. *J. Am. Chem. Soc.* **90**, 5909–5910.
11. Mackor, A., Wajer, Th. A. J. W., and de Boer, Th. J. (1967) C-Nitroso compounds. Part III. Alkoxy-alkyl-nitroxides as intermediates in the reaction of alkoxyl-radicals with nitroso compounds. *Tetrahedron Lett.* 385–390.
12. Iwamura, M. and Inamoto, N. (1970) Reactions of nitrones with free radicals. II. Formation of nitroxides. *Bull. Chem. Soc. Japan* **43**, 860–863.
13. Janzen, E. G. (1971) Spin trapping. *Acct. Chem. Res.* **4**, 31–40.
14. Finkelstein, E., Rosen, G. M., and Rauckman, E. J. (1980) Spin trapping of superoxide and hydroxyl radical: practical aspects. *Arch. Biochem. Biophys.* **200,** 1–16.
15. Janzen, E. G. (1980) A critical review of spin trapping in biological systems, in *Free Radicals in Biology,* (Pryor, W. A., ed.), vol. 4, Academic Press, NY, pp. 116–154.
16. Halpern, H. J., Spencer, D. P., vanPolen, J., Bowman, M. K., Nelson, A. C., Dowey, E. M., and Teicher, B. A. (1989) Imaging radio frequency electron-spin-resonance spectrometer with high resolution and sensitivity for *in vivo* measurements. *Rev. Sci. Instrum.* **60**, 1040–1050.
17. Lai, C.-S. and Komarov, A. M. (1994) Spin trapping of nitric oxide produced *in vivo* in septic-shock in mice. *FEBS Letts.* **345**, 120–124.
18. Halpern, H. J., Yu, C., Barth, E., Peric, M., and Rosen, G. M. (1995) *In situ* detection, by spin trapping of hydroxyl radical markers produced from ionizing radiation in the tumor of a living mouse. *Proc. Natl. Acad. Sci. USA* **92**, 796–800.
19. Jiang, J. J., Liu, K. J., Shi, X., and Swartz, H. M. (1995) Detection of short-lived free radicals by low-frequency electron paramagnetic resonance spin trapping in whole living animals. *Arch. Biochem. Biophys.* **319**, 570–573.
20. Poyer, J. L. and McCay, P. B. (1971) Reduced triphosphopyridine nucleotide oxidase-catalyzed alterations of membrane phospholipids. IV: dependence on Fe^{+3}. *J. Biol. Chem.* **246**, 263–269.
21. Gutteridge, J. M. C. (1987) A method for removal of trace iron contamination from biological buffers. *FEBS Lett.* **214**, 362–364.
22. Buettner, G. R. (1988) In the absence of catalytic metals ascorbate does not autoxidize at pH 7: ascorbate as a test for catalytic metals. *J. Biochem. Biophys. Methods* **16**, 27–40.
23. Buettner, G. R., Oberley, L. W., and Leuthauser, S. W. H. C. (1978) The effect of iron on the distribution of superoxide and hydroxyl radicals as seen by spin trapping and on the superoxide dismutase assay. *Photochem. Photobiol.* **28**, 693–695.

24. Bottomley, P. A., and Andrew, E. R. (1978) RF filed penetration, phase shift, and power dissipation in biological tissues: implications for NMR imaging. *Phys. Med. Biol.* **23**, 630–643.
25. Johnson , C. C. and Guy, A. W. (1972) Nonionizing electromagnetic wave effects in biological materials and systems. *IEEE* **60**, 692–718.
26. Roschmann, P. (1987) Radiofrequency penetration and absorption in the human body: limitations to high-field whole-body nuclear magnetic resonance imaging. *Med. Phys.* **14**, 922–928.
27. Halpern, H. J., Pou, S., Peric, M., Yu, C., Barth, E., and Rosen, G. M. (1993) Detection and imaging of oxygen-centered free radicals with low-frequency electron paramagnetic resonance and signal-enhancing deuterium containing spin traps. *J. Am. Chem. Soc.* **115**, 218–223.
28. Pou, S., Cohen, M. S., Britigan, B. E., and Rosen, G. M. (1989) Spin trapping and human neutrophils: Limits of detection of hydroxyl radical. *J. Biol. Chem.* **264**, 12299–12302.
29. Pou, S., Rosen, G. M., Wu, Y., and Keana, J. F. W. (1990) Synthesis of deuterium- and ^{15}N-containing pyrroline 1-oxides: a spin trapping study. *J. Org. Chem.* **55**, 4438–4443.
30. Pou, S., Ramos, C.L., Gladwell, T., Renks, E., Centra, M., Young, D., Cohen, M. S., and Rosen, G. M. (1994) A kinetic approach to the selection of a sensitive spin trapping system for the detection of hydroxyl radical. *Anal. Biochem.* **217**, 76–83.
31. Ramos, C. L., Pou, S., Britigan, B. E., Cohen, M. S, and Rosen, G. M. (1992) Spin trapping evidence for myeloperoxidase-dependent hydroxyl radical formation by human neutrophils and monocytes. *J. Biol. Chem.* **267**, 8307–8312.

4

Single Photon Counting

Yoshinori Mizuguchi

1. Introduction

Recently the demand for the measurement of very low-light levels has been increasing in the fields of biology, chemistry, medicine, and many others; for instance, analytical chemistry needs to employ fluorescence, luminescence and Raman scattering light for the quantitative operation of substance. Biochemistry calls for the measurement of very low-light level to observe gene expression *(1,2)* and the activity of living cell *(3,4)*, etc. It also calls for radioimmunoassay and fluorescent immunoassay. Medicine asks for a positron computerized tomography (CT) to research a brain physiologically and diagnose cancers. In any case, the absolute levels of light to be detected are very low in every field, so that light that changes with time at a very fast speed can be captured.

The single photon-counting method counts photons of light individually, which gives rise to higher stability and detectability, and also has the advantage of the obtaining a superior signal-to-noise (S/N) ratio than the measurement of direct current (DC) signals from detectors.

The principle of single photon counting is as follows. When incident light enters the detector's sensitive area, i.e., a photocathode, photoelectrons are emitted from photocathode. These photoelectrons then travel through the electron multiplier, where electrons are multiplied by the process of secondary emission, and finally makes a pulse-shape signal as an output. If the intensity of incident light to the detector is changed, the output signal becomes as shown in **Fig. 1**. It is apparent from the figure that the pulse intervals become very narrow in the regions of both high- and low-level light, where the output pulse overlap with the analog pulse, i.e., DC signal is output (*see* **Fig. 1A** and **B**). At this light level, the measurement of the DC-signal level is related to the intensity of the

From: *Methods in Molecular Biology, vol. 108: Free Radical and Antioxidant Protocols*
Edited by: D. Armstrong © Humana Press Inc., Totowa, NJ

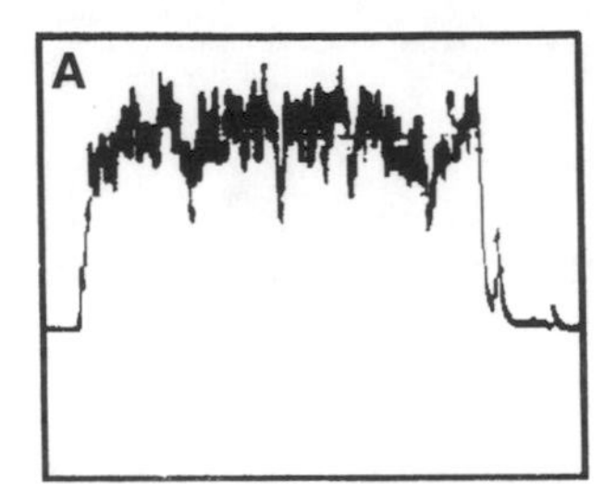

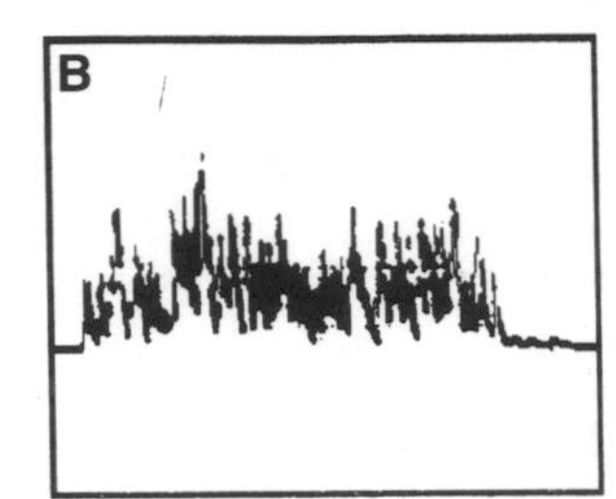

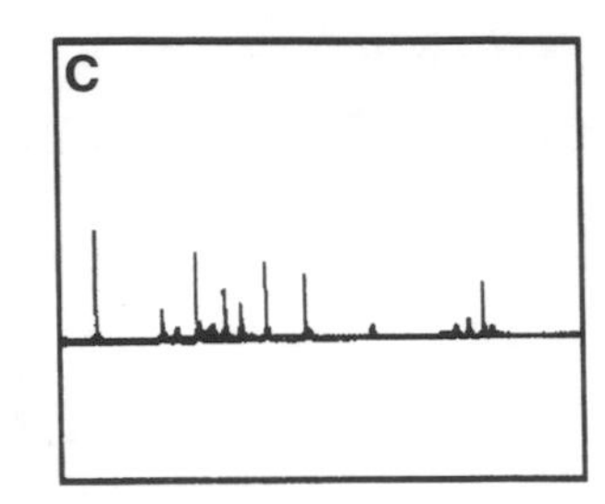

Fig. 1. Output pulses at difference light levels: (**A**) High, (**B**) Low, (**C**) Very Low.

incident light. However, as the intensity of the incident light becomes weak, pulse-shape signal substances (dispersion) increase and finally become discrete pulses (*see* **Fig. 1C**). At this light level, the measurement of the DC-signal level is no longer related to the intensity of the incident light, but the number of the pulses per time is related the intensity of the incident light.

The method of processing output signals can be divided into analog mode and digital mode. The measurement of these discrete pulses is called digital-mode measurement, which is also called the photon-counting mode in general. The process is called single-photon counting because the individual photon signal can be detected. The measurement of DC-signal output at the high-light level is called analog-mode measurement.

In the analog mode, the mean value containing alternating current (AC)-signal substances of **Fig. 1A** becomes a DC signal. On the other hand, the photon-counting mode discriminates a pulse into binary signal, which is then counted. The photon-counting mode therefore has many advantages and it is considerably effective, especially in the region of extremely low-light levels.

In order to perform the single-photon counting, a photomultiplier tube (PMT) is commonly used as a detector. Recently, in order to obtain the image of the photon signal from a sample, the two-dimensional photon-counting tube became usable as a detector. This chapter is intended for those who want to familiarize themselves with how to perform single-photon counting using PMT and the two-dimensional photon-counting tube. In this chapter, the operational principles and the circuit configuration of these devices, as well as some notes, are summarized. In addition, selection guides for these devices are discussed *(5)*.

2. Materials

2.1. PMT

Among the photosensitive devices in use today, a PMT is a versatile device that provides extremely high sensitivity and ultra-fast response. **Figure 2** shows a basic configuration of the photon-counting system. Because signal pulses

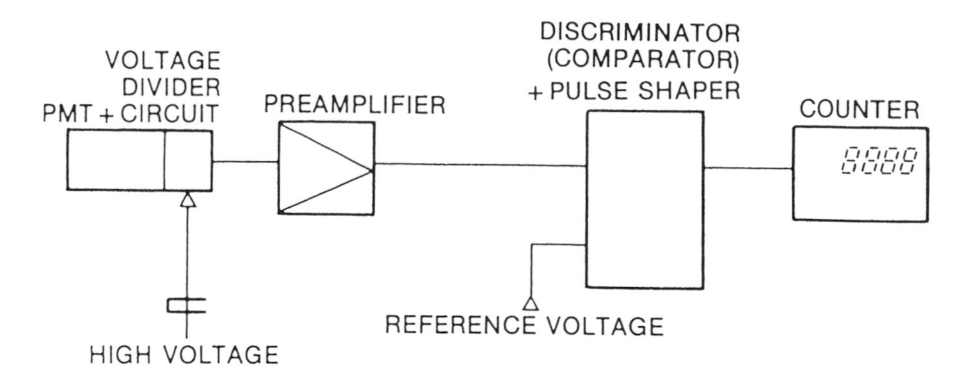

Fig. 2. Basic configuration of photon-counting system.

from PMT are generally small, to amplify those pulses, a preamplifier is usually placed near the PMT for photon counting. Followed by the preamplifier, a circuit discriminator is placed to select out pulses whose height is higher, or lower, than a certain height. Output pulses from the discriminator are then shaped by a pulse shaper into pulses of constant width and height. They are finally counted by a counter.

2.1.1. PMT

A typical PMT consist of a photo-emissive cathode (photocathode) followed by focusing electrodes, an electron multiplier, and an electron collector (anode) in a vacuum tube, as shown in **Fig. 3**.

When light enters the photocathode, the photocathode emits photo-electrons into the vacuum. These photo-electrons are then directed by the focusing electrode voltages towards the electron multiplier, where electrons are multiplied by the process of secondary emission. The multiplied electrons are collected by the anode as an output signal.

Because of secondary-emission multiplication, PMT provides extremely high-sensitivity, exceptionally low noise among the photosensitive devices currently used to detect radiant energy in the ultraviolet (UV) and near infrared (IR) region. PMT also features fast time response, low noise, and choice of large photosensitive areas.

The section describes the prime features of PMT construction and basic operating characteristics *(6)*.

The photocathode of a PMT converts energy of incident light into photoelectrons. The conversion efficiency (photocathode sensitivity) varies with the wavelength of the incident light. This relationship between photocathode sensitivity and the wave-length is called the spectral-response characteristic.

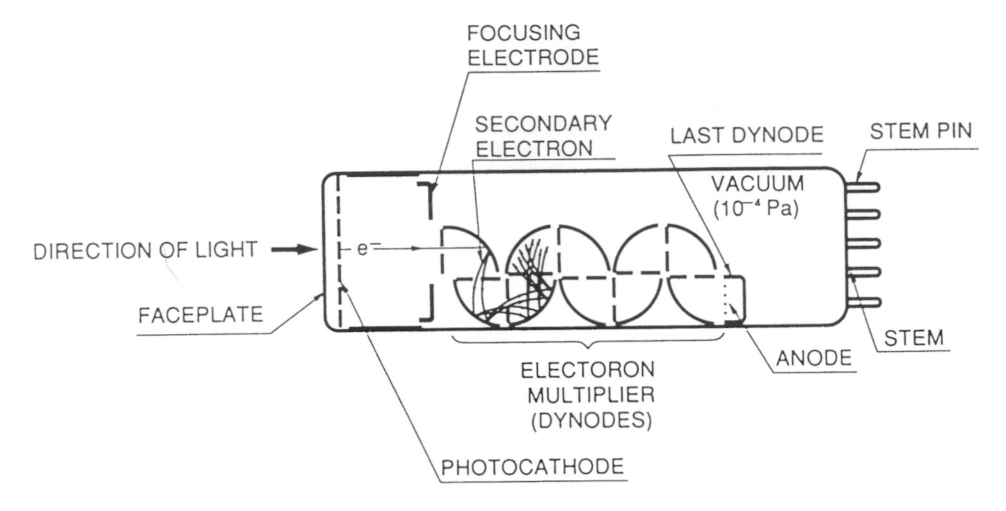

Fig. 3. Cross-section of head-on type PMT.

Figure 4 shows typical spectral response of bialkali PMT. The spectral-response characteristics are determined on the long-wavelength side by the photocathode material and on the short-wavelength side by the window material. Actual data may be different from type to type.

As **Fig. 4** shows, spectral response is usually expressed in terms of radiant sensitivity or quantum efficiency as a function of wavelength. Radiant sensitivity (S) is the photoelectric current from the photocathode, divided by the incident radiant power at a given wavelength, expressed in amperes per watt (A/W). Quantum efficiency (QE) is the number of photo-electrons emitted from the photocathode divided by the number of the incident photons. It is customary to present QE in a percentage. QE and the radiant sensitivity have the following relationship at a given wavelength:

$$QE = (S \times 1240)/(\lambda) \times 100\ (\%)$$

where S is the radiant sensitivity in A/W at the given wavelength, and λ is the wavelength in nm (nano meters).

A low-noise electron multiplier, which amplifies electrons by a cascade secondary electron-emission process provides superior sensitivity (high current amplification and high S/N) of PMTs. The electron multiplier consists of up to about 16 stages of electrodes called dynodes (*see* **Fig. 3**). The photo-electron is multiplied up to 10^6–10^7.

Though secondary electrons at first dynode come as a few discrete pulses with respect to primary electron, they are regarded conjoining to Poisson distribution, and the mean value of this distribution becomes a current amplifica-

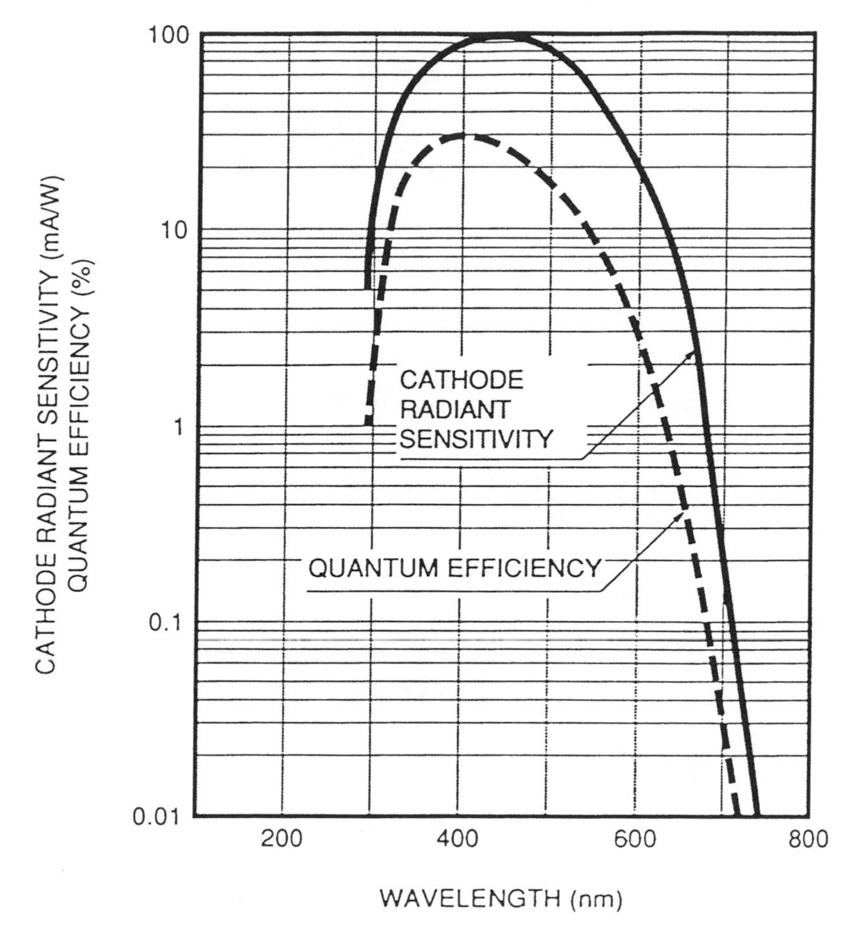

Fig. 4. Typical spectral response of head-on, bialkali photocathode.

tion (gain) σ of PMT. This holds true for further multiplication processes from the second dynode to the others. A set of the last electron group finally appears as an output to the anode with respect to a single photon from the photocathode. In a PMT that has n stage of dynodes, a single photoelectron from the photocathode is multiplied by σ^n as a mean, and the group of electrons makes a pulse-shape output. Because of this, the gain of secondary electrons has variance, i.e., statistical probability, and therefore the value of output pulse height at the anode cannot be aligned (*see* **Fig. 5**).

Furthermore, pulse heights at the anode are non-uniform owing to the non-uniformity of current amplifications caused by position difference of each dynode where electrons strike, and also owing to deviated electrons from the

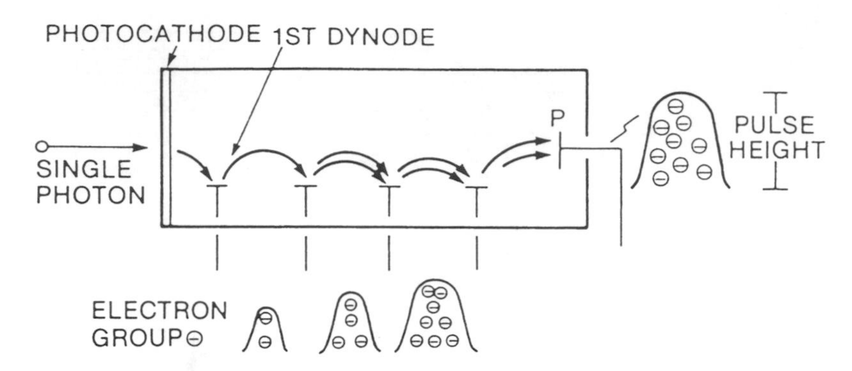

Fig. 5. Electron multiplication process by photon counting mode.

normal multiplication process. Heights of the pulse from the anode plotted in a histogram are called a pulse-height distribution (PHD). Some examples are illustrated in **Fig. 6** which illustrates anode pulse having different height dispersed in time, integrated into a PHD. The abscissa of the figure represents pulse heights, the number of electrons in a group, or the value of the pulse voltage or current produced by that electron group. The PHD curve shows the distribution of the value of voltage or current of the output of PMT.

2.1.2. Power Supply

The output of a PMT is extremely sensitive to the applied voltage by power supply between the photocathode and the anode through the dynodes. Even small variation in applied voltage greatly affect measurement accuracy. Thus, a highly stable source of high voltage is required. (*see* **Fig. 7**).

2.1.3. Preamplifer

Output pulses from a PMT in the single photon-counting mode are very low, and it is therefore necessary to exercise great care in the design of preamplifier and other circuits.

2.1.4. Counter

There are many types of counters, including the power supply, preamplifier, pulse shaper, and discriminator. There are many types of functions from simple counting to time-resolved photometry and computer control, etc. A proper counter can be selected for an application.

2.2. Two-Dimensional Photon-Counting Tubes

Recently, the demand for obtaining an image under very low-light levels has been increasing, especially in the field of biology; for example in gene expression

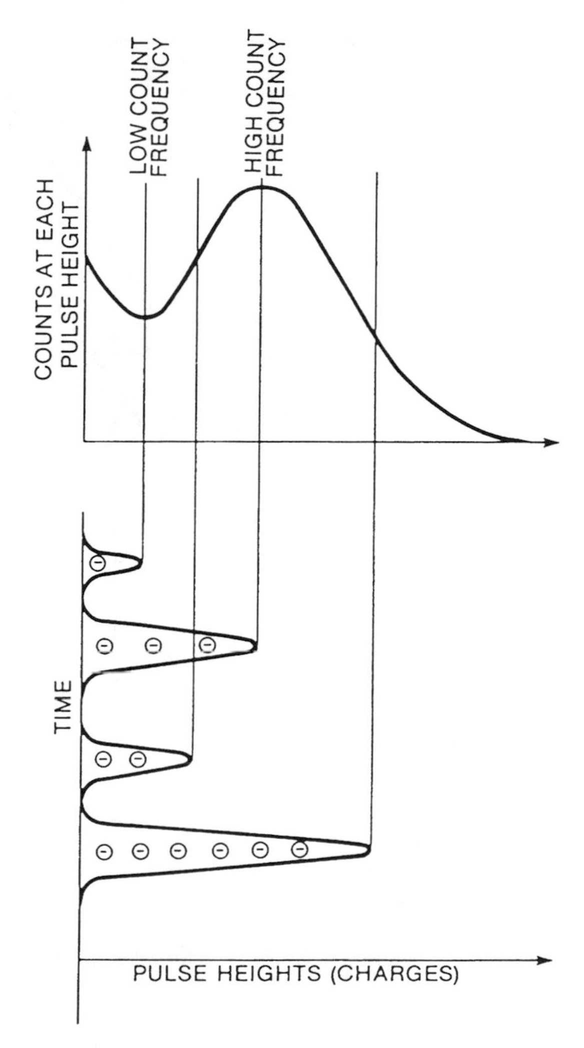

Fig. 6. PMT output and pulse height distribution.

(1,2), and the activity assay of living cell ***(3,4)***, etc. According to such demand, several types of image sensors have been developed, "image intensifier" and "cooled charge coupled device (CCD)." For the purpose of the single-photon counting, the image-intensifier type is more appropriate because of the extremely low noise character compared to the CCD, and such an image intensifier is called a two-dimensional photon-counting tube. It is employed as one part of video camera to get an image under the very low-light level *(7)*.

Just like a PMT, a two-dimensional photon-counting tube consist of a photo-emissive cathode (photocathode) followed by focusing electrodes, and an electron multiplier in a vacuum tube, as shown in **Fig. 8**. The differences are the

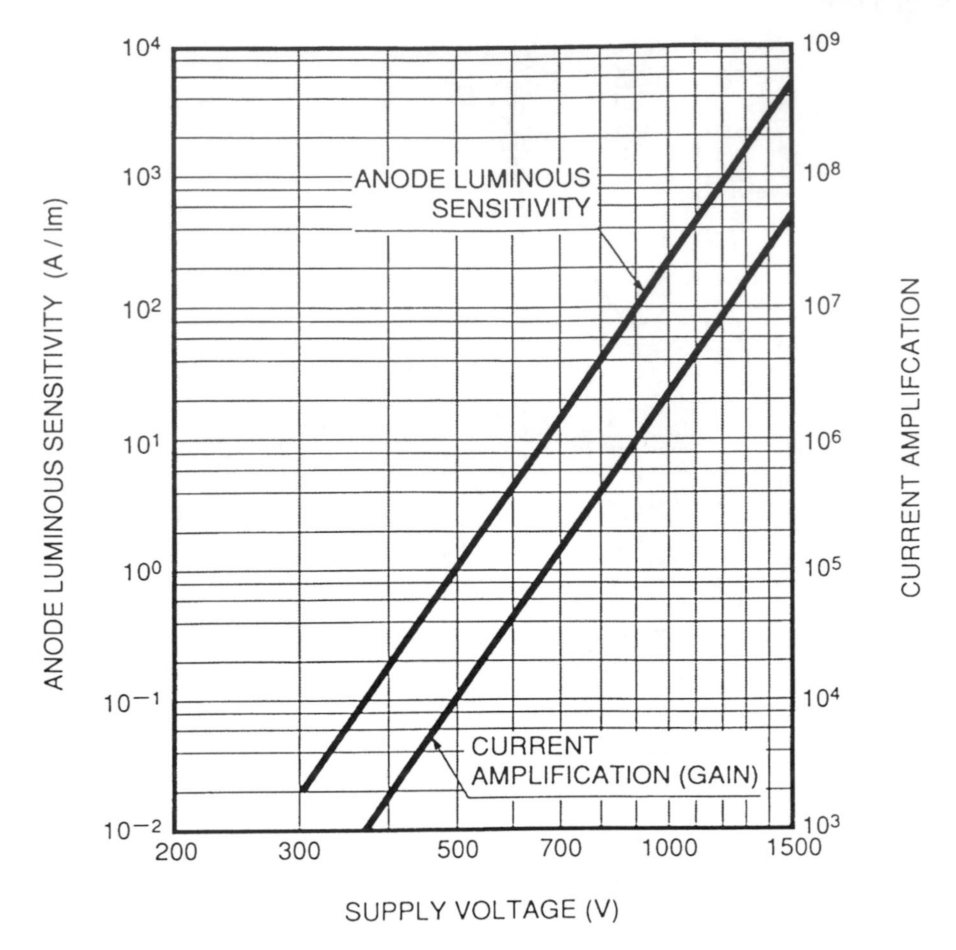

Fig. 7. Example of the current amplification vs supply voltage.

phosphor screen and two or three stages of micro channel plate (MCP) employed in two-dimensional photon-counting tube instead of an electron corrector and dynodes in PMT, respectively.

When incident light enters the photocathode, photo-electrons are emitted from the photocathode. These photo-electrons are then directed by the focusing electrode voltages towards the MCP, where electrons are multiplied to about 10^6 times by the process of secondary emission. The multiplied electrons are bombarded to the phosphor screen and emit light as an output signal. Because the relationship of the positions is maintained from where photo-electrons are first generated to where at last light signal is generated, the light image that is projected on the photocathode is obtained as an intensified image

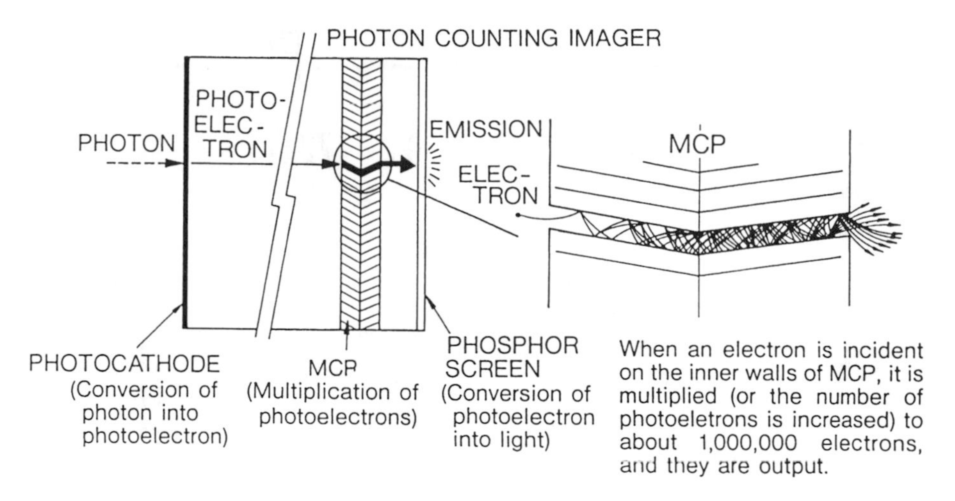

Fig. 8. Operating principle of two-dimensional photon-counting tube.

on the phosphor screen. Therefore, this type of device is called an "image intensifier." The two-dimensional photon-counting tube is a type of the image intensifier that has a high magnitude of multiplication with low noise.

As an important characteristics, a two-dimensional photon-counting tube also has a clear discriminator of the peak and valley in the PHD, the same as in PMT; then it can be used for photon counting (*see* **Fig. 9**). An ordinary two-dimensional photon-counting tube is coupled with a video camera through the optics system, and the output image from the phosphor screen is read by a video camera to get an image.

The video-camera system using two-dimensional photon-counting tube is shown in **Fig. 10**.

3. Methods

3.1. Photon Counting on PMTs

3.1.1. Operation of Photon Counting

Because PMT output contains a variety of noise pulses in addition to the signal pulse representing photo-electron as shown in **Fig. 11**, simply counting pulses without some form of noise elimination will not result in accurate measurement. The most effective approach to noise elimination is to investigate the height of the output pulses.

A typical PHD for the output of PMTs is analyzed as shown in **Fig. 12**. In this PHD, the lower-level discrimination (LLD) is set at the valley through and the upper-level discrimination (ULD) at the foot, where the output pulses are very

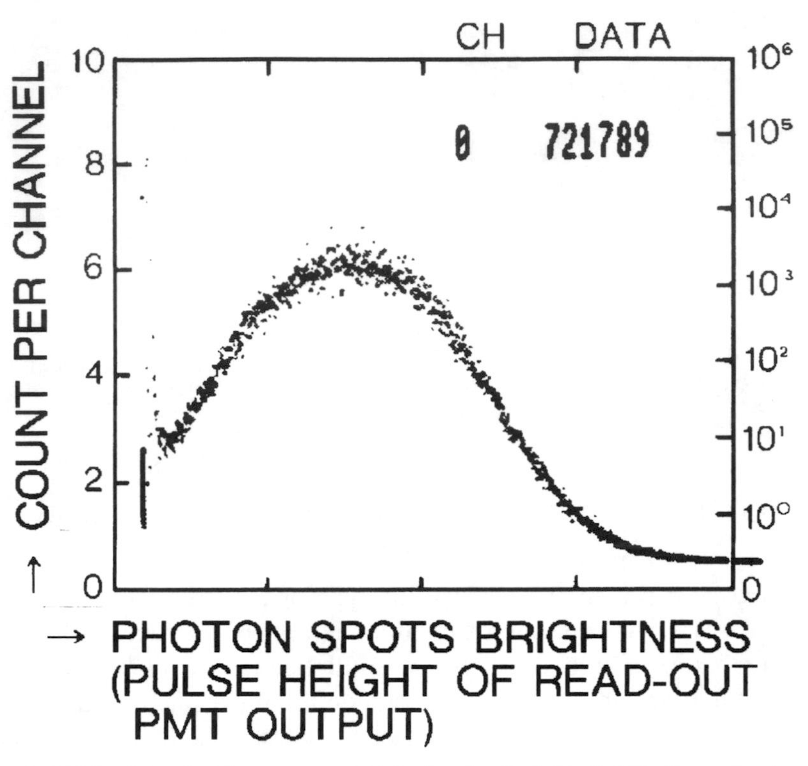

Fig. 9. PHD of single photon spot (typical pulse-height distribution of output spot on output phospho screen).

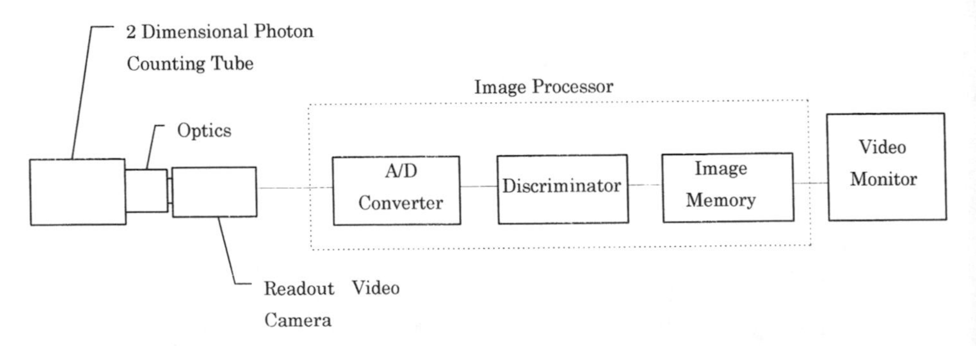

Fig. 10. The block diagram of the video camera system.

few. Most pulses lower than LLD are noise and pulses higher than ULD result from cosmic rays, etc. Therefore, by counting pulses between the LLD and ULD, accurate light measurements become possible. It is recommended that the LLD be set at 1/3 of the mean height of the pulses (Hm) and the ULD at three times Hm. In most cases, however, the ULD setting can be omitted.

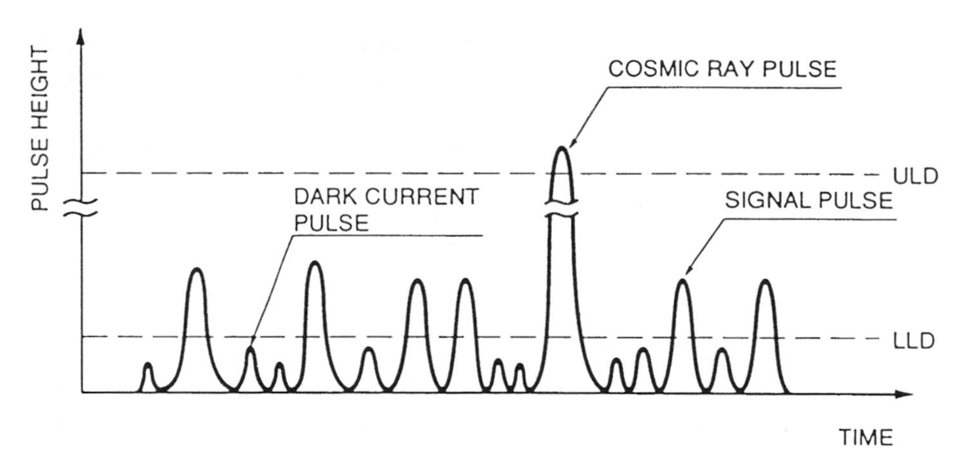

Fig. 11. Output pulse and discrimination level.

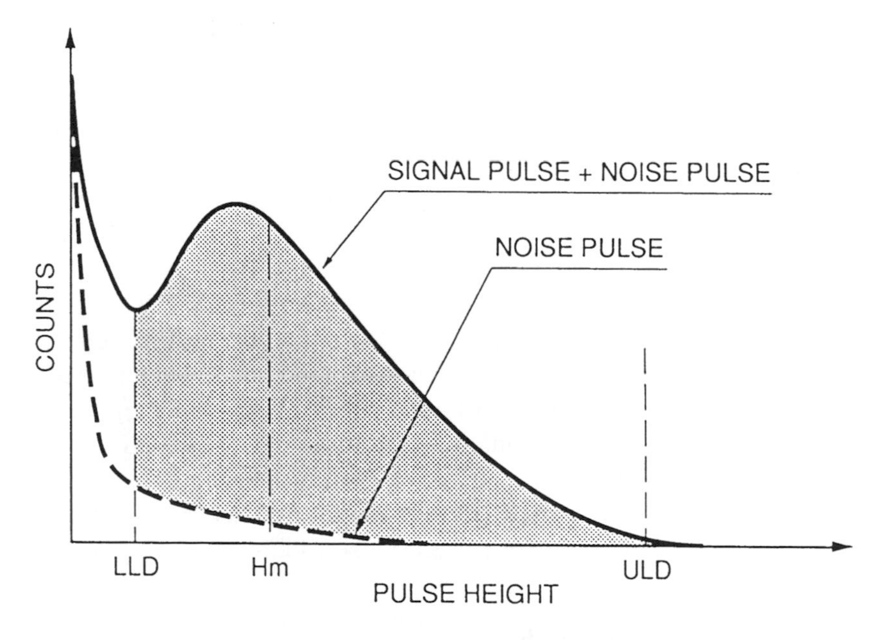

Fig. 12. Typical pulse-height distribution.

Considering the aforementioned, a clear definition of the peak and valley in the PHD is a very significant characteristic for multiplier tubes for use in photon counting. Because of this, such types of PMTs are employed for use in single-photon counting, having a high magnitude of multiplication and low noise among the PMTs.

From the operational principle of photon counting, the photon-counter system and the multichannel pulse height analyzer system have been employed. **Figure 13** shows the actual circuit configurations of each system and the pulse shapes obtainable from circuits of each system. The photon-counter system of **Fig. 13A** shows that output pulse from PMT are amplified by the preamplifier and further amplified by the main amplifier, if necessary. Those amplified pulses are then directed toward the discriminator, for which a comparator integrated circuit (IC) is usually used. They are then compared with the pre-set reference voltage and discriminated into two groups of pulse. One group is lower and the other is higher than the reference voltage. The lower pulses are finally eliminated by the LLD and the higher pulses are eliminated by the ULD. The output of the comparator takes place at a constant level (usually a transistor transistor logic [TTL] level from 0–5 V, and the output of a high-speed comparator takes place at an emitter-coupled logic [ECL] level from –0.9 to –1.7). The counter connected followed by the pulse shaper counts the number of those discriminated pulses only.

In contrast to the photon-counter system, in the multichannel pulse-height analyzer system shown in **Fig. 13B**, output pulses from PMT are generally integrated through a charge-sensitive-type preamplifier, amplified, and shaped by the linear amplifier. These pulses are then discriminated by the discriminator. These discriminated pulses are in turn converted from analog to digital. They are finally accumulated by memory and displayed on the screen.

As shown in the figure, this system is able to output pulse-height information and frequency (the number of counts) simultaneously.

The photon-counter system is used to measure the number of output pulse from PMT corresponding to incident photons, whereas the multichannel pulse-height discriminator system is used to measure the height of each output pulse and the number of output pulses simultaneously. The former system has a superior feature of high-speed counting operation and so is used for general purpose applications. Although the latter system has the disadvantage of obtaining high counts, it is used for the application where a pulse-height analysis is required; for example, radiation scintillation counting and multi-photon analysis.

3.1.2. Setting the Discrimination Level

In the photon-counting mode, it is important to know where to set the discrimination level.

To optimize the S/N and the gain change rate, set a discrimination level near the valley of a PHD curve or somewhere on a plateau of curve. Setting a discrimination level at as low a position as possible provides high counting efficiency, but in reality it should be set near the valley of the PHD curve, somewhere on a plateau of the plateau curve of the PMT. To eliminate noise

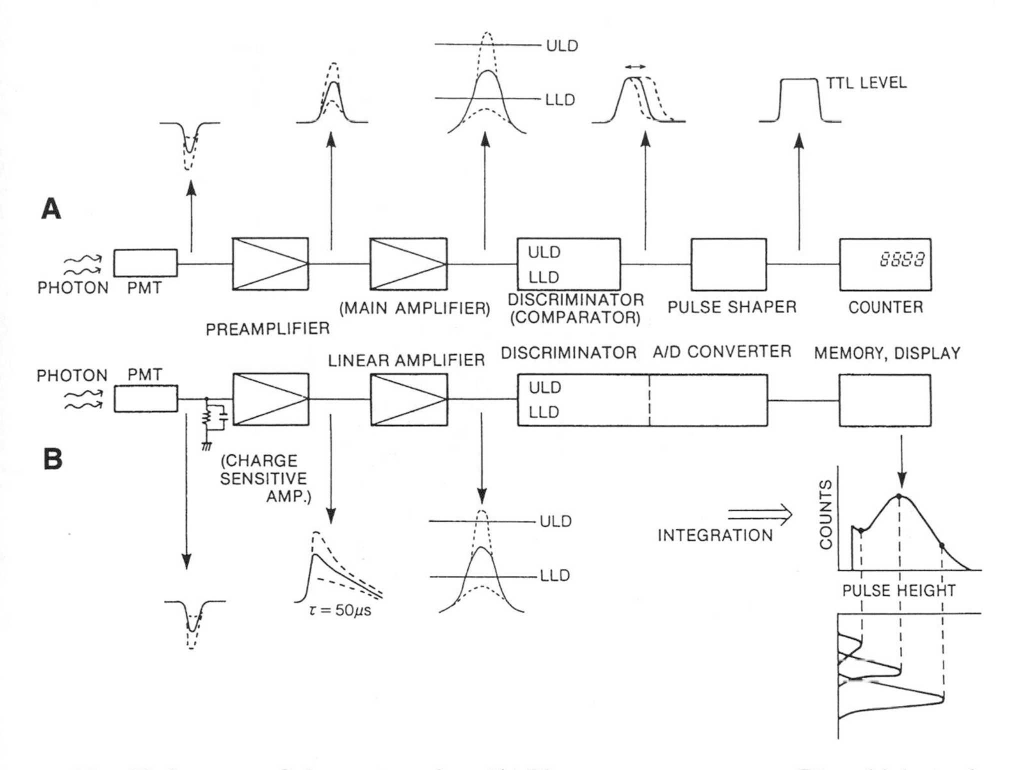

Fig. 13. Systems of photon counting: (**A**) Photon counter system, (**B**) multichannel pulse-height analyzer system.

against signal, set the discrimination level at the middle point between the peak and the valley of a PHD curve.

It is not easy to define a single position discrimination level, so it is necessary to set a discrimination level according to the weight of the aforementioned factors (*see* **Notes 1–4**).

3.1.3. Noise

A very small amount of noise pulses exist in a PMT, even when it is kept in complete darkness. This becomes an important factor in determining the lower detection limit of a PMT. Noises depend upon the supply voltage. Thus, it is suggested that the supply voltage should be as low as possible, if the amplification of the circuits permits. The origin of noise are as follows:

1. Emission of thermal electrons from photocathode and dynodes.
2. Glass scintillations from the glass bulb of a PMT.
3. Leakage current owing to insulation failures of the socket and stem of a PMT.
4. Noise by strong electric field if a PMT is operated at a voltage above the maximum-rated voltage.

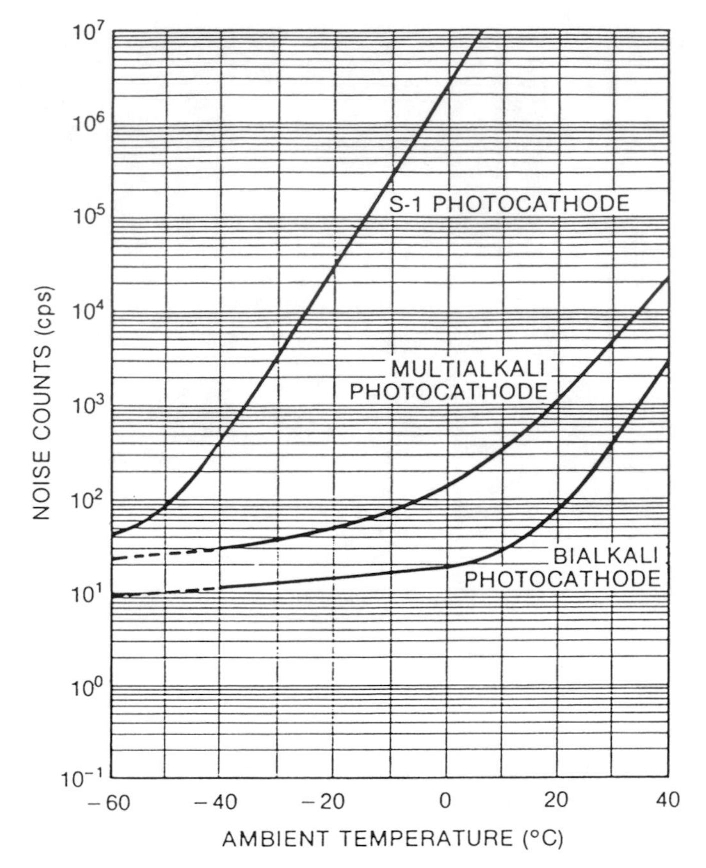

Fig. 14. Temperature characteristics of noise pulse.

5. As an example of external noise, **Fig. 14** shows the temperature characteristics of noise pulse of various photocathodes. The temperature characteristic of PMTs vary with respect to photocathode size and sensitivity, especially red-light sensitivity, etc.

3.2. Photon Counting on Two-Dimensional Photon Counting Tubes

In order to perform photon counting, the same process as with a PMT is employed. The process is shown in **Fig. 15**. The incident photons are multiplied in the tube and turned into bright spots on the phosphor screen (**Fig. 15A**). The intensity of the video signal containing the bright spots are converted to the digital signal and discriminated by a proper discrimination level (**Fig. 15B**). At this time, the spots higher than the discrimination level have the

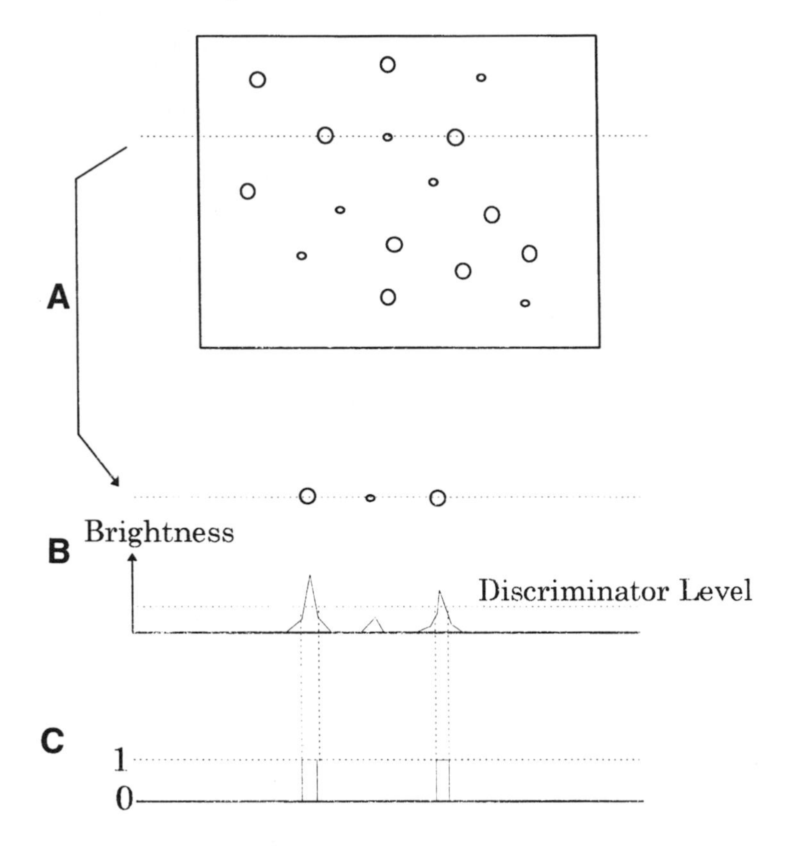

Fig. 15. The process in the video camera system. (**A**) Signal from video camera. (**B**) Discrimination of signal. (**C**) The result of discrimination.

value of 1, whereas the spots lower than the discrimination level have the value of 0 (**Fig. 15C**). Only the positions of the spots that have a value of 1 are stored in the image memory.

3.3. Examples of the Applications

Photomultiplier tubes are widely used as photon-counting detectors; for example, luminometers, spectrometers, and radioactivity detectors (such as scintillation counter and positron CT, etc.).

According to the progress of the image-processing method, two-dimensional photon-counting tubes become useful in biomedical field; for example, gene expression *(1,2)*, the activity assay of living cell *(3,4)*, and the use of a two-dimensional luminometer to analyze samples simultaneously.

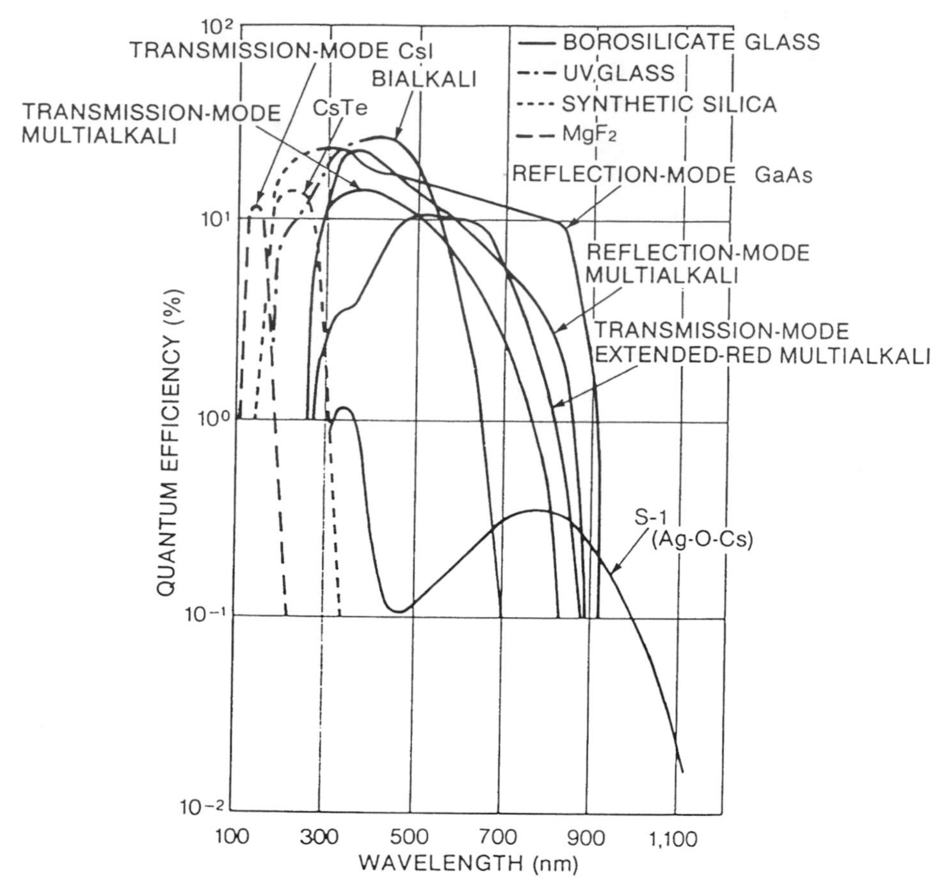

Fig. 16. Spectral response characteristics of typical photocathode.

4. Notes

1. Selection guide: Because there are many types of detectors, it is important to select a proper detector for individual application, based on materials such as data sheets supplied from the manufacturers. The basic characteristics to be considered are the spectral response, the noise characteristics, the magnitude of the multiplication of the signal, etc. The spectral-response characteristics of typical photocathodes are shown in **Fig. 16**.
2. PHD and plateau characteristics: The S/N ratio is one of the most important factors. Here, the S/N is defined as the ratio of the mean value of signal counts to the dispersions of signal and noise counts. The S/N curve shown in **Fig. 17** is the result of how the S/N changes when the supply voltage is changed just as the plateau characteristics are obtained. From a S/N point of view, it is understandable that the supply voltage to the electron multiplier should be set between the voltage (V0), which is the starting point of the plateau region, and the maximum rating of power supply voltage.

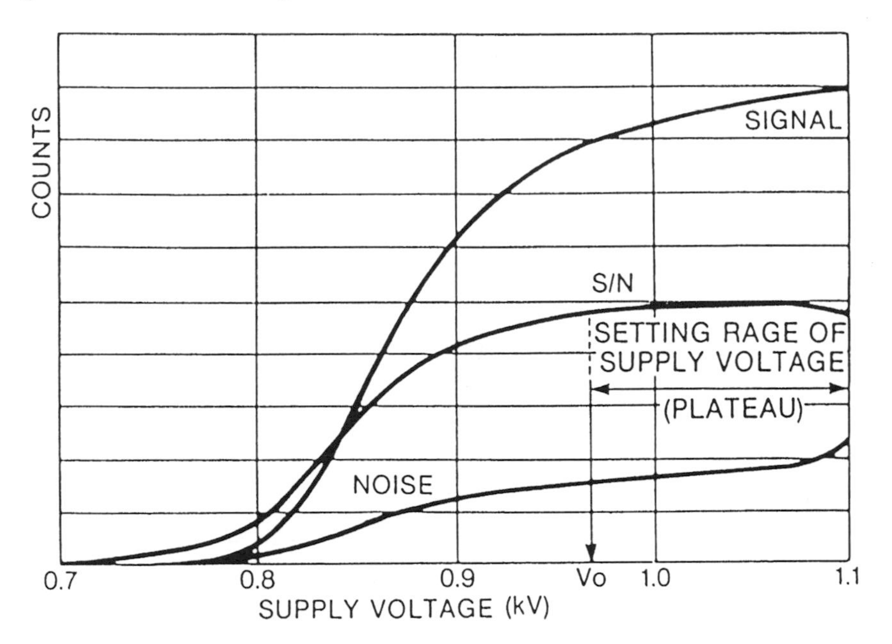

Fig. 17. An example of S/N curve of PMT.

3. Optimizing the S/N: **Fig. 18** shows the signal, the noise counts (integrated value); and the characteristics of the ratio obtained when the discrimination level was changed. This tell us that by setting the discrimination level L2 at about 200ch (which corresponds to near the peak channel), it is possible to optimize S/N ratio. In contrast, the analog mode does not have the capability of pulse-height discrimination and therefore cannot take advantage of such a feature.
4. Linearity of Count Rate: By nature, the photon-counting mode generally used in the region of the very low-light level, and it has an excellent linearity when the count rate is low. However, the count rates becomes nonlinear when the photon-counting mode is performed in the region under the large amounts of incident light. A PMT itself has a wide band width, from 30 up to 300 MHz (which varies depending upon the type of PMT used). This means that the maximum count rate in the photon-counting mode is determined mainly by the time resolution of an electric circuit connected to the PMT.

   Because the time-interval distribution of photons that reach the PMT is a Poisson distribution, the probability of counts (P0) by pulse overlaps, with respect to random events (which corresponds to the normal photon counting), can be obtained using the following formula:

$$\mathrm{P0} = 1 - \exp(-RT)$$

   where T is resolving time and R is average count rate. **Figure 19** shows the comparison between the measured value and the result of the formula above. It can be seen that they match very well.

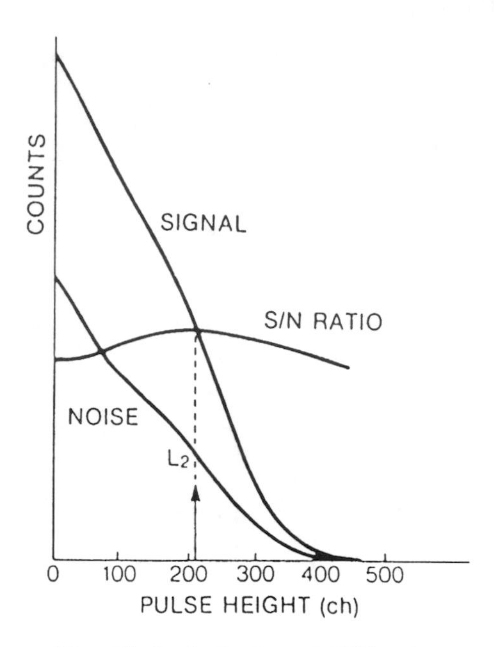

Fig. 18. Optimization of S/N ratio.

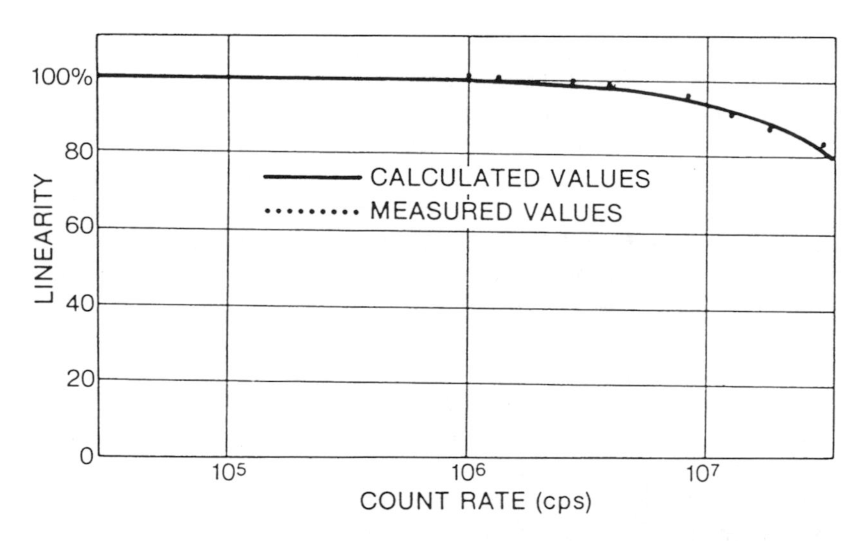

Fig. 19. Linearity of count rate.

Thus, when the count rate is high, it is necessary to reduce them to below the maximum count rate, considering the resolution time of the circuit parameter.

Similarly to PMT, a two-dimensional photon-counting tube has nonlinear characteristics when the count rate is high. Moreover when incident photons are spatially concentrated at a region, a two-dimensional photon-counting tube has nonlinear characteristics, because the output photon spots overlap.

Acknowledgments

I thank Dr. Tadahisa Hiramitsu for reviewing this chapter. I also thank Mr. Kiyoshi Kamiya and Mr. Toshihiro Kume for insight and helpful advice.

References

1. Inoue, S. (1992) Imaging of luciferase secretion from transformed Chinese hamster ovary cells. *Proc. Natl. Acad. Sci. USA* **85,** 9584–9587.
2. Tamiya, E. (1990) Spatial imaging of luciferase gene expression in transgenic fish. *Nucleic Acids Res.* **18(4),** 1072.
3. Masuko, M. (1991) Rapid detection and counting of single bacteria in a wide field using a photon counting TV camera. *FEMS Microbiol. Lett.* **83,** 231–238.
4. Suematsu, M. (1987) Real-time visualization of oxyradical burst from single neutrophil by using ultra-sensitive video intensifier microscopy. *Biochem. Biophys. Res. Comm.* **149(3),** 1106–1110.
5. *How To Perform Photon Counting Using Photomultiplier Tubes* (1993) Hamamatsu Photonics K.K., 325–6, Sunayama-cho, Hamamatsu City, 430, Japan.
6. *Photomultiplier Tubes* (1995) Hamamatsu Photonics K.K., 325-6, Sunayama-cho, Hamamatsu City, 430, Japan.
7. *Image Intensifiers* (1992) Hamamatsu Photonics K.K., 325-6, Sunayama-cho, Hamamatsu City, 430, Japan.

5

Measurement of Reactive Oxygen Species in Whole Blood and Mononuclear Cells Using Chemiluminescence

Kuldip Thusu, Ehad Abdel-Rahman, and Paresh Dandona

1. Introduction

Free radicals generated by a wide variety of processes, such as ionizing radiation *(1)*, toxic xenobiotics *(2)*, inflammation *(3)*, and metabolites of membrane lipid transformation *(4)*, are involved in various diseases *(5–9)*. A number of disorders in cellular and organ systems have been attributed to them. Free radicals damage lipids *(7)*, proteins *(10)* and deoxyribonucleic acid (DNA) *(11,12)* with consequent effects ranging from cell death to neoplasia *(13)*. It is therefore important to measure the amount of free radicals generated by various cell and organ types. Traditionally, one would measure free-radical generation by electron spin resonance, which is costly or complicated, or measure antioxidants that scavenge free radicals to get an estimate of the oxidative state.

We have developed a simple, sensitive, and reproducible method of measuring free radicals in whole blood, mononuclear cells, neutrophils, and a number of cell types using chemiluminescence. Our method also measures free-radical generation in pure chemical systems viz production of superoxide by xanthine/hypoxanthine and xanthine oxidase (XO) in buffer or hydroxyl ion from hydrogen peroxide, and so on.

2. Materials

2.1. Chronolog Aggregometer/Luminometer

A two-channel or a four-channel chronolog aggregometer/ luminometer along with computer interfacing is required. Chronolog (Havertown, PA) also provides agglink software that displays and analyzes the data as the

From: *Methods in Molecular Biology, vol. 108: Free Radical and Antioxidant Protocols*
Edited by: D. Armstrong © Humana Press Inc., Totowa, NJ

experiment is running; this is a great help. Alternately, a graph recorder can be used as well.

2.2. Blood Samples

Ten milliliters of heparinized blood (14.3 USP/mL) is processed for the preparation of mononuclear cells (MNC). The heparinized blood is used for the assay of free radicals in whole blood and suspensions of MNC in buffer or autologous plasma.

2.3. Reagents

1. Luminol: 10 m*M* stock in 0.1 *M* borate buffer.
   a. Borate buffer: 0.1 *M*, pH 9.0 (9.5 g/250 mL is 0.1 *M* stock; adjust pH to 9.0.
   b. Luminol: Dissolve 0.1 77 g luminol in 100 mL of 0.1 *M* borate buffer (0.177 g/l00 mL is 10 m*M*).
2. Zymosan: 10 mg/mL suspension (Stock: 500 mg/50mL).
   a. 500 mg/50mL in saline boiled for 15 min (double boiler) in a small flask placed in a beaker with distilled water and boiled on heating block.
   b. Wash twice with saline by centrifuging at 235*g* and decant. Repeat the same for second wash.
   c. Suspend in saline to 500 mg/50 mL
3. Other agonists for the stimulation of free radicals that can be used are N-formylmethionylleucinylphenylalanine (fMLP) (stock 10 m*M* in DMSO), Phorbol ester (10 m*M* stock), Calcium ionophore (5 m*M* stock).

3. Methods

3.1. Preparation of MNC

1. MNC are isolated (*see* **Note 1**) from heparinized blood (14.3 USP/mL) by a density gradient centrifugation over Ficoll (Organon Teknika Corp; Durham NC) density of 1.077–1.080 g/mL at 20°C. Initially, the blood is centrifuged at 235*g* for 10 min to obtain a buffy coat. The buffy coat is diluted to 3 mL in HBSS and layered on top of 3 mL Ficoll and respun at 235*g* for 30 min at 22°C.
2. The mixed MNC band at the interface is removed with a pasteur pipet and the cells are washed twice in Hank's balanced salt solution (HBSS).
3. The remainder of the blood sample without buffy coat is centrifuged at 1200*g* in order to obtain plasma.
4. MNC are washed in HBSS and resuspended in HBSS or medium M199 (Gibco, Life Technologies Inc., Grand Island, NY) with 10% autologous plasma to make a final concentration of 100,000 cells/mL for free-radical production (*see* **Note 2**).

3.2. Measurement of Oxygen-Free Radicals

1. Heparinized venous blood samples is diluted in HBSS at a dilution of 1:9 and kept at room temperature. One mL sample of diluted blood is pipetted into flat-

bottom plastic tubes and incubated at 37°C in a Chronolog lumi-aggregometer for 3 min.

2. The chemiluminophore luminol (final conc: 300 m*M*) is then added and free-radical generation induced by fMLP (final conc: 20 m*M*) or any other agonist like zymosan, calcium ionophore, phorbol mysteric acid, etc. The dianion of luminol undergoes a reaction with molecular oxygen to form a peroxide. This peroxide is unstable and decomposes to an electronically excited state. This excited dianion emits a photon that is detected by the chemilumeniscent detector.

LUMINOL —$2OH^-$→ DIANION —O_2→ A PEROXIDE

A PEROXIDE → 3-AMINOPTHALATE TRIPLET DIANION (t_1) + N_2 —INTERSYSTEM CROSSING→ 3-AMINOPTHALATE SINGLET DIANION (S_1) —FLUORESCENCE→ 3-AMINOPTHALATE GROUND STATE DIANION, S + hv

Chemiluminescence is recorded for 15 min (a protracted record after 15 min does not alter the relative amounts of chemiluminescence produced by various blood samples). Our method, developed independently, is similar to that published by Tosi and Hamedani ***(14)***.

3. We further established that, in our assay system, there is a dose-dependent inhibition of chemiluminescence by superoxide dismutase and catalase as well as diphenylene iodonium (DPI), a specific inhibitor of nicotinamide adenine dinucleotide phosphate (NADPH) oxidase, the enzyme responsible for the enzymatic production of superoxide ions. Our assay system is exquisitely sensitive to DPI-induced inhibition at nanomolar concentrations ***(15)***. The specific inhibitory effect of DPI on NADPH oxidase has been established by Hancock and Jones ***(16)***.

3.3. Free Radicals in Whole Blood and MNC

1. Heparinized whole blood diluted 1:9 with phenol-free HBSS or MNC suspensions in HBSS or autologous plasma (*see* **Note 2**).
2. Stirrer speed: 1000 rpm, gain; start at 0.01 (to be increased or decreased as required), unit temperature 37°C.
3. Graph: 15 s/division or as required.
4. Test procedure: Run two channels simultaneously.
   a. Pipet 500 µL 1:9 diluted blood into a cuvet containing a stir bar.

b. Place cuvet into the test well and cover the heater block.
c. Display graph on the screen and wait to ensure baseline has stabilized about halfway across the first square.
d. Add 15 µL of 10 m*M* luminol stock solution into each of the cuvets and shut the heating block (*see* **Note 3**).
e. Add 50 µL of (1:19 dilution in HBSS, stock 10 mg/mL), 1 µL of fMLP (stock 10 m*M*), or any other agonist as required, after the baseline has advanced to about half the length of the next square (*see* **Note 3**).
f. Keep the heater block closed and allow the curve to reach the maximum until the end of the last square on the graph.
g. Adjust the gain as required so that the luminescence curves are within the range on the displayed graph.
h. Once the gain is established, the test is performed at that particular gain setting.
i. The peak luminescence is recorded in mV from the point of zymosan or fMLP addition to the end of the graph.
j. The results are determined by computer settings or visual inspection (*see* **Fig. 1**).

3.4. Results

The characterization of active oxygen species using this technique is summarized in **Table 1**.

Luminol-dependent chemiluminescence offers a highly sensitive means of detecting active oxygen, although it is not specific towards any one oxygen species. Superoxide-dismutase (SOD) (10 µg) inhibited 81% of the chemiluminescence, whereas catalase (40 µg) inhibited only 47% of the activity. A catalase concentration (20 µg) that by itself was essentially inactive complemented SOD in eliminating all chemiluminesence. Total inhibition of the response by DPI provides evidence for the detection of plasma membrane-bound NADPH oxidase-mediated superoxide production.

4. Notes

1. MNC should be prepared as soon as possible and the assay completed in less than 2 h.
2 The cell counts should be uniform for all experiments. A differential count (complex blood counts) should be performed on whole blood and the results normalized against either mononuclear or neutrophil cell count. This is particularly important for in vivo experiments in which the administration of a drug or any other intervention may change the cell counts. For example, steroids are known to alter the differential cell counts in whole blood. Therefore, the results should be always expressed per neutrophil or mononuclear cell count.
3. Syringes or pipets should be designated for each reagent. A carryover of an agonist (like fMLP) into luminol will result in erratic baselines or the carryover of DPI/SOD/catalase into luminol may result in an unusual inhibitory response.

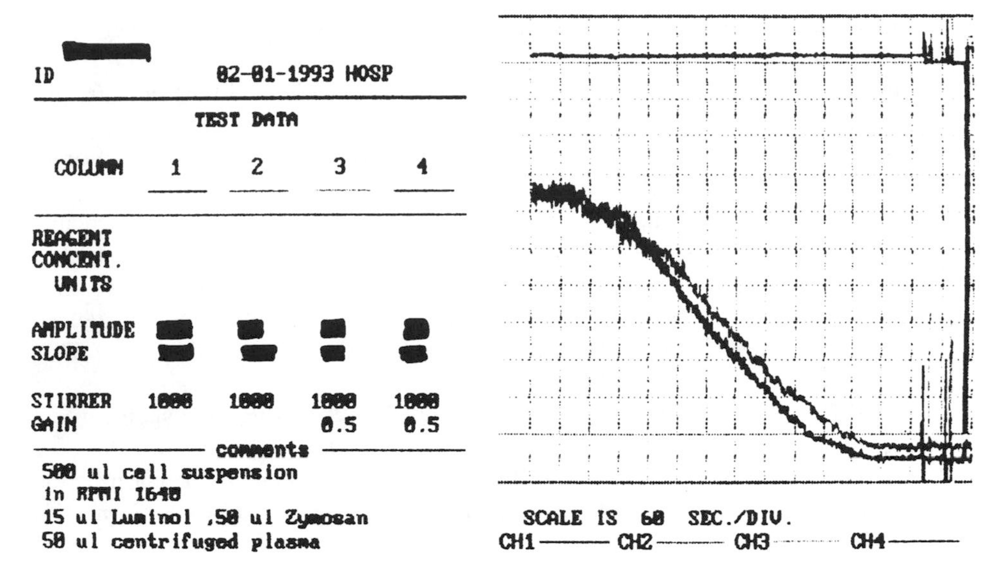

Fig. 1. A representative tracing of zymosan induced free radical generated by MNC suspension diluted in autologous plasma.

Table 1
Characterization of Active Oxygen Species Using Chemiluminescence Technique

Treatment	Inhibition (%) ± S.E
10 µg SOD[a]	81.5 ± 11.5
20 µg Catalase[b]	5.5 ± 13.2
40 µg Catalase	47.4 ± 11.5
10 µg SOD + 20 mg Catalase	100
40 n*M* DPI	100

[a]3100 U/mg protein of SOD.
[b]1500 U/mg protein of catalase.

References

1 Roots, R. and Okada, S. (1975) Estimation of life times and diffusion distances of radicals involved in X-ray induced DNA strand breaks or killing of mammalian cells. *Radiat. Res.* **64,** 306–320.
2. Oyanagi, Y. (1977) Stimulatory effect of platinum (IV) ion on the production of superoxide radical from xanthine oxidase and macrophages. *Biochem. Pharmacol.* **26,** 473–476.
3. Tamura, M., Tamura, T., Tyagi, S. R., and Lambeth, J. D. (1988) The superoxide-generating respiratory burst oxidase of human neutrophil plasma membrane. *J. Biol. Chem.* **263,** 17,621–17,626.

4. Prasad, K. and Kalra, J. (1988) Oxygen free radicals and heart failure. *Angiology* **39,** 417–420.
5. Opie, L. (1989) Reperfusion injury and its pharmacologic modification. *Circulation* **80,** 1049–1062.
6. Floyd, R. A. (1990) Role of oxygen free radicals in carcinogenesis and brain ischemia. *FASEB J.* **4,** 2587–2597.
7. Glaumann, B. and Trump, B. F. (1975) Studies on the pathogenesis of ischemic cell injury. III: morphological changes of the proximal pars recta tubules (P_3) of the rat kidney made ischemic in vivo. *Virchows Arch [B]* **19,** 303–323.
8. Baker, G. L., Corry, R. J., and Autor, A. P. (1995) Oxygen free radical induced damage in kidneys subjected to warm ishcemia and reperfusion. *Ann. Surg.* **202,** 628–641.
9. Parks, D. A., Bulkley, G. B., and Granger, D. N. (1983) Role of oxygen derived free radicals in the digestive tract disease. *Surgery* **94,** 415–422.
10. Freeman, B. A., Rosen, G. M., and Barber, M. J. (1985) Superoxide perturbation of the organization of vascular endothelial cell membranes. *J. Biol. Chem.* **261,** 6590–6593.
11. Dandona, P., Thusu, K., Cook, S., Snyder, B., Makowski, J., Armstrong, D., and Nicotera, T. (1996) Increased oxidative damage of deoxyribonucleic acid (DNA) in dependent Diabetes mellitus. *Lancet* **347,** 444–445.
12. Cantoni, O., Sestili, P., Cattabeni, F., Bellomo, G., Pou, S., Cohen, M., and Cerutti, P. (1989) Calcium chelator Quin 2 prevents hydrogen peroxide induced DNA breakage and cytotoxicity. *Eur. J. Biochem.* **82,** 209–212.
13. Cerutti, P. A. (1985) Prooxidant states and tumor promotion. *Science* **227,** 375–381.
14. Tosi, M. F. and Hamedani, A. (1992) A rapid specific assay for superoxide release from phagocytes in small volumes of whole blood. *Am. J. Clin. Pathol.* **97,** 566–573.
15. Nicotera, T., Thusu, K., Dandona, P. (1993) Elevated production of active oxygen in Bloom's syndrome cell lines. *Cancer Res.* **53,** 5104–5107.
16. Hancock, J. T. and Jones, O. T. G. (1987) Inhibition of diphenylene iodonium and its analogues of superoxide generation by macrophages. *Biochem. J.* **242,** 103–107.

6

Assay of Phospholipid Hydroperoxides by Chemiluminescence-Based High-Performance Liquid Chromatography

Yorihiro Yamamoto, Yasuhiro Kambayashi, and Takako Ueda

1. Introduction

Lipid hydroperoxides (LOOH) are the primary products of lipid peroxidation. Therefore, it would be desirable to detect and identify the lipid hydroperoxides in tissues or blood. One of the most advanced methods for the detection of LOOH is a chemiluminescence-based high-performance liquid chromatography (HPLC) assay *(1–5)* as shown in **Fig. 1**. The reaction sequence leading to the emission of light from isoluminol in the presence of LOOH and microperoxidase, a heme catalyst, is assumed to be as follows:

$$\text{LOOH} + \text{microperoxidase} \rightarrow \text{LO}^{\bullet} \quad (1)$$

$$\text{LO}^{\bullet} + \text{isoluminol (QH}^{-}) \rightarrow \text{LOH} + \text{semiquinone radical (Q}^{\bullet -}) \quad (2)$$

$$\text{Q}^{\bullet -} + \text{O}_2 \rightarrow \text{Q} + \text{O}_2^{\bullet -} \quad (3)$$

$$\text{Q}^{\bullet -} + \text{O}_2^{\bullet -} \rightarrow \text{isoluminol endoperoxide} \rightarrow \text{light } (\lambda_{max} = 430 \text{ nm}) \quad (4)$$

This assay has several advantages:

1. It is very sensitive, with the lower limit of detection being about 0.1 pmol or less;
2. Interference by biological antioxidants is avoided because hydroperoxides and antioxidants can be separated by HPLC; and
3. Information on which lipid class is oxidized can be obtained.

Using this assay, the presence of about 3 n*M* cholesteryl ester hydroperoxide (CE-OOH) in healthy human plasmas has been reported *(6)*. Rats have higher plasma CE-OOH levels (about 10 n*M*) than do humans *(7)*. On the other

From: *Methods in Molecular Biology, vol. 108: Free Radical and Antioxidant Protocols*
Edited by: D. Armstrong © Humana Press Inc., Totowa, NJ

hand, phosphatidylcholine hydroperoxide (PC-OOH) is undetectable in human *(2,3,6)* and rat *(7)* plasma. The reason for the absence of PC-OOH in human and rat plasma are as follows: Plasma glutathione peroxidase can reduce PC-OOH to its alcohol, but not CE-OOH *(8)*; PC-OOH in high density lipoprotein is converted to CE-OOH by lecithin:cholesterol acyltransferase *(9)*.

Although phospholipid hydroperoxides (PL-OOH) are likely to be transient products in vivo, it is a very useful marker of lipid peroxidation at least in vitro and ex vivo systems. Here we describe the method for the detection of PL-OOH at picomole levels and examples of its application.

2. Materials

2.1. Phospholipid Hydroperoxide Standards

1. Phosphatidylcholine, soybean (Sigma, St. Louis, MO).
2. α-Tocopherol.
3. 0.02% Triethylamine in methanol.
4. Octadecylsilyl column (Superiorex ODS, 20 × 250 mm, Shiseido, Tokyo).
5. Phosphatidylethanolamine: Soybean (Sigma).
6. Phosphatidylserine: Bovine brain (Sigma).
7. Phosphatidylinositol: Soybean (Sigma).
8. Spectrometer.

2.2. Tissue Extraction

1. 100 μ*M* 2,6-Di-tert-butyl-4-methylphenol (BHT) in methanol.
2. 100 μ*M* BHT in chloroform/methanol (2/1 [v/v]).
3. Microfuge (Eppendorf).

2.3. Reaction Solution

1. 100 m*M* Borate buffer, pH 10.
2. 6-Amino-2,3-dihydro-1,4-phthalazine-dione (isoluminol; Sigma).
3. Microperoxidase (MP-11; Sigma).

2.4. Chemiluminescence Assay

1. 40 m*M* Sodium phosphate, monobasic.
2. *tert*-Butyl alcohol.
3. 5 μ*M*, 4.6 × 250 mm Silica column (Supelco Japan).
4. 5 μm, 4.6 × 250 mm Aminopropyl column (Supelco).
5. 5 μm, 4.6 × 250 mm Silica gel guard column (Supelco).
6. Mixer: Model 2500-0.22 (Kratos, Westwood, NJ).
7. Chemiluminescence detector: Model 825-CL (Japan Spectroscopics, Tokyo).

3. Method

3.1. Preparation of Phospholipids Hydroperoxides

Phosphatidylcholine was purified on a semipreparative octadecylsilyl column using methanol containing 0.02% triethylamine as the mobile phase at a flow rate of 10 mL/min ***(1)***. PC-OOH were prepared by the aerobic autoxidation of purified phosphatidylcholine at room temperature for several d in the presence of α-tocopherol (VE) and in the absence of radical initiators. VE is added to reduce the formation of *trans,trans*-hydroperoxides, and a high concentration of VE accelerates the autoxidation of lipids ***(10)***. Synthesized PC-OOH was purified with the same HPLC conditions previously described. The concentration of PC-OOH was determined by its absorbance at 234 nm (ε = 28,000/M•cm) ***(11)***. Phosphatidylethanolamine hydroperoxide (PE-OOH), phosphatidylserine hydroperoxide (PS-OOH), and phosphatidylinositol hydroperoxide (PI-OOH) were also prepared by the autoxidation of parent lipids.

3.2. Tissue Extraction

PC-OOH in plasma (or lipoprotein) can be extracted by shaking the sample vigorously with 4 volumes of methanol containing 100 μ*M* BHT (to prevent oxidation). It is necessary to extract PC-OOH in tissue homogenate with two volumes of chloroform/methanol (2/1, v/v) containing 100 μ*M* BHT. After centrifugation at 12,000 rpm for 3 min, aliquots of the either methanol extract from plasma or chloroform/methanol phase are injected onto HPLC (*see* **Notes 1** and **2**).

3.3. Chemiluminescence Assay

Equipment is arranged as outlined in **Fig. 1**. Antioxidants and hydroperoxides in the sample are separated by HPLC, using one of the chromatographic conditions described below, and then mixed with a reaction solution containing isoluminol and microperoxidase in a Kratos special mixer (*see* **Note 3**). The reactions leading to the emission of light take place in a mixing coil made from a piece of HPLC tubing (45 cm, inner volume ~92 μL). The emitted light is measured by a chemiluminescence detector.

The reaction solution for pump B is prepared as follows: 100 m*M* aqueous borate buffer (38.14 g of sodium tetraborate decahydrate per/L) is prepared, and the pH is adjusted to 10 with sodium hydroxide. Isoluminol (177.2 mg, final concentration 1 m*M*) is dissolved in 500 mL of methanol and 500 mL of the aforementioned borate buffer, and then 5 mg of microperoxidase is added. The mobile phase for pump A is methanol/*tert*-butyl alcohol/40 m*M* monoba-

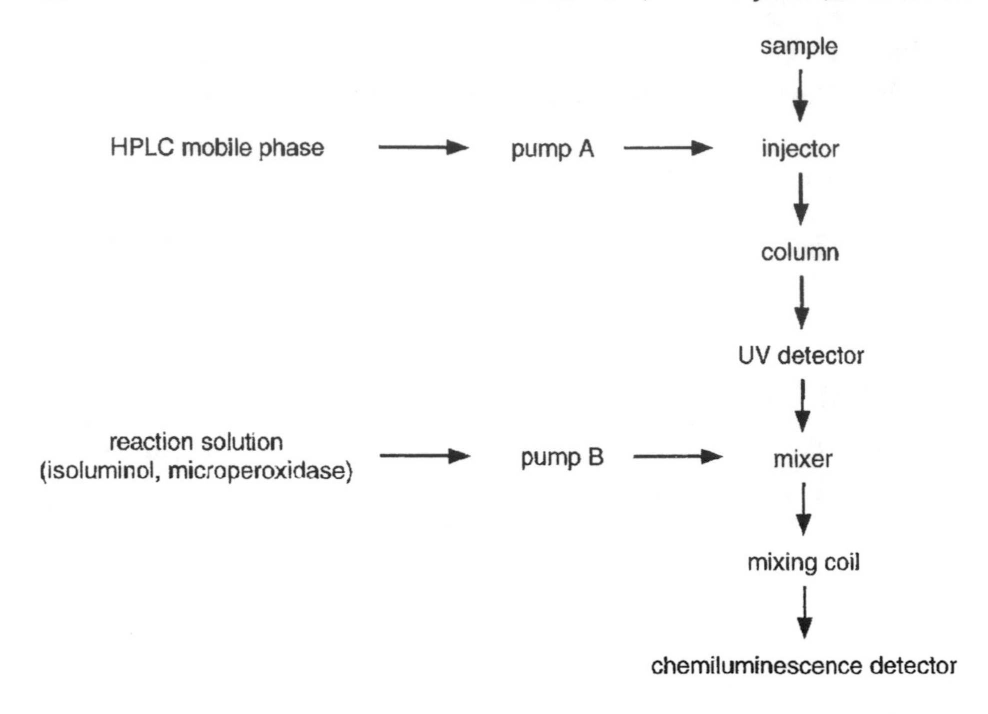

Fig. 1. Schematic diagram of the HPLC-isoluminol chemiluminescence assay.

sic sodium phosphate (6/3/1 [v/v/v]). The flow rates of pump A and B are 1.0 and 1.5 mL/min, respectively (*see* **Notes 4–6**).

3.4. Silica Column for the Assay of PC-OOH

When a silica column is used, PC-OOH elutes at about 12 min. In this condition, ascorbate, urate, and tocopherols elute within 5 min and do not interfere with the PC-OOH detection. **Figure 2** demonstrates the absence of PC-OOH in human plasma (Line 1). Line 2 shows the PC-OOH peak spiked in methanol extract. Line 3 shows the absence of PC-OOH in the monophasic extract of human plasma with four volumes of chloroform/methanol (1/2 [v/v]). Line 4 shows the formation of PC-OOH during the Folch extraction of human plasma. **Figure 3** shows the formation and the decay of PC-OOH during the incubation of isolated hepatocytes with *tert*-butyl hydroperoxide (BOOH) at 37°C under aerobic conditions, suggesting that the cell is capable of decomposing PC-OOH *(12)*.

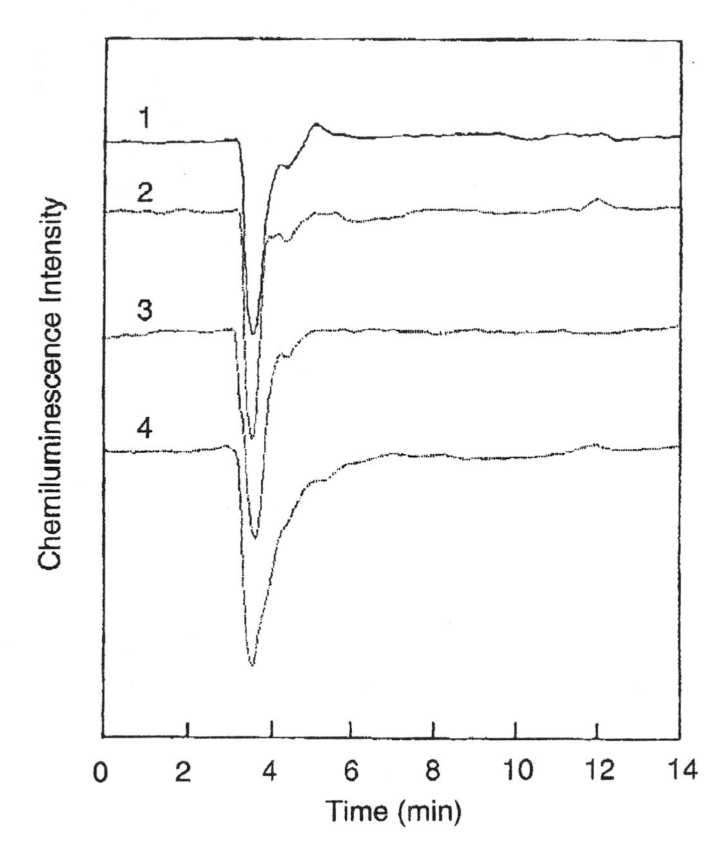

Fig. 2. Chemiluminescence chromatogram of methanol extract from a healthy human plasma analyzed with a silica column using methanol/*tert*-butyl alcohol/40 m*M* monobasic sodium phosphate (6/3/1 [v/v/v]) as the mobile phase. Line 1, injected 50 μL of methanol extract (corresponding to 10 μL of plasma); line 2, injected 50 μL of the same methanol extract spiked with 0.8 pmol (80 n*M*) PC-OOH; line 3, injected 50 μL of chloroform/methanol (1/2 [v/v]) extract (corresponding to 10 μl of plasma); line 4, injected Folch extract (corresponding to 40 μL of plasma).

3.5. Aminopropyl Column for the Assay of PL-OOH

Biological membranes consist of phospholipids, such as phosphatidylcholine, phosphatidylethanolamine, phosphatidylserine, and phosphatidylinositol. The assay for the measurement of hydroperoxides of these phospholipids would be useful for studying the oxidation of biomembranes. **Figure 4** demonstrates

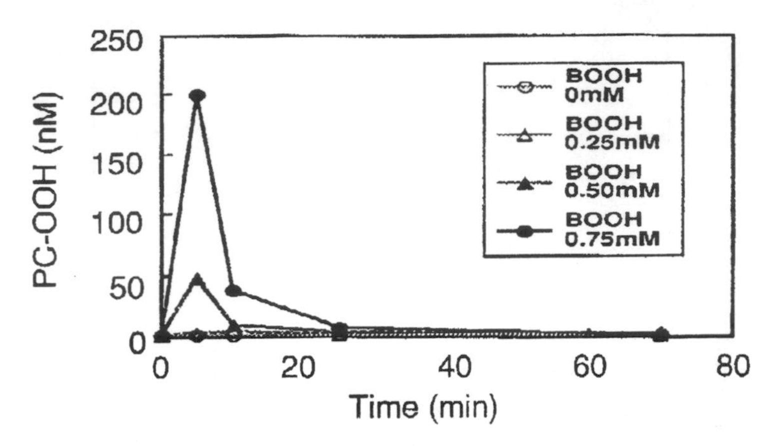

Fig. 3. Formation and the decomposition of PC-OOH during the incubation of isolated hepatocytes with 0, 0.25, 0.50, and 0.75 m*M* *tert*-butyl hydroperoxide (BOOH) at 37°C under aerobic conditions.

the separation of PC-OOH, PE-OOH, PS-OOH, and PI-OOH using two sets of silica-gel guard columns and an aminopropyl column in series. We have observed the formation of PC-OOH and PE-OOH during the oxidation of liver homogenate initiated with free-radical initiators or induced with BOOH ***(13)***.

4. Notes

1. Samples should be kept at –80°C for long storage.
2. Lipids are very susceptible to oxidation when they are extracted from biological samples because peroxidases and other antioxidant enzymes are inactivated by the extraction. α-Tocopherol can work as prooxidant in the organic extract ***(10)***. Evaporation of the extracts usually causes artifactual formation of lipid hydroperoxides as described in the text.
3. To mix an eluant and a chemiluminescence reagent well, the Kratos mixer is used, because it is designed to accomplish a complete mixing of two solutions. Even if the Kratos mixer is replaced by a simple T joint, only a small change was observed in peak areas, suggesting that the rates of reactions of hydroperoxide-induced oxidation of isoluminol are very rapid ***(5)***. Therefore, it is not necessary to use the Kratos mixer.
4. HPLC grade methanol should be used. When 2-propanol is used instead of *tert*-butyl alcohol, the baseline of chemiluminescence detector is high and unstable because of the high content of hydroperoxides in 2-propanol.
5. Methanol/40 m*M* monobasic sodium phosphate (9/1 [v/v]) can be used as a mobile phase for a silica column. This shortens the retention time of PC-OOH from 12 min to 7 min.

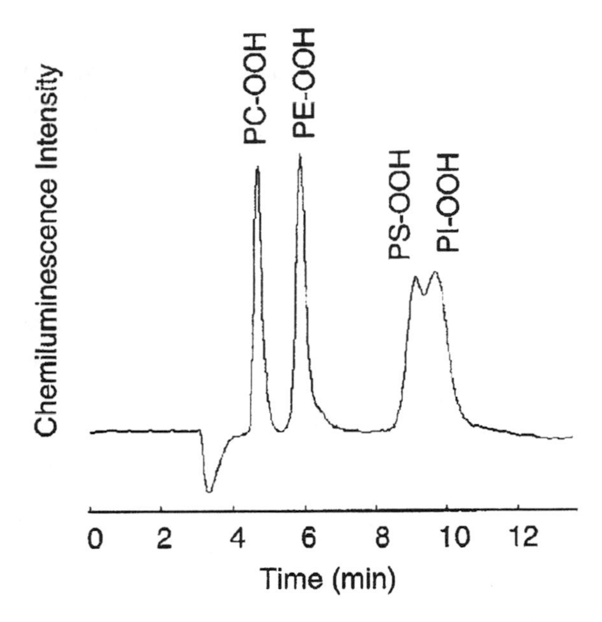

Fig. 4. Separation of 10 pmol of membrane phospholipids hydroperoxides by HPLC monitored by chemiluminescence. Two sets of silica gel guard columns and aminopropyl column in series were used; methanol/*tert*-butyl alcohol/40 m*M* monobasic sodium phosphate (6/3/1 [v/v/v]) was used as the mobile phase.

6. Disappearance of chemiluminescence positive peak by the pretreatment with triphenylphosphine assures that the peak is hydroperoxide, because lipid hydroperoxides are readily reduced by triphenylphosphine at room temperature.

References

1. Yamamoto, Y., Brodsky, M. H., Baker, J. C., and Ames, B. N. (1987) Detection and characterization of lipid hydroperoxides at picomole levels by high-performance liquid chromatography. *Anal. Biochem.* **160,** 7–13.
2. Yamamoto, Y and Ames, B. N. (1987) Detection of lipid hydroperoxides and hydrogen peroxide at picomole levels by an HPLC and isoluminol chemiluminescence assay. *Free Rad. Biol. Med.*. **3,** 359–361.
3. Frei, B., Yamamoto, Y., Niclas, D., and Ames, B. N. (1988) Evaluation of an isoluminol chemiluminescence assay for the detection of hydroperoxides in human blood plasma. *Anal. Biochem.* **175,** 120–130.
4. Yamamoto, Y., Frei, B., and Ames, B. N. (1990) Assay of lipid hydroperoxides using HPLC with post-column chemiluminescence detection. *Methods Enzymol.* **186,** 371–380.
5. Yamamoto, Y. (1994) Chemiluminescence-based high-performance liquid chromatography assay of lipid hydroperoxides. *Methods Enzymol.* **233,** 319–324.

6. Yamamoto, Y. and Niki, E. (1989) Presence of cholesteryl ester hydroperoxide in human blood plasma. *Biochem. Biophys. Res. Commun.* **165,** 988–993.
7. Yamamoto, Y., Wakabayashi, K., Niki, E., and Nagao, M. (1992) Comparison of plasma levels of lipid hydroperoxides and antioxidants in hyperlipidemic Nagase analbuminemic rats, Sprague-Dawley rats, and humans. *Biochem. Biophys. Res. Commun.* **189,** 518–523.
8. Yamamoto, Y., Nagata, Y., Niki, E., Watanabe, K., and Yoshimura, S. (1993) Plasma glutathione peroxidase reduces phosphatidylcholine hydroperoxide. *Biochem. Biophys. Res. Commun.* **193,** 133–138.
9. Nagata, Y., Yamamoto, Y., and Niki, E. (1996) Reaction of phosphatidylcholine hydroperoxide in human plasma: The role of peroxidase and lecithin:cholesterol acyltransferase. Arch. *Biochem. Biophys.* **329,** 24–30.
10. Bowry, V. W. and Stocker, R. (1993) Tocopherol-mediated peroxidation. The prooxidant effect of vitamin E on the radical-initiated oxidation of human low-density lipoprotein. *J. Am. Chem. Soc.* **115,** 6029–6044.
11. Chan, H. W.-S. and Levett, G. (1977) Autoxidation of methyl linoleate. Separation and analysis of isomeric mixtures of methyl linoleate hydroperoxides and methyl hydroxylinoleates. *Lipids* **12,** 99–104.
12. Nakamura, Y,. Ohori, Y., and Yamamoto, Y. (1992) Formation of phosphatidylcholine hydroperoxide in isolated rat hepatocytes treated with *tert*-butyl hydroperoxide, in *Oxygen Radicals: Proceedings of the 5th International Congress on Oxygen Radicals* (Yagi, K., Kondo, M., Niki, E., and Yoshikawa, T., eds.), Excerta Medica, Amsterdam, pp 273–276.
13. Kambayashi, Y., Yamashita, S., Niki, E., and Yamamoto, Y. (1997) Oxidation of rat liver phospholipids: comparison of pathways in homogeneous solution, in liposomal suspension and in whole tissue homogenates. *J. Biochem.* **121,** 425–431.

7

Sensitive and Specific Fluorescent Probing of Oxidative Stress in Different Classes of Membrane Phospholipids in Live Cells Using Metabolically Integrated *cis*-Parinaric Acid

Valerian E. Kagan, Vladimir B. Ritov, Yulia Y. Tyurina, and Vladimir A. Tyurin

1. Introduction

Reactive-oxygen species and organic free radicals are essential metabolic intermediates and have important regulatory functions. Overproduction of these reactive metabolites is implicated in the etiology of a host of degenerative diseases *(1)*, such as cardiovascular disease and neurodegenerative disease, in acute conditions such as trauma and infection, and in aging. Of the various types of oxidative damage that cells undergo, lipid peroxidation is considered to be one of the major contributors to oxidative injury *(2)*; thus, strict regulation of lipid peroxidation is extremely important for normal cell physiology. Quantitative assays of lipid peroxidation in intact cells are essential for evaluating oxidative damage from various sources, and in testing the efficacy of antioxidant interventions.

Although numerous techniques have been developed to quantitate lipid peroxidation in model systems *(3)*, there are only limited applications for assays of lipid peroxidation in live cells or in vivo *(4,5)*. One of the major problems in studies of oxidative stress in membrane phospholipids is that efficient repair mechanism may be activated, e.g., deacylation of oxidatively modified fatty-acid residues and subsequent reacylation of phospholipids with normal, non-oxidized fatty acids *(6,7)*, making accurate determination of lipid oxidation a difficult endeavor in otherwise unperturbed cells.

From: *Methods in Molecular Biology, vol. 108: Free Radical and Antioxidant Protocols*
Edited by: D. Armstrong © Humana Press Inc., Totowa, NJ

In this chapter, we describe a novel method based on the use of *cis*-parinaric acid (PnA) as a reporter molecule for assay of membrane lipid peroxidation in intact mammalian cells. PnA is a natural 18-carbon fatty acid with four conjugated double bonds. The four conjugated double bonds make PnA highly susceptible to peroxidation and also cause it to be fluorescent. Fluorescence is irreversibly lost upon peroxidation thus providing a convenient approach to quantitate lipid peroxidation. PnA has been used as a fluorescent probe in physical-chemical studies of model membranes and as a probe of lipid peroxidation in chemical systems, in lipoproteins, and in simple cell membrane system such as red blood cell ghosts ***(8–10)***. In these model systems, PnA was incorporated into membranes or lipoproteins by physical, nonphysiological means, i.e., by partitioning.

As a natural fatty acid, PnA can be incorporated by metabolism into membrane phospholipids of intact cells ***(11)***. Using metabolically integrated PnA, we developed and optimized procedures for sensitive and specific fluorescent probing of oxidative stress in different classes of membrane phospholipids in live cells ***(12)***. We found that the PnA/living-cell system can be used for testing oxidative stress induced by metabolism or low doses of toxic chemicals and effects of protective agents (antioxidants) in living cells.

2. Materials

2.1. High-Performance Liquid Chromatography (HPLC) Analysis of Cell Lipids

2.1.1. Analytical HPLC System

1. Shimadzu HPLC LC-600 system.
2. Shimadzu fluorescence detector (model RF-551).
3. Shimadzu UV-VIS detector (model SPD-10A V).
4. Shimadzu EZChrom software.
5. A 5 μm Supelcosil LC-Si column (4.6 x 250).

2.1.2. Reagents

All reagents should be of the highest chromatographic quality to ensure accurate and reproducible results.

1. Hexane: Fisher Optima^{TM} grade (Fisher Scientific, Pittsburgh, PA).
2. 2-Propanol: Fisher Optima grade (Fisher Scientific).
3. Water (H_2O) (Aldrich, Milwaukee, WI).
4. Ammonium Acetate (Sigma, St. Louis, MO).

2.1.3. Mobile Phase

1. Solvent A: 2-propanol-hexane-water 57:43:1 (v/v).
2. Solvent B: 2-propanol-hexane-40 m*M* aqueous ammonium acetate 57:43:10 (v/v) pH 6.7.

2.2. High-Performance Thin-Layer Chromatography (HPTLC) Analysis of Cell Lipids

2.2.1. Instruments

1. Heating elements.
2. A Shimadzu UV 160U spectrometer.

2.2.2. Reagents

1. HPLC grade Methanol (CH_3OH) (Sigma-Aldrich).
2. Chloroform ($CHCl_3$) (Fisher Scientific).
3. Acetone (CH_3COCH_3): Aldrich 99.5+%, A.C.S. reagent.
4. Glacial acetic acid (CH_3COOH) (Fisher Scientific).
5. HPLC-grade water (Sigma-Aldrich).
6. Ammonium hydroxide 28% (Fisher Scientific).

2.2.3. Mobile Phase

1. Chloroform:methanol:28% ammonium hydroxide 65:25:5 (v/v).
2. Chloroform:aceton:methanol:glacial acetic acid:water 50:20:10:10:5 (v/v).

2.3. Integration of PnA into Cell Phospholipids

2.3.1. Instruments

1. Constant temperature bath (Fisher Scientific).
2. Centrifuge (Heraeus 28 RS).

2.3.2. Reagents

1. 1.8 m*M* PnA (Z-9, E-11, E-13, Z-15-octadecatetraenoic acid) (Molecular Probes, Eugene, OR).
2. 0.5 mg/mL Human serum albumin (hSA) fatty-acid free.
3. Phosphate buffered saline (PBS): 135 m*M* NaCl, 2.7 m*M* KCl, 1.5 m*M* KH_2PO_4 and 8 m*M* Na_2HPO_4, pH 7.4.
4. L1210 medium: 115 m*M* NaCl, 5 m*M* KCl, 1 m*M* $MgCl_2$, 5 m*M* NaH_2PO_4, 10 m*M* glucose, and 25 m*M* HEPES, pH 7.4.

2.4. Quantitative Analysis of Phospholipid Phosphorus

2.4.1. Reagents

1. 70% Perchloric acid ($HClO_4$) (Fisher Scientific).
2. 12 *N* Hydrochloric acid (Fisher Scientific).
3. Ascorbate, sodium salt, 10% (Fisher Scientific).
4. 2.5% Sodium molibdate (Sigma).
5. Sodium molibdate/malachite-green reagent: 4.2% sodium molybdate in 5 *N* HCl, 0.2% malachite-green base (water solution).
6. Polyoxyethylenesorbitan monolaurate (Tween-20), 1.5% (Sigma).
7. Malachite, green base (Sigma).

3. Methods

3.1. Preparation of Biological Samples

3.1.1. Cell Culture

3.1.1.1. HL-60 Human Leukemia Cells

Human leukemia HL-60 cells are grown in RPMI 1640 medium supplemented with 10% fetal bovine serum (FBS) at 37°C in a 5% CO_2 atmosphere. Cells from passages 25–40 are used for experiments. The density of cells at collection time is 1–2 × 10^6 cells/mL.

3.1.1.2. Aortic Smooth Muscle Cells (SMC)

Aortic SMCs are cultured as explants from the ascending thoracic aortas, obtained from ether-anesthetized Sprague-Dawley male rats (Charles River, Wilmington, MA) weighing 150–200 g, after a midline abdominal incision, including the diaphragm as described ***(13)***. The SMCs grew as explants from the medial tissue and were confluent in 12–14 d. (The cells are available from Dr. R. Dubey, Center for Clinical Pharmacology, University of Pittsburgh.)

Cells are suspended in primary cell culture medium (Dulbecco's modified Eagle's medium [DMEM]/F12 supplemented with penicillin [100 U/mL], streptomycin [100 μg/mL], $NaHCO_3$ [13 mmol/L] and HEPES [25 mmol/L]; Gibco) containing 10% fetal calf serum (FCS), plated in tissue-culture flasks (75 cm^2) and incubated under standard tissue culture conditions (37°C, 5% CO_2-95% air, and 98% humidity. SMCs between the 2nd and 3rd passage were used in experiments.

3.1.1.3. PC12 Rat Pheochromocytoma Cells

1. Control-transfected (pBabe-puro) PC12 cells and PC12 cells transfected with the Bcl-2 gene ligated into the retroviral vector, pBabe, containing a puromycin resistance gene (Bcl-2-pBabe-puro) were provided by Dr. N. Schor (Department of Pediatrics, University of Pittsburgh).
2. Cells are maintained as adherent monolayers in DMEM made with 10% in horse serum, 0.5% in FBS and 1.1% in penicillin/streptomycin. Cells are fed every 3–4 d, and biweekly, 15 μg/mL puromycin was be added to the medium.
3. Cells are examined for Bcl-2 expression by Weatern blotting every 10 passages. Bcl-2 expression did not vary in either cell line over the course of these studies.

3.1.1.4. Sheep Pulmonary Artery Endothelial Cells

1. Sheep pulmonary arteries are obtained from a local slaughterhouse. Sheep pulmonary artery endothelial cells (provided by Dr. B. Pitt, Department of Pharmacology, University of Pittsburgh) are isolated and characterized as described ***(14)***.

2. Cells are harvested by collagenase type I (0.1%; Sigma) digestion and transferred to plastic tissue-culture dishes. Cells are expanded in Opti-MEM (Gibco) with endothelial-cell growth supplement (15 µg/mL; Collaborative Biomedical Products), heparin sulfate (10 U/mL), penicillin (100 U/mL), streptomycin (100 µg/mL), and 10% sheep serum (Sigma).
3. Cells are subcultured by detaching cells with a balanced solution containing trypsin (0.05%) and ethylenediamine tetraacetic acid (EDTA) (0.02%). Cells are used between passages 5 and 8 for all experiments.

3.1.1.5. Cell Viability

Cell viability is determined microscopically on a hemocytometere using the Trypan blue exclusion assay.

3.1.2. Incorporation of PnA into Cell Phospholipids

1. PnA is incorporated into cells by addition of its hSA complex (PnA–hSA) to cell suspensions. The complex is prepared by adding PnA (500 µg, 1.8 µmol in 25 µL of dimethylsilfoxide) to hSA (50 mg, 760 nmol in 1 mL of PBS).
2. Cells in log-phase growth are rinsed twice with L1210 medium. Cells are diluted to a density of 1.0 x 10^6 cells/mL, then incubated with PnA–hSA complex (final concentration of 2.5 µg/mL PnA) in L1210 medium at 37°C in the dark on air for 2 h.
3. After the incubation period, cells are pelleted by centrifugation then washed twice with L1210 medium with and without hSA (0.5 mg/mL). Aliquots are taken for determination of cell viability.

3.1.3. Induction of Lipid Peroxidation in the Cells

Cells loaded with PnA are incubated in the presence of different amounts of oxidants in L1210 medium at 37°C in the dark on air for 2 h. Two different inducers of lipid peroxidation are used: A lipid-soluble initiator of peroxyl radicals, 2,2'-azobis-(2,4-dimethyl-valeronitrile) (AMVN) (0.1–1.0 m*M*) which generates peroxyl radicals in lipid phase of membrane, and H_2O_2 (40 µ*M*).

Hydroperoxide is added four times for 10 µ*M* every 30 min. In experiments with antioxidants, they are added to the incubation medium 15 min before addition of AMVN. After incubation with oxidants aliquots of cell suspension are taken for determination of cell viability and lipids are extracted and resolved by HPLC.

3.2. Extraction Techniques

3.2.1. Extraction of Cell Lipids

1. Total lipids were extracted from cells using a slightly modified Folch procedure ***(15)***. Methanol (2 mL) containing butylated hydroxytoluene (BHT, 0.1mg) is added to the cell suspension (1 × 10^6 cells). The suspension is mixed with

chloroform (4 mL) and, to insure complete extraction, kept for 1 h under a nitrogen atmosphere on ice in the dark.

2. After addition of 0.1 *M* NaCl (2 mL) and vortex mixing (still under a nitrogen atmosphere), the chloroform layer is separated by centrifugation (1500*g*, 5 min).
3. The chloroform is evaporated with a stream of nitrogen and the lipid extract dissolved in 4:3:0.16 (v/v) 2-propanol-hexane-water (0.2 mL). Control experiments demonstrated that the procedure recovered more than 95% of cell phospholipids.

3.3. Assay Conditions

3.3.1. HPLC Analysis of Cell Lipids

1. An HPLC procedure for separation and detection of PnA incorporated into cell phospholipids is used. Total lipid extracts are separated with an ammonium acetate gradient by normal-phase-HPLC as per ***(16)*** with the minor modifications described below.
2. A 5-μm Supelcosil LC-Si column (4.6 x 250 mm) is employed with the following mobile phase flowing at 1 mL/min: solvent A (57:43:1 isopropanol-hexane-H_2O), solvent B (57:43:10 isopropanol-hexane-40 m*M* aqueous ammonium acetate, pH 6.7, 0–3 min linear gradient from 10–37% B, 3–15 min isocratic at 37% B, 15–23 min linear gradient to 100% B, 23–45 min isocratic at 100% B. The elluent is monitored at 205 nm to gauge separation of lipids.
3. Fluorescence of PnA in eluates is monitored by emission at 420 nm after excitation at 324 nm. UV and fluorescence data are processed and stored in digital form with Shimadzu EZChrom software.

3.3.2. Determination of Total Lipid Phosphorus in Lipid Extracts

Total lipid phosphorus in total lipid extracts is estimated spectrophotometrically at 660 nm as described by ***(17)*** with the minor modifications. Aliquot of lipid extract are pipeted into test tubes and solvent evaporated to dryness with a stream of nitrogen. Then, 50 μL of perchloric acid is added to samples and samples incubated for 20 min at 170–180°C on the heating elements.

After the tubes are cool, 0.4 mL of distilled water is added to each tube followed by 2 mL of sodium molybdate/malachite-green reagent (4.2% sodium molybdate in 5.0 *N* HCl - 0.2% malachite green 1:3 [v/v]). Without delay, 80 μl of 1.5% Tween-20 is added, followed by immediate shaking in order to stabilize the color.

3.3.3. HPTLC Analysis of Cell Lipids

1. To identify the HPLC peaks, the fractions of phospholipids are collected and analyzed by two-dimensional HPTLC on silica-gel plates (5 × 5, Whatman, Clifton, NJ). The plates are first developed with a solvent system consisting of chloroform:methanol:28% ammonium hydroxide, (65:25:5 [v/v]). After drying the plates with a forced air blower to remove the solvent, the plates are developed

in the second dimension with a solvent system consisting of chloroform:acetone:methanol:glacial acetic acid:water (50:20:10:10:5 [v/v]).

2. The phospholipids are visualized by exposure to iodine vapor and identified by comparison with migration of authentic phospholipid standards. The identity of each phospholipid is established by comparison with the Rf values measured for authentic standards.
3. The spots are scraped, transferred to tubes, and lipid phosphorus is determined.

3.3.4. Determination of Lipid Phosphorus in Individual Phospholipid Spots

1. Lipid phosphorus in individual phospholipid spots is determined as described ***(18)***. Lipid spots are scraped and transferred to tubes; 125 μL of perchloric acid added and the samples are incubated for 20 min at 170–180°C.
2. After cooling, 825 μL of distilled water is added to each tube followed by 125 μL of 2.5% sodium molybdate and 125 μL of 10% ascorbic acid.
3. Then samples are vortexed and incubated for 10 min at 100°C. After the tubes are cooled again, the color developed is read at 797 nm.

3.4. Results

3.4.1. Integration of PnA into Cell Lipids

Four different cell lines, rat pheochromocytoma PC12, rat aortic smooth muscle cells, sheep pulmonary artery endothelial cells, and human promyelocytic leukemia HL-60 cells were used in the study. Cells are incubated in bovine-serum-free L1210 medium with PnA-hSA (2–5 μg PnA/mL) at 37°C in the dark up to 3 h (to avoid binding of PnA to serum proteins, bovine-serum should be omitted from the incubation medium). PnA may exert cytotoxic effects in prolonged incubations ***(19)***. Under the incubation conditions used (2–5 μg PnA/mL; 1×10^6 cells/mL; 2 h), PnA had no measurable effects on cell viability, which remained almost 100% for all PnA-treated cells in L1210 medium as shown by trypan-blue exclusion. Comparison of the total amounts of unsaturated fatty-acid residues with those of PnA in lipid extracts from cells showed that PnA accounted for less than 1% of membrane phospholipid unsaturated fatty-acid residues ***(12)***. Hence, it is unlikely that membrane structure and characteristics are changed by the presence of such small amounts of incorporated PnA (*see* **Note 1**).

Because PnA is highly oxidizable, it is important to use a delivery system that will prevent its oxidation in the course of incubation. Previous studies ***(12)*** demonstrated that PnA in hSA/PnA complex is sufficiently protected against oxidation during 3 h of incubation. PnA nonspecifically bound to cell surfaces may serve as a pool of free PnA that can be utilized for repair of oxidatively modified PnA-labeled phospholipids. To eliminate this pool as a potential source of error, cells should be incubated with fatty acid-free hSA to remove cell-bound PnA (*see* **Notes 2** and **3**).

3.4.2. HPLC Procedure to Separate and Detect PnA Integrated into Phospholipids

An HPLC procedure to separate and detect PnA integrated into various phospholipid classes in the cells is used. Total lipids extracts from PnA-loaded cells are prepared and the constituent phospholipids separated by HPLC. Mixture of hexane/2-propanol/water is chosen as eluting solvent because lipids are readily dissolved in this system. Prior to HPLC analysis of a sample, the column must be washed for 15–20 min with solvent B (*see* Methods). Then, the column must be equilibrated with the mixture of solvents A and B (1:9) until a stable baseline is obtained. **Figure 1** shows typical fluorescence-detected (**A**) and ultraviolet (UV)-detected (**B**) chromatograms of PnA-labeled phospholipids from sheep pulmonary artery endothelial cells.

Six phospholipid peaks were well-resolved on both fluorescence-detected (**Fig. 1A**) and UV-detected (**Fig. 1B**) chromatograms. Using authentic standards, the peaks are identified as diphosphatidylglycerol (DPG, retention time 13.0 min); phosphatidylinositol (PI, retention time 14.5 min); phosphatidylethanolamine (PEA, retention time 17.3 min); phosphatidylserine (PS, retention time 30.4 min); phosphatidylcholine (PC, retention time 35.6 min); sphingomyelin (SPH, retention time 38.6 min). The identity of phospholipids on HPLC-chromatograms was confirmed by collecting each of the HPLC-fractions and subjecting them to HPTLC analysis (*see* Methods) by comparison with the Rf values measured for authentic standards. Obviously, no fluorescence-detectable HPLC peaks were found in extracts from cells incubated with hSA in the absence of PnA. Whereas quantitatively, the intensities of PnA-labeled phospholipid HPLC-peaks were significantly different in the four cell lines studied, qualitatively, the phospholipid patterns were similar.

For all cell lines used in the study, acylation of different membrane phospholipids by PnA was time-dependent during 1–2 h of incubation with PnA–hSA complex, after which metabolic integration of PnA plateaued. Shown in **Fig. 2** is the time course of PnA incorporation into different phospholipids of rat aortic smooth muscle cells, which was prototypical for the other cell lines used. Based on measurements of total lipid phosphorus in lipid extracts from cells, the amount of PnA integrated into cell lipids was estimated as ng PnA per microgram total lipid phosphorus. Average values for PnA incorporation into five major phospholipid classes of the four cell lines showed that metabolic integration of PnA into membrane phospholipids followed the same order: PC > PEA > PS > SPH (**Table 1**). Quantitatively, the relative amounts

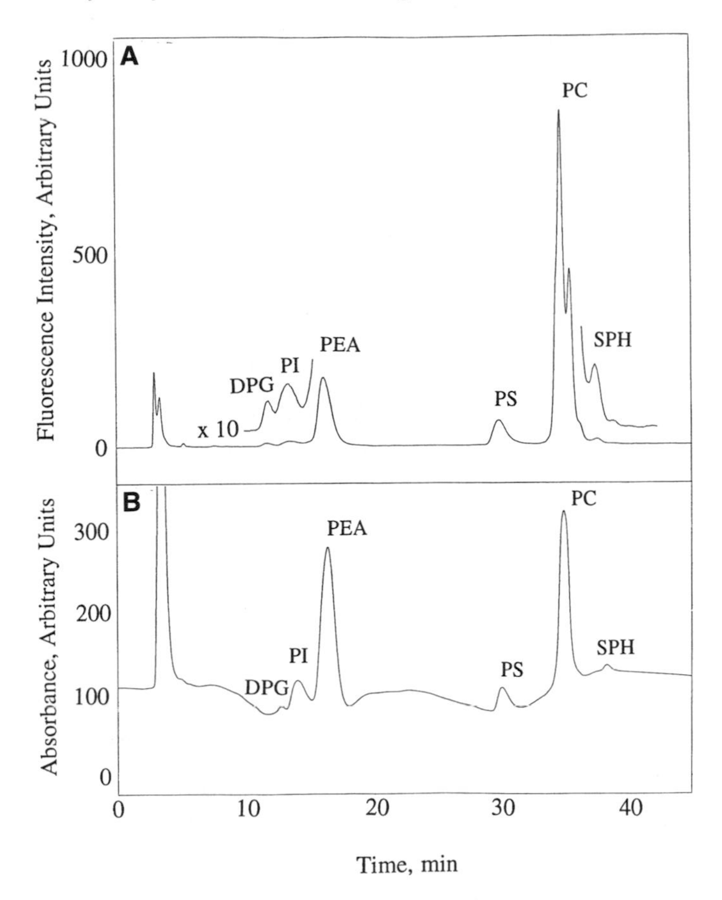

Fig. 1. Normal-phase HPLC chromatograms of total lipids extracted from 1×10^6 sheep pulmonary artery endothelial cells. Cells were incubated with the hSA–PnA complex (5 µg of PnA/0.5 mg hSA/mL of L1210) for 2 h at 37°C then washed with hSA (0.5 mg/mL of L1210 medium) and L1210 medium. Total lipids were extracted and resolved by HPLC. HPLC conditions: mobile phase flowing at 1 mL/min: solvent A (57:43:1 isopropanol-hexane-H_2O), solvent B (57:43:10 isopropanol-hexane-40 m*M* aqueous ammonium acetate, pH 6.7), 0–3 min linear gradient from 10–37% B, 3–15 min isocratic at 37% B, 15–23 min linear gradient to 100% B, 23–45 min isocratic at 100% B. (**A**), Fluorescence-emission intensity, excitation at 324 nm, emission at 420 nm; (**B**), UV absorbance at 205 nm. DPG, diphosphatidylglycerol; PI, phosphatidylinositol; PEA, phosphatidylethanolamine; PS, phosphatidylserine, PC, phosphatidylcholine, SPH, sphingomyelin.

of PnA integrated into phospholipid classes varied significantly for different cell lines.

Based on the measurements of phospholipid profiles, the specific incorporation of PnA into different classes of membrane phospholipids can be estab-

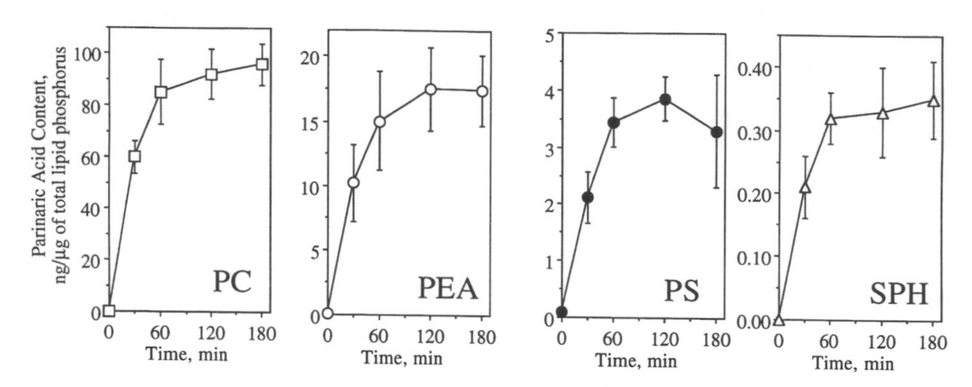

Fig. 2. Time-course of *cis*-parinaric acid integration into major classes of membrane phospholipids in rat aortic smooth muscle cells. Cells (1×10^6) were incubated with 4 μg of PnA/0.5 mg hSA/mL of L1210 for different time intervals at 37°C then washed with hSA (0.5 mg/mL of L1210 medium) and L1210 medium. Then, lipids were extracted and normal-phase HPLC separation with fluorescence detection was performed (for details, *see* legend of **Fig. 1**). PC, phosphatidylcholine, PEA, phosphatidylethanolamine; PS, phosphatidylserine, SPH, sphingomyelin.

lished as shown in **Table 2**. Specific incorporation of PnA followed the same order for the cell lines used: PC > PEA > PS > SPH which was coincident with the relative abundance of the respective phospholipid.

3.4.3. Fluorescent Probing of Oxidative Stress Using PnA Metabolically Incorporated into Phospholipids

Two different kinds of oxidative stressors were chosen for initiation of lipid peroxidation in cells. A lipid-soluble azo-initiator, AMVN generates peroxyl radical at a constant rate at a given temperature. AMVN partitions into lipid domains of membranes and generates radicals within the lipid bilayer ***(20)***. Radical species generated in this way do not escape from the hydrophobic lipid environment ***(21)***. Additionally, hydrogen peroxide was used as an oxidant in HL-60 cells. HL-60 cells contain high myeloperoxidase activity ***(22)***. The myeloperoxidase/H_2O_2/halide system produces potent oxidants such as hypochlorus acid and singlet oxygen ***(23)*** that are known to readily oxidize different lipids ***(24)***.

PnA-labeled rat neuroblastoma PC12 cells, rat smooth muscle cells, and sheep pulmonary-artery endothelial cells were treated with AMVN. HPTLC analysis of phospholipids revealed no significant differences in phospholipid distribution in either cell line following oxidative stress imposed by incubation of the cells with AMVN (**Table 3**). Using HPLC-fluorescence measurements of PnA-labeled cells, we found that all phospholipids in these three different cell lines were affected by AMVN-induced oxidation. Incubation of PnA-labeled rat

Table 1
Incorporation of Acid into Phospholipids of Four Different Mammalian Cell Lines

	PnA content (ng/μg total lipid phosphorus)				
Cell/Line[a]	PC	PEA	PS	PI	SPH
Smooth muscle cells (*n* = 10)	92.2 ± 9.7	17.5 ± 3.2	3.9 ± 0.4	2.7 ± 0.5	0.3 ± 0.1
Pheochromocytoma PC12 (*n* = 7)	105.5 ± 15.0	8.4 ± 2.0	1.7 ± 0.3	0.7 ± 0.1	0.7 ± 0.2
Human leukemia HL-60 cells (*n* = 5)	212.0 ± 39.6	62.7 ± 12.0	6.4 ± 2.3	12.8 ± 3.5	1.3 ± 0.5
Sheep pulmonary-artery endothelial cells (*n* = 3)	141.2 ± 10.0	33.7 ± 4.0	13.3 ± 2.6	1.8 ± 0.3	0.8 ± 0.1

[a]Cells (1×10^6) were incubated for 2 h at 37°C in L1210 medium in the presence of 2–5 μg/mL PnA *cis*-parinaric acid. Cells were washed twice with and without hSA (0.5 mg/mL) in L1210 medium, then total lipids were extracted from cells and HPLC was resolved, as described in Methods section. Values are means ± SD. PI, phosphatidylinositol; PEA, phosphatidylethanolamine; PS, phosphatidylserine; PC, phosphatidylcholine; SPF, sphingomyelin.

Table 2
Specific Incorporation of *cis*-Parinaric Acid into Membrane Phospholipids of Smooth-Muscle Cells (SMC), PC12 Rat Pheochromocytoma Cells, Sheep Pulmonary Artery Endothelial Cells (SPAEC), and Human Leukemia HL-60 Cells

	Specific Incorporation, mol PnA/mol PL			
Phospholipids	SMC	PC12	SPAEC	HL-60
Phosphatidylcholine	1:40	1:67‘	1:30	1:20
Phosphatidylethanolamine	1:133	1:183	1:60	1:30
Phosphatidylserine	1:230	1:744	1:45	1:100
Phosphatidylinositol	1:270	1:1500	1:400	1:50
Sphingomyelin	1:2050	1:1500	1:900	1:600

pheochromocytome PC12 cells with increasing concentrations of AMVN (in the range of concentration from 0.1–0.75 m*M*) ***(25)*** resulted in a dramatic loss of PnA in essentially all classes of phospholipids (**Fig. 3**). The loss of PnA fluorescence, as a result of oxidative damage of its conjugated tetraene system ***(9)***, occurred proportionally to the dose of the oxidant applied. It is important to note that these concentrations of AMVN did not affect cell viability, because >95% cells remained viable after 2 h incubation (*see* **Note 4**). Similarly, AMVN

Table 3
Effect of AMVN on the Phospholipid Composition of Smooth-Muscle Cells, PC12 Rat Pheochromocytoma Cells, and Sheep Pulmonary Artery Endothelial Cells

	Smooth muscle cells		Pheochromocytoma PC12		Sheep pulmonary artery endothelial cells	
Phospholipid	Control	AMVN	Control	AMVN	Control	AMVN
Phosphatidylcholine	44.1 ± 2.2	43.9 ± 2.8	51.9 ± 3.4	49.8 ± 3.2	47.3 ± 2.3	47.2 ± 2.4
Phosphatidylethanolamine	27.5 ± 1.6	27.1 ± 1.5	18.4 ± 1.2	18.1 ± 1.2	24.5 ± 1.5	25.6 ± 1.4
Phosphatidylinositol	8.7 ± 0.6	8.9 ± 0.5	8.5 ± 0.5	8.9 ± 0.5	7.1 ± 0.5	6.6 ± 0.5
Phosphatidylserine	9.7 ± 0.7	8.4 ± 0.8	7.5 ± 0.4	7.6 ± 0.4	9.5 ± 0.6	7.9 ± 0.5
Sphingomyelin	7.3 ± 0.5	8.5 ± 0.8	7.7 ± 0.5	7.8 ± 0.8	8.6 ± 0.5	8.9 ± 0.5
Diphosphatidylglycerol	1.7 ± 0.2	1.3 ± 0.3	3.3 ± 0.3	3.6 ± 0.3	2.3 ± 0.1	2.1 ± 0.1
Lysophosphatidylcholine	0.5 ± 0.1	1.1 ± 0.3	1.5 ± 0.4	2.5 ± 0.3	0.6 ± 0.1	1.6 ± 0.1

[a]All values are mean percent of total phospholids ± SEM (n = 3). The cells were incubated with 0.5 mM AMVN for 2 h at 37°C in L1210 medium.

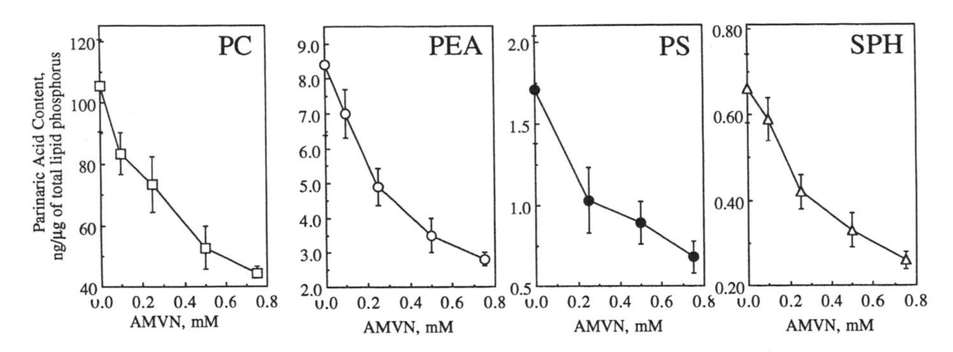

Fig. 3. Peroxidation of PnA-labeled phospholipids induced by a lipid-soluble azo-initiator of peroxyl radicals, AMVN, in PC12 rat pheochromocytoma cells. PnA-labeled PC12 rat pheochromocytoma cells (1×10^6 cells/mL) were incubated with indicated concentrations of AMVN at 37°C for 2 h in L1210 medium. Then, lipids were extracted and normal phase HPLC separation with fluorescence detection was performed (for details, *see* legend of **Fig. 1**). PC, phosphatidylcholine, PEA, phosphatidylethanolamine; PS, phosphatidylserine, SPH, sphingomyelin.

induced a pronounced and uniform oxidation of PnA-acylated phospholipids in sheep pulmonary artery endothelial cells (**Fig. 4**) and human leukemia HL-60 cells. In contrast, incubation of HL-60 cells with H_2O_2 for 2 h (four additions of H_2O_2 10 μ*M* each every 30 min) induced a significant oxidation of PnA in specific phospholipids (**Fig. 5**). The phospholipid pattern of H_2O_2-induced oxidation was different from AMVN-induced oxidation: PEA, PC, and SPH were susceptible to oxidation, whereas PS was not significantly affected (**Fig. 5**). This difference can be attributed to different mechanisms of generation of free radicals in the course of H_2O_2- or AMVN-induced oxidation in the cells.

3.4.4. Quantitation of Antioxidant Effectiveness Using PnA Metabolically Incorporated into Phospholipids

Several natural and synthetic phenolic compounds were used to quantitate their antioxidant effects in cells. In rat aortic smooth-muscle cells, a polar vitamin E homolog, 2,2,5,7,8-pentamethyl-6-hydroxychromane (PMC), 1,3,5[10]-estratrien-2,3-diol-17-one (2-OH-estrone), and 4-OH-tamoxifen produced a concentration-dependent protection against AMVN-induced oxidation *(26)*. The protective effect of antioxidants was phospholipid-specific *(27)*. A complete protection of PEA by antioxidants was achieved at 50 μ*M* PMC and at 13 μ*M* of either 4-OH-tamoxifen or 2-OH-estrone caused (**Fig. 6**) Although 2-OH-estrone at 13 μ*M* was 100% effective in protecting PS, PC, and SPH, 4-OH-tamoxifen's effect was significantly less pronounced towards these phos-

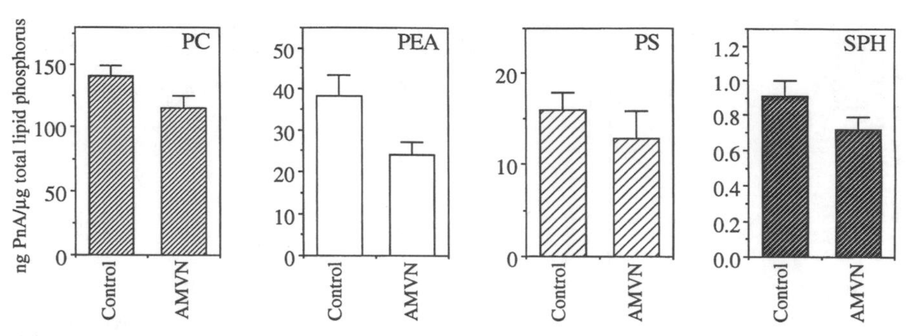

Fig. 4. Peroxidation of PnA-labeled phospholipids induced by a lipid-soluble azo-initiator of peroxyl radicals, AMVN, in sheep pulmonary artery endothelial cells. PnA-labeled PC12 sheep pulmonary artery endothelial cells (1×10^6 cells/mL) were incubated with 0.5 m*M* of AMVN at 37°C for 2 h in L1210 medium. Then, lipids were extracted and normal phase HPLC separation with fluorescence detection was performed (for details, *see* legend of **Fig. 1**). PC, phosphatidylcholine, PEA, phosphatidylethanolamine; PS, phosphatidylserine, SPH, sphingomyelin.

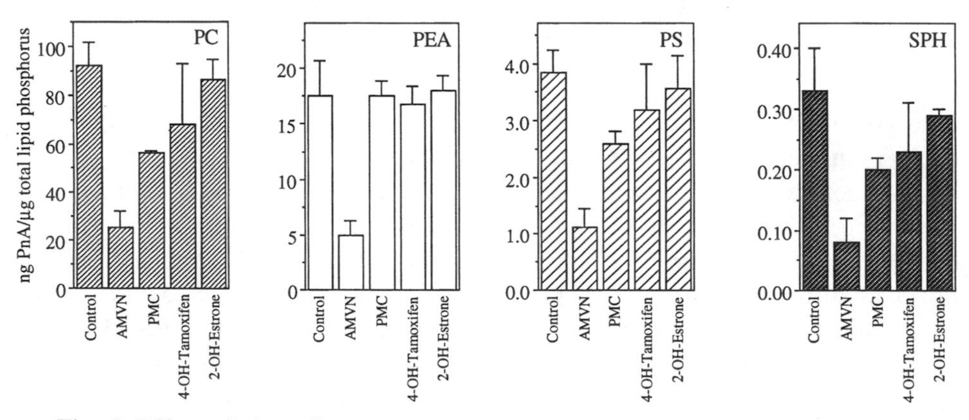

Fig. 5. Effect of phenolic compounds (PMC, 4-OH-tamoxifen, and 2-OH-estrone) on AMVN-induced peroxidation of PnA-labeled phospholipids in rat aortic smooth muscle cells. PnA-labeled rat aortic smooth muscle cells (1×10^6) preincubated with different phenolic compounds for 15 min at 37°C were incubated with AMVN (0.5 m*M*) at 37°C for 2 h in L1210 medium. Then, lipids were extracted and normal-phase HPLC separation with fluorescence detection was performed (for details, *see* legend of **Fig. 1**). Concentrations of PMC, 4-OH-Tamoxifen and 2-OH-estrone were 50 µ*M*, 13 µ*M* and 13 µ*M*, respectively. PC, phosphatidylcholine, PEA, phosphatidylethanolamine; PS, phosphatidylserine, SPH, sphingomyelin.

pholipids. PMC (at 50 µ*M*) was even less effective than 4-OH-tamoxifen. Similarly, H_2O_2-induced peroxidation in HL-60 cells was also inhibited by phenolic antioxidants (PMC, diethylstilbestrol, and a phenolic antitumor drug, VP-16) in a phopspholipid-specific manner (**Fig. 5**).

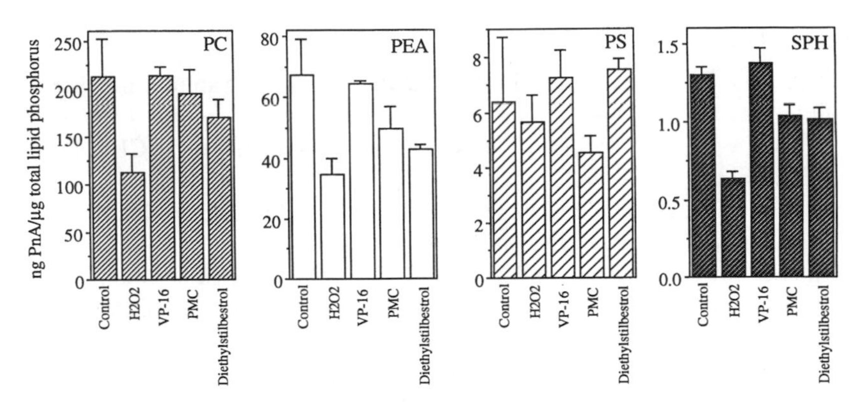

Fig. 6. Effect of phenolic antioxidants (PMC, VP-16 and diethylstilbestrol) on H_2O_2-induced peroxidation of PnA-labeled phospholipids in human leukemia HL-60 cells. PnA-loaded HL-60 (1×10^6) pre-incubated with antioxidants 15 min at 37°C were incubated with 40 µ*M* of H_2O_2 (for additions 10 µ*M* each every 30 min) at 37°C for 2 h in L1210 medium. Then, lipids were extracted and normal-phase HPLC separation with fluorescence detection was performed (for details, *see* legend of **Fig. 1**). Concentrations of PMC, VP-16, and diethylstilbestrol were 50 µ*M*. PC, phosphatidylcholine, PEA, phosphatidylethanolamine; PS, phosphatidylserine, SPH, sphingomyelin.

4. Notes

1. Fluorescent labeling of intracellular phospholipids with metabolically integrated PnA solves the problem of a sensitive probe for lipid peroxidation that is unaffected by cell-repair systems. This model permits an accurate quantification of: The actual degree of phospholipid-specific peroxidation in cells treated with subtoxic levels of oxidants, and the effectiveness of natural and synthetic antioxidants in site-specific protections of phospholipids.
2. To protect PnA against auto-oxidation (during storage, incubation with cells, etc.) the hSA/PnA complex should be used.
3. Excess of free PnA in cell suspensions should be removed after the completion of phospholipid labeling by washing cells with fatty acid free hSA. This is critical because unincorporated PnA can be otherwise utilized for lipid repair (reacylation reactions), thus decreasing (masking) actual amount of oxidative stress.
4. Relatively short incubation periods (not exceeding 3–4 h) are recommended for integration of PnA into cell lipids to avoid its cytotoxicity.

Acknowledgment

This work was supported by grant 96008 from the Johns Hopkins Center for Alternatives to Animal Testing.

References

1. Kehrer, J. P. (1993) Free radical as mediators of tissue injury and disease. *Crit. Rev. Toxicol.* **23,** 21–48.
2. Kagan, V. E. (1988) Lipid Peroxidation in Biomembranes. CRC Press, Boca Raton, FL, pp. 55–146.
3. Draper, H. H. and Hadley, M. (1990) Malondialdehyde determination as index of lipid peroxidation. *Met. Enzymol.* **186**, 421–431.
4. Haklar, G., Yegenada, I., and Yalcin, A. S. (1995) Evaluation of oxidant stress in chronic hemodialysis patients: use of different parameters. *Clin. Chim. Acta* **234**, 109–114.
5. Lazzarino, G., Tavazzi, B., Di Pierro, D., Vagnozzi, R., Penco, M., and Giardina, B. (1995) The relevance of malondialdehyde as a biochemical index of lipid peroxidation of postischemic tissues in the rat and human beings. *Biol. Trace Elem. Res.* **47**, 165–170.
6. Pacifici, E. H., McLeod, L. L., and Sevanian, A. (1994) Lipid hydroperoxide-induced peroxidation and turnover of endothelial cell phospholipids. *Free Radic. Biol. Med.* **17**, 297–309.
7. Salgo, M. G., Corongiu, F. P., and Sevanian, A. (1992) Peroxidation and phospholipase A2 hydrolytic susceptibility of liposomes consisting of mixed species of phosphatidylcholine and phosphatidylethanol-amine. *Biochim. Biophys. Acta* **1127**, 131–140.
8. Kuypers, F. A., Van den Berg, J. J. M., Schalkwijl, C., Roelofsen, B., and Op den Kamp, J. A. F. (1987) Parinaric acid as a sensitive fluorescence probe for the determination of lipid peroxidation. *Biochim. Biophys. Acta* **921**, 266–274.
9. Van den Berg, J. J. M. (1994) Effects of oxidants and antioxidants evaluated using parinaric acid as a sensitive probe for oxidative stress. *Redox Rep.* **1**, 11–21.
10. McKenna, R., Kezdy, F. G., and Epps, D. E. (1991) Kinetic analysis of the free-radical-induced lipid peroxidation in human erythrocyte membranes: evaluation of potential antioxidants using *cis*-parinaric acid to monitor peroxidation. *Anal. Biochem.* **196**, 443–450.
11. Rintoul, D. A. and Simoni, R. D. (1977) Incorporation of a naturally occurring fluorescent fatty acid into lipids of cultured mammalian cells. *J. Biol. Chem.* **252**, 7916–7918.
12. Ritov, V. B., Banni, S., Yalowich, J. C., Day, B. W., Claycamp, H. G., Corongiu, F. P., and Kagan, V. E. (1996) Non-random peroxidation of different classes of membrane phospholipids in live cells detected by metabolically integrated *cis*-parinaric acid. *Biochim. Biophys. Acta* **1283,** 127–140.
13. Dubey, R. K., Gillespie, D. G., Osaka, K., Suzuki, F., and Jackson, E. K. (1996) Adenosine inhibits growth of rat aortic smooth muscle cells: possible role of A2b receptor. *Hypertension* **27,** 786–793.
14. Hoyt, D. G., Mannix, R. J., Rusnak, J. M., Pitt, B. R., and Lazo, J. S. (1995) Collagen is a survival factor against lipopolysaccharide-induced apoptosis in cultured sheep pulmonary artery endothelial cells. *Am. J. Physiol.* **269**, L171–L177.

15. Folch, J., Lees, M., and Sloan-Stanley, G. H. (1957) A simple method for isolation and purification of total lipids from animal tissues. *J. Biol. Chem.* **226**, 497–509.
16. Geurts van Kessel, W. S. M., Hax, W. M. A., Demel, R. A., and Degier, J. (1977) High performance liquid chromatographic separation and direct ultraviolet detection of phospholipids. *Biochim. Biophys. Acta* **486**, 524–530.
17. Chalvardjian, A. and Rubincki, E. (1970) Determination of lipid phosphorus in the nanomolar range. *Anal. Biochem.* **36**, 225–226.
18. Arduini, A., Tyurin, V., Tyurina, Y., Arrigoni-Martelli, E., Malajoni, F., Dottori, S., Federichi, G. (1992) Acyl-trafficking in membrane phospholipid fatty acid turnover: the transfer of fatty acid from the acyl-L-carnitine pool to membrane phospholipids in intact human erythrocytes. *Biochem. Biophys. Res. Commun.* **187**, 313–318.
19. Cornelius, A. S., Yerram, N. R., Kratz, D. A., and Spector, A. A. (1991) Cytotoxic effect of *cis*-parinaric acid in cultured malignant cells. *Cancer Res.* **51**, 6025–6030.1991
20. Niki, E. (1990) Free radical initiators as source of water- or lipid-soluble peroxyl radicals. *Methods Enzymol.* **186**, 100–1108.
21. Krainev, A. G. and Bigelow, D. J. (1996) Comparison of 2,2'-azobis(2-aminopropane) hydrochloride (AAPH) and 2,2'-azobis(2,4-dimethylvalerionitrile) (AMVN) as free radical initiators: a spin-trapping study. *J. Chem. Soc. Perkin Trans.* **2**, 747–754.
22. Kettle, A. J., Gedye, C. A., Hampton, M. B., and Winterbourn, C. C. (1995) Inhibition of myeloperoxidase by benzoic acid hydrazides. *Biochem. J.* **308**, 559–563
23. Steinbecl, M. J., Khan, A. U., and Karnovsky, M. J. (1992) Intracellular singlet oxygen generation by phagocytosing neutrophils in response to particles coated with a chemical trap. *J. Biol. Chem.* **267**, 13,425–13,433.
24. Heinecke, J. W., Li, W., Mueller, D. M., Bohrer, A., and Turk, J. (1994) Cholesterol chlorohydrin systhesis by the myeloperoxidase-hydrogen-chloride system: potential markers for lipoproteins oxdatively damaged by phagocytes. *Biochemistry* **33**, 10,127–10,136.
25. Tyurina, Y. Y., Tyurin, V. A., Carta, G., Quinn, P. J., Schor, N. F., Kagan, V. E. (1997) Direct evidence for antioxidant effect of Bcl-2 in PC12 rat pheochromocytome cells. *Arch. Biochem. Biophys.* **344,** 413–427.
26. Dubey, R. K., Tyurina, Y. Y., Tyurin, V. A., Branch, R. A., Kagan, V. E. (1996) Direct Measurement of Peroxyl radical-induced membrane phospholipids oxidation in cultured vascular smooth muscle cells: evidence for antioxidant protection by 4-Hydroxytamoxifen and estriendiol. *Circulation Res.* (submitted).
27. Pryor, W. A., Cornicelli, J. A., Devall, L. J., Tait, B., Trivedi, D. T., Witial, D. T., and Wu, M. (1993) A rapid screening test to determine the antioxidant potencies of natural and synthetic antioxidants. *J. Org. Chem.* **58**, 3521–3532.

8

Aromatic Hydroxylation

Salicylic Acid as a Probe for Measuring Hydroxyl Radical Production

Andrea Ghiselli

1. Introduction

With the recognition that the imbalance between oxidant and antioxidant compounds is an important causative agent of aging and of various chronic diseases ***(1)***, there has been growing interest in elucidating the pathways involved in oxidative-stress initiation. As oxidative stress is an inescapable repercussion of aerobic life, all organisms that use oxygen protect themselves against free radicals ***(1)***. Although cells are endowed with impressive equipment of antioxidant defenses ***(2)***, they are not completely efficient to resist an increased free-radical formation.

Because of the high reactivity of oxygen free radicals and the lack of specific and sensitive methodologies for their measurements, the assessment of oxidative stress in living systems in vitro and in vivo has been difficult. In particular, hydroxyl radical ($OH^{\bullet}$) has a very short half-life, and it is therefore present in extremely low concentrations.

Aromatic hydroxylation ***(3)*** is one of the most specific methods available for its measure in biological systems and it has been largely used.

The method is based on the ability of $OH^{\bullet}$ to attack the benzene rings of aromatic molecules and produce hydroxylated compounds that may be directly and specifically measured by electrochemical detection. Formation of *o*-tyrosine and *m*-tyrosine from the $OH^{\bullet}$ attack on phenylalanine has been used to study the production of $OH^{\bullet}$ by isolated cells ***(4)*** and by human saliva ***(5)***.

Salicylate (2-hydroxybenzoate) was largely used as a marker of hydroxyl free-radical production both in vitro ***(6,7)*** and in vivo ***(8–10)***. The success of salicylate as a probe of $OH^{\bullet}$ production is mainly owing to three factors:

From: *Methods in Molecular Biology, vol. 108: Free Radical and Antioxidant Protocols*
Edited by: D. Armstrong © Humana Press Inc., Totowa, NJ

1. It has a high reaction rate constant with $OH^{\bullet}$ ($5 \times 10^9/M/s$) *(11)*;
2. It may be used at concentrations sufficient to compete with other possibly present scavenger molecules;
3. Its hydroxylated products are rather stable and some of these are not metabolically produced (*see* **Fig. 1**). In fact, upon salicylate, the attack of $OH^{\bullet}$ produced by different systems such as Fenton reactions and γ-radiolysis, give rise to a set of hydroxylated by-products (2,3-dihydroxybenzoate, 2,5-dihydroxybenzoate, and catechol) *(12–14)*.

Microsomal fractions from mammals treated with inducers of cytochrome P-450, produce 2,5-dihydroxybenzoate (2,5-dHB), but not the 2,3 isomer (2,3-dHB) *(15)*; catechol is only a minor product *(14)*, so that 2,3-dHB appears to be the only reliable marker of $OH^{\bullet}$ production in vivo.

Salicylates are analgesics that are used in the treatment of a variety of pain syndromes. Aspirin is commonly used for the relief of headaches, myalgia, and arthralgia. Once ingested, it is quickly hydrolyzed to salicylate by esterases in the gastrointestinal tract, in the liver, and to a smaller extent in the serum *(16)*; salicylate reaches its peak in plasma about 1.5 h after the oral intake of aspirin *(17)* and is in part metabolized by conjugation with glycine (liver glycine N-acetylase) *(18)*, by conjugation with glucuronic acid to form salicyl acyl and salicylphenolic glucuronide *(19)*, and by hydroxylation (liver microsomalhydroxylases) to form 2,5-dihydroxybenzoic acid *(15)*. About 60% of the salicylate remains unmodified *(19)* and can undergo $OH^{\bullet}$ attack (**Fig. 1**).

The aim of this chapter is to provide a series of validated procedures for detecting $OH^{\bullet}$ production in different models, in vitro and in vivo.

We used salicylate hydroxylation both in simple biochemical systems for assessing the antioxidant capacity of selected compounds, and in cultured cells for evaluating the activity of different antioxidants and to study the mechanisms involved in cell $OH^{\bullet}$ production. Moreover, the versatility of the technique suggests its application also in humans, for evaluating the oxidative stress status both in pathology and in physiology.

2. Materials

2.1. Instruments and Column

2.1.1. High-Pressure Liquid Chromatography (HPLC) System

1. Reverse-phase HPLC analysis is carried out on a Supelcosil™ LC-18, 5 µ*M* (250 × 4.6 mm) analytical column, using a Perkin Elmer series 410 LC pump equipped with a Perkin Elmer SEC-4 solvent environmental control.
2. Perkin Elmer LC-95 UV/Vis Spectrophotometer detector: Set at 300 nm, 0.005 AUFS recorder range.
3. BAS LC-4B amperometric detector (Bio-Analytical Systems, West Lafayette, IN): Equipped with a BAS detector cell. The detector potential is set at +0.76 V versus an Ag/AgCl reference electrode.

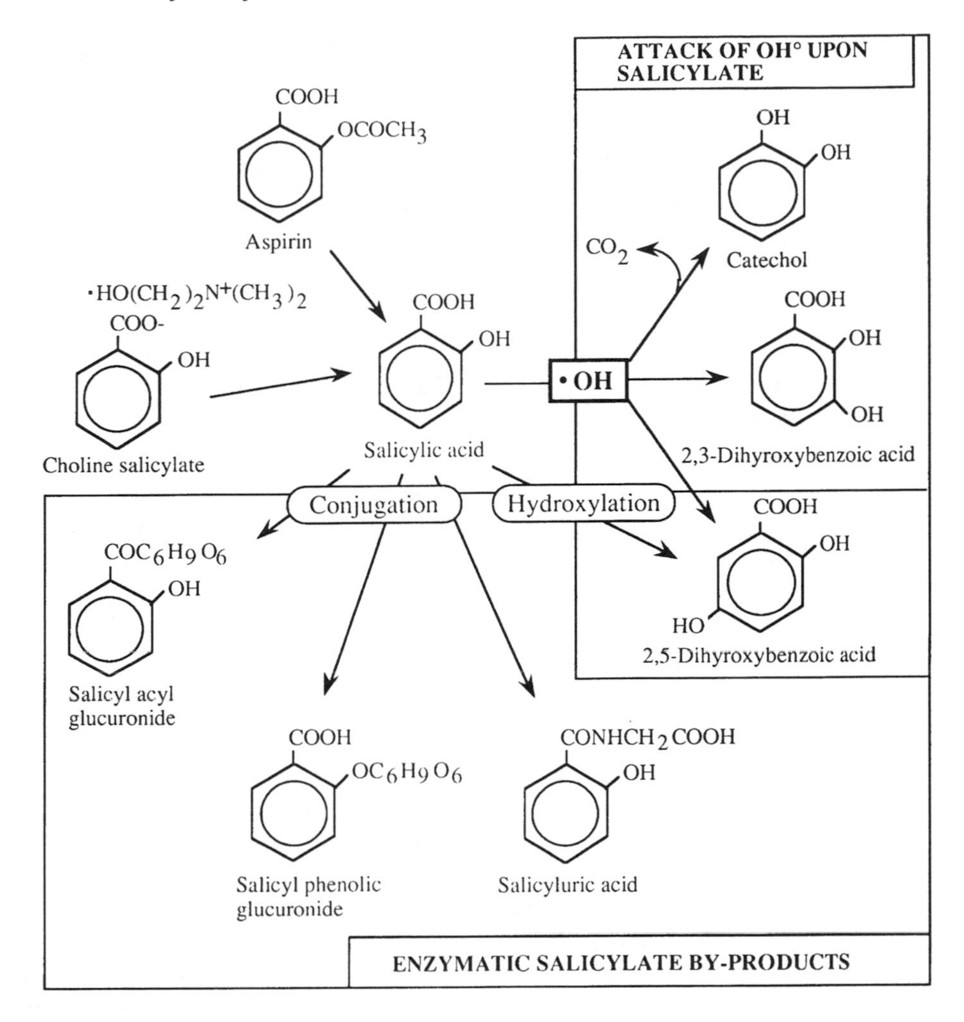

Fig. 1. Diagram showing the main pathways involved in salicylate metabolism. In in vivo studies, this compound may be administered in different forms: aspirin, sodium, or choline salicylate, all releasing salicylate after hydrolysis.

4. ESA (Bedford, MA) Coulochem II detector: Equipped with a conditioning cell (Model 5021) followed by a 2011 analytical cell.

2.2. Reagents

All reagents should be of the highest chromatographic quality to avoid interferences and to ensure reproducibility, and all solutions are made using Milli-Q (Millipore, Bedford, MA) double distilled water (resistance > 18 mΩ/cm^2).

1. Salicylic acid and dihydroxybenzoate isomers (dHBs) (Aldrich, Steinheim, Germany).
2 HPLC-grade solvents (Carlo Erba, Milano, Italy).
3 All other chemicals are purchased from Sigma (St. Louis, MO).
4. The mobile phase (1.0 mL/min flow) consists in 95% (v/v) 30 m*M* sodium citrate/27.7 m*M* acetate buffer and 5% (v/v) methanol, pH 4.75. The solution is filtered through a 0.22 μ*M* pore size Millipore filter and continuously sparged with He gas during elution. The injection loop volume is 100 μL.

3. Methods

3.1. In Vitro Evaluation of the Scavenging Capacity of Antioxidants

1. Salicylate stock solutions (100 m*M*), freshly made in methanol are kept on ice until use. Single antioxidants or mixtures to be analyzed are incubated in 10 m*M* phosphate buffer, pH 7.00, containing 2.0 m*M* salicylate in polypropylene tubes at 37°C (*see* **Note 1**).
2. A flux of $OH^•$ is commonly generated by Fenton's reagent consisting of 100 μ*M* $FeCl_3$, 104 μ*M* EDTA, 100 μ*M* ascorbate and 1.0 m*M* H_2O_2. Iron-EDTA complex is obtained by pre-mixing stock solutions; reaction is started by ascorbate addition *(7)*.
3. After 15 min incubation, the tubes were transferred to an ice-bath and 200 μM deferoxamine and 2.0 mM dimethyl sulfoxide (DMSO) were added. Samples are then extracted into diethyl ether (*see* **Subheading 3.4.1.**). Under these conditions the attack of $OH^•$ on salicylate produces 2,3 and 2,5-dHB to a similar extent (*see* **Note 1**). In the absence of enzymatic systems metabolizing salicylate, the sum of the two dHBs may be used to obtain a measure of the $OH^•$ that has escaped from the antioxidant scavenging. Following this protocol, the anti-hydroxyl radical capacity of single compounds or antioxidant mixtures may be measured. The results may be expressed as the percentage of $OH^•$ trapped, in comparison with the "blank" or standardized by a well-known $OH^•$ scavenger and expressed as mM equivalents of it; thiourea, having a reaction-rate constant with $OH^•$ similar to that of salicylate *(11)* is often used for this purpose.

3.2. Evaluation of $OH^•$ Production by Stimulated Cells

To investigate the capacity of human neutrophils to make $OH^•$ during the "respiratory burst," Halliwell et al. *(19)* used phenylalanine hydroxylation and similar experiments have been conducted with salicylate *(20)*.

We studied the production of $OH^•$ by platelets stimulated to aggregate with arachidonic acid. Blood samples were taken from healthy volunteers who had not ingested salicylate-containing drugs; blood was collected in BD Vacutainer containing 0.13*M* Na-citrate (ratio 9:1). Platelet-rich plasma was separated by centrifugation at 100*g* and platelets were washed three times in phosphate-buffered saline (PBS) as described more in detail elsewhere ***(21)***. Platelets were suspended to a final count of 5 x 10^8/mL in PBS containing 1.5 mg/mL fibrinogen and pre-incubated at 37°C for 5 min with 2.0 m*M* salicylate. Arachidonic acid (10–50 μ*M*) was used as aggregating agent, the reaction allowed for 10 min, then stopped by adding cold 1.0 vol. of 20% trichloroacetic acid. The supernatant was extracted into diethyl ether (*see* **Subheading 3.4.1.**). Also in this system, the rate of dHB hydroxylation is similar for the 2,3- and the 2,5 isomer, thus their sum may be used.

3.3. Evaluation of OH• Production in Humans

The investigation of the rate of in vivo oxidative stress in humans certainly represents the most engaging application of salicylate hydroxylation.

1. We evaluated the in vivo OH• production in two groups of subjects, who were presumably exposed to a high rate of oxidative stress: diabetic patients ***(8)*** and heavy smokers (manuscript in preparation). Diabetics were 20 subjects suffering from insulin-dependent diabetes mellitus in good metabolic control. Smokers (24 healthy subjects, who smoked more than 250 mg tar/d) were studied before and after receiving a daily antioxidant supplementation consisting of ascorbic acid, tocoferol and b-carotene (1000, 600, and 25 mg, respectively) for 28 d (*see* **Note 2**).
2. All subjects suffering from renal and hepatic failure, asthma, bleeding disorders, and peptic diseases were excluded. Each fasting subject took 1.0 g aspirin (Flectadol 1000®, Maggioni, Italy) by mouth and had a breakfast (150 mL whole milk and 50 g carbohydrates) to avoid gastric distress; 3 h after the aspirin intake a blood sample was withdrawn into heparinized BD-Vacutainer and immediately centrifuged at 2000*g* for 10 min at +4°C.
3. Portions of plasma were extracted for benzoates as described below (*see* **Subheading 3.4.1.**).

3.4. Assay Conditions

1. Extraction Procedures: To 500 μL of sample to be extracted, 25 μL 1.0 *N* HCl and 50 n*M* final concentration 3,4-dHB as internal standard are added with mixing. Samples pre-treated with trichloroacetic acid (platelets) are only added with the internal standard. The resulting solution is twice extracted into 5.0 mL of HPLC-grade diethyl ether by thoroughly mixing for 2 min, and subsequently centrifuged for 1 min at 1000*g* to separate the phases. The organic phases are joined, and evaporated to dryness under nitrogen.
2. Owing to the instability of benzoates in biological fluids ***(13)***, samples should be extracted as soon as possible. Evaporated samples may be stored at –80°C until the analysis.
3. Just before the injection into the HPLC system, evaporated samples are reconstituted to 1.0 mL with mobile phase added with 2.0 m*M* DMSO and 200 μ*M*

deferoxamine. These two compounds do not interfere with the chromatographic separation and detection, and by avoiding autooxidation of salicylate and dHBs, ensure accurate and reproducible results.

3.5. Choice of Working Conditions

1. Electrochemical detectors provide significant benefits over other means. There are two approaches to electrochemical detection for HPLC: amperometry and coulometry. Amperometric detectors, such as BAS LC-4B are designed so that the eluent flows by the electrode surface. The fraction of the electroactive compound that reacts is about 10%. On the contrary, coulometric detectors are made so that the eluent flows trough a porous electrode and all the electroactive compounds react. Hence the sensitivity of coulometric detectors is 10 (and more) times higher than that of amperometric detectors. This may be observed in **Fig. 2**, where the hydrodynamic voltammograms relative to the dHB isomers are made with amperometric and coulometric detector. Hydrodynamic voltammetry represents not only the fingerprint of a selected molecule, but it is also used to determine the potential to be applied to the analysis. Coulometric detection provides important advantage respect to amperometry. A sensitivity at least 20 times higher than amperometry, with a significant rise of the signal-to-noise ratio is obtained. According to the results of **Fig. 2**, BAS LC-4B detector was set at 0.76 V in oxidation mode (the potential giving the highest signal-to-noise ratio). To minimize the noise, mobile phase may be prepared the d before the analysis and recycled by placing the outlet tube from the detector into the solvent reservoir. This procedure gives an increase in signal-to-noise ratio without affecting HPLC separation (methanol evaporation is negligible). As ESA Coulochem II oxidizes 100% of analytes the recycle of the mobile phase may persist during working procedures and does not produce interferences with the analysis. Moreover the high voltage necessary to produce detectable peaks in amperometry may shorten the lifetime of electrodes.
2. ESA Coulochem II settings were the following: Conditioning cell and the first electrode of the analytical cell –0.1 V; the second electrode was the analytical one and was set at +0.3 V.
3. The retention times expressed in min, relative to dHBs and salicylate, are 6.8 ± 0.6 for 2,3-dHB, 7.7 ± 0.5 for 2,5-dHB, 9.8 ± 0.7 for 3,4-dHB, and 23.5 ± 1.0 for salicylate. If salicylate values are considered unessential for the experimental protocol, the chromatographic assay is complete within 12 min.
4. Salicylate is not electrochemically active in the conditions described and it needs higher voltage to be detected (1.0 V). Owing to its high concentrations both in vitro (2.0 m*M*) and in plasma samples (about 400 m*M*) it may easily be detected using an ultraviolet (UV) detector set at 300 nm. **Figure 3** reports a typical chromatographic run.

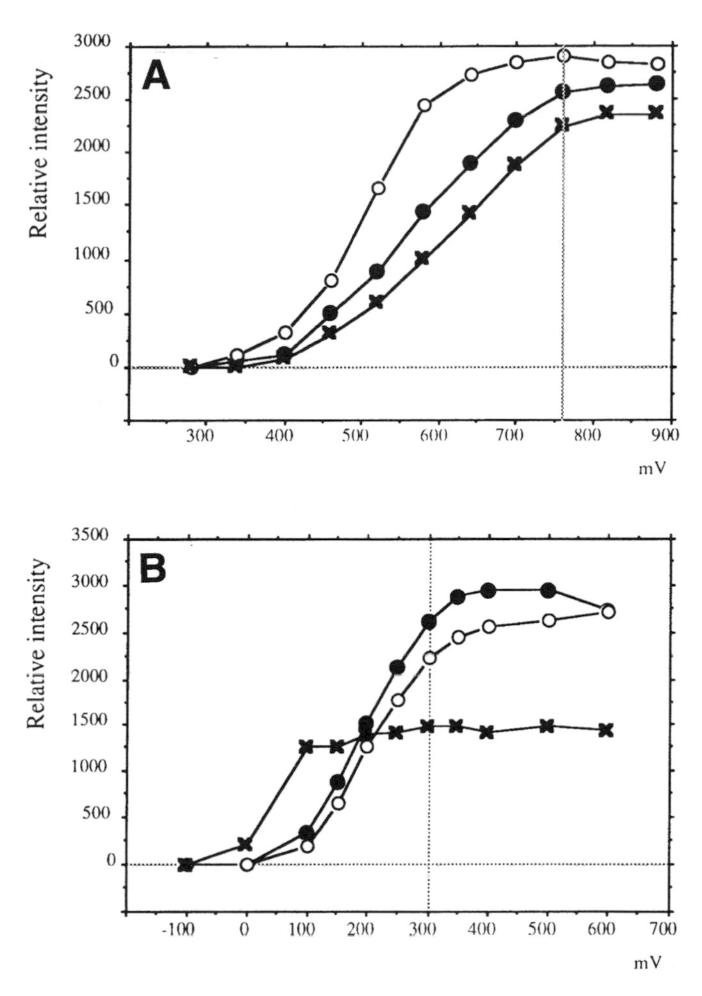

Fig. 2. Hydrodynamic voltammometry of all di-hydroxybenzoate isomers. (**A**) Amperometric detector, 50 pmol injected. Full-scale current was set at 10 nA (**B**) Coulometric detector, 2.5 pmol injected. Full-scale current was set at 100 nA. The dotted lines show the potential used for analytical runs. (o), 2.3-dHB; (•), 2.5-dHB; (X), 3.4-dHB.

3.6. Experimental Results

The results of the analytical procedures are reported in **Subheadings 3.5.1. – 3.5.4.** Here are reported the main results obtained by employing salicylate hydroxylation in proposed experimental protocols for measuring in vitro and in vivo $OH^{\bullet}$ production (*see* Methods).

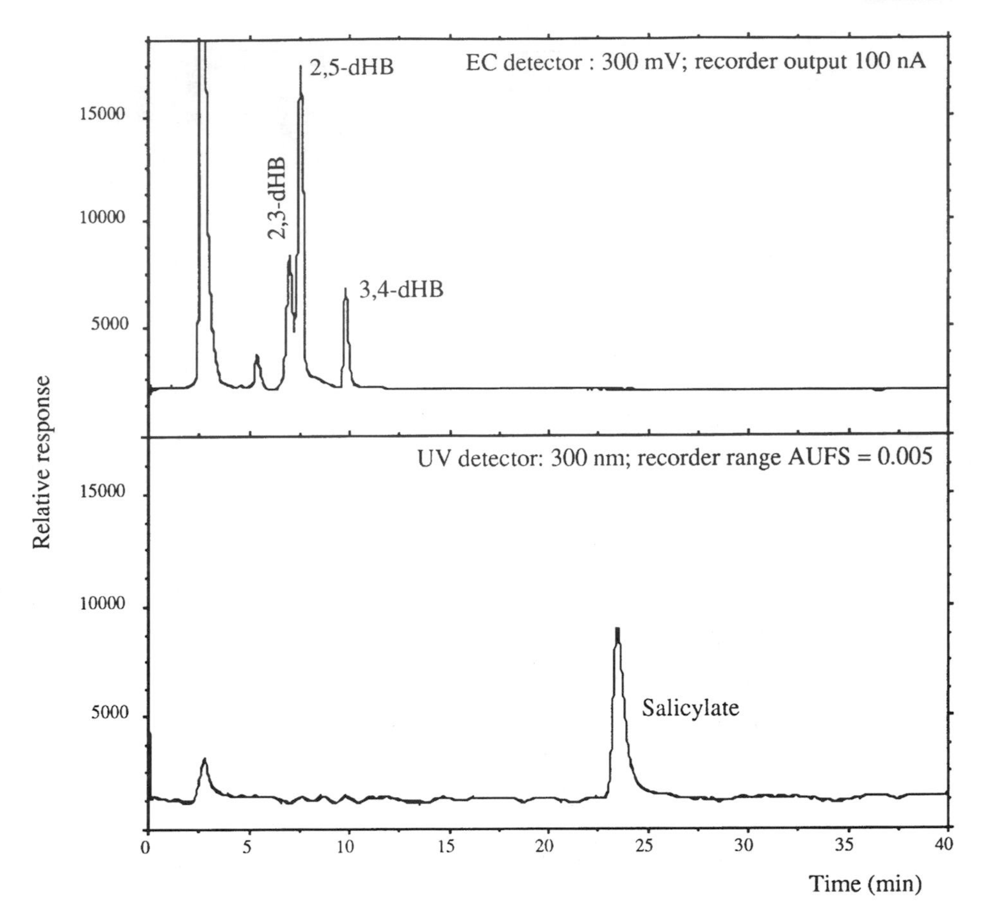

Fig. 3. HPLC-separation of dHB isomers (EC Coulometric detector) and salicylate (UV detector) in human plasma sample of a healthy subject after a single dose of 1.0 g Aspirin.

3.6.1. OH• Production by Stimulated Platelets

A dose-dependent production of hydroxyl radicals by platelets stimulated to aggregate with increasing amounts of arachidonic acid is shown in **Fig. 4**. Salicylic acid at this concentration does not affect cyclooxygenase pathway ***(21)***, consequently it does not interfere with platelet aggregation.

The amount of OH• produced by platelets under these conditions probably originates by Fenton-like reactions triggered by the disruption of cell membranes with subsequent release of metal ions. The production of OH• by platelets would be necessary for aggregation itself ***(22,23)***, and could contribute to the understanding of the relationship between platelets and endothelial damage in atherosclerosis (*see* **Note 3**).

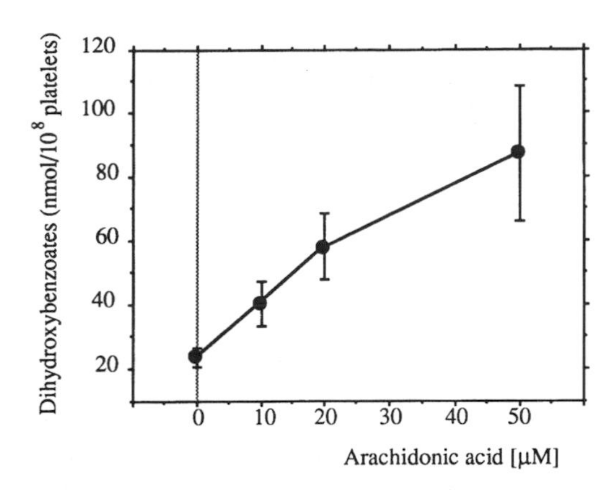

Fig. 4. $OH^\bullet$ production by platelets stimulated to aggregate with increasing amounts of arachidonic acid. Values represent the sum of 2,3 and 2,5-dHB

3.6.2. Assessment of Oxidative Stress in Humans

The results obtained by our in vivo studies are reported in **Table 1**. For the reasons previously discussed, when the experimental protocols are applied in vivo, only 2,3-dHB can be used as marker of $OH^\bullet$ production; the 2,5 isomer is produced by enzymatic pathways.

Diabetic patients (in the left part of the table), even in good metabolic control, are exposed to an hyperproduction of $OH^\bullet$ and at that time, the increase in salicylate hydroxylation is the sole sign that they are exposed to oxidative stress; this observed oxidative-stress status could help to explain why these patients will undergo late complications. Salicylate hydroxylation reveals, therefore, that a certain amount of $OH^\bullet$ is eluding antioxidant defenses before that the phenomenon becomes evident, and produces angiopathy, neuropathy, or cataract.

In contrast heavy smokers (the right hand of the table) did not show any significant difference in comparison with controls, although they showed a trend toward higher values. After 28 d of antioxidant supplementation, however, they showed significantly reduced values of plasma 2,3-dHB. Probably their rate of $OH^\bullet$ production did not change, but the increased antioxidant defenses may have trapped a higher amount of radicals. Controls did not show any significant change (49 ± 6 nmol/l) after antioxidant supplementation.

4. Notes

1. The protection of salicylate and dHBs against auto-oxidation during the extraction and analysis times is extremely important in that unexpected oxidation may occur during these phases giving false results. An antioxidant cocktail containing

Table 1
Values of Hydroxyl Radical Production in Humans Exposed to Two Different Sources of Oxidative Stress[a]

Diabetes mellitus, 2,3 dHB, n*M*		Cigaret smoking, 2,3 dHB, n*M*	
Controls (n = 10)	49 ± 7	Controls (n = 10)	50 ± 17
Diabetics (n = 20)	63 ± 20 p<0.05	Smokers before (n = 24)	56±5
		Smokers after (n = 24)	46 ± 8 p <0.05

[a]Values are expressed mean ± SD and statistical significance.

deferoxamine and DMSO, as previously described, is sufficient in our assay condition to minimize salicylate oxidation. All solution should be made in metal ion-freed distilled water.

2. Plasma-salicylate concentration in humans who ingested 1.0 g aspirin ranges from 200–500 µ*M*. Other salicylate salts (*see* **Fig. 1**) can be administered in higher doses (Na-salicylate or choline salicylate, Arthropan®); they are able to reach higher plasma concentrations, thus trapping a larger amount of OH•.
3. Salicylate salts, moreover, unlike aspirin, do not inhibit platelet function ***(21)***. Because platelets are able to generate OH• (**Fig. 4**) through the cyclooxygenase pathway, their inactivation could give an underestimate of the real amount of OH• produced, especially in patients suffering from degenerative diseases that often are coupled with platelet hyperfunction. The administration of sodium or choline salicylate could alleviate this concern.

References

1. Halliwell, B. and Gutteridge, J. M. C. (1989) Free radicals, aging and disease, in *Free Radicals in Biology and Medicine* (Halliwell, B. and Gutteridge, J. M. C., eds.), Clarendon Press, Oxford, UK, pp. 416–508.
2. Halliwell, B. and Gutteridge, J. M. C. (1990) The antioxidants of human extracellular fluids. *Arch. Biochem. Biophys.* **280,** 1–8.
3. Grootveld, M. and Halliwell, B. (1986) Aromatic hydroxylation as a potential measure of hydroxyl radical formation in vivo. *Biochem. J.* **237,** 499–504.
4. Kaur, H., Fagerheim, I., Grootveld, M., Puppo, A., and Halliwell, B. (1988) Aromatic hydroxylation of phenylalanine as an assay for hydroxyl radicals: application to activated human neutrophils and to the heme protein leghemoglobin. *Anal. Biochem.* **172,** 360–367.
5. Nair, U. J., Nair, J., Friesen, M. D., Bartsch, H., and Ohshima, H. (1995) Otho- and meta-tyrosine formation from phenylalanine in human saliva as a marker of hydroxyl radical generation during betel quid chewing. *Carcinogenesis* **16,** 1195–1198.
6. Floyd, R. A., Watson, J. J., and Wong, P. K. (1984) Sensitive assay of hydroxyl free radicals formation utilizing high pressure liquid chromatography with electrochemical detection of phenol and salicylate hydroxylation products *J. Biochem. Biophys. Methods* **10,** 221–235.

7. Iuliano, L., Pedersen, J. Z., Ghiselli, A,, Praticò, D., Rotilio, G., and Violi F. (1992) Mechanism of reaction of a suggested superoxide-dismutase mimic, Fe(II)-N,N,N′,N′- Tetrakis (2-pyridylmethyl) ethylendiamine. *Arch. Biochem. Biophys.* **293,** 153–157.
8. Ghiselli, A., Laurenti, O., De Mattia, G., Maiani, G., and Ferro-Luzzi, A. (1992) Salicylate hydroxylation as an early marker of oxidative stress in diabetic patients. *Free Rad. Biol. Med.* **13,** 621–626.
9. O'Connell, M. J. and Webster, N. R. (1990) Hyperoxia and salicylate metabolism in rats. *J. Pharm. Pharmacol.* **42,** 205–206.
10. Floyd, R. A., Henderson, R., Watson, J. J., and Wong, P. K. (1986) Use of salicylate with high pressure liquid chromatography and electrochemical detection (LED) as a sensitive measure of hydroxyl free radicals in adriamycin-treated rats. *Free Rad. Biol. Med.* **2,** 13–18.
11. Anbar, M. and Neta, P. (1967) A compilation of specific bimolecular rate constants for the reactions of hydrated electrons, hydrogen atoms, hydroxyl radicals with inorganic and organic compounds in aqueous solutions. *Int. J. Appl. Rad. Isot.* **18,** 493–523.
12. Richmond, R., Halliwell, B., Chauhan, J., and Darbre, A. (1981) Superoxide-dependent formation of hydroxyl radicals: detection of hydroxyl radicals by the hydroxylation of aromatic compounds. *Anal. Biochem.* **118,** 328–335.
13. Maskos, Z., Rush, J. D., and Koppenol, W. H. (1990) The hydroxylation of the salicylate anion by a Fenton reaction and g-radiolysis: a consideration of the respective mechanisms. *Free Rad. Biol. Med.* **8,** 153–162.
14. Grootveld, M. and Halliwell, B. (1986) Aromatic hydroxylation as a potential measure of hydroxyl radical formation in vivo. *Biochem. J.* **237,** 499–504.
15. Halliwell, B., Kaur, H., and Ingelman-Sundberg, M. (1991) Hydroxylation of salicylate as an assay for hydroxyl radicals: a cautionary note. *Free Rad. Biol. Med.* **10,** 439–441.
16. Levy, G. (1965) Pharmacokinetics of salicylate elimination in man. *J. Pharm. Sci.* **54,** 959–967.
17. Rumble, R. H. and Roberts, M. S. (1981) Determination of aspirin and its metabolites in plasma by High Performance Liquid Chromatography without solvent extraction. *J. Chromatog.* **255,** 252–260.
18. Cuny, G., Royer, R. J., Mur, J. M., Serot, J. M., Faure, G., Netter, P., Maillard, A., and Penin, F. (1979) Pharmacokinetics of salicylate in elderly. *Gerontology* **25,** 49–55.
19. Alpen, E. L., Mandel, H. G., Rodwell, V. W., and Smith, P. K. (1951) The metabolism of C^{14} carboxyl salicylic acid in the dog and in man *J. Pharmacol. Exp. Ther.* **102,** 150–156.
20. Davis, W. B., Mohammed, B. S., Mays, D. C., She, Z. W., Mohammed, J. R., Husney, R. M., and Sagone, A. L. (1989) Hydroxylation of salicylate by activated neutrophils. *Biochem. Pharmacol.* **38,** 4013–4019.
21. Hollister, L. and Levy, G. (1965) Some aspects of salicylate metabolism in man. *J. Pharm. Sci.* **54,** 1125–1129.

22. Violi, F., Ghiselli, A., Iuliano, L., Alessandri, C., Cordova, C., and Balsano, F. (1988) Influence of hydroxyl radical scavengers on platelet function. *Haemostasis* **18,** 91–98.
23. Iuliano, L., Pedersen. J. Z., Praticò, D., Rotilio, G., and Violi, F. (1994) Role of hydroxyl radicals in the activation of human platelets. *Eur. J. Biochem.* **221,** 695–704.

9

Simple Assay for the Level of Total Lipid Peroxides in Serum or Plasma

Kunio Yagi

1. Introduction

As is well-recognized, lipid peroxidation is initiated by the abstraction of a hydrogen atom from a polyunsaturated fatty acid, and proceeds via free-radical species to form the primary stable product, lipid hydroperoxide, and then the secondary stable product, lipid aldehyde, as depicted in **Fig. 1.** To check lipid peroxidation in the body, investigate free radical injury, or aid in the diagnosis of lipid peroxide-mediated diseases, one of the most routine test samples is serum or plasma. In serum and plasma, however, free-radical species involved in the process shown in **Fig. 1** all disappear because of their short lifetime, and only stable substances remain. Even such stable substances are liable to decompose to some extent, and, therefore, prolonged treatment or organic solvent-extraction should be avoided. Accordingly, the principle of the procedure presented here is to react these lipid peroxides with a suitable reagent after eliminating other reactive or disturbing substances.

Many years ago, a simple and reliable method for the assay of lipid peroxide level in serum or plasma by use of the thiobarbituric acid (TBA) reaction was devised *(1)*. As was already known, TBA is a sensitive reagent for the detection of lipid peroxides, but its specificity is rather low. Therefore, it was decided to eliminate disturbing substances by a simple procedure. After many trials, a relatively simple procedure was found. For elimination of disturbing substances such as glucose and water-soluble aldehydes, lipids were precipitated along with proteins by use of a phosphotungstic acid-sulfuric acid system, and the reaction of TBA with these lipids was carried out in acetic acid solution to avoid its reaction with other substances such as sialic acid. After the TBA reaction, the product (structure, *see* **Fig. 2**) is measured by fluorometry,

From: *Methods in Molecular Biology, vol. 108: Free Radical and Antioxidant Protocols*
Edited by: D. Armstrong © Humana Press Inc., Totowa, NJ

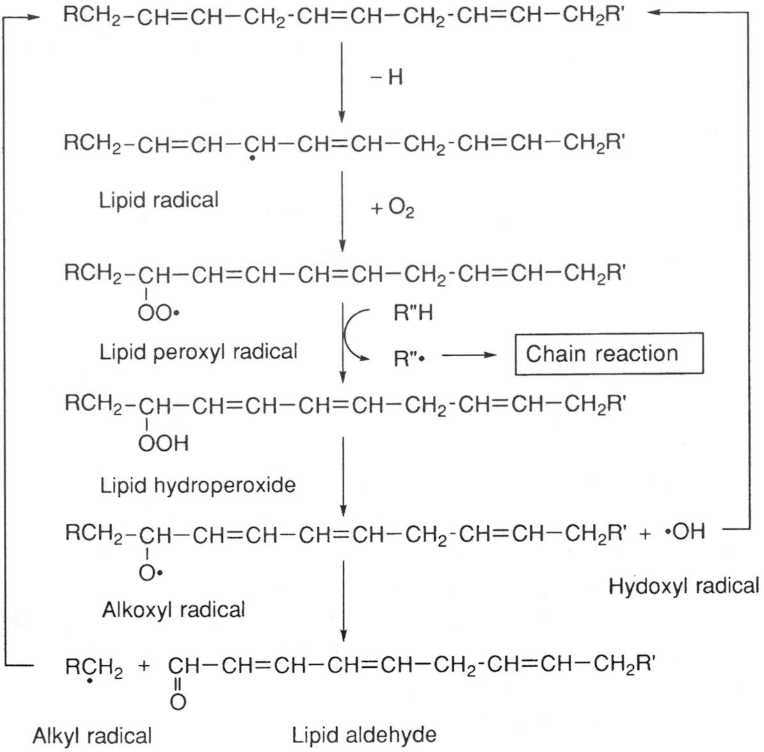

$RCH_2-CH=CH-CH_2-CH=CH-CH_2-CH=CH-CH_2R'$

$-H$

$RCH_2-CH=CH-\dot{C}H-CH=CH-CH_2-CH=CH-CH_2R'$

Lipid radical

$+O_2$

$RCH_2-CH(OO\cdot)-CH=CH-CH=CH-CH_2-CH=CH-CH_2R'$

Lipid peroxyl radical

R"H → R"• → Chain reaction

$RCH_2-CH(OOH)-CH=CH-CH=CH-CH_2-CH=CH-CH_2R'$

Lipid hydroperoxide

$RCH_2-CH(O\cdot)-CH=CH-CH=CH-CH_2-CH=CH-CH_2R'$ + •OH

Alkoxyl radical

Hydoxyl radical

$R\dot{C}H_2$ + $CH(=O)-CH=CH-CH=CH-CH_2-CH=CH-CH_2R'$

Alkyl radical

Lipid aldehyde

Fig. 1. Peroxidation of polyunsaturated fatty acids.

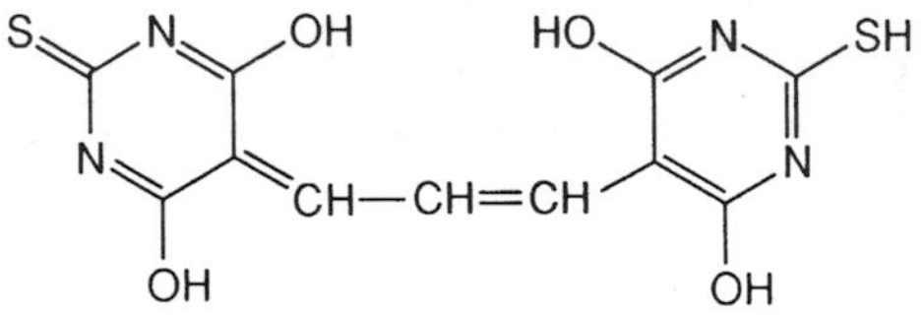

S N OH HO N SH

N CH—CH=CH N

OH OH

Fig. 2. Structure of the product obtained from the reaction of lipid peroxides or malondialdehyde with TBA.

because it was found that the TBA reaction product is a fluorescent substance *(1)* as represented in **Fig. 3**. By use of fluorometry, the specificity of the assay method increased, and, in addition, the assay became applicable to a small volume of blood and, therefore, to infants and small animals.

This method gives the level of the total amount of lipid hydroperoxides and lipid aldehydes derived from them. The lipid peroxide level measured by this

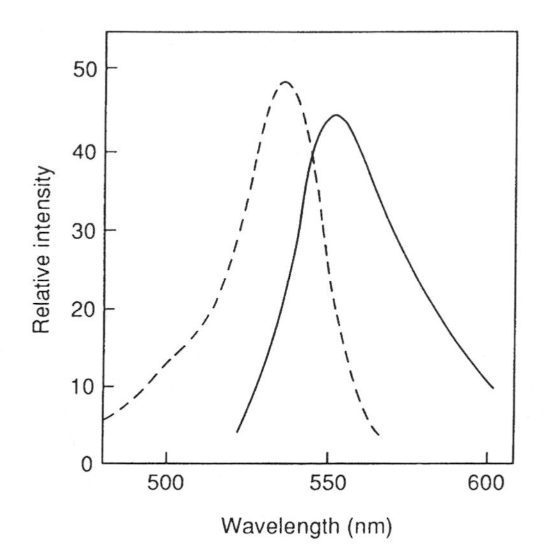

Fig. 3. Excitation and emission spectra of the product obtained from the reaction of lipid peroxides or malondialdehyde with TBA. Dotted line: excitation spectrum monitored at 575 nm; solid line: emission spectrum obtained with excitation at 515 nm.

method was found to be increased in various diseases ***(2)*** such as liver diseases ***(3)***, diabetes ***(4)***, atherosclerosis ***(5)***, and apoplexy ***(6)***.

In this chapter, this simple and reliable method for the assay of the level of the total amount of lipid hydroperoxides and lipid aldehydes in serum or plasma is described.

2. Materials

2.1. Reagents

1. *N*/12 H_2SO_4.
2. 10% Phosphotungstic acid.
3. TBA reagent: mixture of equal volumes of 0.67% TBA aqueous solution and glacial acetic acid.
4. *n*-Butanol, reagent grade.
5. Pure tetramethoxypropane standard.

3. Methods

3.1. Basic Procedure

The standard procedure is presented here. If only a small volume of the blood is available, as in the case of infants, plasma lipid peroxide level/mL of blood can be measured by the next procedure presented (*see* **Notes 1** and **2**).

3.1.1. Standard Procedure for the Assay of Serum or Plasma Lipid Peroxide Level

1. 20 μL of serum or plasma are placed in a glass centrifuge tube.
2. Add 4.0 mL of *N*/12 H_2SO_4 and mix gently.
3. Add 0.5 mL of 10% phosphotungstic acid and mix. After allowing to stand at room temperature for 5 min, the mixture is centrifuged at 1600*g* for 10 min.
4. Discard supernatant and mix sediment with 2.0 mL of *N*/12 H_2SO_4 and 0.3 mL of 10% phosphotungstic acid. The mixture is centrifuged at 1600*g* for 10 min.
5. Sediment is suspended in 4.0 mL of distilled water, and 1.0 mL of TBA reagent is added. Reaction mixture is heated at 95°C for 60 min in an oil bath.
6. After cooling with tap water, 5.0 mL of *n*-butanol is added and the mixture is shaken vigorously.
7. After centrifugation at 1600*g* for 15 min, the *n*-butanol layer is taken for fluorometric measurement at 553 nm with excitation at 515 nm.
8. Taking the fluorescence intensity of the standard solution, which is obtained by reacting 0.5 nmol of tetramethoxypropane in 4.0 mL of water with 1.0 mL of TBA reagent by steps 5–7, as *F* and that of the sample as *f*, the lipid peroxide level can be calculated and expressed in terms of malondialdehyde.

Serum or plasma lipid peroxide level

$$= 0.5 \times \frac{f}{F} \times \frac{1.0}{0.02} = \frac{f}{F} \times 25 \text{ (nmol/mL of serum or plasma)}$$

3.1.2. Procedure for the Assay of Lipid Peroxide Level in a Small Volume of Blood

1. Take 0.05 mL of blood using a micropipet (e.g., from the ear lobe).
2. Put blood into 1.0 mL of physiological saline in a centrifuge tube, and shake gently.
3. After centrifugation at 1600*g* for 10 min, transfer 0.5 mL of the supernatant to another centrifuge tube.
4. Follow **steps 2–8** of the procedure in **Subheading 3.1.1.**

Plasma lipid peroxide level

$$= 0.5 \times \frac{f}{F} \times \frac{1.05}{0.5} \times \frac{1.0}{0.05} = \frac{f}{F} \times 21 \text{ (nmol/mL of blood)}$$

3.2. Results

Since both lipid hydroperoxides and lipid aldehydes are plural substances, and they all react with TBA to give the same product as that of malondialdehyde (**Fig. 2**), malondialdehyde is appropriate as standard. Practically, however, tetramethoxypropane is used, because it is stable for storage and quantitatively

Table 1
Serum Lipid Peroxide Levels of Normal Subjects[a]

	Lipid peroxide level	
Age (years)	Male	Female
≤10	1.86 ± 0.60 (10)	2.08 ± 0.48 (8)
11–20	2.64 ± 0.60 (10)[b]	2.64 ± 0.54 (9)
21–30	3.14 ± 0.56 (10)	2.98 ± 0.50 (9)
31–40	3.76 ± 0.52 (11)[c]	3.06 ± 0.50 (9)[d]
41–50	3.94 ± 0.60 (11)	3.16 ± 0.54 (10)[d]
51–60	3.92 ± 0.92 (8)	3.30 ± 0.74 (10)
61–70	3.94 ± 0.70 (10)	3.46 ± 0.72 (10)
≥71	3.76 ± 0.76 (12)	3.30 ± 0.78 (10)
Mean	3.42 ± 0.94 (82)	3.10 ± 0.62 (75)[d]

[a]Lipid peroxide levels were measured according to this method and expressed in terms of malondialdehyde (nmol/mL serum). Mean ± SD is given. The number of subjects is given in parentheses. Significant difference:
[b]$p < 0.05$ vs the age group ≤10;
[c]$p < 0.05$ vs the age group 21–30;
[d]$p < 0.05$ vs the corresponding group of males.

converted to malondialdehyde upon being heated in acidic solution. Thus, the value obtained by this method gives relative level of lipid hydroperoxides and/or lipid aldehydes expressed in terms of malondiadehyde, but not their absolute quantity. Therefore, it is recommended to describe the value as "lipid-peroxide level expressed in terms of malondialdehyde." The term "TBA-reactive substances (TBARS)" is often used by some researchers. In the case of the measurement by the present method, TBARS that are not related to lipid peroxides are eliminated; therefore, the obtained value should better be described as "lipid peroxide level," and not as "TBARS."

Normal values are summarized in **Table 1** *(3)*. Values tend to increase with age, but do not exceed 4 nmol/mL in terms of malondialdehyde. Between 30–50 years of age, the value of women is significantly lower than that of men. This explains why women have a lower incidence of atherosclerotic heart disease than men *(7)*.

4. Notes

1. Effect of anti-coagulants: Anti-coagulants used to obtain plasma, *viz*, heparin, citrate, oxalate, and ethylenediaminetetraacetic acid, have no effect on the reac-

tion involved in this method. The values obtained for serum and plasma separated from the same blood are essentially the same.

2. Throughout this method, distilled water should be free of transient metals, especially iron. The contamination of a minute amount of iron in the water will give inaccurate values for lipid peroxides.

References

1. Yagi, K. (1976) A simple fluorometric assay for lipoperoxide in blood plasma. *Biochem. Med.* **15,** 212–216.
2. Yagi, K. (1987) Lipid peroxides and human diseases. *Chem. Phys. Lipids* **45,** 337–351.
3. Suematsu, T., Kamada, T., Abe, H., Kikuchi, S., and Yagi, K. (1977) Serum lipoperoxide level in patients suffering from liver diseases. *Clin. Chim. Acta* **79,** 269–270.
4. Sato, Y., Hotta, N., Sakamoto, N., Matsuoka, S., Ohishi, N., and Yagi, K. (1979) Lipid peroxide level in plasma of diabetic patients. *Biochem. Med.* **21,** 104–107.
5. Plachta, H., Bartnikowska, E., and Obara, A. (1992) Lipid peroxides in blood from patients with atherosclerosis of coronary and peripheral arteries. *Clin. Chim. Acta* **211,** 101–112.
6. Kawamoto, M., Kagami, M., and Terashi, A. (1986) Serum lipid peroxide level in apoplexia. *J. Clin. Biochem. Nutr.* **1,** 1–3.
7. Castelli, W. P. (1984) Epidemiology of coronary heart disease: the Framingham study. *Am. J. Med.* **76(2A),** 4–12.

10

Simple Procedure for Specific Assay of Lipid Hydroperoxides in Serum or Plasma

Kunio Yagi

1. Introduction

The level of lipid peroxides in serum or plasma has been found to be increased in various diseases *(1)*, especially in vascular disorders such as angiopathy in diabetes *(2)*, atherosclerosis *(3)*, and apoplexy *(4)*.

For verification of the hypothesis that increased lipid peroxides are causative of vascular disorders, experiments were conducted on the effect of linoleic acid hydroperoxide on blood vessels, and injury to the endothelial cells was detected both in vivo *(5)* and in vitro *(6)*. Recently, the generation of hydroxyl radicals from lipid hydroperoxides contained in low-density lipoprotein by the addition of ferrous iron or epinephrine-iron complexes was demonstrated *(7)*. This result would explain, at least in part, the pathogenesis of catecholamine-induced atherogenesis *(8)* and stress-induced atherosclerosis *(9)*. In addition, linoleic acid hydroperoxide was found to inhibit the replication of DNA in the nucleus *(10)*. This induces the inability of cell proliferation, and results in the atrophy of organs and tissues.

From these points of view, special attention has been focused on the level of lipid hydroperoxides in serum or plasma. As in the case of the method described in Chapter 9, a method to determine lipid hydroperoxides by their reaction with a suitable reagent without the disturbance of any other substances has been searched. It has already been reported from our laboratory that lipid hydroperoxides react with a methylene-blue derivative by the catalysis of hemoglobin to yield methylene blue *(11)*. Later, 10-(N-methylcarbamoyl)-3,7-(dimethylamino)-phenothiazine (MCDP) was used for the measurement of lipid hydroperoxides contained in lipids extracted from foods *(12)*. We called this method the hemoglobin-methylene blue (Hb-MB) method, which is based on the reaction of lipid

From: *Methods in Molecular Biology, vol. 108: Free Radical and Antioxidant Protocols*
Edited by: D. Armstrong © Humana Press Inc., Totowa, NJ

hydroperoxides with MCDP in the presence of a catalyzer hemoglobin, as depicted in **Fig. 1**. For direct application of this method to serum or plasma, suitable conditions were investigated, and it was found that dissociation of lipid hydroperoxides from serum or plasma proteins and elimination of the effects of holotransferrin and ascorbic acid are necessary. The dissociation of lipid hydroperoxides from proteins can be attained by lipoprotein lipase in the presence of the detergent Triton X-100. The elimination of the effect of holotransferrin can be made by the addition of triethylenetetraminehexaacetic acid, which forms a chelate with holotransferrin. And ascorbic acid can be eliminated by ascorbate oxidase. Thus, pretreatment of serum or plasma with these agents makes it possible to measure the amount of lipid hydroperoxides in serum or plasma by the application of the Hb-HB method to the sample ***(13)***.

In this chapter, a simple procedure for the specific assay of the total amount of lipid hydroperoxides in serum or plasma is described.

2. Materials

2.1. Assay Using Linoleic Acid Hydroperoxide as External Standard

1. Pretreatment reagent: 0.1 *M* 3-(N-morpholino) propanesulfonic acid (MOPS) buffer, pH 5.8, containing 0.1% Triton X-100, 1.5 units/mL lipoprotein lipase, 9 m*M* trimethylenetetraminehexaacetic acid, and 15 units/mL ascorbate oxidase.
2. MCDP reagent is 0.1 *M* MOPS buffer (pH 5.8) containing 0.1% Triton X-100, 0.04 m*M* MCDP, and 60 µg/mL hemoglobin.
3. Linoleic acid hydroperoxide standard is prepared by the reaction of linoleic acid with soybean lipoxygenase as described by Ohkawa et al. ***(14)***. Briefly, linoleic acid (1 g in ethanol) is dissolved in 1000 mL of 50 m*M* borate buffer, pH 9.0, mixed with 2 mg of lipoxygenase (130,000 U/mg protein), and incubated at 23°C. When the increase in absorbance at 233 nm reaches maximum, the reaction is stopped by the addition of a sufficient amount of 1 *N* HCl to adjust the pH to 4.0. Linoleic acid hydroperoxide is extracted twice with about 500 mL of diethyl ether each time. The extract is then dehydrated with solid Na_2SO_4 and concentrated under vacuum. The hydroperoxide thus obtained is purified by high-performance thin-layer chromatography on a Kiesel Gel_{60} F_{254} plate with a mixture of hexane: diethyl ether (8:7, v/v) as the mobile phase (Rf = 0.27). The purified hydroperoxide is dissolved in ethanol and stored at –20°C. As the external standard, 0.1 mL of 50 nmol/mL linoleic acid hydroperoxide in 10% ethanol is used.

2.2. Assay Using Cumene Hydroperoxide as External Standard

1. Pretreatment reagent (*see* **Subheading 2.1., item 1**).
2. MDCP reagent (*see* **Subheading 2.1., item 2**). The reagent described in 2.1.2 is supplemented with 6 mg/mL 2-methylbenzimidazole and 0.75 mg/mL Sarcosinate LN.
3. Pure cumene hydroperoxide standard.

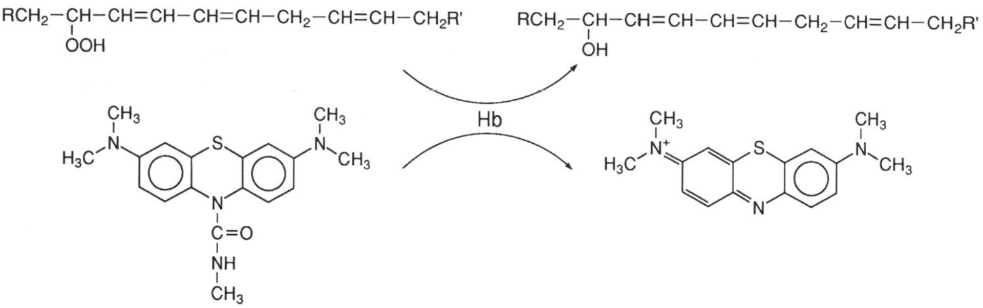

Fig. 1. Hemoglobin-methylene blue (Hb-MB) method for the measurement of lipid hydroperoxides.

3. Method

The standard procedure using linoleic acid hydroperoxide as the external standard is presented here.

1. 100 µL Serum or plasma arc placcd in a glass tube.
2. Add 1.0 mL of the pretreatment reagent and mix.
3. Incubate at 30°C for 5 min.
4. Add 2.0 mL of the MCDP reagent and mix.
5. Incubate at 30°C for 10 min; measure absorbance at 665 nm.
6. Taking the absorbance of the standard solution, which is obtained by treating 0.1 mL of 50 nmol/mL linoleic acid hydroperoxide in 10% ethanol by **steps 2–5**, as A and that of the sample as a, the amount of lipid hydroperoxides is calculated.

$$\text{Serum or plasma lipid hydroperoxides} = 5 \times \frac{a}{A} \times 10 \text{ (nmol/mL of serum or plasma)}$$

The standard procedure using cumene hydroperoxide as the external standard is the same as that shown in the above procedure, except that the supplemented MCDP reagent is used and 0.1 mL of 50 nmol/mL of cumene hydroperoxide in aqueous solution is employed as the external standard (*see* **Notes 1–3**).

4. Notes

1. The value obtained by this method represents the amount of lipid hydroperoxyl groups in a sample, not the amount of lipids containing lipid hydroperoxyl groups. In normal serum or plasma, the value is approximately 1 nmol/mL or less ***(13)***.
2. None of the anti-coagulants mentioned in Chapter 9 have an effect on the reaction involved in this method. The values obtained for serum and plasma separated from the same blood are essentially the same.

3. Although linoleic acid hydroperoxide is a suitable standard, cumene hydroperoxide can be used as a convenient one because of its greater stability. The reaction of cumene hydroperoxide with MCDP is slow, but the reactivity is increased in the presence of 2-methylbenzimidazole and Sarcosinate LN *(13)*.

References

1. Yagi, K. (1987) Lipid peroxides and human diseases. *Chem. Phys. Lipids* **45,** 337–351.
2. Sato, Y., Hotta, N., Sakamoto, N., Matsuoka, S., Ohishi, N., and Yagi, K. (1979) Lipid peroxide level in plasma of diabetic patients. *Biochem. Med.* **21,** 104–107.
3. Plachta, H., Bartnikowska, E., and Obara, A. (1992) Lipid peroxides in blood from patients with atherosclerosis of coronary and peripheral arteries. *Clin. Chim. Acta* **211,** 101–112.
4. Kawamoto, M., Kagami, M., and Terashi, A. (1986) Serum lipid peroxide level in apoplexia. *J. Clin. Biochem. Nutr.* **1,** 1–3.
5. Yagi, K., Ohkawa, H., Ohishi, N., Yamashita, M., and Nakashima, T. (1981) Lesion of aortic intima caused by intravenous administration of linoleic acid hydroperoxide. *J. Appl. Biochem.* **3,** 58–65.
6. Sasaguri, Y., Nakashima, T., Morimatsu, M., and Yagi, K. (1984) Injury to cultured endothelial cells from human umbilical vein by linoleic acid hydroperoxide. *J. Appl. Biochem.* **6,** 144–150.
7. Yagi, K., Komura, S., Ishida, N., Nagata, N., Kohno, M., and Ohishi, N. (1993) Generation of hydroxyl radical from lipid hydroperoxides contained in oxidatively modified low-density lipoprotein. *Biochem. Biophys. Res. Commun.* **190,** 386–390.
8. Kukreja, R. S., Datta, B. N., and Chakravarti, R. N. (1981) Catecholamine-induced aggravation of aortic and coronary atherosclerosis in monkeys. *Atherosclerosis* **40,** 291–298.
9. Kaplan, J. R., Manuck, S. B., Clarkson, T. B., Lusso, F. M., Taub, D. M., and Miller, E. W. (1983) Social stress and atherosclerosis in normocholesterolemic monkeys. *Science* **220,** 733–735.
10. Fukuda, S., Sasaguri, Y., Yanagi, H., Ohuchida, M., Morimatsu, M., and Yagi, K. (1991) Effect of linoleic acid hydroperoxide on replication of adenovirus DNA in endothelial cells of bovine aorta. *Atherosclerosis* **89,** 143–149.
11. Ohishi, N., Ohkawa, H., Miike, A., Tatano, T., and Yagi, K. (1985) A new assay method for lipid peroxides using a methylene blue derivative. *Biochem. Int.* **10,** 205–211.
12. Yagi, K., Kiuchi, K., Saito, Y., Miike, A., Kayahara, N., Tatano, T., and Ohishi, N. (1986) Use of a new methylene blue derivative for determination of lipid peroxides in foods. *Biochem. Int.* **2,** 367–371.
13. Yagi, K., Komura, S., Kayahara, N., Tatano, T., and Ohishi, N. (1996) A simple assay for lipid hydroperoxides in serum or plasma. *J. Clin. Biochem. Nutr.* **20,** 181–193.
14. Ohkawa, H., Ohishi, N., and Yagi, K. (1978) Reaction of linoleic acid hydroperoxide with thiobarbituric acid. *J. Lipid Res.* **19,** 1053–1057.

11

Antibodies Against Malondialdehyde-Modified Proteins

Induction and ELISA Measurement of Specific Antibodies

Yves Chancerelle, J. Mathieu, and J. F. Kergonou

1. Introduction

Polyunsaturated fatty acids undergo in vivo oxidative damage called lipid peroxidation, breaking down to lipid hydroperoxides and various secondary products *(1)*. Increase in lipid peroxides have been reported, for example, in radiation damage, diabetes, hyperlipidemia, and vascular diseases, and there is considerable interest in the relation between oxidative stress and development of atherosclerosis *(2,3)*.

Malondialdehyde (MDA) is a highly reactive dialdehyde generated during arachidonic-acid catabolism in thrombocytes *(4)*, but is also produced during nonenzymatic lipid peroxidation of unsaturated fatty acids *(5)*. MDA reacts in vivo and in vitro with free amino groups of proteins, phospholipids, or nucleic acids, producing inter- or intra-molecular 1-amino-3-iminopropene (AIP) bridges, leading to structural modifications of biological molecules *(6)*. The structural changes produced by MDA lead the modified molecules to be recognized as nonself, inducing immune response against the novel structures, especially AIP bridges. Thus, MDA confers antigenic properties to MDA-conjugated proteins *(7)*, analogous to those reported for glutaldehyde- or acetaldehyde-conjugated proteins *(8,9)*. It is now well-established that sera of healthy mammals contain antibodies reacting specifically with MDA-modified proteins (AbAIP), and recognizing an epitope consisting in AIP bridge and possibly the two surrounding lysyl residues *(10–12)*. The fact MDA, AIP bridges, and AbAIP are produced not only in disease but also under healthy condition, leading one to question their possible physiological role. Immunization of

From: *Methods in Molecular Biology, vol. 108: Free Radical and Antioxidant Protocols*
Edited by: D. Armstrong © Humana Press Inc., Totowa, NJ

rabbits with MDA-modified proteins results in an increased production of AbAIP, with an enhancement factor of 400–1600, depending on the animals *(7)*. Such antibodies react with various MDA-modified proteins, whereas they do not react with the corresponding native proteins; this indicates that the recognized epitope is directly related to MDA modifications. Confirmation of the nature of this epitope was obtained by Western blot ***(13)***. Immunoblot of all MDA-modified proteins and of MDA-modified polylysine reacted with antibodies obtained in immunized rabbits, whereas native proteins did not react.

This chapter will describe the procedure to obtain AbAIP by rabbit immunization and the ELISA technique for AbAIP evaluation in sera.

2. Materials

2.1. Antibodies Induction

1. Male white New Zealand rabbits, weighing about 3 kg.
2. Complete Freund's adjuvant (Sigma Chemical Corp., St. Louis, MO, F5881).
3. Incomplete Freund's adjuvant (Sigma, F5506).
4. Rabbit serum albumin (RSA) (Sigma, A0764).
5. Tetramethoxypropane (TMP), i.e., MDA-bis-(dimethyl-acetal) (Merck, Munchen, Germany, 805797).
6. Isotonic NaCl.

2.2. Enzyme-Linked Immunosorbent Assay (ELISA)

1. Polystyrene 96-wells microplates (Nunc, Wiesbaden-Biebrich, Germany, Maxisorp F96).
2. Chicken egg-white lysozyme (Sigma, L6876).
3. TMP.
4. Goat IgG anti-rabbit IgG, peroxidase-labeled (Cappel, furnished by Organon Teknika Corp., Turnhout, Belgium, 55679).
5. Goat IgG anti human IgG, peroxidase-labeled (Cappel, 55226).
6. H_2O_2 30 volumes (Sigma, H1009).
7. Orthophenylene-diamine (OPD), i.e., peroxidase substrate (Sigma, P1526).
8. Gelatin from cold water fish skin, i.e., teleostean gelatin (Sigma, G7765).
9. 375 m*M* Phosphate buffer, pH 7.0, 35.5 g anhydrous Na_2HPO_4, 15.0 g anhydrous NaH_2PO_4, distilled water up to 1000 mL, pH adjusted to 7.0.
10. Phosphate buffered saline (PBS) buffer, pH 7.4, containing 0.05% Tween-20 (PBST), 8.0 g NaCl, 0.2 g anhydrous KH_2PO_4, 1.15 g anhydrous Na_2HPO_4, 0.2 g KCl, 0.5 mL Tween-20, distilled water up to 1000 mL, pH adjusted to 7.4.
11. Acidic phosphate buffer, pH 6.2, containing 0.05% Tween-20 (ABT), 8.7 g NaCl, 8.97 g anhydrous NaH_2PO_4, 6.25 g anhydrous Na_2HPO_4, 0.5 mL Tween-20, distilled water up to 1000 mL, pH adjusted to 6.2.
12. 0.15 *M* Phosphate-citrate buffer, 4.7 g citric acid monohydrate, 7.3 g anhydrous Na_2HPO_4, distilled water up to 1000 mL, pH adjusted to 5.0 just before use. Store at 4°C.

13. PBST-gelatin: 12g teleostean gelatin; PBST up to 1000 mL. Prepared just before use.
14. OPD solution: 625 mg OPD; phosphate-citrate buffer up to 50 mL. Aliquots of 3.5 mL are stored at –30°C and defrost light-sheltered just before use.
15. Peroxidase substate solution: 20 µL H_2O_2 30 volumes; 3.2 mL OPD solution; 0.15 *M* phosphate-citrate buffer up to 100 mL. This solution must be prepared just before use and is light-sensitive.
16. Automatic microplate reader with a 492 nm filter.
17. Microplate washer.
18. Multiwell distributor.

3. Methods

3.1. Antibodies Induction

3.1.1. Preparation of MA

1. MA is prepared according to Allen ***(14)***. Briefly, a 1 m*M* solution of TMP in 10 m*M* HCl is allowed to hydrolize to MDA for 2 h in the dark at 25°C.
2. Then, 1 mL of the resulting solution is added to 9 mL of a 1 mg/mL solution of RSA in 37.5 m*M* phosphate buffer (rabbit serum albumin: 10 mg; 375 m*M* phosphate buffer: 1 mL; distilled water up to 10 mL).
3. After incubation for 72 h at 25°C, the mixture is stored at 4°C until use.

3.1.2. Rabbit Immunization Procedure

1. Rabbits (male white New Zealand, 6–8-wk-old, 3 kg) are injected subcutaneously in four different sites with MA (500 µL of the above MA solution mixed with 500 µL of isotonic NaCl and then emulsified with 150 µL of complete Freund's adjuvant), and subsequently once a mo during 3 mo with the same dose in incomplete Freund's adjuvant.
2. Blood is collected before each injection and IgG antibody titers are determined by ELISA.
3. It should be noted that MDA-modified heterologous proteins may also be used for immunization of rabbits (e.g., MDA-modified bovine casein), but in such cases they also develop antibodies against the native protein. These antibodies do not interfere in AbAIP determination if the MDA-modified protein used for coating of ELISA microplates is different from that used for immunization (*see* **Fig. 1**, for example).

3.2. ELISA

3.2.1. Preparation of ML

ML is prepared according to Kergonou ***(11)***. 360 µL of TMP corresponding to 1.5 mmol of MDA is suspended in 50 mL of an aqueous solution of acetic acid (1% v/v). This suspension is allowed to stand overnight at 4°C in order to achieve its hydrolysis to free MDA. The resulting solution is then mixed with an equal volume of 5 m*M* chicken egg-white lysozyme in 1% (v/v) aqueous acetic acid.

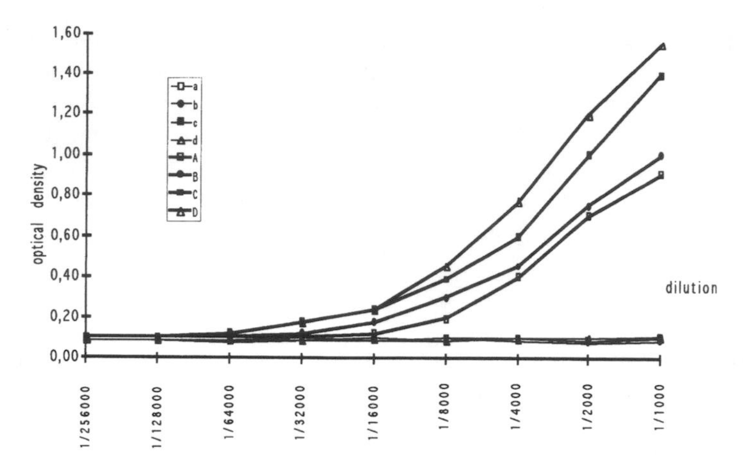

Fig. 1. Evolution, in function of immunizations, of AcAIP in rabbit immunized with malondialdehyde-modified caseine. A,B,C,D: reaction in ML coated wells. a,b,c,d: reaction in native lysozyme coated wells.

Concentrations in the final solution are: 15 m*M* MDA, 5 m*M* lysozyme 2, 1% acetic acid. This solution is allowed to stand at 37°C in a screw-capped flask for 5 wk and then stored at 4°C until use.

3.2.2. Preparation and Storage of Coated Microplates

Coating (50 microplates) is performed with 200 µL per well of a 5 µ*M* solution of ML in 75 m*M* phosphate buffer (aforementioned 2 mL ML solution; 200 mL 375 m*M* phosphate buffer; distilled water; up to 1000 mL). After coating, microplates are allowed to stand for 2 h at 37°C and for 22 h at 4°C, then washed three times PBST, pH 7.4, and stored covered at 4°C until use.

3.2.3. Determination of Human Seric AbAIP

Just before use, microplates are washed three times with ABT, pH 6.2 (*see* **Note 1**).

The first reagents (150 µL/well) are distributed as follows on a microplate:

1. All wells of lines 1 and 2: PBST-gelatin.
2. All wells of line 3: reference human serum diluted to 1/400 or 1/800 in PBST-gelatin.
3. The mean optical density (OD) of this line will be used as corrective factor for mean OD of other diluted sera.
4. Lines 4–12: quadruplicate samples of human sera diluted to 1/400 or 1/800 in PBST-gelatin; thus, 18 human sera are determined in these lines.

Microplates are incubated for 2 h at 37°C in the dark, then washed five times with PBST. The second reagents (200 µL/well) are distributed as follows on each microplate: All wells of line 1: PBST-gelatin. All wells of lines 2–12: peroxidase-labeled antibody to human IgG, diluted to 1/10000 in PBST-gelatin (*see* **Note 2**).

Microplates are incubated for 2 h at 37°C in the dark, and washed five times with PBST. 200 µL/well of peroxidase substrate solution are then distributed in every well of lines 1–12. Reactions are stopped 5 min later with 50 µL/well of 2.5 *M* H_2SO_4, and OD are measured at 492 nm.

Calculations are performed as follows:

1. Line 1 receives only the peroxidase substrate solution in order to detect a possible spontaneous reaction of H_2O_2 with OPD (*see* **Note 3**).
2. Line 2 receives all reagents except serum, and is an indication of nonspecific binding of peroxidase-labeled antibody.
3. Line 3 receives a reference human serum diluted to 1/400 or 1/800 in PBST-gelatin, and all other reagents. The mean OD of this line will be used as corrective factor for mean OD of other diluted sera (*see* **Note 4**).

Results are expressed as corrected relative OD units, calculated with the following formula (*see* **Note 5**):

$$\text{Corrected OD} = \frac{\text{Mean of quadruplicate determination on the serum} - \text{Mean of line 2}}{\text{Mean of line 3} - \text{Mean of line 2}}$$

3.2.4. Determination of Rabbit Seric AbAIP

Rabbit seric AbAIP are determined like human ones, with two differences. First, sera from immunized rabbit are serially diluted in PBST-gelatin, from 1/50 to 1/25600. Quadruplicate 150 µL samples of each dilution are distributed in lines 3–12, while all wells of line 1 receive 150 µL of PBST-gelatin; thus titrations of 2 rabbit sera are obtained on each microplate. Second, a peroxidase-labeled antibody to rabbit IgG is used instead of the labeled antibody to human IgG.

Calculations are performed as for human sera, except that means of quadruplicate determinations on each serum are not corrected vs a standard serum:

$$\text{Corrected OD} = \text{Mean of quadruplicate determination on the serum} - \text{Mean of line 2}$$

3.3. Results

Shown in **Fig. 1** is the evolution of antibody level in rabbit immunized with malondialdehyde-modified caseine. Sera are tested on ML coated and lysozyme-coated plates. The reactivity toward ML increase from A to D corresponding to

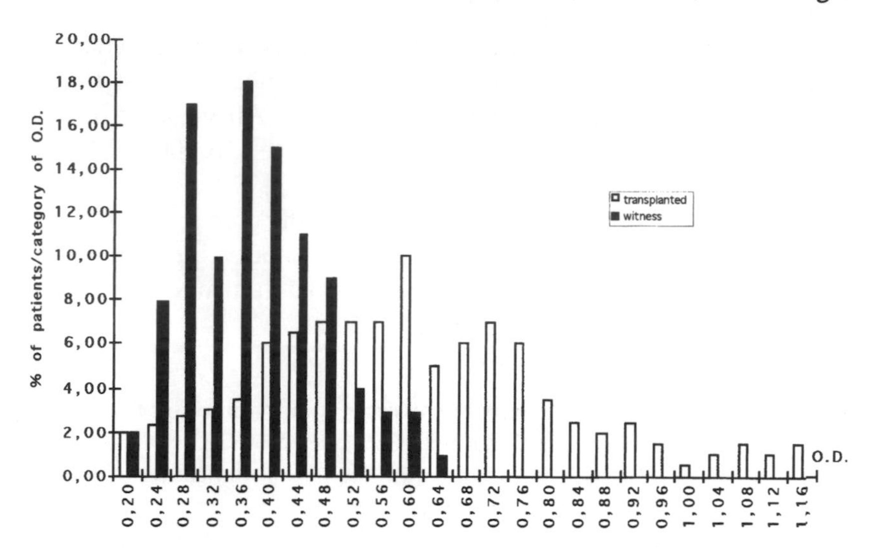

Fig. 2. Evaluation by ELISA of AbAIP in sera of heart transplant recipients compared to witness. Results are expressed as the mean of corrected optical density measured in 4 wells for each tested serum.

repeated immunizations. On the opposite side, any variation is observed for native lysozyme (lines a–d). These results confirm that antibodies recognizing a chemical structure, including AIP bridges, are produced with MDA-modified proteins and that the antibody production is a physiological process.

Figure 2 presents the distribution of antibody level in sera of 25 heart-transplant recipients compared to 40 witness patients. Results are expressed in terms of statistical repartition in classes of OD (from 0.2 to 1.16 OD Units). The mean of optical density for witness patients is 0.36 and 0.60 for heart-transplant recipients, and the maximum is 0.64 for witness compared to 1.16 for transplanted. These results demonstrate that everyone has a physiological antibody level, and that the antibody level is statistically higher in transplanted patients, with a higher range of repartition. Such an antibody production in transplanted patients is related to a process that involves oxygen-free radicals and increased lipid peroxidation, possibly related to cyclosporin nephrotoxicity.

4. Notes

1. All buffers may be at room temperature (climatized at 20–22°C) before use in order to minimize fluctuations.
2. Choice of the peroxidase-labeled Ig is very important, because sera of mammals naturally contain AbAIP, which reacts with antigenic sites on coated wells, lead-

ing to an increased response for tested sera. Nonspecific fixation of peroxidase-labeled Ig is determined on line 2 of each microplate. Peroxidase-labeled Ig must be selected in order to obtain a minimal response in line 2.

3. H_2O_2 concentration must be checked at regular intervals by spectrophotometric measurement of OD at 230 nm. For 50 µL of 1/10 diluted H_2O_2 solution added to 3 mL distilled water, OD at 230 nm must be from 1.1 to 1.2. If OD is less than 0.9, H_2O_2 must be changed.
4. Because it is now impossible to quantify AbAIP with a standard, and because reaction rates are dependent on temperature, we use a pool of sera in column 3 to minimize fluctuations from plate to plate. With this correction, fluctuations from plate to plate are less than 10%.
5. Plates are coated by adsorbtion and we advise working with the same lot of plates for a whole experimentation. For the same reason, some plates may be tested with the same serum at the time of change of lot.

Acknowledgments

We thank Roseline Soropogui for insight and passion for this work.

References

1. Esterbauer, H. (1985) Lipid peroxidation products: formation, chemical properties and biological activities., in *Free Radicals in Liver Injury* (Poli, G., Cheeseman, K. H., Dianzani, M. U., and Slater, T., eds.), IRL Press, Washington, DC, 29–47.
2. Esterbauer, H., Dieber-Rotheneder, M., Waeg, G., Srtiegl, G., and Jurgens G. (1990) Biochemical, structural and functional properties of oxidized low-density lipoprotein. *Chem. Res. Toxicol.* **3**, 77–92.
3. Wojcicki, J., Dutkiewicz, T., Gieldanowski, J., Samochowiec, L., Barcew-Wiszniewska, B., Rozewicka, L., Wira, D., Wesolowska ,T., Torbus-Lisiecka, B., and Gonet, B. (1992) Essential phospholipids modify immunological functions and reduce experimental atherosclerosis in rabbits. *Atherosclerosis* **93**, 7–16.
4. Hecker, M. and Ulrich, V. (1989) On the mechanism of prostacyclin and thromboxane A_2 biosynthesis. *J. Biol. Chem.* **264**, 141–150.
5. Draper, H. H. and Hadley, M. (1990) A review of recent studies on the metabolism of exogenous and endogenous malondialdehyde. *Xenobiotica* **20**, 901–907.
6. Halliwell, B. and Gutteridge, J. M. C. (1989) *Free Radicals in Biology and Medicine.* Oxford Science Publications, Clarendon Press, Oxford, UK.
7. Kergonou, J. F., Bruna, E., Pennacino, I., and Ducousso, R. (1988) Immunogenicity of proteins reacted with malonic dialdehyde (MDA). *Adv. Biosci.* **71**, 121–124.
8. Onica, D., Dobre, M., and Lenkei, R. (1978) Immunogenicity of glutaraldehyde treated homologous monomeric albumin in rabbits. *Immunochemistry* **15**, 941–947.
9. Israel, Y., Hurwitz, E., Niemela, O., and Arnon, R. (1986) Monoclonal and polyclonal antibodies against acetaldehyde-containing epitopes in acetaldehyde-protein adducts. *Proc. Natl. Acad. Sci. USA* **83**, 7923–7927.

10. Kergonou, J. F., Pennacino, I., Lafite, C., and Ducousso, R. (1988) Immunological relevance of malonic dialdehyde (MDA): II: precipitation reactions of human sera with MDA-crosslinked proteins. *Biochem. Int.* **16**, 835–843.
11. Kergonou, J. F., Pennacino, I., Lafite, C., and Ducousso, R. (1988) Immunological relevance of malonic dialdehyde (MDA): III. immunoenzymatic determination of human immunoglobulin binding to MDA-cross-linked proteins. *Biochem. Int.* **16**, 845–852.
12. Chancerelle, Y., De Lorgeril, M., Viret, R., Chiron, B., Renaud, S., and Kergonou J. F. (1991) Increased lipid peroxidation in cyclosporin-treated heart recipients. *Am. J. Cardiol.* **68**, 813–816.
13. Chancerelle, Y., Alban, C., Viret, R., Tosetti, F., and Kergonou, J. F. (1993) Immunization of rabbits with proteins reacted with malonic dialdehyde (MDA): kinetics and specificity of the immune response. *Biochem. Int.* **29,** 141–148.
14. Allen, D. W., Burgoyne, C. F., Groat, J. D., Smith, C. M., and White, J. G. (1984) Comparison of hemoglobin Köln erythrocyte membranes with malondialdehyde-reacted normal erythrocyte membranes. *Blood* **64**, 1263–1269.

12

Oxidized LDL and Lp(a)

Preparation, Modification, and Analysis

Jan Galle and Christoph Wanner

1. Introduction

During the last several years, it has become increasingly apparent that oxidation of low density lipoproteins (LDL) takes place in vivo as a key event in the development of human atherosclerosis ***(1,2)*** and vascular dysfunction ***(3,4)***. Thus, many laboratories have focused their interest on the in vitro effects of isolated LDLs on a variety of biological systems, including cultured vascular endothelial or smooth-muscle cells, intact arterial preparations, blood cells, and whole organ systems. LDLs were studied in their native form and after in vitro oxidative modification, which generates an LDL with similar characteristics as LDL isolated from advanced atherosclerotic lesions. Most frequently, ultracentrifugation methods were used to isolate LDL from human plasma, and lipoprotein oxidation was induced by metal ions (Cu^{2+} or Fe^{2+}). However, the extent of lipoprotein oxidation encircles a wide range and depends on several variables, and one should be aware of the fact that the results obtained with oxidized LDL described in the literature are achieved with differing preparations, ranging from minimally modified to extensively oxidized LDL. It is therefore the purpose of this chapter to describe methods for the preparation and the oxidative modification of LDL with special attendance on the degree of lipoprotein oxidation, and to explain different ways to analyze the oxidative modification.

More recently, lipoprotein (a) [Lp(a)], a lipoprotein with partially similar characteristics as LDL but even higher atherogenic potential, has been identified. Lp(a) is a plasma lipoprotein composed of apolipoprotein B and a large glycoprotein termed apolipoprotein(a) [apo(a)]. Lp(a) deposits have—like

From: *Methods in Molecular Biology, vol. 108: Free Radical and Antioxidant Protocols*
Edited by: D. Armstrong © Humana Press Inc., Totowa, NJ

LDL—been identified in atherosclerotic arteries *(5)* and other tissues *(6)*, and Lp(a) is likely to undergo oxidative modification in vivo like LDL. In this chapter, we therefore also describe methods for isolation, oxidative modification, and analysis of Lp(a).

2. Materials

2.1. Lipoprotein Source

For LDL preparation: fresh pooled human plasma, obtained from the blood bank before freezing. For Lp(a) preparation: LDL apheresis filters after treatment of patients rich in Lp(a); if not available, use fresh plasma from individuals rich in Lp(a).

2.2. Instruments

1. Ultracentrifuge.
2. Ultracentrifugation rotors: For LDL preparation: Beckman (Fullerton, CA) VTI 50; for Lp(a) preparation: Beckman 50.2 Ti.
3. Ultracentrifugation tubes: Beckman Quick Seal™ Tubes with 39 mL.
4. Spectrophotometer: e.g., Perkin Elmer (Norwalk, CT) Spectrometer Lambda 12.
5. Agarose gel electrophoresis kit (Lipidophor™ All In 12, Immuno).
6. Dialysis tubes: e.g., Spectra Por™ (Los Angeles, CA), Spectrum, pore size 50,000 Dalton.
7. Ultrafree™ centrifugal ultrafilters from Millipore (Bedford, MA): pore size 30,000 Dalton.
8. Sterile filters: pore size 0.2 µm.
9. Cutting system for Quick Seal™ Tubes from Beckman.

2.3. Reagents

2.3.1. LDL Preparation and Oxidation

1. Phosphate buffered saline (PBS): Stock solution: 400 g NaCl, 10 g KCl, 10 g KH_2PO_4, 72 g $Na_2HPO_4 \cdot 2H_2O$; and 5 mL H_2O; for preparation, we use sterile and pyrogen-free AMPUVA™ from Fresenius (Bad Hamburg, Germany).
2. $CuSO_4$: Stock solution 10 m*M*; can be kept at 4°C.
3. Preservative cocktail to avoid autooxidation, proteolytic digestion, and bacterial growth: Consists of ethylenediamine tetraacetic acid (EDTA) (0.2 *M* stock solution), aprotinin, chloramphenicol, and phenyl methyl sulfonyl fuoride (PMSF; from Sigma, St. Louis, MO), prepared fresh each time.
4. Potassium bromide (KBr) for density adjustment.

2.3.2. For Lp(a) Preparation

1. Wash-out buffer for LDL apheresis filters: 0.9% NaCl, 0.1 *M* Tris-HCl, 1.0 *M* EDTA, 0.02 NaN_3; adjust to a pH of 8.5 with HCl.

2. Preservative cocktail.
3. Lysin-Sepharose (Lysin-Sepharose 4 B, Pharmacia, Upsala, Sweden), columns with 20 mL; loading buffer: PBS + Tween-20 1%; elution buffer: eta-amino-n-caproic acid (eACA, Sigma) + PBS + Tween-20 1% (0.66 g eACA per 100 mL); regeneration buffer: PBS + Tween-20 1% + thimerosal 0.02%.
4. Potassium bromide.
5. Lp(a) analysis kits: Lp(a) Phenotyping™ (Immuno, Vienna, Austria); Vectastain ABC antimouse IgG test (Vector Laboratories, Burligame, CA).

3. Methods

3.1. LDL Preparation

1. An important issue when starting a LDL preparation is to obtain fresh human plasma. In untreated plasma, there may be a rapid proteolytic digestion and auto-oxidation of LDL, which can be demonstrated with the use of sodium dodecyl sulfate-polyacrylamide gel electrophoresis (SDS-PAGE) by the partial degradation of Apo B 100, the LDL-specific apoprotein. We therefore put effort into an as early as possible addition of the "preservative cocktail" *(7)* to human plasma obtained from the local blood bank.
2. For a typical preparation, we pool three units of fresh plasma, which yields a total volume of approximately 600 mL (*see* **Note 1**). Per 100 mL of plasma, we add 1.6 mL of 0.2 *M* EDTA, 2.8 mL of 0.3 *M* NaCl, 1000 IE aprotinin (Trasylol™), 10 mg chloramphenicol (*see* **Note 2**), and 40 mg PMSF (*see* **Note 3**). We add the preservative cocktail to the plasma in a glass cylinder under continuous stirring, strictly avoiding foam formation, which denaturates LDL.
3. The principal of the LDL preparation is the separation of the lipoprotein subclasses by density ultracentrifugation in several steps. We start with the separation of chylomicrons, very low density lipoprotein (VLDL) and LDL from HDL and other serum proteins in one step, and then separate VLDL/chylomicrones from LDL. For this purpose, we use a vertical rotor, which significantly decreases the centrifugation time necessary for an adequate separation, in comparison to fixed angle rotors.
4. To separate VLDL, chylomicrons, and LDL from HDL, the density of the plasma must be adjusted to 1.063 g/mL. A density of 1.063 g/mL is achieved by adding approximately 0.0834 g KBr/mL plasma under continuous stirring, according to approximately 50 gr KBr/600 mL plasma. Accuracy of the density adjustment can be checked with a spindle or with exact weighing of 1 mL of fluid. The plasma will then be distributed into 16 quick-seal centrifugation tubes with 39 mL (*see* **Note 4**). Eight tubes will be placed in the Beckman VTI 50 rotor, the remaining eight are kept at 4°C. The first centrifugation run can now be started, at 4°C and 200,000*g*, for 8 h. The brake function of the ultracentrifuge can be used. The second run of the other eight tubes can be started immediately thereafter. After centrifugation, one will see in the tubes on top an orange layer, consisting of LDL, covered by the creamy VLDL and chylomicrones. Below this layer,

there will be clear fluid, consisting of approximately 20% of the whole volume in the tube. Below this clear fluid are HDL and plasma proteins. We now sting into the centrifugation tube with a strong syringe needle and carefully aspirate the top layer. The total volume aspirated from 16 tubes will account to approximately 150 mL.

5. In the next step, LDL will be separated from VLDL and chylomicrones by density-gradient ultracentrifugation. For this purpose, the fluid will be readjusted to a density of 1.063 mg/mL by adding solid KBr according to the following formula:

$$\text{KBR [g]} = \frac{\text{Volume} \times (\text{density A} - \text{density B})}{1 - 0.267 \times \dfrac{(\text{density A} + \text{density B})}{2}}$$

where Volume is the volume of the fluid [mL], density A is the desired density [g/mL], and density B is the actual density (usually approximately 1.040 mg/mL) (*see* **Note 5**).

We first fill the quick seal tubes with 19 mL of PBS, and then gently underlay the buffer with 20 mL of the VLDL/chylomicrones/LDL mixture (*see* **Note 6**). After closure of the tubes, we start a centrifugation run in the Beckman VTI 50 rotor (200,000*g,* at 4°C, for 3 h). Again, the brake function of the ultracentrifuge can be used. After the run, there will be three bands to see: on top, a creamy layer, consisting of VLDL/chylomicrones, in the middle, a broad, orange band consisting of LDL, and on the bottom, contaminating plasma proteins and HDL. Between these bands, there is clear fluid. The LDL band can be aspirated using the same technique as for the VLDL/chylomicrone/LDL collection (*see* **Note 7**).

Thus, when starting a LDL preparation with 600 mL plasma in the morning, the first centrifugation run will be finished in the evening, the second run can be done during the night, and the separation of VLDL/chylomicrones/LDL from HDL can be finished 24 h after the start. Accordingly, the final run, separating LDL from VLDL/chylomicrones, can be done during the second day of the preparation.

6. Collection of the orange middle band after the final ultracentrifugation run yields 30–40 mL of fluid containing LDL in a concentration of approximately 5 mg/mL. This solution can be further concentrated in Millipore UltraFree™ centrifugal ultrafilters. To gain a final concentration of 15–20 mg/mL, we centrifuge the LDL-solution for approximately 2 h at 1,500 rpm at 4°C in centrifugation filters, which can be filled with maximally 19 mL. Protein content of LDL can be measured according to the method described by Lowry *(8)*.
7. Finally, we extensively dialyze the LDL stock solution in dialysis tubes with a pore size of 50,000 Dalton against PBS supplemented with 0.2 m*M* EDTA in the dark at 4°C (10 mL of stock solution against 3 L PBS over 3 d, two changes of buffer). When LDL will be oxidized, we dialyze the stock solution against PBS in the absence of EDTA.

3.2. Analysis of LDL Purity

LDL purity can be evaluated by various methods, including spectrophotometry, agarose-gel electrophoresis, apoprotein detection, and SDS-PAGE. We normally use agarose-gel electrophoresis and spectrophotometry because these methods are simple to perform and also valuable for the detection of LDL oxidation.

3.2.1. Agarose-Gel Electrophoresis

Agarose-gel electrophoresis, performed with a commercially available kit (Lipidophor, Immuno), separates lipoproteins according to their molecule size and electronegativity (*see* **Note 8**). During a normal 80-min run, native LDL migrates approximately 11 mm from the starting point, HDL 20 mm, and VLDL 14 mm. The bands can be made visible after the run with the developer of the electrophoresis kit in a cell-culture dish. For native LDL, only one sharp band should be visible at approximately 11 mm.

3.2.2. Spectrophotometry

Spectrophotometry can be used to identify the typical spectrum of LDL and to monitor the oxidative modification *(9)*. **Figure 1** shows the spectrum of native LDL in the range between 300 and 600 nm, with two absorption maxima at approximately 465 and 485 nm. Absorption increases in the ultraviolet (UV) range owing to the protein content of LDL. With ongoing, increasing LDL oxidation, these absorption maxima gradually disappear.

3.3. LDL Oxidation

LDL oxidation can easily be induced by incubation of the lipoprotein solution (filled in dialysis tubes) with metal ions, which catalyze a lipid-peroxidation process *(10)*. Most laboratories use Cu^{2+} to start the reaction. However, the extent of the resulting oxidation is a function of multiple parameters, including composition of the buffer, temperature, duration of Cu^{2+} exposition, the concentration of Cu^{2+}, and the concentration of LDL during Cu^{2+} exposition. The influence of all these parameters on LDL oxidation can be checked for example by examining the electrophoretic mobility of oxidized LDL on agarose-gel electrophoresis. Increasing the temperature from 21°C to 37°C increases the rate of oxidation. Increased rates of oxidation are also observed with a higher Cu^{2+} concentration or a lower LDL concentration in relation to the surface of the dialysis tube. **Figure 2** gives an example of how variation of the incubation time at a temperature of 37 °C influences LDL oxidation.

Thus, to achieve reproducible results, it is extremely important to standardize the conditions under which LDL shall be oxidized. We normally fill 6 mL of LDL at a concentration of 10 mg/mL in a dialysis tube of 10-cm length under sterile conditions. We use sterile filters with a pore size of 0.3 μM. Before the start of the oxidation procedure, LDL is dialyzed for 3 d at 4°C in the dark against 3 L EDTA-free PBS (*see* **Subheading 3.1.7.**). The dialysis tube is then

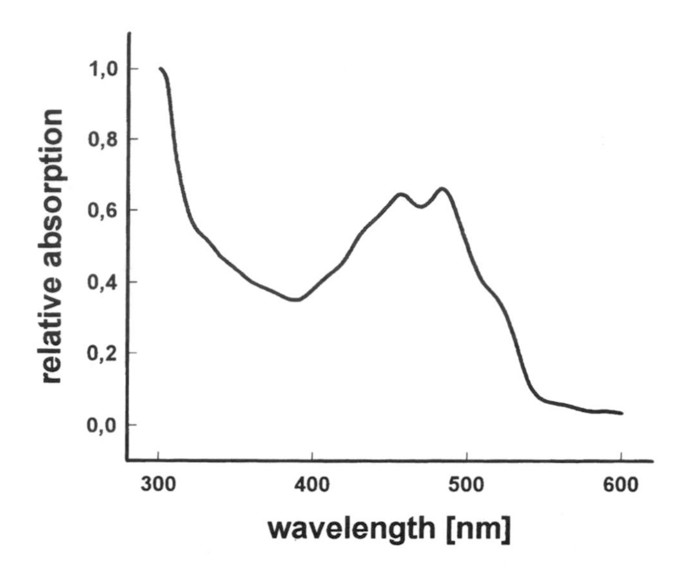

Fig. 1. Absorption spectrum of native LDL in the range between 300 and 600 nm. There are two absorptions maxima at 460 and 485 nm characteristic for native LDL. After oxidative modification, these absorptions maxima disappear.

placed in 3L PBS containing 10 µ*M* Cu^{2+} at 21°C. After approx 5–8 h, one can note a change in color from orange to yellow. This time point corresponds with the end of the lag phase of the lipid peroxidation (*see* **Fig. 3**). We leave LDL exposed to Cu^{2+} for 24 h, and than change the buffer against PBS without Cu^{2+} but containing EDTA (200 µ*M*). Oxidized LDL will be dialyzed for 3 d against three changes of this buffer (each 3 L) at 4°C in the dark. After a final sterile filtration, it is ready to use.

3.4. Analysis of LDL Oxidation

Various methods can be used to analyze LDL oxidation. Here, we describe only two simple techniques that in our opinion are sufficient to assess the extent of LDL oxidation. Other methods (SDS-PAGE, tests for thiobarbituric acid-reactive substances, analysis of lipid composition) have been described elsewhere in detail ***(11–13)***.

3.4.1. Photometric Detection

Photometric analysis is a simple method to study the individual resistance of an LDL preparation against oxidation. The absorption at 234 nm depicts the formation of conjugated dienes, which are formed during the early phase of LDL oxidation ***(9)***. For this purpose, we add 50 µg of native LDL to 1 mL PBS

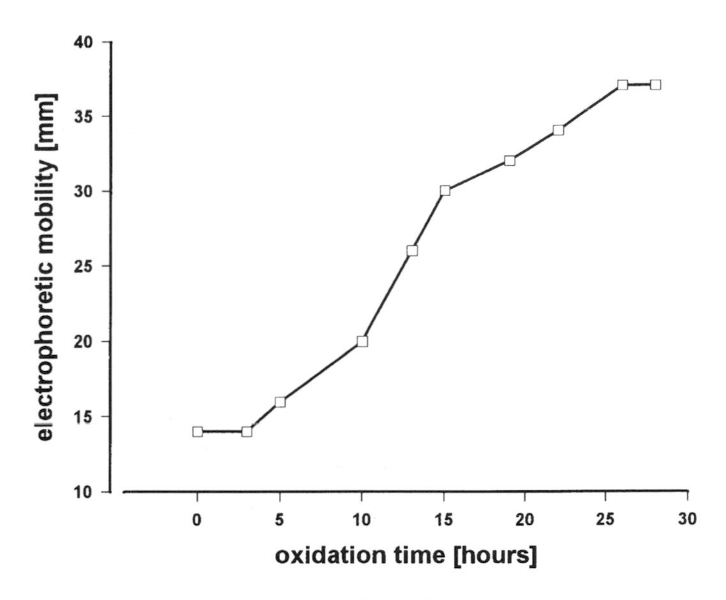

Fig. 2. Electrophoretic mobility [mm] of oxidized LDL on agarose gel in relation to the oxidation time [h]. With increasing duration of exposition to oxidative conditions, the electrophoretic mobility of LDL increases.

in a quarz cuvet. **Figure 1** shows the absorption spectrum of native LDL between 300 and 600 nm. We then add 1.66 μM Cu^{2+} to the cuvet and set the spectrophotometer to the time-drive program at 234 nm. Absorption will automatically be measured every 2 min, and will increase only slowly during the first 2–3 h, representing the lag phase where the antioxidative defense of LDL is consumed (**Fig. 2**). After this lag phase, diene formation and absorption at 234 nm will rapidly increase and reach their maximum after approximately 4–6 h (**Fig. 3**). The lag phase of an individual LDL preparation shows great variability and depends, e.g., on the nutrition, smoke habits, medication, and other factors of an individuum. The lag-phase interval depends also on the oxidative conditions (temperature, Cu^{2+} concentration, LDL concentration, buffer), and its determination is therefore a useful tool for analysis and comparison of different conditions.

3.4.2. Agarose-Gel Mobility

During oxidation of LDL, its electronegativity increases, and therefore electrophoretic mobility on agarose gel is a good indicator of the extent of oxidation *(**10**)*. Under the conditions as described in the Methods section, native LDL migrates approximately 11 mm. Electrophoretic mobility of oxidized

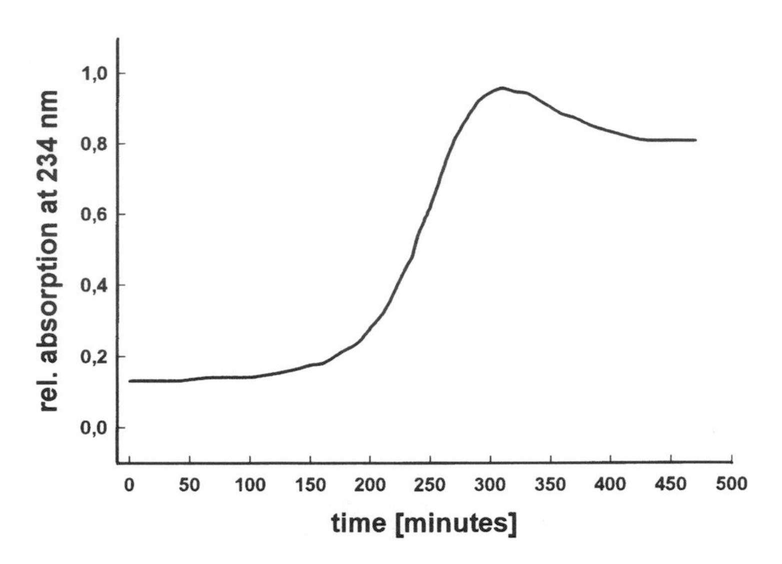

Fig. 3. Continuous detection of diene formation as a parameter for lipid peroxidation of LDL by measurement of absorption at 234 nm after addition of Cu^{2+}. Following an initial lag phase of variable duration, diene formation rapidly increases and reaches a maximum after 4–6 h.

LDL varies between 12 and 30 mm, depending on the oxidative conditions (*see* **Fig. 4**). When LDL is oxidized as described in **Subheading 3.3.**, its electrophoretic mobility will be approximately 16 mm.

3.5. Lp(a) Preparation

Lp(a) is prepared in several steps including ultracentrifugation and affinity gel chromatography. Lp(a) source can be the plasma of individuals rich in Lp(a) or the wash out fluid of LDL-apheresis filters after treatment of a patient with high Lp(a) levels. We prefer the use of apheresis filters because the preparation is repeatable from an identical Lp(a) source.

The preparation starts with the wash out of the apheresis filter after the treatment session with 500 mL of buffer (composition *see* **Subheading 2.3.2.**), followed by the addition of the preservative cocktail (*see* **Subheading 3.1.**). This fluid will then be adjusted with KBr to a density of 1.12 g/mL according to the formula as described in **Subheading 3.1.**, be filled in Quick Seal ultracentrifugation tubes with 39 mL, and be centrifuged for 24 h in a Beckman 50.2 Ti rotor at 150,000*g* and 16°C. With this centrifugation step, VLDL, LDL, Lp(a), and part of HDL will be separated from more dense HDL particles. After this run, one can see a dark-yellow pellet in the top region of the tubes. The tubes will now be cut

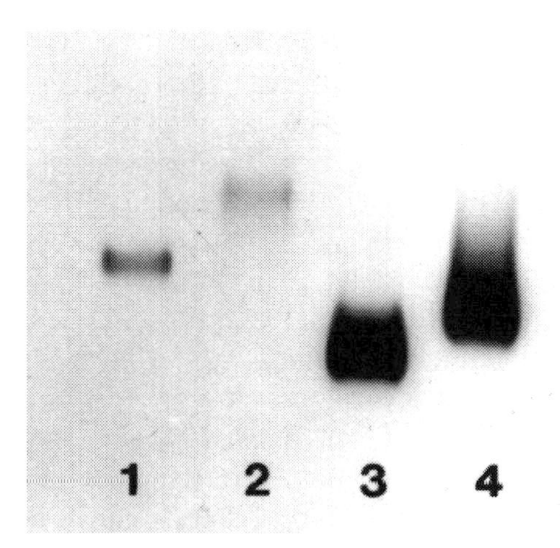

Fig. 4. Agarose-gel electrophoresis of native Lp(a) (line 1), oxidized Lp(a) (line 2), native LDL (line 3), and oxidized LDL (line 4). Electrophoretic mobility increases significantly after oxidative modification of the lipoproteins, indicating enhanced negative charge.

just below this pellet, and the pellets will be dissolved in 50 mL PBS. The purpose of the next step is to separate VLDL/LDL from Lp(a) and contaminating HDL. Therefore, the fluid must be adjusted to a density of 1.065 g/mL, and be centrifuged for 24 h in a 50.2 Ti rotor at 150,000*g* and 16°C. After centrifugation, one will see in these tubes on top an orange/yellow layer consisting of LDL and VLDL. In the middle, there will be clear fluid, and in the lower part one can identify a yellow layer consisting of Lp(a) and HDL. This fluid can be aspirated in a way as described above (*see* **Subheading 3.1.** and **Note 9**). Starting with 500 mL, we usually obtain 100–150 mL fluid at this point, enriched in Lp(a) but still containing HDL. Now HDL has to be separated from Lp(a). This can be achieved by affinity chromatography using lysine sepharose columns. The principal of this method is that Lp(a) binds to the gel while HDL runs through when the columns are filled with PBS-Tween buffer. Using an elution buffer, Lp(a) can be regenerated. In a standard preparation, we use 5 columns with 20 mL lysine sepharose in parallel. First, the lysine sepharose columns must be washed extensively with PBS-Tween buffer to clean them from thimerosal. Then the columns can slowly be filled with 10 mL of the Lp(a)/HDL fluid, followed by washing with PBS-Tween buffer. The first 10 mL of the effluant contain still a significant amount of unbound Lp(a) and can be collected for immediate reloading of the columns. Now the columns will be washed with 80 mL PBS-Tween buffer each. In the final step, Lp(a) will be eluted with 15 mL elution buffer per column (*see*

Note 9). The columns can be regenerated with PBS-Tween buffer and be reused. The preparation of Lp(a) as described here yields a solution with a concentration of approximately 0.5–1 mg/mL. Lp(a) can be further concentrated with centrifugation filters as described in **Subheading 3.1.**, with the difference that Tween should be used to prevent sticking to the filters. Like LDL, Lp(a) must now extensively be dialyzed. There are, however, two important differences to the preparation of LDL: Lp(a) must not be dialyzed at 4°C, because it tends to agglutinate in the cold, and it should be kept in glass vials because it binds to several nonglass surfaces.

3.6. Analysis of Lp(a) Purity

Standard methodology consists of SDS-PAGE 5–10 % (**Fig. 4**). If available 3.5–13% gels give slightly higher resolution. Apo(a) can be distinguished from apoB-100 by Coomassie-blue conventional staining. Using high and low molecular-weight standards, the purity can further be verified by the absence of other apolipoproteins such as apoE and apoCI-III. Identification of apo(a) should be confirmed by immunoblotting using a monoclonal antibody (MAB) (4F3, 1:2000; Cappel, Durham NC) or polyclonal antibody (PAB) (Immuno) against apo(a). A concise description of the methods comes with the commercially available kits. Immunoblotting can also be used to distinguish the amount of native Lp(a) from that of oxidized Lp(a) using SDS-agarose-gel electrophoresis. Samples are mixed with SDS (final concentration, 10 g/L), 0.5% bromophenol blue and 75% glycerol. After electrophoresis, proteins are transferred to a nitrocellulose membrane (immobilion-polyvinylidene-difluoride membrane, Millipore, Bedford, MA) by electroblotting or thermoblotting (75°C for 45 min). Membranes must be blocked for 60 min in 5% (wt/vol) powdered skim milk, incubated for 2 h with the anti-apo(a) antibody previously described. Thereafter, extensive washing in Tris-buffered saline, pH 7.4, containing 0.05% Tween-20 is necessary. For detection of the bands the horseradish peroxidase-labeled protein A (Vectastain ABC antimouse IgG test Kit; Vector Laboratories) or the alcaline phosphatase method (Lp(a) Phenotyping Kit; Immuno) should be used according to the first antibody used. The Immuno Kit includes a defined apo(a) isoform-standard that consists of a mixture of five distinct isoforms (Kringle IV repeats 14, 19, 23, 27, 35).

3.7. Lp(a) Oxidation

Oxidation of Lp(a) is carried out essentially in the same way as oxidation of LDL in **Subheading 3.3**. One difference may be a longer lag phase of Lp(a) oxidation, compared to LDL. In preparations from our laboratory, the duration was significantly longer until a change in color of Lp(a) within the dialysis tube was noteable (*see* **Subheading 3.3.**).

3.8. Analysis of Lp(a) Oxidation

For analysis of Lp(a) oxidation, the same tests as described in **Subheading 3.4.** can be applied.

3.9. Results

3.9.1. Photometric Detection

The spectrum of native Lp(a) in the range between 300 and 600 nm also shows two absorption maxima at 460 and 480 nm, like LDL **(Fig. 1)**. These absorption maxima also gradually disappear with ongoing, increasing Lp(a) oxidation. However, spectrophotometry is not a suitable method for the detection of increased formation of conjugated dienes as an online measurement of Lp(a) oxidation **(Fig. 2)**.

3.9.2. Agarose-Gel Mobility

For analysis of Lp(a) oxidation, electrophoretic mobility on agarose gel is a useful tool and is carried out as previously described with the commercially available kits (**Fig. 3**). Native Lp(a) migrates approximately 20 mm during a 80 min run, and oxidized Lp(a) up to two times faster, depending on the degree of oxidation (**Fig. 4**).

4. Notes

1. Fresh plasma must not be frozen. Freezing the plasma results in degradation of Apo B 100.
2. Use always cloves and work with great care with fresh human plasma, because the results of the ongoing routine tests for hepatotrop viruses or HIV infections are usually not yet available when starting the preparation.
3. Dissolve 250 mg PMSF in 6.5 mL ethanol and add 3 mL of the solution after density adjustment. Caution: PMSF is toxic; work under a flue.
4. Fill the centrifugation tubes with a 50-mL syringe and a 15-cm long steel cannula, connected to the syringe with an elastic silicon tube. This allows you to fill gently without foam formation.
5. We measure the actual density by weighing 1 mL of fluid in a precision scale.
6. Take care to seal the quick-seal centrifugation tubes strictly vertically; otherwise, they might leak during the centrifugation run and interrupt the procedure.
7. To underlayer PBS buffer with the lipoprotein mixture adjusted to a density of 1.063 g/mL, we again use a 50-mL syringe and a 15-cm long steel cannula, connected to the syringe with an elastic silicon tube. Take care not to mix the gradient.
8. Lipidophor agarose-gel electrophoresis kits are commercially available and are ready to use. We dilute 10 µL of LDL stock solution in 100 µL PBS, and mix 50 µL of this dilution with 50 µL of prewarmed (50°C) agar. We then fill 20 µL of this mixture into the prepared cavities of the gel plates. The electrophoresis run lasts for 80 min at 70 Volt and 15 mAmp.

9. Before elution of Lp(a) from the lysine sepharose columns, one can see a yellow band in the upper half of the column, consisting of bound Lp(a). During elution with eACA, one can see how this band gradually disappears.

References

1. Steinberg, D., Parthasarathy, S., Carew, T. E., Khoo, J. C, .and Witztum, J. L. (1989) Beyond cholesterol: modifications of low-density lipoprotein that increase its atherogenicity. *N. Engl. J. Med.* **320,** 915–924
2. Ylä-Herttuala, S., Palinski, W., Rosenfeld, M. E., Parthasarathy, S., Carew, T. E., Butler, S., Witzum, J. L., and Steinberg, D. (1989) Evidence for the presence of oxidatively modified low density lipoprotein in atherosclerotic lesions of rabbit and man. *J. Clin. Invest.* **84,** 1086–1095
3. Flavahan, N. A. (1992) Atherosclerosis or lipoprotein-induced endothelial dysfunction-potential mechanisms underlying reduction in EDRF/nitric oxide activity. *Circulation* **85,** 1927–1938
4. Galle, J., Bengen, J., Schollmeyer, P. and Wanner, C. (1995) Impairment of endothelium-dependent dilation in rabbit renal arteries by oxidized lipoprotein(a): role of oxygen-derived radicals. *Circulation* **92,** 1582–1589
5. Rath, M., Niendorf, A., Reblin, T., Dietel, M., Krebber, H. J., and Beisiegel, U. (1989) Detection and quantification of lipoprotein(A) in the arterial wall of 107 coronary bypass patients. *Arteriosclerosis* **9,** 579–592
6. Sato, H., Suzuki, S., Ueno, M., Shimada, H., Karasawa, R., Nishi, S.-I., and Arakawa, M. (1993) Localization of apolipoprotein(a) and B-100 in various renal diseases. *Kidney Int.* **43,** 430–435
7. Edelstein, C. and Scanu, A. M. (1986) Precautionary measures for collecting blood destinated for lipoprotein isolation. *Methods Enzymol.* **128,** 151–155.
8. Lowry, O. H., Rosebrough, N. J., Farr, A. L., and Randall, R. J. (1951) Protein measurement with phenol reagent. *J. Biol. Chem.* **193,** 265–275
9. Esterbauer, H., Striegel, G., Puhl, H., and Rotheneder, M. (1989) Continuous monitoring of in vitro oxidation of human low density lipoprotein. *Free Rad. Res. Comm.* **6,** 67–75
10. Steinbrecher, U. P., Witztum, J. L., Parthasarathy, S., and Steinberg, D. (1987) Decrease in reactive amino groups during oxidation or endothelial cell modification of LDL. correlation with changes in receptor-mediated catabolism. *Arteriosclerosis* **7,** 135–143
11. Laemmli, U. K. (1970) Cleavage of structural proteins during the assembly of the head of bacteriophage T4. *Nature* **227,** 680–685
12. Janero, D. R. (1990) Malondialdehyde and thiobarbituric acid-reactivity as diagnostic indices of lipid peroxidation and peroxidative tissue injury. *Free Rad. Biol. Med.* **9,** 515–540
13. Esterbauer, H., Jürgens, G., Quehenberger, O., and Koller, E. (1987) Autoxidation of human low density lipoprotein: loss of polyunsaturated fatty acids and vitamin E and generation of aldehydes. *J. Lipid. Res.* **28,** 495–509

13

Oxidized and Unoxidized Fatty Acyl Esters

Odile Sergent and Josiane Cillard

1. Introduction

Membrane phospholipids contain polyunsaturated fatty-acid side chains esterified to phosphoglycerol. These polyunsaturated fatty acyl esters are easily oxidable lipids. A sensitive method for determination of lipid peroxidation is of considerable interest. Usually lipid peroxidation is determined by evaluation of a by-product, malondialdehyde, which is produced by lipid-peroxide breakdown. Malondialdehyde, however, does not represent the early stage of membrane lipid peroxidation nor the true amount of oxidized lipids. Moreover, most techniques for malondialdehyde use thiobarbituric acid-reactants evaluation *(1)*, which is an indirect measurement.

The first stage of the peroxidation process consists of the molecular rearrangement of the double bonds originally present in the polyunsaturated fatty acyl esters, which leads to the formation of conjugated dienes. Measurement of the ultraviolet (UV) absorbance of conjugated dienes in the wavelength range from 230 to 235 nm cannot always be used because the shoulder of the UV absorbance owing to the conjugated diene formation is partially masked by the unoxidized fatty acids themselves, or extracted contaminants that absorb at wavelengths nearby that of conjugated dienes *(2,3)*. A new technique developed by Corongiu et al. *(4,5)* improved the accuracy of the measurement by using second derivative spectroscopy. This method changes the large absorbance shoulder to a sharp minimum peak. Minima at 233 and 242 nm, ascribed to the *trans,trans* and *cis,trans* conjugated diene isomers, respectively, are quantified in arbitrary units as $d^2A/d\lambda^2$ *(5)*. A linear regression curve between arbitrary units and absorbance at 234 nm has been established using autoxidized linoleic-acid micelles *(6)*. Then, using an average extinction coefficient of 27,000 $M^{-1}cm^{-1}$ for hydroperoxides, absorbance could be converted to oxidized fatty acyl ester concentration.

From: *Methods in Molecular Biology, vol. 108: Free Radical and Antioxidant Protocols*
Edited by: D. Armstrong © Humana Press Inc., Totowa, NJ

Because the quantity of extracted lipids can vary from one sample to another, conjugated dienes should be reported to the amount of total fatty acyl esters present in the extract. Most of the methods described for lipid estimation in biological samples use colorimetric techniques. Some only quantify total lipids *(7)*, others, which measured the fatty acyl ester group, lack sensitivity and specificity for cell samples *(8)*. In this chapter, we present a method for determination of total (oxidized and unoxidized) fatty acyl esters by Fourier transform infrared (FTIR) spectrocopy using the specific carbonyl ester absorption band at 1740 cm^{-1} *(9)*. FTIR instruments are well-known to improve the signal-to-noise ratio and allow better analysis of smaller samples than dispersive instruments.

This chapter will describe accurate and sensitive methods for oxidized and unoxidized lipid determination in very small samples, such as cells. Three separate procedures are necessary: lipid extraction, diene conjugation signal determination, and measurement of total fatty acyl esters.

2. Materials

2.1. Lipid Extraction

1. Chloroform.
2. Methanol.
3. 0.88% Potassium chloride in aqueous solution.
4. 0.*01 M* Phosphate buffer, pH 7.45.

2.2. Conjugated Dienes Measurement by Second Derivative Spectroscopy

1. Spectrophotometer managed by a computer, which allows the automatic calculation of second derivative spectra (Secomam, S1000, Sarcelles, France), scanning speed = 100 nm/min; interval = 0.25 nm.
2. Cyclohexane spectrophotometric grade.

2.3. Total Fatty Acyl Esters Determination by FTIR

1. 16 PC Perkin-Elmer FTIR: Sensitivity can be improved by a special cell with a path length of 10 mm (Eurolabo-Hellmat, Paris, France).
2. Glycerol trioleate.
3. Anhydrous chloroform.

3. Methods

1. Measurement of oxidized and unoxidized fatty acyl esters results from three steps: lipid extraction, evaluation of oxidized fatty acyl esters by the determination of diene conjugation signal, and assay for total (oxidized and unoxidized) fatty acyl esters.
2. These methods are suitable for very small samples. For this reason, we describe the techniques by considering measurement from cells.

3. The low quantity of sample will lead to only a slight production of conjugated dienes. Consequently, the presence, even in low concentration, of interfering products may prevent detection of the conjugated diene spectrum (*see* **Note 1**). Many precautions are required, especially during lipid extraction.

3.1. Lipid Extraction

The method used is that of Folch et al. ***(10)***. Cells are washed twice with 0.01 *M* phosphate buffer, pH 7.45, and then resuspended in 300 µL of methanol (*see* **Note 2**). Cells are lysed by ultrasonication for 1 min in ice bath and then vortexed for another 1 min, after which 300 µL of chloroform is added. Cells are vortexed for 2 min, and then another 600 µL of chloroform is added and cells are vortexed again for another 2 min. Potassium chloride in aqueous solution (0.88%) is added to the previous mixture (100 µL/tube), which is vortexed again for 1 min. The mixture is centrifuged at 4°C for 10 min at 3000*g*. The lower chloroform lipid layer is removed and saved at –20°C for subsequent quantitative determination of oxidized and total fatty acyl esters (*see* **Note 3**).

3.2. Conjugated Diene-Signal Determination

1. Conjugated dienes are measured by second derivative spectroscopy according to the method of Corongiu et al. ***(4,5)***.
2. Chloroform lipid solutions are dried under nitrogen at 40°C (*see* **Note 4**). Lipids are then dissolved in 500 µL of spectrophotometric-grade cyclohexane and vortexed for 30 s. Lipid solutions are immediately scanned from 200 to 300 nm (*see* **Note 5**).
3. Before the scanning operation, a background-correction-memorized scan is performed with a control composed of 200 µL phosphate buffer (0.01 *M*, pH = 7.45) extracted and treated as a sample, in order to correct any minor differences related to the extraction conditions. Samples are scanned for absorbance and the second derivative spectra are obtained. The minima at 233 and 242 nm are quantified in arbitrary units as $d^2A/d\lambda^2$, representing the measurement for minima to adjacent maxima at the higher wavelength (*see* **Note 6**). The sum of both peak heights is calculated (**Fig. 1**).

A linear regression curve is established between arbitrary units (x) and absorbance (y) at 234 nm using oxidized linoleic-acid micelles ($y = 63x + 0.03$; $r = 0.9989$) ***(6)***. Then absorbance at 234 nm can be converted to hydroperoxide concentration using an average extinction coefficient of 27,000 M^{-1} cm^{-1} ***(11)***.

3.3. Assay of Total Fatty Acyl Esters

Cyclohexane solutions must be kept at –20°C for analysis of total fatty acyl esters (oxidized and unoxidized fatty acyl esters) (*see* **Note 7**). Samples are placed overnight in a vacuum dessicator with phosphoric anhydride and, just

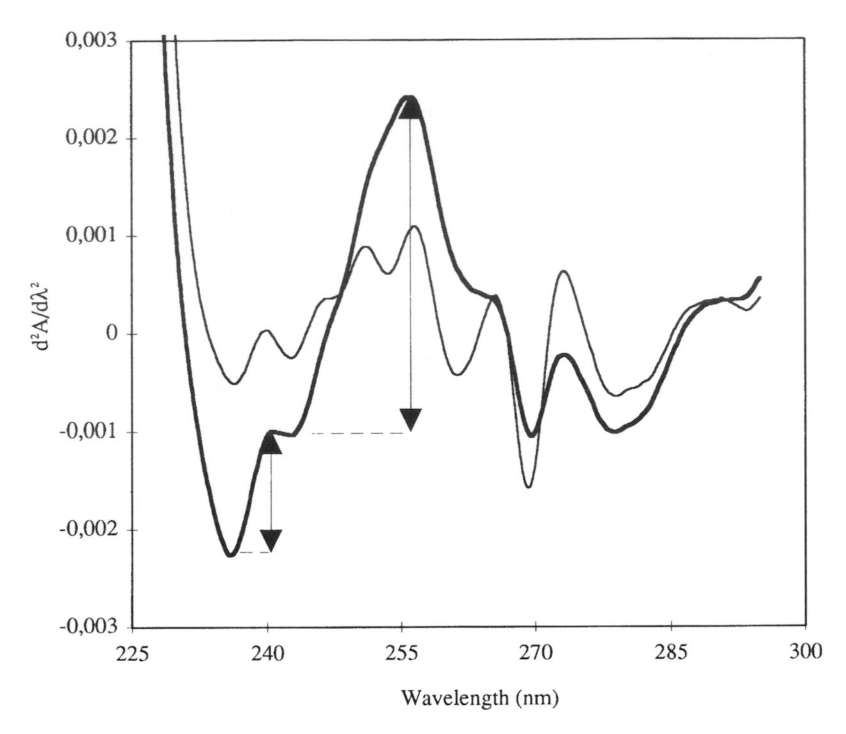

Fig. 1. Second derivative of the UV spectra of lipids extracted from control rat hepatocytes and from hepatocytes for which an oxidative stress is induced by addition of 100 μ*M* iron-nitrilotriacetic acid (NTA) for 5 h *(6)*. The arrows show the measured heights for minima to adjacent maxima at the higher wavelength in the spectrum of oxidized cells. (—) Control; (**—**) iron.

before analysis, are dissolved in a known volume of anhydrous chloroform (*see* **Note 8**).

Infrared analysis is carried out using FTIR spectrophotometer. A blank composed of anhydrous chloroform is scanned over the wavenumber range of interest in the study (1640–1800 cm^{-1}). The total acquisition number is 20 scans. The areas under the carbonyl ester band at 1740 cm^{-1} of the lipids are determined by a computer (**Fig. 2**).

In cell membranes, fatty acyl esters are represented by many lipids such as cholesteryl esters, phospholipids, and triglycerides. Consequently, the concentration of fatty acyl esters is expressed as nanomoles fatty acid equivalent per mL (*see* **Note 9**). A standard curve expressed as carbonyl-ester band area in function of nanomoles fatty-acid equivalent is obtained using 1–50 μL of glycerol trioleate in anhydrous chloroform (1 μL = 1.82 μg glycerol trioleate = 6.17 nanoequivalents of fatty acids) (*see* **Note 10**).

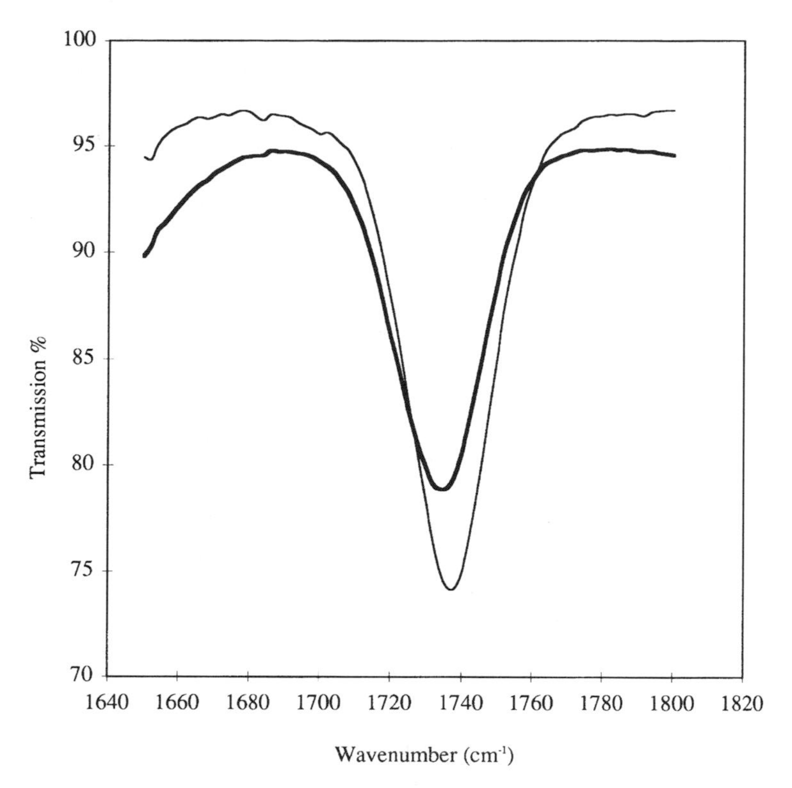

Fig. 2. Infrared absorption spectra of fatty acyl esters extracted from rat hepatocytes and of glycerol trioleate standard (247 nanomoles fatty acid equivalent/mL) ***(6)***. (——) Control; (**▬▬**) standard.

Then, the hydroperoxide levels are related to the amount of total fatty acyl esters. Oxidized fatty-acyl esters can be expressed as picomoles of hydroperoxides per nanomole fatty-acid equivalent (*see* **Note 11**).

4. Notes

1. A minimum of 2.5 million cells could be used if precautions are taken to avoid interfering substances.
2. Plastic should be avoided because the chloroform/methanol mixture is a very strong solvent. For instance, automatic pipets with plastic tips, plastic tops, and also tops of agglomerated cork are to be avoided. Glass syringes and glass tubes are preferred.
3. If tubes used for lipid extraction are kept from one experiment to another, they should be carefully washed with chloroform, then dried and washed again with distilled water.
4. Be careful not to use a plastic system for providing nitrogen just above chloroform extracts, because even chloroform vapor can dissolve plastic.

5. Quartz cuvets for UV-spectra obtention should be carefully washed especially if used in another experimentation. We recommend washing it with chloroform, then with ethanol, and finally with distilled water.
6. Minima at 233 and 242 nm were ascribed to the *trans,trans* and *cis,trans* conjugated diene isomers, respectively ***(5)***. A high ratio of *cis,trans* to *trans,trans* owing to inhibition of *trans,trans* isomers production by hydrogen donors would reflect an efficient activity of radical termination by antioxidants.
7. Cyclohexane lipid solutions can be kept for many weeks at –20°C in order to analyze together a large number of samples by FTIR spectroscopy.
8. It is of great importance to eliminate traces of water in order to prevent opacity of the FTIR cell windows.
9. For lipid samples extracted from 2.5 million rat hepatocytes, an average of 160 nanomole fatty-acid equivalent could be found ***(6)***.
10. In 2.5 million rat hepatocytes incubated for 5 h with iron-NTA (100 μ*M*), an average of 32 picomole hydroperoxides per nanomole fatty-acid equivalent was found ***(6)***.
11. Other fatty acyl esters, such as cholesteryl oleate or phosphatidylcholine dipalmitoyl, can be used as standards. But be careful about the number of carbonyl ester groups by molecule to plot the standard curve (for instance, 1, 2, or 3 for cholesteryl oleate, phosphatidylcholine dipalmitoyl or glycerol trioleate, respectively).

References

1. Janero, D. R. (1990) Malondialdehyde and thiobarbituric acid-reactivity as diagnostic indices of lipid peroxidation and peroxidative tissue injury. *Free Rad. Biol. Med.* **9,** 515–540.
2. Buege, J. A. and Aust, S. D. (1978) Microsomal lipid peroxidation. *Methods Enzymol.* **52,** 302–310.
3. Recknagel, R. O. and Glende, E. A. (1984) Spectrophotometric detection of lipid conjugated dienes. *Methods Enzymol.* **105,** 331–337.
4. Corongiu, F. P. and Milia, A. (1983) An improved and simple method for determining diene conjugation in autoxidized polyunsaturated fatty acids. *Chem. Biol. Interact.* **44,** 289–297
5. Corongiu, F. P., Poli, G., Dianzani, M. U., Cheeseman, K. H., and Slater, T. F. (1986) Lipid peroxidation and molecular damage to polyunsaturated fatty acids in rat liver: recognition of two classes of hydroperoxides formed under conditions in vivo. *Chem. Biol. Interact.* **59,** 147–155
6. Sergent, O., Morel, I., Cogrel, P., Chevanne, M., Beaugendre, M., Cillard, P., and Cillard, J. (1993) Ultraviolet and infrared spectroscopy for microdetermination of oxidized and unoxidized fatty acyl esters in cells. *Anal. Biochem.* **211,** 219–223
7. Chiang, S. P., Gessert, C. F. and Lowry, O. H. (1957) Colorimetric determination of extracted lipids, in *Air Force School of Aviation Medicine-Research Report 56–113.* Randolph Field, Texas, pp. 1–4.
8. Snyder, T. and Stephens, N. A. (1959) A simplified spectrophotometric determination of ester groups in lipids. *Biochim. Biophys. Acta* **34,** 244–245.

9. Cronin, D. A. and McKenzie, K. (1990) A rapid method for the determination of fat in foodstuffs by infrared spectrometry. *Food Chemistry* **35,** 39–49
10. Folch, J., Lees, M., and Sloane-Stanley, G. H. (1957) A simple method for the isolation and purification of total lipids from animal tissues. *J. Biol. Chem.* **226,** 497–509
11. Pryor, W. A. and Castle, L. (1984) Chemical methods for the detection of lipid hydroperoxides. *Methods Enzymol.* **105,** 293–299.

14

Synthesis of Lipid and Cholesterol Hydroperoxide Standards

Donald Armstrong and Richard W. Browne

1. Introduction

Phospholipids and cholesterol are the lipid component of cellular and subcellular membranes and are transported primarily (>65%) in low-density lipoproteins *(1)*. The fatty acids of these compounds are easily attacked by reactive oxygen species leading to the formation of their respective peroxides *(2–4)*. Cholesterol can be oxidized directly with hydroperoxide formation at the 5, 7, and 25 positions *(5)*. The ratio of oxidized phospholipid to cholesterol markedly increases membrane fluidity and deformability characteristics *(6)*.

Like the diseases associated with lipid peroxides described in Chapter 15, hydroperoxides of cholesterol and cholesterol esters have also been linked to a number of pathological conditions, especially atherosclerosis *(7)* and hemolytic disorders *(8)*, which were recognized over 20 yr ago. The detection of both lipid and sterol hydroperoxides provides valuable information about radical injury by reactive oxygen species during oxidative stress. These oxidized lipids are also useful in determining peroxidation of liposome formulations *(9)*. Since the methods employed to study peroxidation rely upon chromatographic identification, synthesis of appropriate standards is essential. An alternative method to identify lipid hyperperoxides (LHP) and cholesterol hydroperoxides is by their reduced hydroxy product, which is the stable metabolite formed in vivo *(9,10)*. This may be done by gas chromatography *(11)*, mass spectrometry (GC-MS) *(12)*, or high-pressure liquid chromatography (HPLC) (*see* Chapter 15).

The goal of this chapter is to present methods for preparing pure standards of LPH, cholesterol hydroperoxides, and cholesterol ester hydroperoxide. Identification by HPLC is discussed in Chapter 15.

From: *Methods in Molecular Biology, vol. 108: Free Radical and Antioxidant Protocols*
Edited by: D. Armstrong © Humana Press Inc., Totowa, NJ

2. Materials

2.1. LHP

1. Standards: Linoleic (18:2), linolenic (18:3), arachidonic (20:4), and docosahexaneoic acid (22:6) are synthesized with soybean lipoxidase (Sigma Chem. Co., St. Louis, MO) in 0.1 *M* borate buffer, pH 9.
2, Chromatography: Thin layer silica-gel plates (0.25-mm thickness) are used to separate unoxidized fatty-acid standards from LHP (EM Science, Gibbstown, NJ). The mobile phase is heptane/ethyl acetate/acetic acid (95:5:0.1, v/v/v).

2.2. Cholesterol Hydroperoxides

1, Standards: Cholesterol and cholesterol linoleate are synthesized using hematoporphyrin and pyridine (Sigma).
2. Chromatography: Silica-gel plates described in **Subheading 2.1.2.** are used to separate the peroxides from unreacted cholesterol. The mobile phase is heptane/ethyl acetate (1:1, v/v). Thin-layer chromatography (TLC) chambers are equilibrated for 1 h to allow the mobile-phase vapors to saturate the chamber.

3. Methods

3.1. Synthesis of LHP

1. The procedure has been reported previously by our laboratory ***(14)*** and uses 250 mg of the fatty acid with 10 mg of soybean lipoxygenase in a total volume of 100 mL of 0.1 *M* borate buffer, pH 9.
2. Oxygen is continuously perfused over the solution and the initial incubation period is 2 h at 4°C. A second 10-mg aliquot of lipoxygenase is added and incubation continued for an additional 2 h.
3. Five grams of sodium chloride are dissolved in the solution and the LHP are extracted three times with 100 mL of anhydrous diethyl ether (*see* **Note 1**). Fifty grams of oven-dried sodium sulfate crystals are added to remove extraneous water. The ether extract is filtered through Whatman No. 1 paper to trap the crystals, then aliquots are evaporated to dryness under nitrogen, and the resulting yellow LHP oil stored at –80°C. Stability of LHP standards is 1 yr.
4. Conversion is >95% as determined by HPLC or TLC

3.2. TLC

1. The residue from **Subheading 3.1.3.** is resuspended in methanol and continuously applied along the length of the plate 2 cm from the bottom.
2. Plates are developed by placing in a chamber presaturated with mobile phase and chromatographed to a pre-marked distance of 17 cm. Plates are air-dried, re-chromatographed, dried again, the plate covered with another clean glass plate, and sandwiched between cold packs so that only 1 cm is exposed at either edge, sprayed with 0.1 *M* sulfuric acid, and the edges only heated to 150°C for 5 min on a hot plate to visualize the LHP and unoxidized fatty acid.

3. Lines are drawn across the plate to outline the LHP area, the gel scraped off, resuspended in methanol, centrifuged at 1000*g* for 5 min to sediment the silica gel, the supernatant decanted and dried under nitrogen. and resuspended in 300 µL chloroform. A molar-extinction coefficient of 25,000 is used to determine concentration.

3.3. Synthesis of Cholesterol and Cholesterol Ester Hydroperoxides

1. Photosensitized oxidation of cholesterol is accomplished by adding 16 mg of cholesterol and 14.8 mg of hematoporphyrin dissolved in 200 mL of dry pyridine, which is irradiated with a 250 W lamp for 48 h while a slow stream of oxygen is bubbled through the solution. At the end of this time, the volume is reduced to about 80 mL.
2. The standard is extracted with an equal volume of diethyl ether and 1.5 g of activated charcoal is added to stop the reaction. After filtration through Celite, the clear yellow solution is dried under nitrogen.
3. Cholesterol linoleate (16 mg) is dissolved in 200 µL of chloroform-methanol (2:1, v/v), evaporated to dryness under nitrogen, resuspended in 3 mL of aqueous 70% t butyl peroxide with 100 µ*M* ferrous sulfate, placed in the dark, and mechanically shaken for 48 h at 37°C. The reaction is stopped by diluting the mixture with 15 mL of chloroform-methanol (2:1).
4. The solution is washed three times with 5 mL of Type 1 water and dried under nitrogen. The residue is dissolved in 300 µL of chloroform.

3.4. TLC

Cholesterol hydroperoxide and the ester hydroperoxide solutions are applied to a silica-gel plate as described in **Subheading 3.2.**, separated in ethyl acetate/heptane (1:1 v/v) mobile phase, removed and air dried, the edges sprayed first with either 1% N,N-dimethyl 1- phyenylenediamine in 50% methanol/1% acetic acid, or with 0.5 *M* sulfuric acid, and the plate heated at 120°C for 15 min on a hot plate.

The characteristic color of the spots are recorded, the Rf values calculated, the appropriate bands scraped off, recovered as previously described, and concentration determined by weight or spectrophotometrically.

3.5. Preparation of Hydroxy Derivatives

One hundred mg of LHP standard is dissolved in 2 mL of methanol, and 20 mg of sodium borohydride added. The mixture is allowed to stand for 30 min at 4°C with occasional stirring, 1 mL of water is added, and the mixture extracted three times with 2 mL of chloroform, which forms the lower phase, evaporated to dryness under nitrogen, and stored at –80°C (*see* **Note 2**).

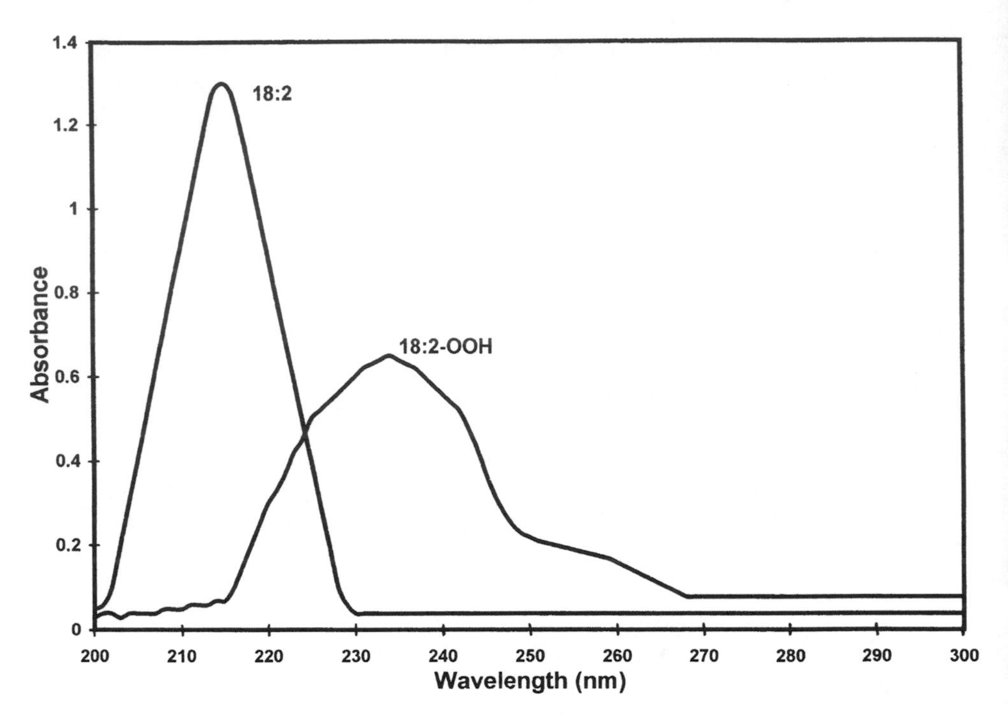

Fig. 1. Ultraviolet spectrum of linoleic acid (18:2ω3λ_{max} = 209 nm) and its hydroperoxide (λ_{max} = 234 nm).

Cholesterol hydroperoxide standards are treated with 1% sodium borohydride in methanol according to the method of Lier and Smith ***(15)***.

3.6. Results

3.6.1. Identification

3.6.1.1. Spectrophotometry of LHP

The absorption maxima at 234 nm, corresponding to the conjugated-diene system of LHP peroxides is used to compare against a sharp peak at 209–215 nm (**Fig. 1**) which represents the native, unoxidized fatty acid. Under the conditions described in **Subheading 3.1.**, conversion is >95%.

3.6.1.2. TLC of LHP

This technique is confirmation for **Subheading 4.1.1.** by establishing the purity of the standard (**Fig. 2**) and provides a means for calculating the relative amounts of free fatty-acid LHP in biological samples. The RF (mobility relative to solvent front) values for peroxide standards of 18:2, 18:3, 20:4, and

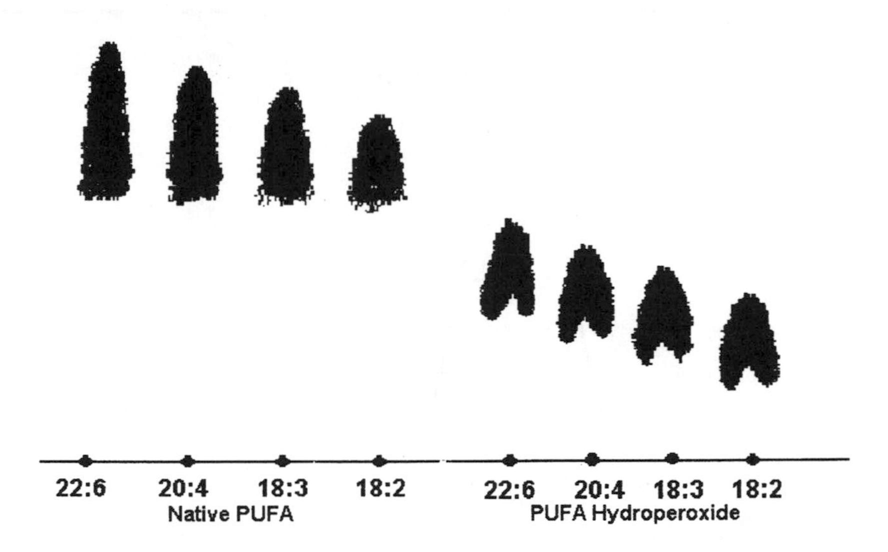

Fig. 2. Thin-layer chromatogram of polyunsaturated fatty acids and the corresponding hydroperoxides developed two times on silica-gel G60 using mobile phase of heptane/ethyl acetate/acetic acid (95:5:1). Spots are visualized by charring at 150°C after spraying the plate with 0.1 *M* sulfuric acid.

22:6 are 0.37, 0.36, 0.35, and 0.33, respectively. The hydroxy derivative co-chromatographs with LHP, but can be resolved by HPLC analysis described in Chapter 15.

3.6.1.3. Spectrophotometry of Cholesterol Peroxide

The absorption maxima is at 238 nm (**Fig. 3**).

3.6.1.4. TLC of Cholesterol Peroxide

Charring with H_2SO_4 gives differential colors for cholesterol (red), peroxides (blue), and hydroxides (purple). The RF values are 0.92, 0.68 and 0.24, respectively.

4. Notes

1. A strong emulsion tends to form when separatory funnels are shaken vigorously. Adding several drops of ethanol will break the emulsion, leading to a well-separated phase. The ethanol has no effect on the recovery of the hydroperoxides.
2. When collecting the lower phase it is preferable to leave some of the chloroform phase behind rather than contaminate this phase with any of the aqueous methanol/water phase, which contains the borohydried.

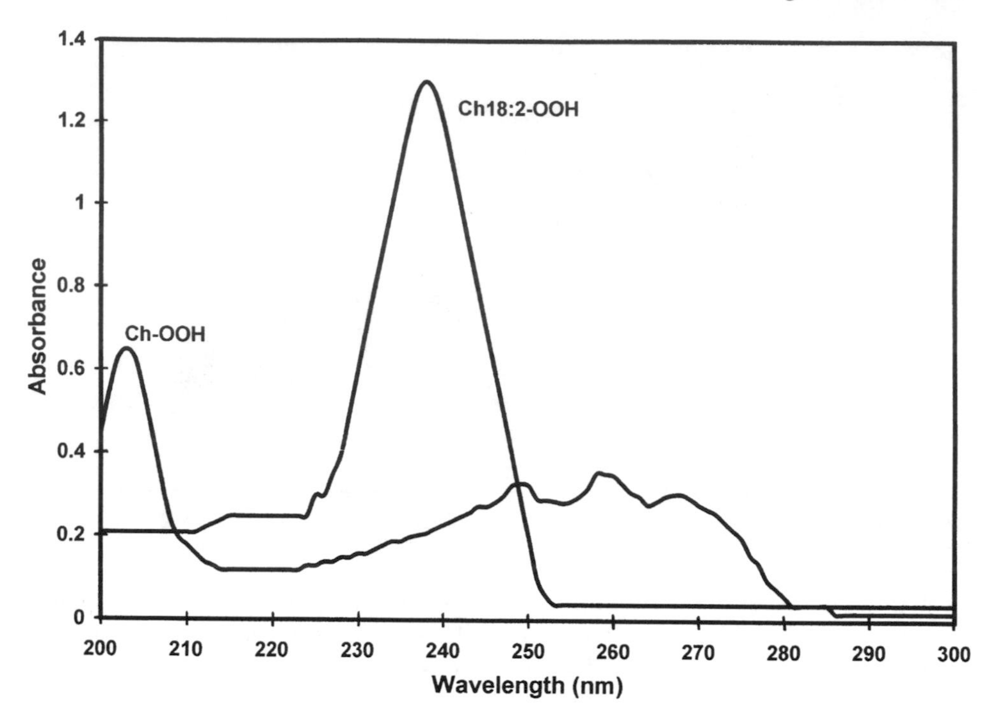

Fig. 3. Ultraviolet spectrum of cholesterol ester hydroperoxide (CH18:2-OOH λ_{max} = 238 nm) and cholesterol hydroperoxide (Ch-OOH λ_{max} = 203 nm) in methanol. The peaks at 251–257 nm belong to other oxides of cholesterol.

References

1. Nourooz-Zadeh, J., Tajaddini-Sarmadi, J., Ling, K. L., and Wolff, S. P. (1996) Low density lipoprotein is the major carrier of lipid hydroperoxides in plasma: relevance to determination of total plasma lipid hydroperoxide concentrations. *Biochem. J.* **313,** 781–786.
2. Esterbauer, H., Gebricki, J., Puhl, H., and Jurgens, G. (1992) The role of lipid peroxidation and antioxidants in oxidative modification of LDL. *Free Rad. Biol. Med.* **13,** 341–390.
3. Janero, D. R. (1990) Malondialdehyde and thiobarbituric acid-reactivity as diagnostic indices of lipid peroxidation and peroxidative tissue injury. *Free Rad. Biol. Med.* **9,** 515–540.
4. Tien, M. and Aust, S. D. (1982) Comparative aspects of several model lipid peroxidation systems, in *Lipid Peroxides in Biology and Medicine* (Yagi, K., ed.), Academic Press, Orlando, FL, pp. 23–29.
5. Smith, L. (1990) Cholesterol oxidation, in *Membrane Lipid Oxidation (vol. 1)* (Vigo-Pelfrey, C., ed.), CRC Press, Boca Raton, FL, pp. 129–154.

6. Hockstein, P. and Rice-Evans, C. (1982) Lipid peroxidation and membrane alterations in erythrocyte survival in, *Lipid Peroxides in Biology and Medicine* (Yagi, K., ed.), Academic Press, Orlando, FL, pp. 81–88.
7. Harlan, W., Gilbert, J., and Brooks, C. (1973) Lipids of human atheroma. VIII: oxidized derivatives of cholesterol linoleate. *Biochem. Biophyis. Acta* **316,** 378–385.
8. Lamola, A., Yamane, T., and Trozzolo, A. (1993) Cholesterol hydroperoxide formation in red cell membranes and photochemolysis in erythropoietic protoporphyria. *Science* **179,** 1131–1133.
9. Lang, J. K. and Vigo-Pelfrey, C. (1993) Quality control of liposomal lipids with special emphasis on peroxidation of phospholipids and cholesterol. *Chem. Phys. Lipids* **64,** 19–29.
10. Miyazawa, T., Fujimoto, K., and Oikawa, S. (1990) Determination of lipid hydroperoxides in low density lipoprotein from human plasma using HPLC with chemiluminescence detection. *Biomed. Chromatog.* **4,** 131–134.
11. Peng, S-K. (ed.) (1991) *Biological Effects of Cholesterol Oxides.* CRC Press, Orlando, FL.
12. Nikkari, T., Malalo-Ranta, U., Hiltunen, T., Jaakkola, O., and Yla-Herthuala, S. (1995) Monitoring of lipoprotein oxidation by gas chromatographic analysis of hydroxy fatty acids. *J. Lipid Res.* **36,** 200–207.
13. Thomas, D. W., van Kuijk, F. J., and Stephens, R. J. (1992) Quantitative determination of hydroxyl fatty acids as an indicator of an vivo lipid peroxidation: oxidation products of arachidonic and docosapentaenoic acids in rat liver after exposure to carbon tetrachloride. *Anal. Biochem.* **206,** 353–358.
14. Armstrong, D., Hiramitsu, T., Gutteridge, J. M. C., and Nilssen, S. E. (1982) Studies on experimentally induced retinal degeneration I: effect of lipid peroxides on electroretinographic activity in the albino rabbit. *Exp. Eye Res.* **35,** 157–171.
15. Lier, J. and Smith, L. (1968) Sterol metabolism II: gas chromatographic recognition of cholesterol metabolites and artifacts. *Anal. Biochem.* **24,** 419–430.

15

Separation of Hydroxy and Hydroperoxy Polyunsaturated Fatty Acids By High-Pressure Liquid Chromatography

Richard W. Browne and Donald Armstrong

1. Introduction

Lipid peroxidation is an autocatalytic free radical-mediated chemical mechanism in which polyunsaturated fatty acids (PUFA) undergo oxidation to form lipid hydroperoxide (LHP). Increased lipid peroxidation has been implicated in many disease conditions including acute myocardial infarction *(1)*, stroke *(2)*, diabetes mellitus *(3)*, hepatic disorders *(4)*, and toxicity *(5)* by certain drugs, pesticides, and metals. Study of lipid peroxidation in disease requires simple, discriminative, and sensitive methods for LHP measurement because of their low concentration and because of the multitude of isomeric and esterified forms. Methods have been developed that use spectrophotometry, fluorometry, chemiluminescence, and chromatography as well as enzymatic and non-enzymatic techniques *(6–8)*. These methods determine total hydroperoxides as well as interfering substances particular to each assay. High pressure liquid chromatography (HPLC) techniques remove interfering substances from the analysis of LHP as well as separate hydroperoxides of different lipid classes. Because lipid peroxidation occurs predominantly in PUFA, esterified to cholesterol and glycerol-based lipids, assessment of the extent of lipid peroxidation can be made through direct measurement of total PUFA hydroperoxides and indirectly by measurement of their hydroxy derivative. HPLC-ultraviolet (UV) detection of PUFA hydroperoxides, based upon the absorbance of the conjugated diene at 234 nm, has been criticized owing to its inability to differentiate LHP from the corresponding alcohol-reduction products that co-elute in procedures reported to date *(9,10)*. To achieve specificity,

From: *Methods in Molecular Biology, vol. 108: Free Radical and Antioxidant Protocols*
Edited by: D. Armstrong © Humana Press Inc., Totowa, NJ

HPLC has been used with several different detection schemes including electrochemistry and post-column chemiluminescence ***(9,10)***. We have developed an HPLC technique that separates four major hydroperoxides and hydroxy PUFA as well as the parent, unoxidized fatty acids, on a single chromatogram, thereby eliminating the need for specialized detection schemes. The combination of HPLC separation and UV detection provides a simple and comprehensive analysis of lipid peroxidation, applicable to blood and tissue samples.

2. Materials

2.1. Analytical HPLC System

1. Shimadzu (Columbia, MD) LC-6A pump System (max flow rate = 9.99 mL/min.).
2. Shimadzu LPM-600 low-pressure mixing/proportioning manifold.
3. Shimadzu SIL-7A autosampler/injector.
4. Shimadzu SPD-M6A W/VIS photodiode array.
5. Shimadzu ECD-6A electrochemical detector.
6. Kontes (Vineland, NJ) 3-PORT helium-degassing manifold.
7. IBM 486Dx on-line computer.
8. Shimadzu CLASS-VP chromatography software.

2.2. HPLC Columns

1. Supelco (Bellefonte, PA) LC-18: 4.6 × 150 mm, 5 micron particle size, 100 Å pore.
2. Supelco LC-18: 4.6 × 250 mm, 5 micron particle size, 100 Å pore).
3. Supelcoguard C-18 guard column: 4.6 × 20 mm, 5 micron particle size, 100 Å pore.
4. Supelcoguard guard column cartridge holder.

2 .3. Reagents

All chemicals are obtained from Sigma Chemical Company (St. Louis, MO) in the highest available purity. All solvents are HPLC grade and filtered through 0.22-µ*M* nylon membranes prior to use.

1. Acetonitrile (ACN).
2. Tetrahydrofuran (THF).
3. Butylated hydroxy toluene (BHT) (30 mg/L) in HPLC-grade ethanol.
4. Sodium hydroxide: 10*N* in HPLC-grade water.
5. Acetic acid(0.5%): in HPLC-grade water.
6. Anhydrous diethyl ether containing 0.001% BHT.
7. Anhydrous sodium sulfate.
8. Lipoxidase (Type 1-B, from soybean).
9. *cis*-9,*cis*-12-Linoleic (18:2ω6).
10. 9, 12, 15-Linolenic (18:3ω3).
11. 5,8,11,14-Arachidonic (20:4ω6).
12. 4,7,10,13,16,19-Docosahexasnoic acid (22:6ω6).

3. Methods

3.1. Samples

3.1.1. Serum

Collect whole blood into a Vacutainer-type tube (Baxter Scientific, McGraw Park, IL) containing no anticoagulant. Let the specimen stand at room temperature for 15 min to allow clotting, and centrifuge at 3000*g* for 10 min at 4°C. Decant serum and place on ice for immediate analysis or store in a freezer-safe tube at –80°C.

3.1.2. Plasma

Collect whole blood into a Vacutainer-type tube containing ethylenediamine tetraacetic acid (EDTA) as the anticoagulant. Mix by inversion for at least 1 min to ensure even distribution of the EDTA. Centrifuge the specimen at 3000*g* for 10 min at 4°C, remove plasma, and place on ice for immediate analysis or store in a freezer-safe tube at –80°C.

3.1.3. Tissue Homogenate

Tissues for lipid-peroxide analysis should be as fresh as possible. Weigh approximately 0.1 g of tissue, trimmed of fat and connective tissue, and dilute with 10 volumes of deionized water. Grind tissue in a clean Potter-Elvejhem glass homogenizer and transfer to a sonication-safe tube. Sonicate specimen for 15 s at a setting of 40 Hz using a probe-type ultrasonicator. Use whole uncentrifuged homogenate for lipid-peroxide analysis.

3.2. Standards

3.2.1. PUFA Hydroperoxides

1. The synthesis of LHP compounds is covered in much greater detail in **Chapter 14** so only a brief description is given here. The peroxides of linoleic (18:2ω6), linolenic (18:3ω3), arachidonic (20:4ω6), and docosahexaenoic (22: 6ω6) acids are prepared by a modification of Armstrong et al. ***(6)***, using the enzyme Lipoxidase.
2. Between 250 and 300 mg of each fatty acid are prepared in 100 mL of 0.1 *M* sodium borate, pH 9.0, and saturated with oxygen. Ten mg of soybean lipoxidase is added and the mixture is incubated at 4°C for 2 h. A second 10-mg aliquot of lipoxidase is added and the mixture is incubated at 4°C for another 2 h. Oxygen is continuously perfused over the mixture during incubations.
3. After incubation, 5 g of sodium chloride is dissolved in the incubation medium and the lipid peroxides are extracted three times with 100 mL of anhydrous diethyl ether. 50 g of oven-dried sodium sulfate is added to the 300 mL of pooled ether extract to remove any water, and then filtered through Whatman Type 1 paper. The ether extract is then evaporated to dryness under nitrogen.

4. The dry extract (consisting of a yellow oil) is resuspended in 5.0 mL of methanol and stored at –80°C prior to HPLC (*see* **Note 1**).

3.2.3. PUFA Hydroxy Fatty Acids

1. The hydroxy derivatives of the four LHP standards are prepared by the method described by Terao, Asano, and Matsushita ***(10)***, for the preparation of hydroxy derivatives of phospholipid hydroperoxides.
2. One mL of the hydroperoxide stock containing the equivalent of approximately 100 mg/mL of the corresponding LHP is dissolved in 2 mL of methanol. Twenty milligrams of sodium borohydried is added and the solution is vortexed vigorously.
3. The mixture is incubated for 30 min at 4°C with occasional agitation. Products are extracted three times with 2 mL of chloroform and evaporated to dryness under nitrogen.
4. The derivatives are filtered through 0.22-µm nylon membranes and stored at –80°C prior to use on HPLC.

3.3. HPLC Sample Preparation

3.3.1. Total-Lipid Extraction

Total lipids are extracted from serum, plasma, or tissue homogenate by adding 1 volume of sample to 1 volume of absolute ethanol containing 30 mg/L BHT to precipitate proteins and protect the sample from in vitro peroxidation. The sample is vortexed, extracted 2 times with four volumes anhydrous diethyl ether, and the upper phases pooled. Fifty microliters of 5% acetic acid is added to the remaining lower phase and again extracted two times with diethyl ether. All four extracts are pooled and evaporated to dryness under nitrogen.

3.3.2. Saponification

The dried total-lipid extract from **Subheading 3.3.1.** is resuspended in 1.0 volume of ethanol, which has been purged with helium gas. Fifty microliters of 10 *N* sodium hydroxide is added and the sample is then vortexed and heated to 60°C for 20 min. The solution is neutralized by addition of 30 µL glacial acetic acid and evaporated to dryness under nitrogen. The mixture of hydrolyzed lipids and free fatty acids is reextracted according to **Subheading 3.3.1.**, resuspended in 0.5 volumes of mobile phase, and analyzed by HPLC.

3.4. HPLC Separation

3.4.1. Standards and Calibration

1. Working calibrators of LHP and hydroxy derivatives are prepared fresh prior to analysis. Stock solutions of LHP and hydroxy derivatives from **Subheading 3.2.2.**, and **3.2.3.** are diluted in ethanol to approximately 20 µmol/L by applying Beer's law using the molar-extinction coefficients given in **Table 1** (*see* Cayman Chemical Co. Catalog, Ann Arbor, MI).

Table 1
Molar-Extinction Coefficients (Σ) of Polyunsaturated Fatty Acid Hydroperoxide andHydroxy-Derivative Standards

Standard	Σ (L/M/cm)
Hydroperoxy 18:2	23,000
Hydroperoxy 18:3	23,000
Hydroperoxy 20:4	27,000
Hydroperoxy 22:6	27,000
Hydroxy 18:2	23,000
Hydroxy 18:3	23,000
Hydroxy 20:4	27,000
Hydroxy 22:6	27,000

2. Fifty microliters of this solution is injected onto a Supelcosil LC-8 analytical column (4.6 mm × 15 cm, 5 micron particle size, 100Å pore) using an isocratic mobile phase of water/ACN/THF/AcAc (30.8:44.4:24.8:0.025) at a flow rate of 1.0 mL/min.
3. The percent purity of the standard is determined as the percent of the total chromatogram area attributed to the standard. The percent purity of the standard times the concentration of standard obtained by spectrophotometry yields the true calibrator concentration. Specific dilutions of the calibrators are analyzed according to the conditions described for the analysis of samples in the following section.
4. Calibration curves are generated and stored by the on-line PC.

3.4.2. Sample Analysis

1. Reconstituted lipid extracts or specific calibration solutions are loaded into 300 µL conical-glass inserts and placed into spring loaded, amber-glass autosampler vials equipped with Teflon septa/seals. Typically, between 25 and 150 µL is injected onto the analytical column and separated isocratically on a Supelcosil LC-18 analytical HPLC column, using an isocratic mobile phase of water/ACN/THF/AcAc (50:32.7:18.3:0.025) at a flow rate of 1.3 mL/min.
2. The column eluate is monitored at 236 nm. Results are collected at 0.64-s intervals by the PC and integrated by area.

3.5. Analysis of Method

3.5.1. Imprecision

A series of five identical samples are analyzed each day and the mean, standard deviation (SD), and coefficient of variation (%CV) are determined by computer analysis. Between-run imprecision is calculated using the values from two within-run tests. Similarly, between-d imprecision is tested by comparing the within-run imprecision value from four consecutive days.

3.5.2. Linearity

The upper limit of linearity is determined as the point where the resolution of the peaks is lost owing to the broadening of peaks with larger sample concentrations. Because the concentration of LHP is normally very small, the lower limit of linearity or minimum detectable quantity (MDQ) is the more likely encountered problem (*see* **Note 2**).

3.5.3. MDQ

The MDQ is the concentration of analyte, which gave a signal three times greater than the baseline SD. The MDQ is established by each individual laboratory because different HPLC systems will vary in their ability to measure LHP (*see* **Note 3**).

3.5.4. Recovery

Method recovery is determined by standard-additions methodology ***(11)***. The total-lipid extract of a control sera is separated into two aliquots. One is spiked with a 2.0 nmol/mL concentration of PUFA hydroxy or hydroperoxy standard and the other with an equal volume of deionized water. Using the concentration of the spiking solution and the dilution of the control sera, the added concentration of analyte is determined. The amount of LHP recovered is expressed as a percentage of the amount added. These samples are assayed in triplicate and the average value is used for calculations (*see* **Note 4**).

3.6. Results

3.6.1. Separation of Hydroperoxides from Native Unoxidized Fatty Acid

PUFA hydroperoxides are separated from their unoxidized precursors by reverse-phase (RP) HPLC. Both C8 and C18 columns of various dimensions are used successfully. A Supelcosil C8 column (4.6 mm × 15 cm, 5 micron particle size, 100Å pore) is ideal for separating solutions of pure fatty acids from their oxidized products during lipoxidase-catalyzed synthesis of the hydroperoxides. The C8 column provides resolution of these species in short run times. These separations require a two channel or a programmable UV detector in order to monitor both the native fatty acid (215 nm) and the hydroperoxides (236 nm). **Figure 1** shows a chromatogram obtained after 15 min incubation of unoxidized linoleic acid with, soybean lipoxidase (*see* **Note 5**).

3.6.2. Separation of Hydroperoxide Standards

Complete resolution of the eight oxidized PUFA standards requires a C-18 column and is accomplished by progressively increasing mobile-phase polar-

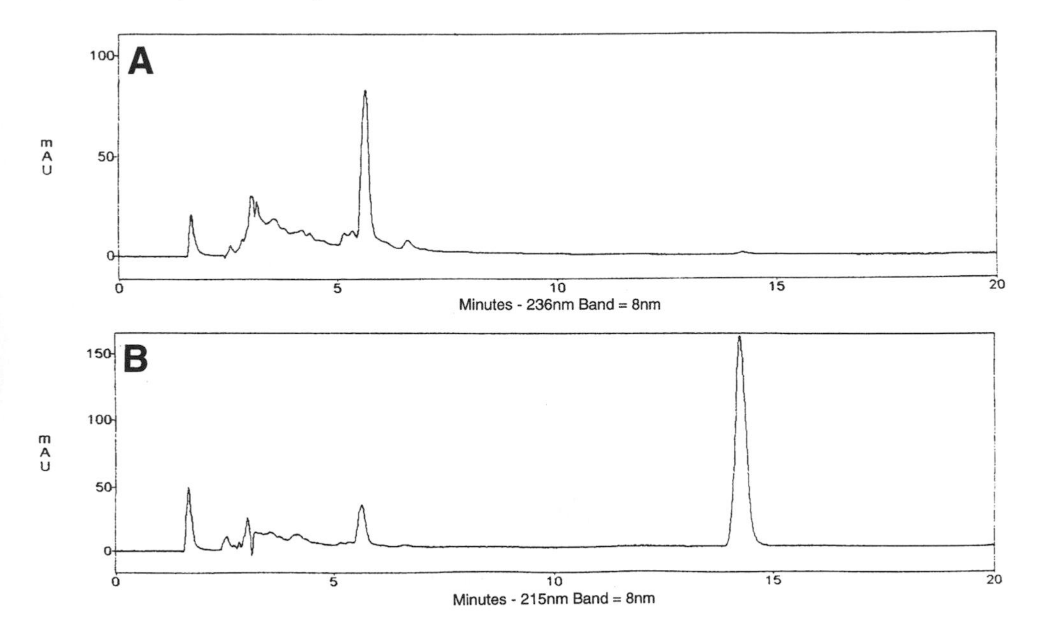

Fig. 1. Chromatogram of the simultaneous separation of hydroperoxy 18:2 at 234 nm (**A**) from unoxidized 18:2 at 215 nm (**B**).

ity, analytical column length, and system flow rate. Baseline resolution of all eight oxygenated PUFA standards is obtained within 30 min by the conditions described in **Subheading 3.4.1**. The analytical HPLC column was a Supelcosil LC18 (4.6 mm × 25 cm, 5 µ*M* particle size, 100Å pore). Peak identification is made by absolute retention time of 50 nmol of each standard spiked into an equimolar mixture of the eight standards. This is required because the retention time of an individual standard does not match its retention time when it is injected as part of a mixture. Results of a separation of the eight standards is shown in **Fig. 2**. The performance characteristics of this method are outlined in **Table 2**.

4. Notes

1. Synthesis of PUFA hydroperoxides by this method produces predominantly one isomeric form depending on the enzyme used. Synthesis of standards using chemical means such as cupric sulfate and hydrogen peroxide will give a mixture of isomers. For specific isomer identification, it is necessary to purchase pure standards available from commercial sources such as Caymen Chemical Co. (Ann Arbor, MI) or Cascade Biochemicals Ltd. (Berkshire, UK).

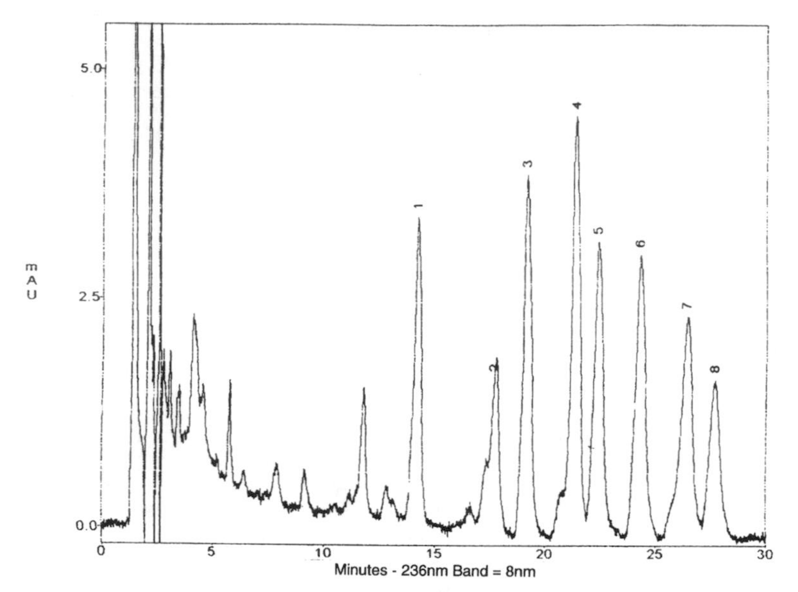

Fig. 2. Chromatogram of the separation of hydroperoxy and hydroxy polyunsaturated fatty acids by UV absorbance at 234 nm.1, hydroxy 18:3; 2, hydroperoxy 18:3; 3, hydroxy 18:2; 4, hydroxy 20:4; 5, hydroxy 22:6; 6, hydroperoxy 18:2; 7, hydroperoxy 20:4; 8, hydroperoxy 22:6. *See* text for mobile phase and flow rate.

Table 2
Performance Characteristics

Reproducibility	CV = 4.6%
Linear range	5–250 pmol
Minimum detectable quantity	5 pmol
Recovery	96–102%

2. The upper limit of linearity for individual standards is considerably higher than standards in a mixture. At levels above 250 pmol, the broadening of peaks associated with increasing concentration obscures adjacent peaks. Most modern chromatography-integration software has the capability to compensate for a limited loss of resolution in samples with high concentrations, but for practical purposes, these samples are diluted or less sample is injected.
3. The MDQ reported in **Table 2** is for an HPLC system using a photodiode array UV-VIS detector. This amount should be less than 5 pmol on a system using a dedicated UV detector, owing to its inherent higher sensitivity.
4. It is necessary to perform recovery studies by spiking the PUFA hydroperoxides into total-lipid extracts. When spiked into fresh serum, plasma, or tissue homogenate, rapid reduction of these compounds occurs, presumably owing to the action of endogenous peroxidase. In this case, the majority of the spiked hydro-

peroxide (approx 70%), is recovered as the corresponding hydroxy derivative depending upon the length of time used to make the preparation.

5. Auto-oxidation of native PUFA occurs rapidly at room temperature and any preparation will always contain some oxidation products. To minimize this, keep standards cold and store them under nitrogen or argon and dilute in solvents that have been purged with helium to remove dissolved oxygen.

Acknowledgments

This research was accomplished with support from The Mark Diamond Graduate Student Research Fund of The State University of New York at Buffalo.

References

1. Anzar, J., Santos, M., and Valles, J. (1983) Serum malondialdehyde-like material (MDA-LM) in acute myocardial infarction. *J. Clin. Pathol.* **36,** 312–715.
2. Satoh, K. (1987) Serum lipid peroxide in cerebrovascular disorders determined by a new colorimetric method. *Clin. Chem. Acta.* **90,** 37–43.
3. Armstrong, D., Abdella, N., Salman, A., Miller, N., Abdel Rahman, E., and Bojancyzk, M. (1991) Relationship of lipid peroxides to diabetic complications. *J. Diab. Comp.* **6,** 116–122.
4. Suematsu, T., Kamada, T., and Abe, H. (1977) Serum lipid peroxide level in patients suffering from liver disease. *Clin. Chem. Acta.* **79,** 267–270.
5. Comporti, M. (1985) Biology of disease: lipid peroxidation and cellular damage in toxic liver injury. *Lab. Invest.* **53(6),** 599–618.
6. Armstrong, D. and Browne, R. (1994) The analysis of free radicals, lipid peroxides, antioxidant enzymes and compounds related to oxidative stress as applied to the clinical chemistry laboratory, in *Free Radicals in Diagnostic Medicine* (Armstrong, D., ed.), Plenum Press, NY, pp. 43–58.
7. Yagi, K. (1984) A simple fluorometric assay for the lipidperoxide in blood plasma. *Biochem. Res.* **15,** 212–216
8. Ohkawa, H., Ohoshi, N., and Yagi, K. (1979) Assay for lipid peroxides in animal tissues by thiobarturic acid reaction. *Anal. Biochem.* **95,** 351–358.
9. Yamamoto, Y. and Frei, B. (1990) Evaluation of an isoluminol chemiluminescence assay for the detection of hydroperoxides in human blood plasma. *Methods Enzymol.* **186,** 371– 380.
10. Matsushita, S., Teraro, J., Yamada, K., and Shibata, S. (1988) Specific detection of lipid hydroperoxides using HPLC-EC method, in *Oxygen Radicals in Biology and Medicine* (Simic, M. and Taylor, K., eds.), Plenum Press, NY, pp.164–168.
11. Strobel, H. A. and Heineman, W. R. (eds.) (1989) Standard additions method, in *Chemical Instrumentation: A Systematic Approach* (3rd ed.). John Wiley and Sons, NY, pp. 391–397.

16

Products of Creatinine with Hydroxyl Radical as a Useful Marker of Oxidative Stress In Vivo

Kazumasa Aoyagi, Sohji Nagase, Akio Koyama, Mitsuharu Narita, and Shizuo Tojo

1. Introduction

Methylguanidine (MG) was first discovered in the serum and urine of uremic patients *(1)*. Increased synthesis of MG and its toxicity has been implicated as a potent uremic toxin *(2)*.

We have clarified that MG was synthesized from creatinine by hydroxyl radical and proposed the pathway as shown in **Fig. 1** *(3,4)*. We also have reported MG synthesis by activated human leukocytes *(5)*, isolated rat hepatocyte *(4,6)*, and micro-organelles such as microsomes *(7)*, which generate hydroxyl radicals. We also reported that MG/creatinine excreted in urine was significantly increased in humans after 1 h of treatment in the hyperbaric oxygen chamber *(8)*.

Recently, a direct product of creatinine with hydroxyl radical was identified as creatol (CTL) *(9)*. MG was synthesized via creatol, creaton A, and creaton B, as shown **Fig. 2**.

Creatinine is an intrinsic and relatively low toxic substance that is distributed equally in the body as the substrate. Both MG and CTL are easily excreted into the urine. Conversion of CTL to creaton A is carried out by both enzymatic and non-enzymatic reactions *(10)*. The enzymes that forms creaton A from creatol were purified from rat liver *(11)* and kidney *(12)*. Non-enzymatic conversion of CTL to creaton A was performed by oxidative or non-oxidative reactions; however, the rate of chemical reaction is lower than that of enzymatic conversion *(10)*. The rapid conversion of creaton A to creaton B has been reported *(10)*.

MG is an end product of creatinine reacted with hydroxyl radical. Therefore, MG synthetic rate from creatinine indicates the amount of hydroxyl-radical

From: *Methods in Molecular Biology, vol. 108: Free Radical and Antioxidant Protocols*
Edited by: D. Armstrong © Humana Press Inc., Totowa, NJ

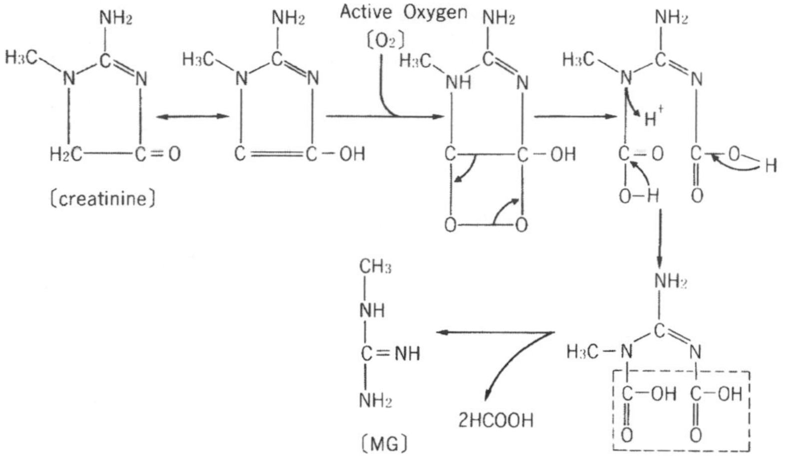

Fig. 1. Proposed MG synthetic pathway from creatinine by reactive oxygen species.

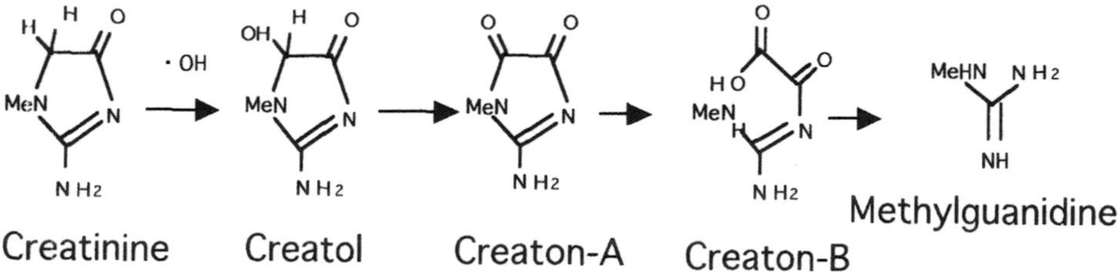

Fig. 2. Production of MG from creatinine via creatol *(10)*.

generation in most cases. CTL, a direct hydroxyl-radical product of creatinine, is also a useful marker of hydroxyl-radical generation in certain conditions.

Cretol+MG/creatinine is a more accurate indicator of hydroxyl-radical generation. Biosyntheses of MG and creatol by isolated rat hepatocytes are shown in **Fig. 3**. Urinary excretion of creatol and MG reflects hydroxyl-radical generation in the whole body. For example, we have reported that puromycin aminonucleoside, which causes heavy proteinuria, increased MG synthesis in isolated rat hepatocytes *(4)*. Urinary excretion of MG is increased in the urine treated by puromycin aminonucleoside *(13)*.

MG and creatol is determined by high performance liquid chromatography (HPLC) *(14)*, using post-column derivertization and 9,10-phenanthrequinone (PQ) as fluorogenic reagent *(15)*. In the case of CTL, CTL separated by analytical column was changed to MG by heating in 2 *N* NaOH in a coil before reacting with fluorescent reagent for mono-substituted guanidino compounds.

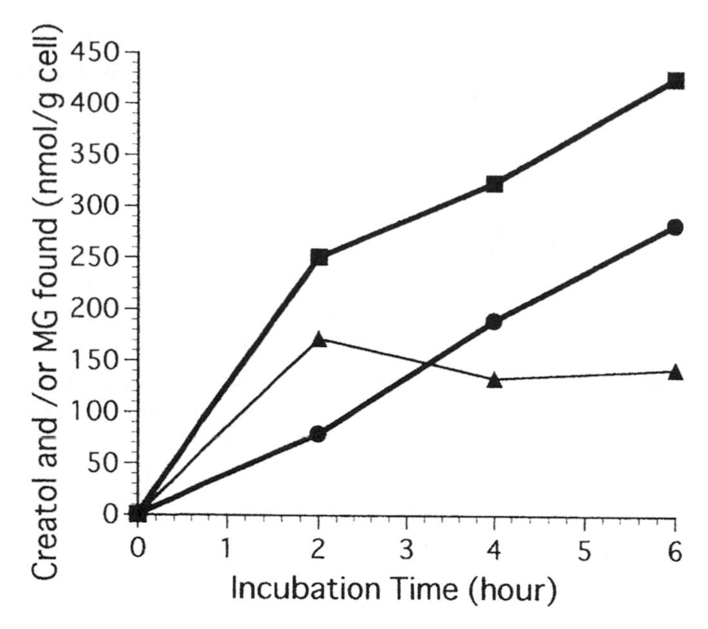

Fig. 3. Biosynthesis of MG or creatol by isolated rat hepatocytes in the presence of 16 m*M* creatinine. ● MG, ▲ Creatol, ■ MG+creatol.

2. Materials

2.1. Determination of MG by HPLC

1. 500 mg of PQ (Wako Pure Chemical Industries, LTD, Osaka, Japan, Cat. No. 166-11301, for biochemistry use) is dissolved in 1000 mL of dimethylformamide (DMF) (Wako Cat. No. 042-20621 for HPLC) before use under dark conditions. PQ is a highly light-sensitive reagent.
2. Standard Solution: 2.5 nmol of MG (Sigma Chem. Co., St. Louis, MO) dissolved in 1 mL of 10% TCA is stored at –20°C.
3. Mobile phase: First buffer: 0.4 *N* Na-Citrate buffer pH 2.8. Second buffer: 1 *N* NaOH.
4. JASCO (Hachioji, Japan) 865-10 column oven set at 65°C
5. Auto sampler (Kyowa Seimitu Co., Japan, KSST-60), samples are chilled at 4°C.
6. Elusion buffer exchanger (Kyowa Seimitu, KTC-8A).
7. Reaction coil (at 65°C): 1/16 inch stainless-steel tube for HPLC (5 mm i.d. × 5 m).
8. Analytical column: 2.1 mm × 100 mm stainless-steel column packed with TSK gel SCX (10 µm) (Tosoh, Tokyo, Japan, cat. no. 07154).
9. Fluorescence spectrometer (JASCO, FP-110).
10. 3 HPLC pumps: (JASCO, 880-PU).

2.2. Determination of Creatol by HPLC

1. Standard solutions of CTL (100 µ*M*) are kept at –20°C (*see* **Note 1**).
2. JASCO RO-961 oven at 125°C.

3. An additional coil (6a in **Fig. 4**; 0.5 mm i.d. × 5 m) is inserted in place of a short tube for hydrolysis of CTL in a reaction bath (7 in **Fig. 4**; JASCO, RO-961) after the column separation and before reaction with PQ.
4. TSK gel SCX (10 μm) 6 mm i.d. × 150 mm.

3. Methods

3.1. Preparation of Samples

1. An equal volume of cold 20% TCA is added to serum or plasma and kept at 4°C for 30 min and centrifuged at 1100*g* for 15 min at 4°C. 100 μL of the supernatant is used for analysis.
2. 10 Volumes of cold 10% TCA is added to the urine and centrifuged at 1100*g* 15 min at 4°C. The supernatant (100 μL) is used directly for MG or CTL analysis.
3. CTL is stable at –20°C (*see* **Note 2**), however, at 4°C (*see* **Note 3**), slow conversion to MG *(9)* is not negligible for long-term storage, because CTL concentration in the serum is 10 times that of MG *(16)*.

3.2. Determination of MG by HPLC

1. From biological material, MG is separated by HPLC (**Fig. 4**) using a column packed with cationic ion-exchange resin and reacted with phenanthrequinone under alkaline condition for the determination (*see* **Note 4**).
2. Flow diagram of the new analytical system (**Fig. 4**): The automated guanidino compounds analyzer, includes HPLC with PQ as a fluorogenic reagent: pumps, column oven, fluorometer, autosampler, eluent exchanger, cation exchange column, TSK gel SCX (2.1 × 100 mm stainless-steel column packed with TSK gel SCX; 10 μm: one coil (6b in **Fig. 4**; 0.5 mm i.d. × 5 m) for the reaction with PQ is placed inside the column oven, along with the separation column. (The second coil 6a in **Fig. 4** for the conversion of creatol is replaced by a short tube in the case of MG assay.)
3. Analytical method for MG: The first eluent is 1 *M* NaOH. For the labeling reaction, 2 *M*-NaOH solution and PQ solution (500 mg of PQ in 1 L of DMF) are prepared and the stock vessel for PQ and its subsequent flow pathway are completely shielded from light.
4. HPLC conditions are as follows: temperature of column oven, 65°C; flow rate for eluents, 1.4 mL/min; equilibrated by the first buffer: 0.4 *N* Na-Citrate buffer (0.4 *M*: Na concentration), pH 2.8, for 3 min and eluted by the second solution: 1 *M* NaOH for 25 min; flow rate for 2 *M* NaOH solution, 0.7 mL/min; flow rate for PQ solution, 0.7 mL/min; fluorescence, λmax (em.) 495 nm and λmax (ex.) 340 nm. Each sample is injected 3 min after the first eluent has started. Effluent is subsequently reacted with PQ in another reaction coil (6b in **Fig. 4**) inside the column oven.
5. Chromatograms of standards and urine from a normal subject are shown in **Fig. 5**.

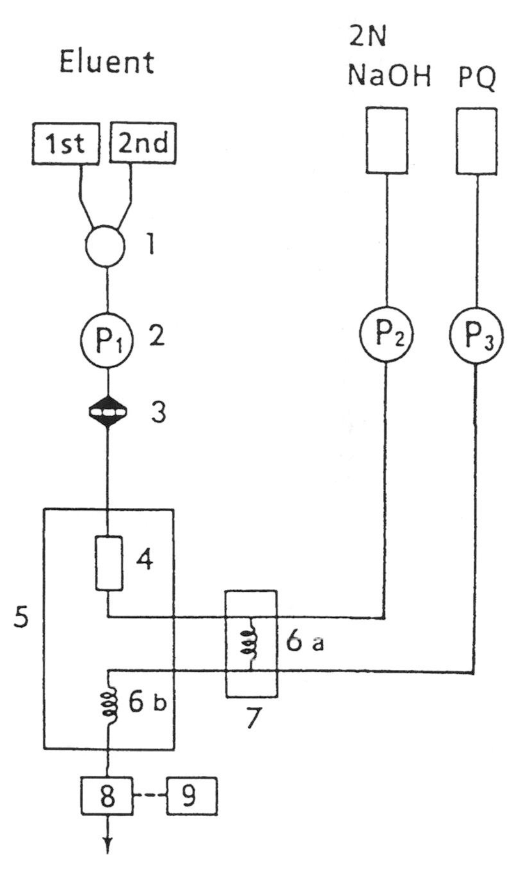

Fig. 4. Schematic flow diagram for the analytical system of MG or creatol. Components: 1, eluent exchanger; 2, HPLC pump; 3, autosampler; 4, column; 5, column oven (65°C); 6a, reaction coil for hydrosis; 6b, coil for reaction with PQ; 7, reaction bath (125°C); 8, fluorometer; 9, data processor.

3.3. Determination of Creatol by HPLC

1. Analytical method for CTL: The first eluent is 0.4 *M* (Na concentration) trisodium citrate (9 parts) + dimethyl sulfoxide (DMSO) (1 part), subsequently adjusted to pH 5.0 with conc. HCl; the second is 1 *M* NaOH. For the labeling reaction, 2 *M*-NaOH solution and PQ solution (500 mg of PQ in 1 L of DMF) are prepared, and the stock vessel for PQ and its subsequent flow pathway are completely shielded from light.
2. HPLC conditions are as follows: temperature of column oven, 65°C; flow rate for eluents, 1.0 mL/min (first eluent [pH 5.0], 20 min; second eluent [1 *M* NaOH],10 min); flow rate for 2 *M* NaOH solution, 0.5 mL/min; flow rate for PQ

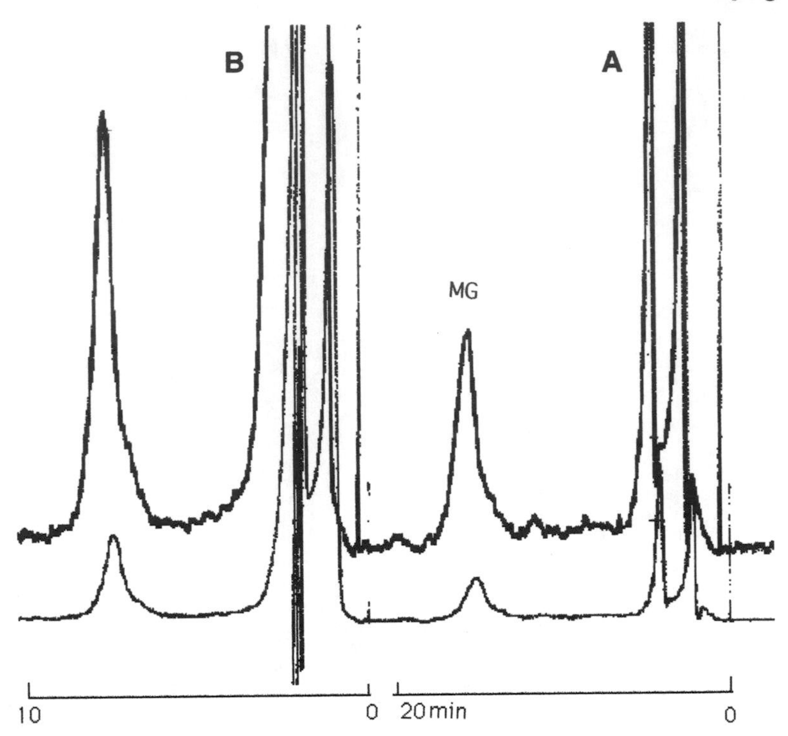

Fig. 5. Chromatogram of MG analysis. (**A**) Standard MG (0.25 nmol). (**B**) MG in urine from normal human.

solution, 0.5 mL/min; fluorescence, λmax (em.) 495 nm and λmax (ex.) 340 nm fluorescence. Each sample is injected 11 min after the first eluent has started. The post-column alkaline hydrolysis of CTL is carried out in the reaction coil immersed in the reaction bath at 125°C (6a in **Fig. 4**), and the hydrolysate is subsequently reacted with PQ in another reaction coil (6b in **Fig. 4**) inside the column oven.

3. Chromatograms of standards and urine are shown in **Fig. 6**.

4. Notes

1. Forty percent of CTL is degraded at 4°C in 16 wk *(15)*.
2. CTL is less stable at around pH 5, but stable at pH 1.0.
3. It is preferable to cool urine during collection. Using fresh urine and MG/creatinine will provide useful information.
4. To estimate urinary MG, MG contained in food must be considered. Roasted meat contains a large amount of MG. Sometime, in normal subjects, MG from food is about 50% of the MG in urine. We used a noncreatinine diet in such experiments.

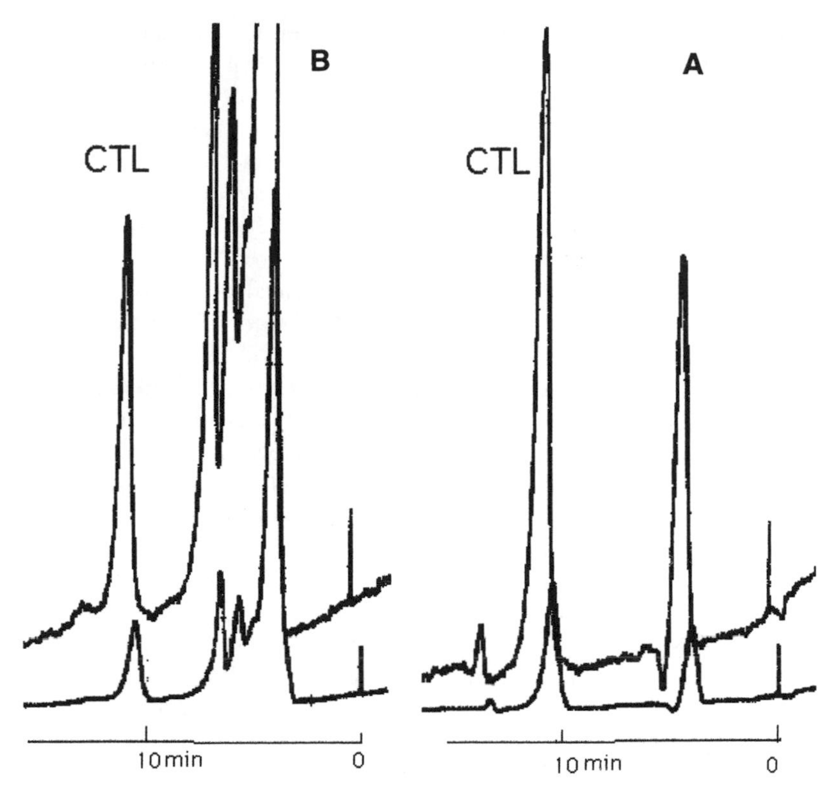

Fig. 6. Chromatogram of creatol analysis. (**A**) Standard CTL (0.2 nmol). (**B**) Serum of patient under hemodialysis.

References

1. Pfiffiner, J. J. and Meyers, V. C. (1930) On the calolimetric estimation of guanidine base in blood. *J. Biol. Chem.* **87,** 345–353.
2. Giovannetti, S., Biagini, M., Balestri P. L., Navalesi, R., Giagnoni, P., de Matleis, A., Ferro-Milone, P., and Perfetti, C. (1969) Uremia-like syndrome in dogs chronically intoxicated with methylguanidine and creatinine. *Clin. Sci.* **36,** 445–452.
3. Nagase, S., Aoyagi, K., Narita, M., and Tojo, S. (1986) Active oxygen in methylguanidine synthesis. *Nephron* **44,** 299–303.
4. Aoyagi, K., Nagase, S., Narita, M., and Tojo, S. (1987) Role of active oxygen on methylguanidine synthesis in isolated rat hepatocytes. *Kidney Int* **22,** s229–s233.
5. Sakamoto, M., Aoyagi, K., Nagase, S., Ishikawa, T., Takemura, K., and Narita, M. (1989) Methylguanidine synthesis by reactive oxygen species from human leukocytes (in Japanese). *Jpn. J. Nephrol.* **31,** 851–858.
6. Aoyagi, K., Nagase, S., Sakamoto, M., Narita, M., and Tojo, S. (1989) Active oxygen in methylguanidine synthesis by isolated rat hepatocytes, in *Guanidines,* vol. 2 (Mori, A. et al., eds.), Plenum, New York, pp. 79–85.

7. Nagase, S., Aoyagi, K., Sakamoto, M., Takemura, K., Ishikawa, T., and Narita, M. (1992) Biosynthesis of methylguanidine in the hepatic microsomal fraction. *Nephron* **62,** 182–186.
8. Takemura, K., Aoyagi, K., Nagase, S., Sakamoto, M., Ishikawa, K., and Narita, M. (1992) Effect of hyperbaric therapy on urinary methylguanidine excretion in normal human and patients with renal failure, in *Guanidino Compounds in Biology and Medicine* (Marescau, B., ed.), John Libbey & Co., London, pp. 301–307.
9. Nakamura, K. and Lenaga, K. (1992) Stability of creatol, an intermediate of methylguanidine production from creatinine, and analysis in physiological fluids, in *Guanidino Compounds in Biology and Medicine* (Marescau, B. ed.), John Libbey, London, pp. 329–331.
10. Nakamura, K., Lenaga, K., Yokozawa, T., Fujitsuka, N., and Oura, H. (1991) Production of methylguanidine from creatinine via creatol by active oxygen species: analyses of the catabolism in vitro. *Nephron* **58,** 42–46
11. Yokozawa, T., Fujitsuka, N., Oura, H., Akao, T., Kobashi, K., Lenaga, K., Nakamura, K., and Hattori, M. (1993) Purification of methylguanidine synthase from rat kidney. *Nephron* **63,** 452–457.
12. Ozasa, H., Horikawa, S., and Ota, K. (1994) Methylguanidine synthetase from rat kidney is identical to long-chain L-2-hydroxy acid oxidase. *Nephron* **68,** 279.
13. Aoyagi, K., Nagase, S., Takemura, K., Ohba, S., and Narita, M. (1992) Dipyridamole decreased urinary excretion of methylguanidine increased by puromycin aminonucleoside in vivo, in *Guanidino Compounds in Biology and Medicine* (Marescau, B., ed.), John Libbey, London, pp. 309–313.
14. Yamamoto, Y., Manji, T., Saito, A., Maeda, K., and Ohta, K. (1979) Ion exchange chromatographic separation and fluorometric determination of guanidino compounds in physiologic fluids. *J. Chromatogr.* **162,** 327–340.
15. Yamada, S. and Itano, H. A. (1966) Phenanthrequinone as an analytical reagent for arginine and other monosubstituted guanidines. *Biochem. Biophys. Acta* **130,** 538–540..
16. Ienaga, K., Nakamura, K., Fukunaga, Y., Nakano, K., and Kanatuna, T. (1994) Creatol and chronic renal failure. *Kidney Int.* **46(Suppl. 47),** S22–S24

17

Quantitative Analysis of Peptide and Protein Changes in Ischemic Hippocampal Tissue by HPLC

Thomas M. Wengenack, J. Randall Slemmon, and J. Mark Ordy

1. Introduction

Profiling of hippocampal peptides by reversed-phase HPLC (RP-HPLC) was used as a tool for identifying molecular events associated with neurodegeneration following global cerebral ischemia. Global cerebral ischemia is known to cause impairment of working memory, but not reference memory in humans and rats *(1,2)*. Histologically, global cerebral ischemia results in the selective, delayed neurodegeneration of hippocampal CA1 pyramidal neurons in humans and rats *(1,3)*. Quantitatively, there is no significant decrease in the number of CA1 pyramidal cells on post-ischemic d 1, but highly significant decreases on d 3 and 7 *(3)*. Consequently, the unique, delayed-onset, and selective vulnerability of post-ischemic hippocampal CA1 neurodegeneration in the four-vessel occlusion (4-VO) rat model of transient, global cerebral ischemia have become of intense interest for clarification of the mechanisms of neurodegeneration and pathophysiology of amnesia after global ischemia, as well as identification of molecular targets for pharmacological intervention and treatment *(4)*.

After transient ischemic attacks, global cerebral ischemia, or cardiac arrest, altered integrity of the blood-brain barrier (BBB) may have important consequences for post-ischemic pathophysiology, efficacy of different types of drug treatment, and duration of a "therapeutic window" *(5)*. Several recent studies have reported increased BBB permeability following 4-VO global ischemia *(6–8)*. Increased permeability of the BBB could result in leakage of proteins or smaller molecules, which normally do not cross the BBB, into the parenchyma, and contribute to pathogenic mechanisms *(5)*. For example, increased levels of hemoglobin fragments have been reported in Alzheimer's disease brain using

From: *Methods in Molecular Biology, vol. 108: Free Radical and Antioxidant Protocols*
Edited by: D. Armstrong © Humana Press Inc., Totowa, NJ

the present RP-HPLC methods *(9)*. Few studies, however, have assessed the BBB permeability of hemoglobin following global cerebral ischemia. This is especially relevant because hemoglobin has been reported to be neurotoxic in vitro and in vivo *(10,11)*. It has been well-documented that blood-borne molecules, particularly iron-containing molecules such as hemoglobin, generate free radicals that cause lipid peroxidation and cellular-membrane damage *(12,13)*. Free radicals, or reactive oxygen species, have been proposed to be pathogenic in several neurodegenerative disorders, including ischemia *(14,15)*. Therefore, identification of changes in specific peptide and protein fragments following global cerebral ischemia might identify proteins involved in the neurodegenerative process that may be used as targets for therapeutic agents.

The specific aims of this study were to provide a quantitative and statistical analysis of RP-HPLC hippocampal peptide profiles across post-ischemic d 1, 3, and 7 and identify specific peptide fragments of cellular and vascular origin that might be involved in initiation and propagation of hippocampal CA1 neurodegeneration. Global ischemia was produced in adult rats for 30 min with the 4-VO rat model of transient, global cerebral ischemia. Post-ischemic changes in peptides were studied with RP-HPLC. Comparisons of post-ischemic differences in hippocampal peptides indicated significant changes in peptides of cellular and vascular origin that occurred on post-ischemic d 1, before neurodegeneration was histologically apparent. There were also significant changes on post-ischemic d 3 and 7. The post-ischemic CA1 neurodegeneration and pattern of concomitant changes in hippocampal peptides demonstrated a temporal dissociation between the histologically demonstrable delayed-onset CA1 neurodegeneration and RP-HPLC profiles of peptide changes, suggesting complex interactions among pyramidal-cell degeneration and BBB permeability as part of the early hippocampal response to global ischemia *(3)*. Preliminary reports of these data have been presented previously *(16,17)*.

2. Materials

2.1. Instruments and Columns

1. Peptide extraction and fractionation: Peptides are extracted from tissue supernatants using Waters C_{18} Sep-Pak Environmental cartridges (Millipore Corp., Milford, MA), fractionated using Sephadex G-50 fine (Pharmacia/LKB, Princeton, NJ) in Kontes columns (1.5 × 20 cm; Vineland, NJ), and further separated by ion-exchange using Waters Accell QMA and CM columns (Millipore).
2. RP HPLC peptide separation: Analytical peptide separation is performed using a Vydac C_{18}, small-bore column (5 µm, 0.46 × 25 cm; Separations Group, Hesperia, CA, #218TP54). The HPLC consisted of LDC pumps and automated sample injector (Laboratory Data Control, Riviera Beach, FL) with a Waters 441

(Millipore) ultraviolet absorbance detector and controlled by a PC-compatible computer running Axxiom 727 software (Axxiom Chromatography, Moorpark, CA) (*see* **Note 1**).

3. Peptide identification: Tryptic fragments are separated by RP-HPLC using a Vydac C_{18}, small-bore column, 5 µm, 0.46 × 25 cm (Separations Group, #218TP54) on a Beckman System Gold HPLC with Pump Module 126 and Detector Module 166 (San Ramon, CA) (*see* **Note 1**). Tryptic fragments were sequenced with an Applied Biosystems (Foster City, CA) model 477/120 automated, pulsed-liquid protein sequencer.

2.2. Reagents

1. Acetonitrile (ACN), trifluoroacetic acid (TFA), water: (J. T. Baker, Phillipsburg, NJ), HPLC grade.
2. Glycerol, guanidine hydrochloride, methanol, 2-[N-morpholino] ethanesulfonic acid (MES), phosphoric acid, sodium chloride, sodium hydroxide, tris(hydroxymethyl)aminomethane-trifluoroacetic acid, urea: (VWR Scientific, Rochester, NY), ACS grade.
3. Cytochrome C (horse heart), trypsin (bovine, EC 3.4.21.4): (Sigma, St. Louis, MO).

2.3. Mobile Phase

The buffers of the mobile phase should be sparged with helium for 5 min and kept under pressure with helium in solvent delivery bottles. Aqueous phase (Buffer A): 0.1% TFA/99.9% HPLC water (v/v). Organic phase (Buffer B): 70% ACN/0.1% TFA/29.9% HPLC water (v/v).

3. Methods

3.1. Subjects

Forty-five male Sprague-Dawley rats, 2–3-mo-old and weighing 250–300 gm, were used for studying the effects of 4-VO ischemia on hippocampal RP-HPLC peptide profiles (*see* **Note 2**). Animals were housed individually under a 12-h light/dark cycle with *ad lib* access to food and water. All procedures performed were in accordance with institutional guidelines using protocols approved by the Institutional Animal Care and Use Committee.

3.2. 4-VO Global Ischemia

1. Transient, global cerebral ischemia is produced using the 4-VO rat model *(18)*, as modified for complete occlusion of the vertebral arteries and control of collateral circulation *(19)*. Recognized advantages of this model include: definable 4-VO procedures for reproducible, selective, graded, and bilaterally symmetrical CA1 hippocampal pyramidal cell damage, low incidence of seizures, and ease of preparation of a significantly large number of animals for statistical evaluations *(20)*. Such neurophysiological criteria as binocular corneal and pupillary reflexes,

respiration rate, and duration of post-ischemic recovery, were monitored as *a priori* confirmatory measures of complete ("grade 1") occlusion of vertebral arteries, control of collateral circulation, and temporary occlusion of forebrain circulation via the carotid arteries, to ensure bilaterally symmetrical pyramidal-cell degeneration *(3)*.

2. Rats are anesthetized with halothane (1.5%) in O_2/N_2O (1:1, v/v) and both vertebral arteries are permanently occluded by cauterization, with visualization of the full perimeter of the alar foramina. The common carotid arteries are then exposed and isolated. A small loop of silicone tubing (Silastic, Dow Corning, Midland, MI) is placed around each vessel to allow easy access to them for occlusion on the following day. Next, a 30-cm length of #1 surgical suture is passed through the rat's neck using an 18-gauge arterial needle so as to lie dorsal to the trachea, esophagus, external jugular veins, common carotid arteries, and the vagus nerves, but ventral to the cervical muscles.
3. The rats are allowed to recover from anesthesia and surgery overnight, during which time they are fasted, but have free access to water. After brief exposure to halothane, global ischemia is produced on the next day by re-exposure of the carotid arteries and clamping them closed with atraumatic aneurysm clips. When the rat does not regain consciousness, the neck suture is tightened around the cervical muscles to control collateral circulation.
4. The carotid-artery clips are released after 30 min and restoration of blood flow through the arteries is verified by direct visual inspection. The ventral neck incision is closed with wound clips and the neck suture is carefully removed.
5. During ischemia and reperfusion, body temperature is maintained at 37.5 ± 0.5°C. with a rectal thermistor connected to a heating lamp until the rats recover thermal homeostasis. Owing to the severe ischemia produced by 30 min of occlusion used in this study, only 15 of 30 ischemic rats survived or achieved a "grade 1" criterion and were used for subsequent evaluations (*see* **Note 3**). Fifteen control animals received sham surgery, which included all surgical manipulations except occlusion.

3.3. RP HPLC Peptide Profiling

1. The specific aim of this study is to quantitatively analyze endogenous hippocampal peptides using RP-HPLC in order to identify post-ischemic peptide changes and relate them to histologically observed changes in hippocampal pyramidal-cell degeneration and loss.
2. A total of 30 rats were used for peptide profiling. Five rats were used for each of three control groups and three 4-VO groups, at d 1, 3, and 7 after sham surgery or ischemia to obtain adequate amounts of tissue for the analyses (0.5–1.0 gm/group).
3. Each animal is decapitated and the brain removed after the blood had drained (*see* **Note 4**). The brain is then rinsed with saline to remove any residual blood. The left and right hippocampi are quickly dissected, frozen on dry ice, and stored at –80°C.

3.3.1. Peptide Extraction and Fractionation

1. The hippocampi are fractionated into four crude pools of acid-soluble peptides using conventional methods of size exclusion and ion-exchange chromatography:
   a. High molecular-weight peptides;
   b. Neutral peptides;
   c. Anionic peptides; and
   d. Cationic peptides. These methods are similar to those described previously for profiling brain peptides in human tissue with Alzheimer's disease ***(9,21)***.
2. Frozen hippocampi are ground with a mortar and pestle chilled on dry ice. The tissue is then homogenized in 50 mL of 250 m*M* phosphoric acid, pH 2.5, with a Tekmar Tissumizer at 20,000 rpm for 1 min and then boiled for 5 min to halt proteolysis. The homogenates are centrifuged at 35,000*g* for 20 min at 4°C. The supernatants are concentrated using two Waters C_{18} Sep-Pak Environmental cartridges in series that have been equilibrated with 10 mL of 0.1% TFA:99.9% water.
3. The cartridges are mounted on a vacuum manifold (Alltech, Deerfield, IL) with stopcocks and run at one drop per s. Dry cartridges are first wetted with 5 mL of methanol and then washed with 10 mL of distilled water (*see* **Note 5**). After the supernatants are loaded, the cartridges are washed with 5 mL of 0.1% TFA:99.9% water. The peptides are eluted with 8 mL of 70% ACN:0.1% TFA:29.9% water and dried in the presence of 120 µL of 6 *M* urea per tube in a Speed-Vac (Savant Instr., Farmingdale, NY) (*see* **Note 6**).
4. The samples are resuspended in 1 mL of 0.1% TFA:99.9% water with 50 µL of glycerol and layered on top of a bed of Sephadex G-50 fine (1.5 × 20 cm) in 0.1% TFA:99.9% water using a Pasteur pipet (*see* **Note 7**). The first 8 mL is discarded as the dead volume. The next 7 mL is collected as the void volume and is considered the high molecular-weight fraction. These void volumes are concentrated to 3 mL with a Speed-Vac before RP-HPLC profiling. The next 25 mL is taken as the included and salt volumes and contained the low molecular-weight fraction. This low molecular-weight fraction is concentrated with the C_{18} Sep-Pak cartridges, as previously described, and dried with 50 µL of 6 *M* urea per tube in a Speed-Vac.
5. The low molecular-weight fraction is separated by ion-exchange chromatography into anionic, cationic, and neutral peptide fractions using Waters Accell QMA and CM columns. The columns are first wetted with 10 mL of methanol and washed with 10 mL distilled water, followed by 10 mL of 500 m*M* MES, pH 6.0, and then 10 mL of 10 m*M* MES, pH 6.0. The columns are mounted on a vacuum manifold (Alltech, Deerfield, IL) with stopcocks and run at one drop per second. The dried samples are resuspended in 10 mL of 10 m*M* MES, pH 6.0, and the pH adjusted to 6.0 with 40 m*M* sodium hydroxide to compensate for the residual TFA salts.
6. The samples are first passed over the QMA columns and washed with 5 mL of 10 m*M* MES, pH 6.0. All of this flowthrough is collected, passed over the CM

columns, and then washed with 5 mL of 10 m*M* MES, pH 6.0. The total flowthrough is taken as the neutral peptide fraction.

7. The QMA and CM columns are eluted with 8 mL of 10 m*M* MES, pH 6.0, containing 500 m*M* sodium chloride (*see* **Note 8**). The QMA eluent is taken as the anionic fraction and the CM eluent taken as the cationic fraction. The pH of the three fractions is adjusted to 2.5 with phosphoric acid.
8. The fractions are then concentrated with the C_{18} Sep-Pak cartridges as previously noted, dried in the presence of 15 µL of 6 *M* guanidine hydrochloride per tube with a Speed-Vac, and stored at –20°C (*see* **Note 9**).

3.3.2. RP HPLC Peptide Separation

1. The four crude peptide fractions are then profiled and separated into component peptides by RP-HPLC. A C_{18} column (0.46 × 25 cm, 1 mL/min, Vydac) is used with a binary mobile-phase system of 0.1% TFA:99.9% water (Buffer A) and 70% ACN:0.1% TFA:29.9% water (Buffer B). Samples are resuspended in 1.7 mL of 0.1% TFA:99.9% water and filtered with 0.45 µm polyvinylidene fluoride (PVDF) syringe filters (Acrodisc, Gelman Sciences, Ann Arbor, MI) before RP-HPLC profiling. Sample volumes are adjusted to contain equal tissue concentrations based on the original weight of the frozen tissue.
2. One milliliter of sample is injected and separation is performed by a gradient method with increasing concentration of Buffer B. The gradient of Buffer B is indicated by the secondary vertical axis of the chromatograms in **Fig. 1**. The HPLC (Laboratory Data Control) is equipped with an automatic sample injector (Laboratory Data Control) and ultraviolet absorbance detector (Waters 441), and controlled by a PC-compatible computer running Axxiom 727 software (Axxiom Chromatography). The resulting chromatograms contained approximately 200 peptide peaks per sample per fraction. The peaks are integrated using the Axxiom 727 software to determine retention time, peak height, and peak area, and then analyzed by programs using SAS software (SAS Institute, Cary, NC).
3. These programs sorted homologous peaks into corresponding observations across samples within each fraction and then compared them for post-ischemic differences. A minimum post-ischemic change of 25% is used as a criterion for indicating peaks that increased or decreased in the post-ischemic profiles. This criterion of 25% is used because it is greater than an estimate of the 95% confidence interval, defined as 1.96 times the experimental error. The mean experimental error is calculated from the differences between homologous peaks in the control groups and found to be 11.1%. Changes are verified by visual inspection of the chromatograms. Peaks of interest are collected during RP-HPLC of a reserve sample and then sequenced for peptide identification.

3.4. Peptide Identification

1. Isolated peptide peaks are identified by amino-terminal sequencing of tryptic fragments. Because the amino terminal of endogenous peptides is often blocked

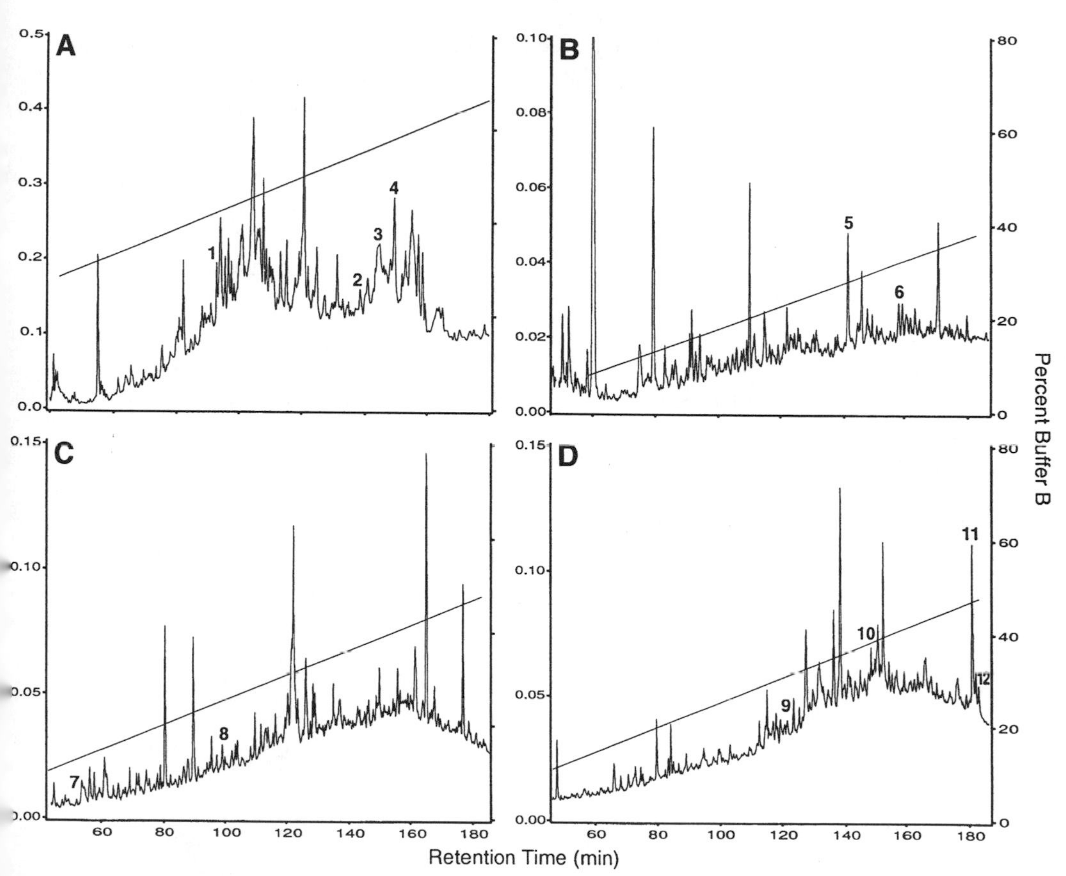

Fig. 1. RP-HPLC separation of acid-soluble hippocampal peptides in the: (**A**) high molecular weight fraction, (**B**) neutral fraction, (**C**) anionic fraction, and (**D**) cationic fraction from the sham d 1 control sample. In each panel, the vertical axis plots the absorbance by ultraviolet (UV) detection at 214 nm vs the retention time in min on the horizontal axis. The secondary vertical axis plots the mobile-phase gradient as a function of percent Buffer B (70% ACN:0.1% TFA:29.9% water, 1.0 mL/min). In each panel, peaks that exhibited significant post-ischemic changes are indicated by numbers. These numbered peaks are identified in **Tables 1** and **2** (*see* pp. 176–177).

(J. R. Slemmon, unpublished observations), and owing to the limited amount of material, which did not allow multiple samples for sequencing, the peptides are first trypsinized.

2. Bovine trypsin (1 mg, EC 3.4.21.4; Sigma) is purified by RP-HPLC using a C_{18} column (0.46 × 25 cm, 1 mL/min; Vydac) and a 1-h gradient of 0–50% Buffer B. Purified trypsin (140 µg) is collected as the large, defined peak emerging at approximately 38% Buffer B (*see* **Note 10**). One microgram of trypsin is added

to each peptide sample resuspended in 50 μL of tris(hydroxymethyl)amino-methane-trifluoroacetic acid, pH 8.5. The samples are incubated at 37°C. for 18–20 h. A sample of cytochrome C (1 mg, horse heart; Sigma) is also trypsinized as a positive control.
3. Digestion is halted by adding 1 mL of 0.1% TFA:99.9% water to the samples. Tryptic fragments are collected during RP-HPLC separation with a C_{18} column (0.46 × 25 cm, 1 mL/min, Vydac) on a Beckman System Gold HPLC using 0–50% Buffer B gradient. The tryptic fragments are dried with a Speed-Vac and stored at –80°C until sequencing.
4. Tryptic fragments are sequenced with an Applied Biosystems model 477/120 automated, pulsed-liquid phase protein sequencer with RP-HPLC module. The dried tryptic fragments are resuspended in 90 μl of 25% TFA:75% water. The sample is spotted onto a glass-fiber disc that has been treated with 1.5 mg of Polybrene (Biobreen, Applied Biosystems) to prevent sample washout in the reaction chamber.
5. The peptide is sequenced automatically using Edman degradation chemistry after a sample of amino-acid standards is run through the HPLC module of the sequencer. The sequence is reconstructed by comparing the location of the peak in each cycle with the standard chromatogram (*see* **Note 11**). The peptide fragments are then identified by comparing the amino-acid sequence of the tryptic fragment to the GenBank and EMBL sequence data bases using the FASTA program and the GCG program package (Genetics Computer Group, Madison, WI) ***(22,23)***. Sequencing of multiple tryptic fragments of each peptide confirmed the identification of each peptide.

3.5. Statistical Analyses

1. For the statistical analysis of control-ischemia differences in peak heights of the identified hippocampal peptides, three-way, mixed-model, between and within-subject analyses of variance (ANOVA: 2 × 3 × number of peptides; control vs ischemia × d 1, 3, 7 × peptides) are first performed to test for overall ischemia effects, treating peptide fragments as a within-subject measure.
2. Overall significant ischemia effects on fragmentation of proteins are not obtained because of increases offset by decreases across post-ischemic days. Therefore, in order to test for ischemia effects on specific days across fragments, and because fragmentation of proteins generally increased on specific days and decreased on others following global ischemia, nonparametric, one-tailed, Wilcoxon matched-pairs signed-ranks tests are performed at each post-ischemic day vs the mean of the control values. Then, in order to test for individual day effects on specific protein fragments, because there is only one value for each peptide at each day, each of the three post-ischemic days is compared with the mean of the three control values using a Studentized test for statistical outliers ***(24)***.
3. Individual post-ischemic d 1, 3, and 7 values are considered to be significantly different at $p < 0.05$ only when they differed from the control mean by 4.30 rather than 3.00 standard errors, which normally defines statistical outliers. There are large variances

among the control values of some protein fragment peaks, which could not be ruled out as being sham-surgery effects. Although this variance may have masked statistically significant post-ischemic differences across days, only the peaks that exhibited a significant post-ischemic change are included in the results.

3.6. Results

1. The major goal of this study was to examine the fragmentation of hemoglobin and other hippocampal proteins after 4-VO global cerebral ischemia. The more traditional molecular techniques such as differential cDNA library analysis or immunohistochemical methods are of limited use because they are not able to demonstrate the generation of specific peptide fragments. Two-dimensional gel electrophoresis is very useful for proteins, but peptide studies using this approach can be very problematic. Peptides are often poorly resolved on sodium dodecyl sulfate-polyacrimide gel electrophoresis (SDS-PAGE) and quantification of such species using electrotransfer can be complicated by low recovery. In order to circumvent these problems, we chose to use an RP-HPLC-based approach. HPLC offers several important advantages when studying peptides because it separates peptides with high resolution and the technique is inherently quantitative. It also lends itself to automation, thereby simplifying the analysis and making it less prone to subjective comparisons. HPLC-based peptide analyses also offer the advantage that many peptides other than the ones being directly studied are automatically made part of the study. This offers both an excellent baseline for determining the extent to which the tissue was altered after insult. The peptide profiles developed in the present study contained a total of approximately 5000 peptides that were analyzed from six experimental pools of tissue. This represented about 800 peptides from each tissue sample. Of this number, the levels of only 12 peptide species changed significantly. Those peaks are indicated by numbers in **Fig. 1**, which illustrates the RP-HPLC chromatograms of the d 1 control sample fractionated by gel-sizing and ion-exchange chromatography. It may be important to note that these were only the acid-soluble peptides. There may have been additional changes in lipid soluble and other peptide fractions.
2. The first peptides characterized as showing major changes in their post-ischemic levels are from serum albumin (peaks 3 and 10, **Fig. 1**). Peak #3 appears to be intact serum albumin based on its elution time from the RP-HPLC column (J. R. Slemmon, unpublished observation). These peptides began increasing in the 4-VO animals at the earliest time point of 1 d. Fragmentation of serum albumin is greatest on d 3 and 7 post-ischemia. Further examination of peptides that displayed changes in amount as a function of 4-VO insult identified four peptide fragments from hemoglobin. Fragmentation of alpha and beta hemoglobin is also greatest on d 3 post-ischemia. All four of these peptide species (peaks 2, 4, 5 and 6, **Fig. 1**) are not intact alpha or beta hemoglobin based on their elution time (*see* **Fig. 2**). Intact alpha and beta hemoglobin did not change significantly across samples, indicating no apparent gross changes in blood volume in the tissue. Significant increases in fragmentation of blood-derived proteins, including hemoglobins, are observed particularly on d 3 and 7 post-ischemia.

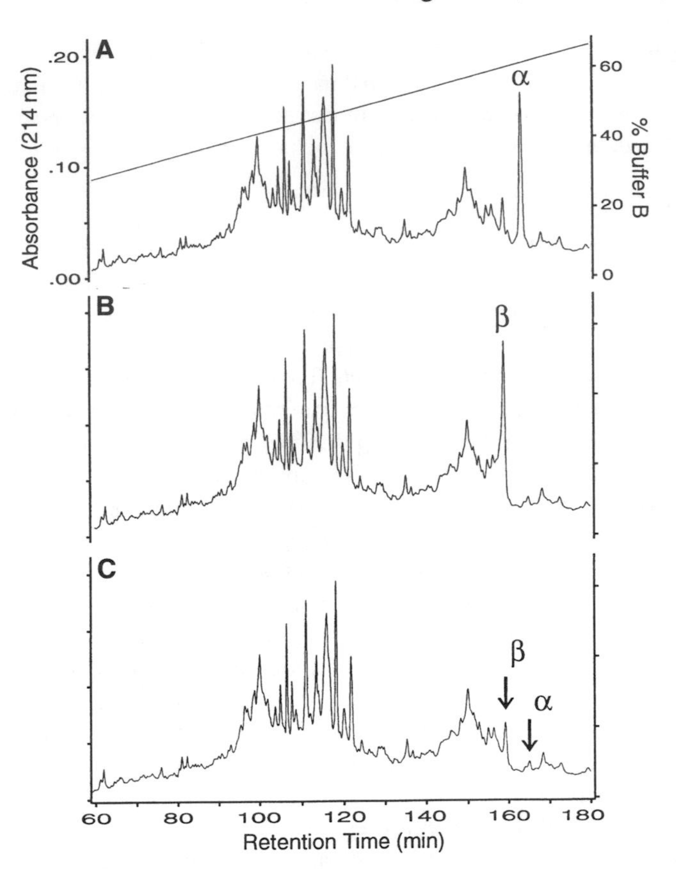

Fig. 2. Intact hemoglobin subunits elute in the high molecular weight fraction of rat brain. RP-HPLC chromatograms of samples spiked with purified, intact subunits of hemoglobin: (**A**) alpha hemoglobin-spiked sample, (**B**) beta hemoglobin-spiked sample, (**C**) control sample of rat brain. In each panel, the vertical axis plots the absorbance by UV detection at 214 nm vs the retention time in min on the horizontal axis. The secondary vertical axis plots the mobile-phase gradient as a function of percent Buffer B (70% ACN:0.1% TFA:29.9% water, 1.0 mL/min).

3. To test for overall ischemia effects on the fragmentation of blood-derived proteins, a mixed-model, between and within-subject ANOVA is performed and showed that there is no overall effect owing to ischemia, across days, or an interaction of peptides by days, owing to increases and decreases in different peptides on different days. There is, however, an overall significant difference among the peak heights across protein fragments owing to differences in relative amounts of the various peptides. Although there is no significant overall ischemia effect, post-ischemic fragmentation of blood-derived proteins is generally increased, particularly on d 3 post-ischemia. One-tailed, nonparametric, Wilcoxon matched-

pairs signed-ranks tests, performed at each post-ischemic day vs the mean of the control values, indicated that there is a significant number of increases on post-ischemic d 3, but not on d 1 or 7 post-ischemia. **Table 1** contains a summary of the identity, tryptic sequence, peak height, and statistical comparisons of significant control-ischemia differences in fragmentation of blood-derived hippocampal proteins at d 1, 3, and 7 post-ischemia.

4. Coincidental with the elevation in specific peptide fragments from serum albumin and hemoglobin are changes in the fragmentation or expression of peptides, which represent proteins that are not normally restricted to the vasculature. Based on partial peptide sequence from trypsin-generated samples, two of these peptides are fragments of calmodulin (peaks 7 and 8, **Fig. 1**). Interestingly, there is also a decrease in the neuron-specific peptide neurogranin (peak 1, **Fig. 1**). That this peptide decreased in parallel with calmodulin may not be coincidental because neurogranin binds calmodulin and has been observed to regulate calmodulin activation of targets in vitro *(25)*. Not surprisingly, a major and minor form of ubiquitin is also seen to increase on d 3 and persist through d 7 post-ischemia. After the initial, general decline in peptide levels seen among the non-albumin or hemoglobin peptides at post-ischemic d 1, ubiquitin showed the largest subsequent increase. This probably reflects the large amount of protein degradation, which should accompany the neurodegenerative processes that follow ischemia. An early decrease in a peptide from adenylate kinase is also observed on post-ischemic d 1, and this species did not increase until d 7 in a manner similar to calmodulin. This may reflect, along with calmodulin, a decrease in general metabolic activity early after the insult, which is then followed by a late increase around d 7 post-ischemia.
5. To test for overall effects of ischemia on fragmentation of these proteins, an ANOVA is performed and shows that there are no significant overall effects owing to ischemia, across days, or interactions of peptides across days, owing to increases and decreases in different peptides on different days. There is an overall significant difference among the peak heights across protein fragments, which would be expected owing to differences in relative amounts of the various peptides. For these other hippocampal proteins, there appeared to be an overall trend of decreased fragmentation on post-ischemic d 1 and increased fragmentation on d 7 post-ischemia. One-tailed, nonparametric, Wilcoxon matched-pairs signed-ranks tests, performed at each post-ischemic day vs the mean of the control values, indicated that there is a significant number of decreases on post-ischemic d 1, but not on d 3 post-ischemia. Although five of the six peptides increased on post-ischemic d 7, it is not significant owing to the small sample size. **Table 2** contains a summary of the identity, tryptic sequence, peak height, and statistical comparisons of significant control-ischemia differences in the fragmentation of other hippocampal proteins across d 1, 3, and 7 post-ischemia.
6. Studies have shown that excessive activation of glutaminergic N-methyl-D-aspartate (NMDA) and amino-3-hydroxy-5-methylisoxazole-4-propionic acid (AMPA) receptors, Ca^{2+} entry into CA1 neurons, and stimulation of calmodulin-

Table 1
Summary of Significant Changes in Blood-Derived Hippocampal Protein Fragments

	Peak	Tryptic	Control Peak Height (μV)			4-VO Peak Height (μV)			% Change[b]		
Peak Identity	Number[a]	Sequence	D 1	D 3	D 7	D 1	D 3	D 7	D 1	D 3	D 7
Alpha Hemoglobin	2	MFAAFPTTK	45297	50271	40390	57280	75222	40703	+26	+66[c]	−10
Alpha Hemoglobin	4	TYFSHIDVSP	140943	169213	132844	204628	131413	65814	+39	−11	−55[c]
Alpha Hemoglobin	5	LASVSTVLT	31061	31038	33880	27187	41905	41263	−15	+31[c]	+29[c]
Beta Hemoglobin	6	VVYPWTQRY	9530	13522	14392	11680	24695	19728	−6	+98[c]	+58
Serum Albumin	3	TPVSEK	87021	118251	78004	113599	186589	83630	+20	+98[c]	−11
Serum Albumin	10	DLGEQHFK	19014	13911	16463	16609	54670	36452	+1	+232[d]	+121[d]

[a]Refer to peak numbers in **Fig. 1**.

[b]Relative post-ischemic change (%); individual 4-VO (d 1, 3, or 7) vs control mean. Overall ischemia effects—ANOVA ($2 \times 3 \times 6$; control vs ischemia × d 1, 3, 7 × 6 peptides): ischemia vs control [$F(1,2) = 1.86$; NS]; days [$F(2,2) = 3.73$; NS]; peptides [$F(5,10) = 28.89$; $p < 0.001$]; peptides × day [$F(10,10) = 1.52$; NS]. Wilcoxon matched-pairs signed-ranks one-tailed test (each post-ischemic day vs the mean of the control values): d 1 ($N = 6$, $T = 5$, NS); d 3 ($N = 6$, $T = 3$, $p < 0.05$); d 7 ($N = 6$, $T = 10$, NS).

[c]Studentized outlier test, df = 2, $p < 0.05$; individual 4-VO value (d 1, 3, or 7) vs control mean.

[d]Studentized outlier test, df = 2, $p < 0.01$; individual 4-VO (d 1, 3, or 7) vs control mean.

Table 2
Summary of Significant Changes in Other Hippocampal Protein Fragments

	Peak	Tryptic	Control Peak Height (µV)			4-VO Peak Height (µV)			% Change[b]		
Peak Identity	Number[a]	Sequence	D 1	D 3	D 7	D 1	D 3	D 7	D 1	D 3	D 7
Adenylate Kinase	9	KVNAEGSVD	6596	8385	7491	3034	6921	11408	–60[c]	–8	+52[c]
Calmodulin	7	ADIDGDGQV	2603	2804	2503	1502	2002	4005	–43[c]	–24	+52[c]
Calmodulin	8	GERLTDEEV	10041	10566	9943	5731	8965	13874	–44[d]	–12	+36[d]
Neurogranin	1	IQASFRGHMAR	81615	88381	91346	76250	54413	66110	–13	–38[c]	–24
Ubiquitin	11	TLSDYNIQK	65035	76119	70577	32508	74537	95425	–54[c]	+6	+35[c]
Ubiquitin	12	MQIFVK	7870	9838	8854	4319	12600	17741	–51[c]	+41	+100[c]

[a]Refer to peak numbers in **Fig. 1**.

[b]Relative post-ischemic change (%); individual 4-VO (d 1, 3, or 7) vs control mean. Overall ischemia effects—ANOVA ($2 \times 3 \times 6$; control vs ischemia × d 1, 3, 7×6 peptides): ischemia vs control [$F(1,2) = 1.23$; NS]; day [$F(2,2) = 2.02$; NS]; peptides [$F(5,10) = 76.83$; $p < 0.001$]; peptides × day [$F(10,10) = 1.14$; NS]. Wilcoxon matched-pairs signed-ranks one-tailed test (each post-ischemic day vs the mean of the control values): d 1 ($N = 6$, $T = 0$, $p < 0.02$); d 3 ($N = 6$, $T = 9$, NS); d 7 ($N = 6$, $T = 5$, NS).

[c]Studentized outlier test, df = 2, $p < 0.05$; individual 4-VO value (d 1, 3, or 7) vs control mean.

[d]Studentized outlier test, df = 2, $p < 0.01$; individual 4-VO (d 1, 3, or 7) vs control mean.

dependent kinase may all be involved in initiation of damage in global ischemia and other neuropathologies ***(3,16,21)***. The findings of the present study using quantitative RP-HPLC analysis suggest that post-ischemic alterations in cellular peptides are initiated preceding demonstrable neurodegeneration, whereas the appearance of fragments of blood-borne proteins accompanies or follows neurodegeneration, as detected by quantitative histology ***(3)***. The present findings also suggest that changes in BBB permeability may be involved in initiation and propagation of selective CA1 vulnerability and neurodegeneration. Significant increases in hemoglobin fragments coincident with hippocampal CA1 pyramidal-cell neurodegeneration also have serious implications for recognition of a role in free radical-mediated ischemic neurotoxicity and neurodegeneration. Consequently, early post-ischemic changes in cellular peptides observed with RP-HPLC may provide new molecular targets for therapeutic intervention while neurotoxicity and neurodegeneration are still reversible. Furthermore, post-ischemic increases in hemoglobin fragmentation observed with RP-HPLC analysis in the present study suggest that free-radical production could be more extensive than previously assumed in global ischemia and other neurodegenerative disorders ***(14)***.

4. Notes

1. The HPLC systems and columns should be flushed and stored with 70% ACN:0.1% TFA: 29.9% (Buffer B) water to prevent bacterial growth. Flush the system and column with 0.1% TFA:99.9% water (Buffer A) thoroughly just prior to use.
2. Wistar rats may be used to achieve a higher percentage of "grade 1" criterion animals. Sprague-Dawley rats are used when behavioral studies are conducted because they are less neophobic.
3. Post-ischemic rats may be fed ground food pellets mixed with water until they recover fully.
4. Alternately, the rats may be flushed transcardially with 50 mL of heparinized saline quickly after euthanasia with CO_2 to profile the peptides and proteins only within the parenchyma.
5. The Waters C_{18} Sep-Pak Environmental cartridges may be stored in methanol and reused.
6. Each sample is collected in two tubes of 4 mL each and then recombined later during resuspension.
7. The Sephadex G-50 fine was swelled overnight in 0.1% TFA:99.9% water. It was then poured into the columns and washed with 3 bed volumes of 0.1% TFA:99.9% water to pack the columns. The void volume was determined by running a sample of bovine serum albumin (BSA) through a column and detecting its elution at 280 nm with a Pharmacia UV1 detector (Pharmacia/LKB).
8. The Waters Accell QMA and CM ion-exchange columns were discarded after each use.
9. Dried samples may be stored at –20ºC. and remain stable for several months.

10. The protein concentration of the purified trypsin was estimated with a photometer by its absorbance at 280 nm using an extinction coefficient of 1.0 AU/mg protein.
11. The appearance and position of the amino-acid standard peaks in the first cycle must be inspected after each sequence in order to verify and maintain correct sequence reconstruction. Changes can be rectified by adjusting the pH or salt concentration of Buffer B for the HPLC module.

Acknowledgments

This work was supported by NSF grant BNS-9021042 and NIA grant AG00107. The authors would like to thank Greg Campbell for his expert technical assistance.

References

1. Zola-Morgan, S., Squire, L. R., and Amaral, D. G. (1986) Human amnesia and the medial temporal region: enduring memory impairment following a bilateral lesion limited to field CA1 of the hippocampus. *J. Neurosci.* **6,** 2950–2967.
2. Ordy, J. M., Thomas, G. J., Volpe, B., Dunlap, W. P., and Colombo, P. M. (1988) An animal model of human-type memory loss based on aging, lesion, forebrain ischemia, and drug studies with the rat. *Neurobiol. Aging* **9,** 667–683.
3. Ordy, J. M., Wengenack, T. M., Bialobok, P., Coleman, P. D., Rodier, P., Baggs, R. B., Dunlap, W. P., and Kates, B. (1993) Selective vulnerability and early progression of hippocampal CA1 pyramidal cell degeneration and GFAP-positive astrocyte reactivity in the rat four-vessel occlusion model of transient global ischemia. *Exp. Neurol.* **119,** 128–139.
4. Ordy, J. M., Volpe, B., Murray, R., Thomas, G., Bialobok, P., Wengenack, T. M., and Dunlap, W. (1992) Pharmacological effects of remacemide and MK-801 on memory and hippocampal CA1 damage in the rat four-vessel occlusion (4-VO) model of global ischemia, in *The Role of Neurotransmitters in Brain Injury* (Globus, M. Y.-T. and Dietrich, W. D., eds.), Plenum Press, NY, pp. 83–92.
5. Wahl, M., Unterberg, A., Baethmann, A., and Schilling, L. (1988) Mediators of blood-brain barrier dysfunction and formation of vasogenic brain edema. *J. Cereb. Blood Flow Metab.* **8,** 621–634.
6. Dobbin, J., Crockard, H. A., and Ross-Russell, R. (1989) Transient blood-brain barrier permeability following profound temporary global ischaemia: an experimental study using 14C-AIB. *J. Cereb. Blood Flow Metab.* **9,** 71–78.
7. Schmidt-Kastner, R., Szymas, J., and Hossmann, K.-A. (1990) Immunohistochemical study of glial reaction and serum-protein extravasation in relation to neuronal damage in rat hippocampus after ischemia. *Neurosci.* **38,** 527–540.
8. Preston, E., Sutherland, G., and Finsten, A. (1993) Three openings of the blood-brain barrier produced by forebrain ischemia in the rat. *Neurosci. Lett.* **149,** 75–78.
9. Slemmon, J. R., Hughes, C. M., Campbell, G. A., and Flood, D. G. (1994) Increased levels of hemoglobin-derived and other peptides in Alzheimer's disease cerebellum. *J. Neurosci.* **14,** 2225–2235.

10. Sadrzadeh, S. M. H., Anderson, D. K., Panter, S. S., Hallaway, P. E., and Eaton, J. W. (1987) Hemoglobin potentiates central nervous system damage. *J. Clin. Invest.* **79,** 662–664.
11. Regan, R. F. and Panter, S. S. (1993) Neurotoxicity of hemoglobin in cortical cell culture. *Neurosci. Lett.* **153,** 219–222.
12. Sadrzadeh, S. M. H., Graf, E., Panter, S. S., Hallaway, P. E., and Eaton, J. W. (1984) Hemoglobin: a biologic Fenton reagent. *J. Biol. Chem.* **259,** 14354–14356.
13. Winterbourn, C. C. (1985) Free-radical production and oxidative reactions of hemoglobin. *Envir. Health. Persp.* **64,** 321–330.
14. Halliwell, B. (1992) Oxygen radicals as key mediators in neurological disease: fact or fiction? *Ann. Neurol.* **32,** S10–S15.
15. Schurr, A. and Rigor, B. M. (1992) The mechanism of cerebral hypoxic-ischemic damage. *Hippocampus* **2,** 221–228.
16. Wengenack, T. M., Slemmon, J. R., and Coleman, P. D. (1992a) Effects of transient forebrain ischemia on high molecular weight fraction peptides in the rat hippocampus, in *Molecular Biology of Aging*, Cold Spring Harbor Laboratory, Cold Spring Harbor, NY, pp. 28.
17. Wengenack, T. M., Slemmon, J. R., Ordy, J. M., Bialobok, P., Dunlap, W. P., and Coleman, P. D. (1992b) Global cerebral ischemia effects on high molecular weight peptides and CA1 pyramidal neurons in the rat hippocampus. *Soc. Neurosci. Abstr.* **18,** 1261.
18. Pulsinelli, W. A. and Brierley, J. B. (1979) A new model of bilateral hemispheric ischemia in the unanesthetized rat. *Stroke* **10,** 267–271.
19. Pulsinelli, W. A. and Buchan, A. M. (1988) The four-vessel occlusion rat model: method for complete occlusion of vertebral arteries and control of collateral circulation. *Stroke* **19,** 913–914.
20. Pulsinelli, W. A. (1985) Selective neuronal vulnerability: morphological and molecular characteristics. *Prog. Brain Res.* **63,** 29–37.
21. Slemmon, J. R. and Flood, D. G. (1992) Profiling of endogenous brain peptides and small proteins: methodology, computer-assisted analysis, and application to aging and lesion models. *Neurobiol. Aging* **13,** 649–660.
22. Devereux, J., Haeberli, P., and Smithies, O. (1984) A comprehensive set of sequence analysis programs for the VAX. *Nucleic Acids Res.* **12,** 387–395.
23. Pearson, W. R. and Lipman, D. J. (1988) Improved tools for biological sequence comparison. *Proc. Natl. Acad. Sci. USA* **85,** 2444–2448.
24. Barnett, V. and Lewis, T. (1984) *Outliers in Statistical Data.* John Wiley & Sons, NY.
25. Martzen, M. R. and Slemmon, J. R. (1995) The dendritic peptide neurogranin can regulate a calmodulin-dependent target. *J. Neurochem.* **64,** 92–100.

18

Electrochemical Detection of 8-Hydroxy-2-Deoxyguanosine Levels in Cellular DNA

Thomas M. Nicotera and Sofia Bardin

1. Introduction

Oxidative stress has been implicated in the etiology of many pathological states and known to result in DNA damage. Oxidative DNA damage can lead to mutagenesis *(1–3)* and has been associated with aging *(4)*, diabetes mellitus *(5)*, inflammatory disease *(1)* and carcinogenesis *(1–3,6)*. The 8-hydroxy-2-deoxyguanosine lesion (8-OHdG) is often used in the assessment of oxidative DNA damage and has become the de facto marker for oxidative damage to DNA. 8-OHdG is one of the most prominent lesions observed following exposure to ionizing radiation *(7)* but also results from treatment with many xenobiotics *(8)* as well as by endogenous mechanisms. Thus, normal endogenous levels becomes a critical issue in the assessment of cellular 8-OHdG levels in pathological states. Furthermore, 8-OHdG is efficiently repaired by a DNA glycosylase specific for this lesion *(9)* and its contribution to mutagenesis is relatively weak *(10–12)*. However, it is representative of approximately 20 additional oxidatively-derived lesions known to result from radiation damage *(13)* and whose mutagenic outcome are largely unknown in mammalian-cell systems.

Recent analysis of 8-OHdG in tumor tissue has found this lesion to be consistently elevated in tumor tissue as compared to surrounding tissue *(14–17)*. However, the relatively large variations in both basal levels and in tumor tissue prevents the comparison of data among these studies. It is likely that the numerous techniques employed to measure the levels of DNA damage contributes to these large variations. These methods include gas chromatography-mass spectrometry (GC-MS) *(16,17)*, ^{32}P-postlabeling *(18)*, post-labeling with ^{3}H-acetic anhydride *(19)*, electrochemical detection *(7,8,10,14)* and antibody

From: *Methods in Molecular Biology, vol. 108: Free Radical and Antioxidant Protocols*
Edited by: D. Armstrong © Humana Press Inc., Totowa, NJ

detection *(20)*. Each of these methods provides certain advantages and disadvantages and have been discussed elsewhere. Also, various methods of DNA extraction, digestion, and storage conditions likely contribute to these variations.

In this chapter, we present a method to accurately quantitate 8-OHdG levels in human tissue using high performance liquid chromatography (HPLC) followed by electrochemical (EC) detection. The electrochemical method for the detection of modified nucleosides, coupled with the separation of enzymatically digested DNA components by HPLC has become the method of choice owing to its relative ease of use, its high selectivity and high sensitivity of detection. Typically, femtomole amounts of 8-OHdG can be detected in enzymatic digests of DNA following HPLC separation. This technique incorporates the relatively mild enzymatic digestion of DNA rather than the harsh acidic digestion and the high sensitivity of detection obviates the need for radioactive post-labeling of DNA components. Furthermore, newer refinements of this technique permit the analysis of a wide array of nucleosides adducts *(21)*.

2. Materials

2.1. Instruments and Columns

2.1.1. HPLC System

1. Perkin Elmer (Norwalk, CT) Model 250 pump.
2. Altex (a subsidiary of Beckman Instruments, Fullerton, CA) model 153 UV detector with a 254 nm lamp.
3. Supelco (Bellefonte, PA) LC-18-S Reverse Phase C-18 column (25 cm × 4.6 mm, for EC detection).
4. Rainin (Woburn, MA) Dynamax Macro C-18 Column (10 mm by 250 mm, used for isolation of 8-OHdG).
5. Shimadzu (Columbia, MD) model CR501 integrator was used to record and calculate the area under the curve.
6. Low speed, tabletop centrifuge.

2.1.2. Electrochemical Detector

Bioanalytical Systems (West Lafayette, IN) model LC-4B amperometric detector with a dual glassy carbon electrode.

2.2. Reagents

1. Calf thymus DNA, purchased from Cooper Biomedical.
2. 50 m*M* Tris buffer, pH 8.0.
3. 0.05 *M* potassium phosphate buffer: (J. T. Baker Analyzed): Prepared fresh each day, adjust to pH 5.5, and filter through a 0.45-micron filter by suction.
4. Adenosine, Guanosine triphosphate (Pharmacia, Piscataway, NJ).

5. Methanol, HPLC grade (J. T. Baker, Phillispburg, NJ).
6. Acetonitrile, HPLC grade (J. T. Baker).
7. Ammonium acetate, sodium acetate (Sigma).
8. Chloroform/Isoamyl alcohol (24:1) (J. T. Baker Analyzed).

2.2.1. Glassware

All laboratory glassware used for electrochemical detection was passivated using 3*N* nitric acid and washed copiously with water (*see* **Note 1**).

2.2.2. Enzymes

1. Deoxyribonuclease I (bovine pancreas) (Sigma).
2. Snake venom phosphodiesterase (*crotalus atrox venom*) (Pharmacia).
3. Alkaline phosphatase (*E. Coli*, Type IIIN) (Sigma).
4. Ribonuclese (DNAse free) (Pharmacia).
5. Proteinase K (Pharmacia).

2.2.3. Mobile Phase

1. 0.1 *M* ammonium acetate pH 7.0 with a 0–10% acetonitrile gradient is used for separation of 8-OHdG standard.
2. 0.05 *M* potassium phosphate buffer, pH 5.5, containing acetonitrile (97:3, v/v) filtered through a 0.2 µ*M* Rainin filter.

3. Methods

3.1. Preparation of 8-OHdG Standard

1. The 8-OHdG standard is prepared by the X-irradiation of dGMP (15 mg in 60 mL H_2O) in the presence of O_2 and 1.0 µ*M* Cu(II) using a Maxitron 250 (General Electric).
2, Products are isolated by HPLC using a Dynamax Macro reverse phase C-18 column (10 mm × 250 mm) in a solvent system consisting of 0.1 *M* NH_4Ac, pH 7.0, with a 0–10% acetonitrile gradient and monitored at 254 nm wavelength.
3. Peaks are collected, desalted, and purified to homogeneity by a second HPLC run using H_2O/methanol (0–50% gradient). 8-OHdGMP is identified by NMR and mass spectrometry *(22)*. The nucleoside is prepared by incubating with alkaline phosphatase for 1 h in 50 m*M* Tris buffer, at pH 8.0.

3.2. Preparation of Mononuclear Cells (MNC)

1. Blood samples are obtained from subjects from the antecubital vein. Twenty mL of heparinized blood (14.3 USP units/mL) is processed for the preparation of MNC by density gradient centrifugation over Ficoll (Organon Teknika Corp; Durham NC; density of 1.077–1.080 g/mL at 20°C).
2. Initially, the blood is centrifuged at 235*g* for 10 min to obtain a buffy coat. The buffy coat is diluted to 3 mL in Hank's Balanced Salt Solution (HBSS) and lay-

ered on top of 3 mL Ficoll and respun at 235*g* for 30 min at 22°C. The mixed MNC band at the interface is removed with a pasteur pipet and the cells washed twice in HBSS.

3. The remainder of the blood sample without buffy coat are centrifuged at 1200*g* in order to obtain plasma. MNC are washed in HBSS and resuspended in autologous plasma for DNA extraction.

3.3. Cell Lysis

Cells are washed twice with phosphate-buffered saline (PBS) and centrifuged for 5 min at 250*g* in a tabletop centrifuge. The pellet is taken up in 0.25 mL of 0.5 *M* disodium ethylenediamine tetraacetic acid (EDTA), pH 8.0 and the pellet broken up by pipetting.

The cells are lysed overnight (18 h) with the addition of 0.35 mL of proteinase K (3.0 U/mL), 0.35 mL of a 10% SDS, and taken up in a total of 2.5 mL with water in a conical centrifuge tube.

3.4. DNA Extraction

The following day, 0.2 mL of prewarmed 3.0 *M* sodium acetate is added to the lysis mixture, followed by the addition of an equal volume of chloroform/isoamyl alcohol (24:1). This suspension is centrifuged for 5 min at 1600*g*. The resulting solution is separated into three layers and the upper layer containing the DNA is carefully pipetted off with a wide-mouth pipet and this entire procedure is repeated once. The DNA is precipitated with the slow addition of two volumes of cold 100% ethanol. When these layers are gently mixed, the DNA begins to precipitate and is removed with a glass rod. The DNA is washed twice, first in 70% ethanol, then in 100% ethanol, and solubilized in 0.5 mL of prewarmed Tris-EDTA, pH 8.0 (50 m*M* and 10 m*M*, respectively).

3.5. Enzymatic Hydrolyses of Calf Thymus

1. The DNA in Tris-EDTA buffer is treated with Dnase-free Rnase (2 Units) and incubated for 3 h at 37°C with gentle mixing throughout the incubation. At the end of the incubation. 0.2 mL of a 3.0 *M* sodium acetate is added and gently swirled. To this solution, add an equal volume of isopropanol and gently swirl by inverting the tube until the two phases are mixed. The strands of DNA will slowly begin to form. Remove the DNA with a glass rod and wash consecutively with 70% and 100% ethanol. Place the DNA into a 1.0 mL eppendorf tube containing 0.5 mL of 50 m*M* Tris, pH 8.0, and gently dissolve.
2. DNA samples are adjusted to a concentration of 1.0 mg/mL in Tris buffer (50 m*M*, pH 8.0). The DNA is digested to the nucleoside level by incubation for 2 h at 37°C with a mixture of DNAseI (120 U/mg DNA), phosphodiesterase I (1.25 U/mg DNA) and alkaline phosphatase (2.0 U/mg DNA).

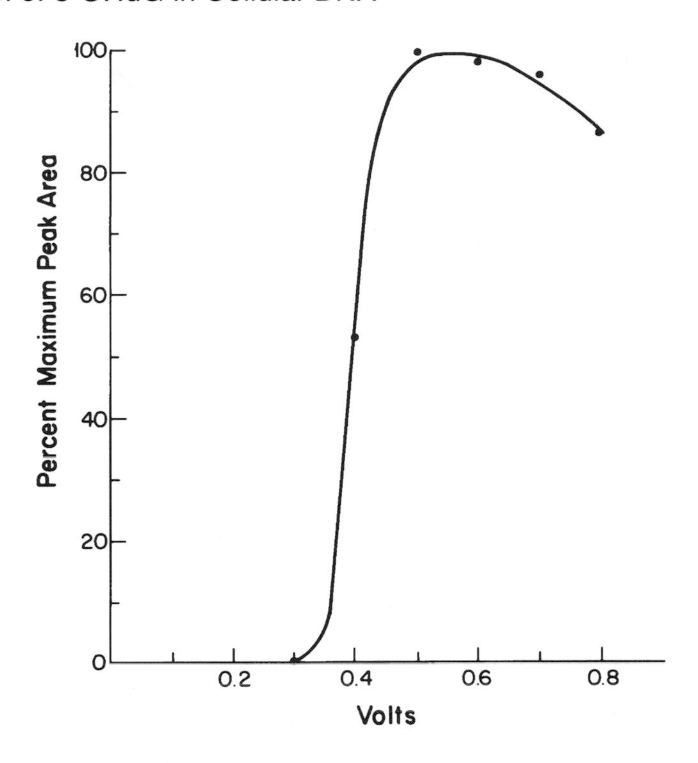

Fig. 1. The hydrodynamic voltammogram of 8-OHdG. The area under each peak is determined by means of an integrator following elution of 8-OHdG from an HPLC column in 0.05 *M* KH_2PO_4, pH 5.5, containing 2.5% acetonitrile. Each point is an average of 3 determinations.

3. Following digestion, the hydrolysate is filtered through a 10,000 MW filter (Millipore) and aliquots of the desired working concentrations are lyophilized to dryness (**see Note 2**).

3.6. HPLC

1. The HPLC system consisted of a Perkin Elmer model 250 pump with a Supelco LC-18-S reverse-phase column (25 cm × 4.6 mm), an Altex model 153 UV detector with a 254 nm lamp, and a Bioanalytical Systems model LC-4B amperometric detector with a dual glassy carbon electrode.
2. A Shimadzu model CR501 integrator is used to record and calculate the area under the curve. For amperometric detection, a potential of 600 mV and a current of 2 nA is used. The mobile phase consisted of potassium phosphate buffer (0.05 *M*, pH 5.5) containing acetonitrile (97.5:2.5 v/v) filtered through a 0.2 µm Rainin filter (*see* **Note 3**).

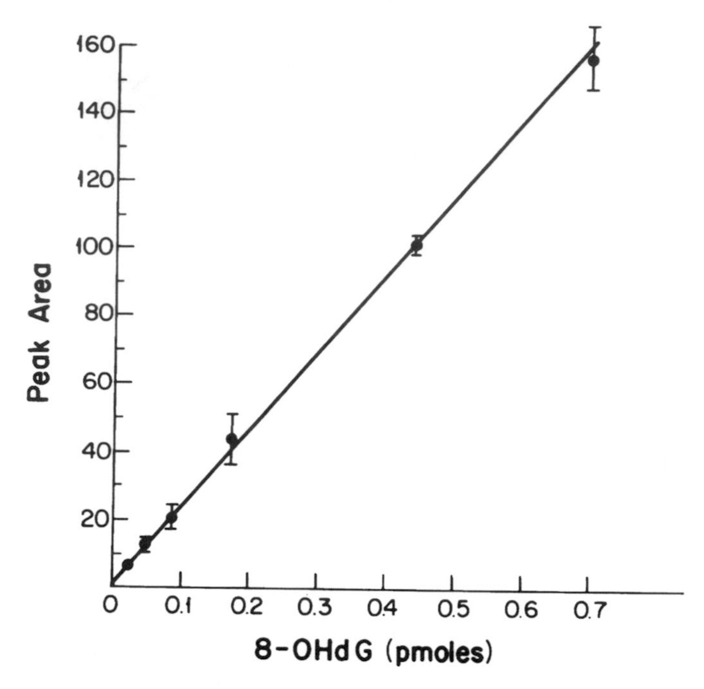

Fig. 2. The 8-OHdG standard curve. HPLC-EC response is a function of the amount of 8-OHdG injected. Each data point is the average of three separate determinations. The signal to noise ratio is approximately 5:1 for the most sensitive point, the 20 femtomole standard and >10:1 for the remaining points.

3. The flow rate is 1.0 mL/min. Known amounts of 8-OHdG standard are run at the beginning and end of each experiment to ensure against loss of electrode sensitivity. Standard curves for electrochemical responses of 8-OHdG and for for the ultraviolet (UV) detection of adenosine are constructed and used for quantitation of unknown samples.

3.7. Voltammogram and Standard Curves

Known amounts of characterized standard 8-OHdG are injected into the HPLC according to protocol and the relative output is measured using an integrator. The voltage is increased with each injection of standard until the maximum output is obtained (*see* **Fig. 1**). For 8-OHdG, the voltage selected for analysis (600 nA) is sufficient for the complete ionization of the sample and below which the sample begins to degrade. A unique hydrodynamic voltammogram is obtained for each compound of interest and can be used to characterize unknown compounds.

Once the conditions for detection of 8-OHdG are set, an EC standard curve can be generated (*see* **Fig. 2**). The level of sensitivity selected (i.e., the current applied)

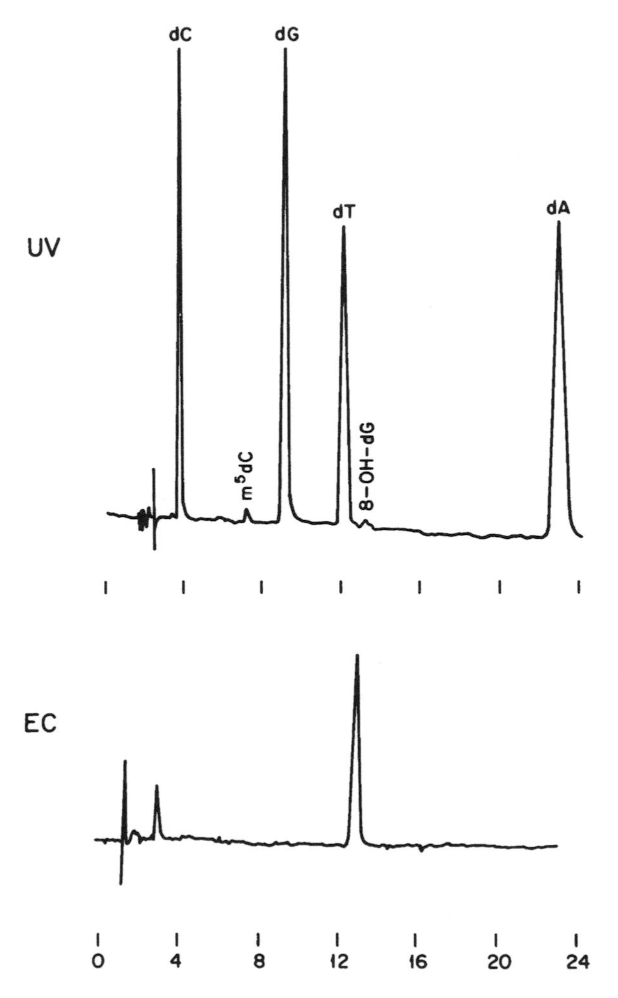

Fig. 3. Comparison of optical detection (upper trace) with electrochemical detection (lower trace) of a 175 ng sample of O_2 saturated, 30 Gy irradiated calf thymus DNA. The optical detector (254 nm) is set at a range of 0.005. The electrochemical detector (0.6V) is set at the 1 nA range.

determines the limit of detection and the duration of the equilibration time required. Selecting too high a sensitivity increases the equilibration dramatically.

A UV standard curve is also necessary (not shown). In our case, we selected adenosine since it is the most stable nucleoside. Dual UV and EC analysis provides several advantages which are the capacity to relate the amount of 8-OHdG to amount of DNA in the sample of interest, and the use of calf thymus DNA as a standard both as an absolute standard (*see* **Fig. 3**). Use of calf thymus

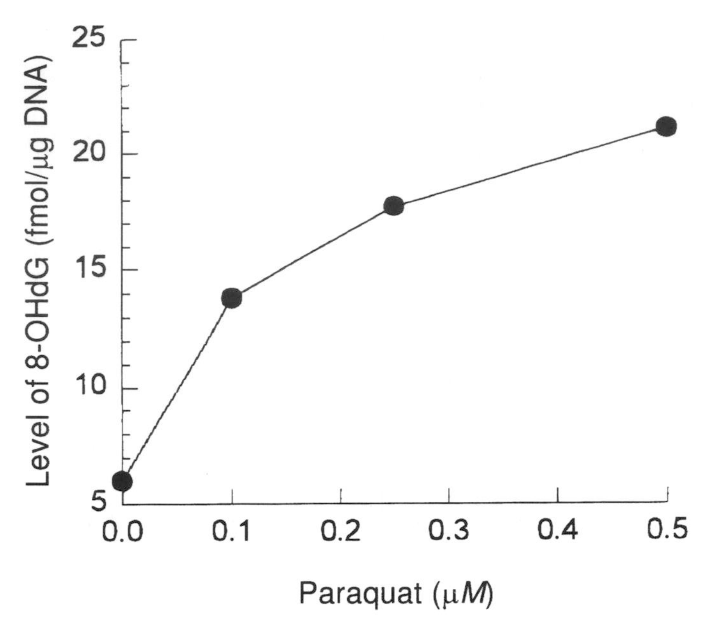

Fig. 4. Levels of 8-OHdG in the B-lymphblastoid cell line GM3299 treated with increasing concentrations of the superoxide generating compound paraquat. Log phase cells (10^7) are treated overnight with paraquat, after which the cells are collected, washed twice in PBS, and lysed overnight with SDS, proteinase K, and EDTA as described in Methods section.

DNA also provides a cost-effective means to check the electrode for potential loss of sensitivity.

3.8. Analysis of Cellular DNA

A representative analysis of 8-OHdG in DNA from B-lymphoblastoid cells is provided (*see* **Fig. 4**) following overnight incubation (18 h) with paraquat (0, 0.1, 0.25, and 0.5 m*M*). The DNA is extracted, digested, and analyzed according to the protocol provided. Duplicate analysis of the above graph can easily be accomplished in 1 d.

4. Notes

1. All laboratory glassware should be passivated with 3.0*N* and rinsed extensively with distilled/deionized water. This is also true for all metal components such as HPLC tubing, injectors etc. (do not passivate electrodes or columns). Alternatively, Teflon or metal-free components should be substituted wherever possible. It is highly recommended that these materials be dedicated for electrochemical detection in order to avoid potential contamination with metals and other oxidants.
2. The running buffer should be degassed for approximately 20 min prior to turning on the EC electrode in order to remove dissolve oxygen and other gasses. This

will avoid trapping gas bubbles in the working electrode. After the initial period of degassing, a small positive pressure is sufficient to prevent oxygen from re-entering into the mobile phase. Other methods such as in-line gas exchange membranes work equally well.

3. In our experience, DNA samples degrade quickly after enzymatic digestion. It is therefore recommended that DNA samples be prepared on 1 d then digested and immediately analyzed the next day. Undigested DNA samples can better withstand degradation of 8-OHdG and can be stored for short periods of time. Depending on the purity of the DNA sample preparations, it may be possible to freeze digested DNA samples overnight without significant loss of 8-OHdG. The effect of storage conditions on samples stability should be verified experimentally.

References

1. Ames, B. N. (1983) Dietary carcinogens and anticarcinigens. *Science* **221,** 1256–1264.
2. Cerutti, P. A. (1985) Prooxidant states and tumor promotion. *Science* **225**, 375–381.
3. Breimer, L. H. (1990) Molecular mechanisms of oxygen radical carcinogenesis and mutagenesis: the role of DNA base damage. *Mol. Carcinogenesis* **3,** 188–197.
4. Fraga, C. G., Shigenaga, M. K., Park, J-W., Deagn, P., and Ames, B. N. (1990) Oxidative damage to DNA during aging: 8-hydroxy-2'-deoxyguanosine in rat organ DNA and urine. *Proc. Natl. Acad. Sci. USA* **87,** 4533–4537.
5. Dandona, P., Thusu, K., Cook, S., Snyder, B., and Nicotera, T. (1996) Oxidative DNA damage diabetes mellitus. *Lancet* **347,** 444–445.
6. Floyd, R. A. (1990) The role of 8-hydroxyguanosine in carcinogenesis. *Carcinogenesis* **11,** 1447–1450.
7. Kasai, H., Crain, P. F., Kuchino, Y., Nishimura, S., Ootsuyama, A., and Tanooka, H. (1986) Formation of 8-hydrodroxyguanine in cellular DNA by agents producing oxygen radicals and evidence for its repair. *Carcinogenesis* **8,** 1849–1851.
8. Fiala, E. S., Conaway, C. C., and Mathis, J. E. (1989) Oxidative DNA and RNA damage in the livers of Sprague-Dawley rats treated with the hepatocarcinigen 2-nitropropane. *Cancer Res.* **49,** 5518–5522.
9. Chung, M. H., Kasai, H., Jones, D. S., Inoue, H., Ishikawa, H., Ohtsuka E., and Nishimura, S. (1991) An endonuclease activity of *Escherichia Coli* that specifically removes 8-hydroxyguanine residues from DNA. *Mutat. Res.* **254,** 1–12.
10. Kuchino, T., Mori, F., Kasai, H., Inoue, H., Iwai, S., Miura, K., Ohtsuka E., and Nishimura, S. (1987) Misreading of DNA templates containing 8-hydroxydeoxuguanosine at the modified base and at adjacent residues. *Nature (Lond.)* **327,** 77–79.
11. Shibutani, S., Takeshita, M., and Grollman, A. P. (1991) Insertion of specific bases during DNA sysnthesis past the oxidation-damaged base 8-oxodG. *Nature (Lond.)* **349,** 431–434.
12. Wood, M. L., Dizdaroglu, M., Gajewski, E., and Essigmann, J. M. (1990) Mechanistic studies of ionizing radiation and oxidative mutagenesis: genetic effects of a

single 8-hydroxyguanine (7-hydro-8-oxoguanine) residue inserted at a unique site in a viral genome. *Biochem.* **29,** 7024–7032.
13. Cadet, J. (1994) DNA repair caused by oxidation, deamination, ultraviolet radiation and photoexcited psoralens in DNA adducts, in *Identification and Biological Significance*, (Hemminki, K., Dipple, A., Shuker, D. E. P., Kadlubar, F. F., Segerback, D., and Bartsch, H., eds.), IARC Scientific Publication No. 125. International Agency for Research on Cancer, Lyon, pp. 245–276.
14. Roy, D., Floyd, R. A., and Liehr, J. (1991) Elevated 8-hydroxyguanosine levels in DNA of diethylstilbesterol-treated Syrian hamsters: covalent DNA damage by free radicals generated by redox cycling of ditheylstilbesterol. *Cancer Res.* **51,** 3882–3885.
15. Wei, H. and Frenkel, K. (1991) In vivo formation of oxidized DNA bases in tumor promoter-treated mouse skin. *Cancer Res.* **51,** 4113–4449.
16. Malins, D. and Haimanot, R. (1991) Major alterations in the nucleotide structure of DNA in cancer of the female breast. *Cancer Res.* **51,** 5430–5432.
17. Olinski, R., Zastawny, T., Budzbon, J., Skowkowski, J., Zegarski, W., and Dizdaroglu, M. (1992) DNA base modifications in chromatin of human cancerous tissues. *FEBS* **309,** 193–198.
18. Maccubbin, A. E., Przybyszewski, J., Evans, M., Budzinski, E. E., Patryzc, H. B., Kulesz-Martin, M., and Box, H. B. (1995) DNA damage in UVB-irradiated keratinicytes. *Carcinogenesis* **16,** 1659–1660.
19. Frenkel, K., Zhong, Z., Wei, H., Karkoszka, J., Patel, U., Rashid, K., Georgescu, M., and Solomon, J. J. (1991) Quantitative high-performance chromatography analysis of DNA oxidized in vitro and in vivo. *Anal. Biochem.* **196,** 126–136.
20. Degan, P., Shinegawa, M. K., Park, E-M., Alperin, P. E., and Ames, B. N. (1991) Immunoaffinity isolation of urinary 8-hydroxy-2'-deoxyguanosine and 8-hydroxyguanine and quantitation of 8-hydroxy-2'-deoxyguanosine in DNA by polyclonal antibodies. *Carcinogenesis* **12,** 865–871.
21. Park, J-U., Cundy, K. C., and Ames, B. N. (1989) Detection of DNA adducts by high performance liquid chromatography with electrochemical detection. *Carcinogenesis* **10,** 827–832.
22. Paul, C. R., Wallace, J. C., Alderfer, J. A., and Box, H. C. (1988) Radiation chemistry of d(TpApCpG) in oxygenated solution. *Int. J. Radiat. Biol.* **54,** 403–415.
23. Marmur, J. (1961) A procedure for the isolation of deoxyribonucleic acid from micro-organisms. *J. Mol. Biol.* **3,** 208–218.

19

Nitric Oxide Synthase

Ahmad Aljada and Paresh Dandona

1. Introduction

Nitric oxide (NO) is an important bio-regulatory molecule in the nervous, immune and cardiovascular systems. The physiological involvement of NO in neuronal transmission, control of vascular tone, and immune response-induced cytostasis as well as the deleterious effects associated with altered levels of NO synthesis *(1,2)* culminated in the selection of the free-radical gas as "1992 Molecule of the Year" *(3)*.

NO is synthesized from one of the guanidine nitrogens of L-arginine by the enzyme nitric oxide synthase (NOS). The reaction represents a novel enzymatic process that involves a five-electron oxidation of L-arginine to NO, together with the stoichiometric production of L-citrulline (*see* **Fig. 1**). The reaction utilizes reduced nicotinamide adenine dinucleotide phosphate (NADPH) as an electron donor and also requires molecular oxygen, calcium, NOS-bound FAD, flavin mononucleotide and tetrahydrobiopterin *(4)*. NOS also binds calmodulin and contains heme.

There are three NOS isozymes. Isozyme I is constitutively present in central and peripheral neuronal cells and certain epithelial cells. Its activity is regulated by Ca^{2+} and calmodulin. Its function includes long-term regulation of synaptic transmission in the central nervous system, central regulation of blood pressure, and smooth-muscle relaxation. It has also been implicated in neuronal death in cerebrovascular stroke *(1,5)*. Isozyme II (in cytokine-induced cells) is usually not constitutively expressed but can be induced in macrophages and many other cells *(6)*. The activity of isoform II is not regulated by Ca^{2+}. A similar isozyme can be induced in other cell types such as smooth-muscle cells, fibroblasts, hepatocytes, etc. Induced NOS II is involved in the pathophysiology of autoimmune diseases and septic shock *(7)*. Isoform III has

From: *Methods in Molecular Biology, vol. 108: Free Radical and Antioxidant Protocols*
Edited by: D. Armstrong © Humana Press Inc., Totowa, NJ

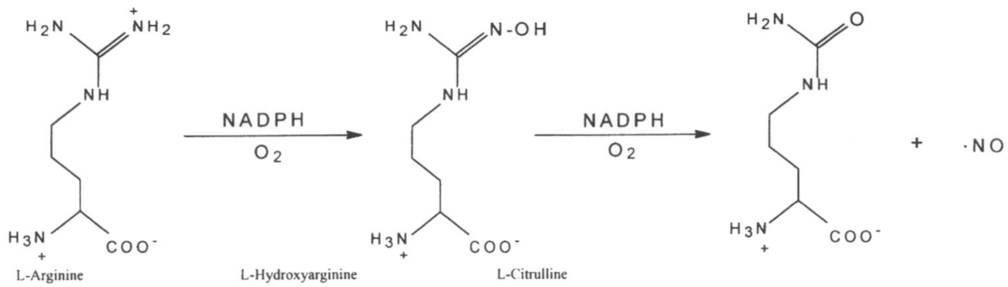

Fig. 1. Reaction catalyzed by nitric oxide synthase.

been found mostly in endothelial cells. It is constitutively expressed and regulated by Ca^{2+} and calmodulin. NO from endothelial cells keeps blood vessels dilated, and prevents the adhesion of platelets and white cells *(8)*. All the NOS enzymes characterized thus far are dimers comprised of subunits of 130,000–150,000 Dalton. The quantity of each isoform can be determined by Western blotting. An antibody for each isoform can be purchased from Transduction Laboratories (Lexington, KY).

Total NOS activity can easily be measured in three ways:

1. Spectrophotometrically, by exploiting the reaction of NO with oxyhemoglobin to form methemoglobin, which absorbs at 401 nm.
2. Quantification of [^{14}C] or [^{3}H] citrulline from the respective radiolabeled L-arginine.
3. Measurement of nitrite (NO_2^-) and nitrate (NO_3^-), the solution decomposition products of NO.

Measuring NOS activity by monitoring the conversion of arginine to citrulline is currently the standard assay for NOS activity. This sensitive and specific assay involves the addition of [^{14}C] or [^{3}H] arginine to intact tissue or protein extract. After incubation, the reactions are stopped with a buffer containing EDTA, which chelates the calcium required by NOS, and consequently, inactivates the NOS. The sample reactions are then applied to equilibrated cation-exchange resin to which the arginine binds. Citrulline flows through the column completely because it has a neutral charge at pH 5.5. The NOS activity is then quantitated by counting the radioactivity in the eluate.

2. Materials

2.1. Reagents

1. L-[2,3,4,5-^{3}H] or L-[U-^{14}C] Arginine is purchased from Amersham (Arlington Heights, IL) as the monohydrochloride salt with a specific activity of approx 300

mCi/mmol. A working solution of 3 μCi/μmol is prepared by diluting the stock with unlabeled arginine.

2. 10 m*M* Arginine (Sigma, St. Louis, MO).
3. 20 μ*M* Phosphodiesterase 3′:5′-Cyclic Nucleotide-activator (Calmodulin) (Sigma). When testing NOS activity from tissue extracts, addition of calmodulin to the reaction is not required. However, when testing purified NOS, the addition of calmodulin is required.
4. 40 μ*M* Flavin Adenine Dinucleotide (FAD) (Sigma). This solution should be kept out of direct light because FAD is light sensitive.
5. 5 m*M* (6R)-5,6,7,8-Tetrahydro-L-Biopterin Dihydrochloride (H_4B) (Sigma). Tetrahydropterins are readily oxidized in air and must therefore be made up as anaerobic solutions under argon or nitrogen or prepared in buffer containing dithiothreitol (DTT). A 5 m*M* H_4B solution prepared in 10 m*M* Tris-HCl, pH 7.4, and 100 m*M* DTT can be frozen at –80°C for over 3 mo without any observable oxidation. Tetrahydropterins are light sensitive and should be kept out of direct sunlight.
6. 10 m*M* β-Nicotinamide Adenine Dinucleotide Phosphate, reduced form (β-NADPH) (Sigma). This solution should be freshly prepared in 10 m*M* Tris-HCl, pH 7.4.
7. 2X Reaction Buffer (*see* **Note 1**): 100 m*M* Tris-HCl, pH 7.4, 280 m*M* NaCl, 1.0 m*M* $CaCl_2$, 10.8 m*M* KCl.
8. "Stop" Buffer: 50 m*M* Tris-HCl, pH 5.5, 4.0 m*M* EDTA.

2.2. Cation-Exchange Resin

1. AG 50W-8 cation-exchange resin can be purchased from Bio-Rad (Hercules, CA). It is available with H^+ , Na^+, or NH_3^+ counterions.
2. Resin is converted from one ionic form to another. AG 50W resin is available in several particle-size ranges. The flow rate in a chromatographic column increases with increasing particle size. However, the attainable resolution increases with decreasing particle size and narrower size-distribution ranges. Particle size is given either in mesh size or micron size. The larger the mesh-size number, the smaller the particle size. AG 50W is used in either a batch method or a column method.
3. A batch method consists of adding the resin directly to the sample and stirring. The column method requires preparing a column filled with resin, and passing the sample through to achieve the separation. In the citrulline assay, the H^+ form is converted to Na^+ form with 1 *M* NaOH and 200–400 mesh size is used in the column method (*see* **Note 2**).

2.2.1. AG 50WX-8 Resin Conversion

The resin is converted from H^+ to Na^+ with 1 M NaOH. A 2 volumes of solution/volume of resin with a flow rate of 2 mL/min/cm^2 is used for conversion. A pH value of 9 is used as a test for completeness of conversion.

The resin is then rinsed with 4 volumes of distilled water and a pH<9 is used as an indicator for the completion of rinsing. The resin is stable in acid, base, and organic solvents, and may be autoclaved. To prevent bacterial growth during prolonged storage, use a preservative such as 0.05% sodium azide (*see* **Note 3**).

3. Methods

3.1. Sample Preparation

1. The citrulline assay can be used with a variety of tissue homogenate from numerous sources such as blood vessels, visceral organs, nervous tissue, immune cells, and cultured cells.
2. NOS activity is relatively unstable. It is best to freeze intact tissues or harvested cultured cells prior to homogenization. Flash-freeze the tissues in liquid nitrogen and then store at –80°C.
3. Add 5 volumes (volume of buffer/weight of tissue) of ice-cold homogenization buffer (0.1 *M* NaCl, 10 m*M* Tris.Cl, –pH 7.4), to a tissue sample.
4. Homogenize the tissue with a tissue grinder. A high-intensity ultrasonic liquid processor can be used as well. The tissue homogenate should be kept on ice.
5. Transfer 1 mL of the tissue homogenate into a microcentrifuge tube and spin at 14,000*g* for 10 min at 4°C.
6. Transfer the supernatant to another microcentrifuge tube and keep the tube on ice. (Save a small aliquot from each sample for total protein measurements) (*see* **Note 4**).

3.2. Procedure

1. The volumes given here yield sufficient reaction mix for 10 reactions. If more or less reaction mix is needed, adjust the volumes proportionally (*see* **Note 5**): 500 µL of 2X reaction buffer, 100 µL of 40 µ*M* FAD, 10 µL of 5 m*M* H_4B, 50 µL of 20 µ*M* calmodulin, 100 µL of 10 m*M* NADPH (freshly prepared), 20 µL of 10 m*M* arginine (*see* **Note 6**), 20 µL of 300 mCi/mmol ^{14}C or ^{3}H-arginine.
2. Start the reaction by combining 20 µL of tissue homogenate with 80 µL of reaction mix in 5 mL tube.
3. Incubate the reaction at 30°C for 10–60 min depending on the tissue used. High levels of neuronal NOS in the nervous tissues and skeletal muscle permit brief assays (10–15 min) of NOS. Lower levels of endothelial NOS in vascular tissues require prolonged periods (60 min).
4. Stop the reaction by adding 2 mL of "stop" buffer to the reaction, and incubate at 4°C.
5. Equilibrate 2.5 mL of Dowex AG 50WX-8 (Na^+ form) with a minimum of 2.0 mL "stop" buffer (Pasteur pipets can be used as columns after blocking the tapered end with glass fibers).
6. Allow the excess buffer to pass through the column, leaving enough buffer to just cover the top of the resin bed.
7. Apply individual incubation mixtures (2.1 mL) dropwise to the top of the column without disturbing the resin bed. Drain the sample into the top of the bed and

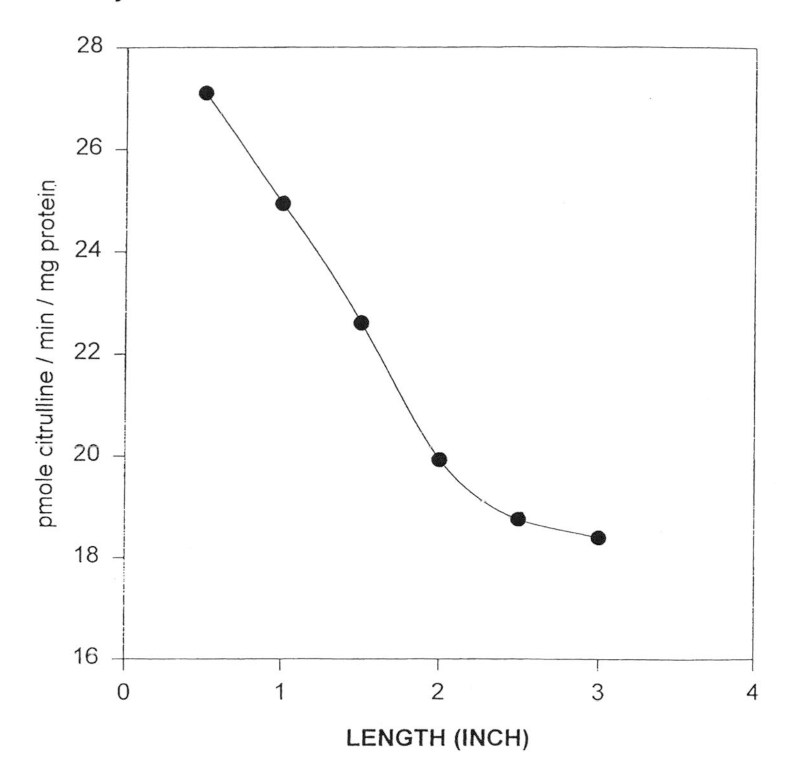

Fig. 2. Column length vs NOS activity. The high activity at column length <2 inches is owing to radiolabeled arginine that passed through the column.

wash the column by applying 1 mL of "stop" buffer, being very careful to rinse down the sides of the column and to avoid stirring up the bed. Drain each portion to the level of the resin bed before the next portion is added. Never allow the liquid level to drain below the top of the resin.

8. Collect the eluate into scintillation vials and add 18 mL of liquid-scintillation fluid.
9. Quantitate the radioactivity in a liquid-scintillation counter (*see* **Note 7**). Each sample must be corrected for the amount of arginine that flows through the column and for citrulline recovery. This can be achieved by inhibiting a reaction with 1 m*M* N_{ω}-nitro-L-arginine methyl ester HCl (a competitive NOS inhibitor) or by boiling the sample or by omitting NADPH or calcium from the reaction. A high blank value will greatly reduce the assay sensitivity (*see* **Note 8**).

3.3. Results

1. Different resin column lengths are tested using homogenized rabbit brain and 2.5 inches of cation-exchange resin is found to be efficient to trap ~95% of the arginine (**Fig. 2**).

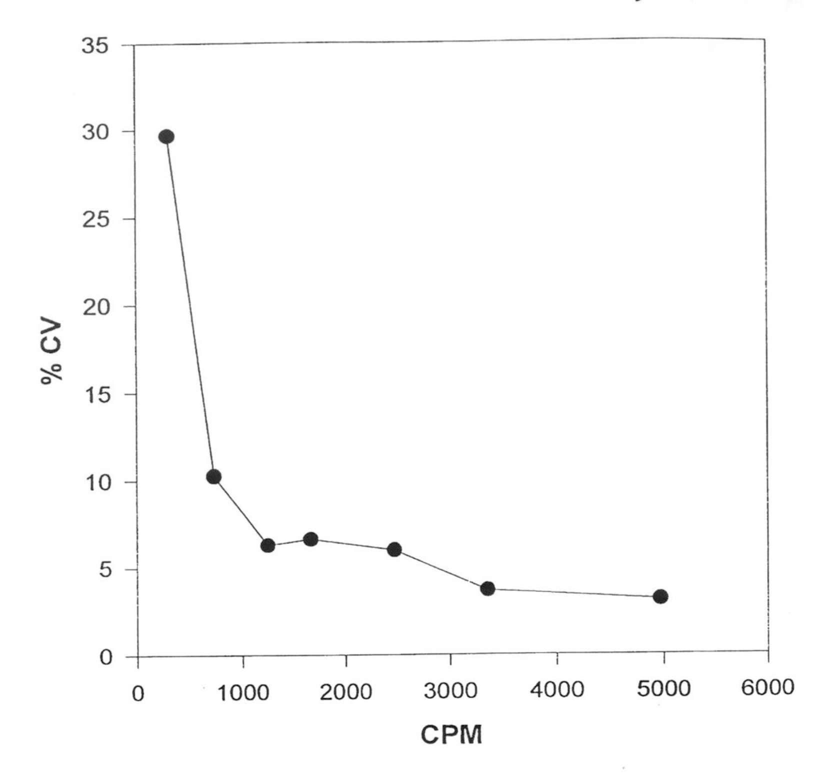

Fig. 3. %CV vs CPM of NOS activity.

2. To determine the coefficient of variation at different CPM, homogenized rabbit-brain cerebellum with a protein concentration of 10 mg/mL is serially diluted and each dilution is analyzed 5 times, (**Fig. 3**).
3. Homogenized rabbit-brain cerebellum with protein concentration of 10 mg/mL are aliquoted and frozen at –70°C for reproducibility and linearity studies. NOS activity is measured in these aliquots four times and the variation between the runs is less than 10% (**Fig. 4**).

4. Notes

1. A final free calcium concentration of 75 μ*M* is required for optimal NOS activity *(9)*. Ca^{2+} is not required for the inducible form of NOS.
2. A mesh size of 200–400 produces the lowest background and using different mesh size might lead to low percentage recovery.
3. The cation-exchange separation is affected by changes in salt concentration and pH. Therefore, the recovery of citrulline and the amount arginine that does not bind to the resin should always be reassessed if the reaction conditions are changed.

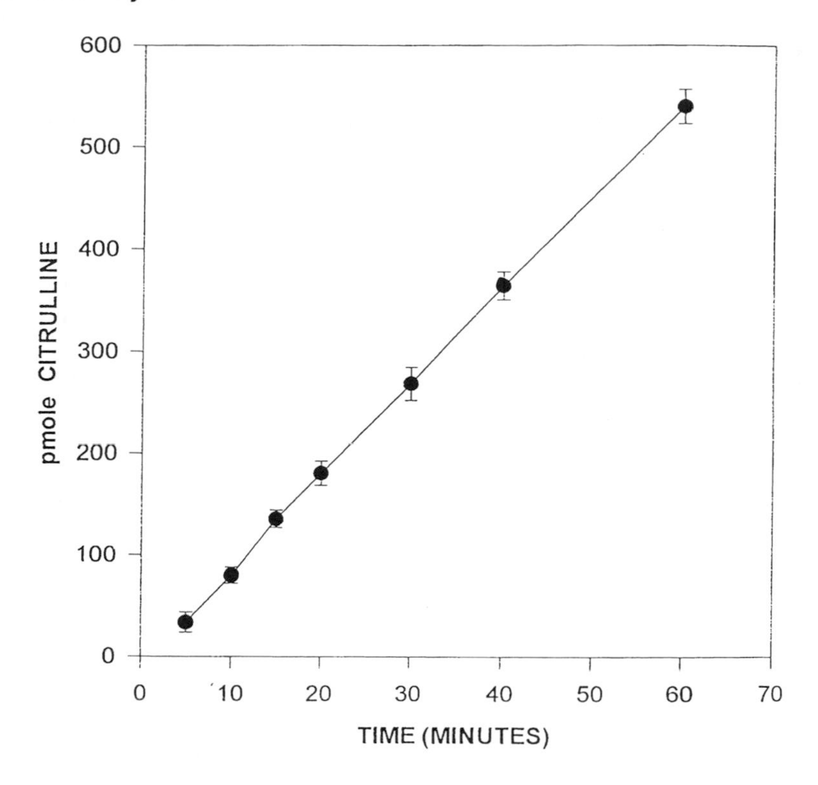

Fig. 4. Time vs arginine conversion.

4. NOS activity in soluble and membrane-associated fractions is discriminated by centrifuging the homogenized tissue at 100,000*g* for 60 min and assaying the supernatant activity. The supernatant contains the soluble form of NOS and the pellet contains the membrane-associated form.
5. Crude extracts of the macrophage NOS require the presence of Mg^{2+} for maximum activity, whereas the purified NOS shows no such dependence. A final concentration of 1 m*M* magnesium diacetate is used only with crude and semipurified macrophage NOS.
6. The K_m (Michaelis constant) of NOS is in the range of 2–20 μ*M* ***(6)*** and an arginine concentration of 200 μ*M* is enough to ensure maximum NOS activity.
7. The resin should retain >95% of the applied radioactivity. If more than 5% of the activity flows through, the arginine should be purified prior to conducting the assay.
8. In this assay, the major radiolabeled compound in the eluate is citrulline. This is verified using thin-layer chromatography (TLC). Using silica-gel chromatography plates with $CH_3OH:NH_4OH$ (6:1), arginine migrates at R_f (the ratio of the distance traveled by a compound to the distance traveled by the solvent) of ~0.1, whereas citrulline migrates at an R_f of ~0.5.

References

1. Moncada, S., Palmer, R. M. J., and Higgs, E. A. (1991) Nitric oxide: physiology, pathophysiology, and pharmacology (review). *Pharmacol. Rev.* **43(2),** 109–142.
2. Nathan, C. (1992) Nitric oxide as a secretory product of mammalian cells (review). *FASEB J.* **6(12),** 3051–3064.
3. Culotta, E. and Koshland, D. J.(1992) NO news is good news. *Science* **258,** 1862–1863.
4. Stuehr, D. J., Kwon, N. S., Nathan, C., Griffith, O., Feldman, P., and Wiseman, J. (1991) N^{φ}-hydroxy-L-arginine is an intermediate in the biosynthesis of nitric oxide from L-arginine. *J. Biol. Chem.* **266,** 6259–6263.
5. Bredt, D. S. and Snyder, S. H. (1990) Isolation of nitric oxide synthase, a calmodulin requiring enzyme. *Proc. Natl. Acad. Sci. USA* **87(2),** 682–685.
6. Stuehr, D. J., Cho, H. J., Kwon, N. S., Weise, M. F., and Nathan, C. F. (1991) Purification and characterization of the cytokine induced macrophage nitric oxide synthase: an FAD and FMN containing flavoprotein. *Proc. Natl. Acad. Sci. USA* **88(17),** 7773–7777.
7. Forstermann, U. (1994) Biochemistry and molecular biology of nitric oxide synthases (review). *Arzneimittel-Forschung* **44(3A),** 402–407.
8. Sessa, W. C. (1994) The nitric oxide synthase family of proteins. *J. Vas. Res.* **31(3),** 131–143.
9. Knowles, R. G., Palacios, M., Palmer, R. M., and Moncada, S. (1989) Formation of nitric oxide from L-arginine in the central nervous system: a transduction mechanism for stimulation of the soluble guanylate cyclase. *Proc. Natl. Acad. Sci. USA* **86(13),** 5159–5162.

20

Quantitation of Heme Oxygenase (HO-1) Copies in Human Tissues by Competitive RT/PCR

Nader G. Abraham

1. Introduction

A variety of oxidative stress-inducing agents, such as metals, ultra-violet (UV) light, heme, and hemoglobin have been implicated in the pathogenesis of the inflammatory process. The cellular response to such agents involves the production of a number of soluble mediators including acute phase proteins, cicosanoids, and various cytokines.

The rate-limiting enzyme in heme catabolism, heme oxygenase (HO), is a stress-response protein and its induction has been suggested to represent an important cellular-protective response against oxidative damage produced by free heme and hemoglobin ***(1–5)***. Induction of HO may specifically decrease cellular heme (pro-oxidant) and clcvatc bilirubin (anti-oxidant) levels ***(5–9)***. Two HO isozymes, the products of distinct genes, have been described ***(10,11)***. HO-1, which is ubiquitously distributed in mammalian tissues, is strongly and rapidly induced by many compounds that elicit cell injury; the natural substrate of HO, heme, is itself a potent inducer of the enzyme ***(10,11)***. HO-2, which is believed to be constitutively expressed, is present in high concentrations in such tissues as the brain and testis, and is believed to be non-inducible ***(10)***.

Endotoxin, IL-1, and other stress agents cause a rapid (within 5–10 min) activation of the HO gene and a subsequent accumulation of HO mRNA ***(12–14)***. This process involves transcriptional activation of several regulatory sites in the HO promoter region. AP-1 and IL-6 responsive elements are found in the promoter region of this gene ***(13–15)***. A recent study from this laboratory demonstrated that the proximal promoter region of the human HO gene also contains NF-kB and AP-2 binding sequences ***(16)***. The finding of AP-2 and NF-kB binding sites on the HO promoter suggests the importance of HO in

From: *Methods in Molecular Biology, vol. 108: Free Radical and Antioxidant Protocols*
Edited by: D. Armstrong © Humana Press Inc., Totowa, NJ

stress/injury responses, when these transcriptional factors are known to be activated ***(16)***.

Because the samples under investigation are of limited cell number and/or copy number of mRNAs, conventional methods of mRNA analysis such as Northern hybridization, nuclease-protection mapping, and *in situ* hybridization are insufficient in sensitivity and/or accuracy for our purpose. The polymerase chain reaction (PCR) has been widely used to amplify cDNA copies of low levels of mRNAs after reverse transcription (RT)***(17–19)***. We achieved reliable quantitation by including an internal standard of equivalent amplification efficiency in the reaction mixture. Several strategies are available to ensure that RT-PCR is quantitative ***(20)***. We chose competitive RT-PCR, which has become the most widely used form of quantitative RT-PCR, because the greatest advantage of using competitive PCR is that it is not necessary to assay PCR products exclusively during the exponential phase of the amplification ***(21)***. Quantitation of target transcripts was achieved by inclusion of a competitor template in the PCR reaction; PCR products derived from the competitor can be distinguished from the products derived from the sample by a difference in size. In competitive PCR, serial dilutions of the sample are used with a fixed amount of competitor template (internal standard) cDNA and this mixture was subjected to amplification. We have applied this approach in an effort to determine the levels of HO-1 in human tissues.

2. Materials

Oligonucleotides: Two previously published oligonucleotides ***(22)*** are used as primers for the RT-PCR reaction: CAG GCA GAG AAT GCT GAG TTC (sense) and GAT GTT GAG CAG GAA CGC AGT (antisense) (NBI, Plymouth, MN). These primers are designed to amplify a 555 bp stretch (79 bp to 633 bp) of the human HO-1 ***(23)***.

3. Methods

3.1. Internal Standard

The plasmid pCMV-HHO1 is linearized with *Eco*47 III (a single restriction site in human HO-1 cDNA at position 429 bp). The linearized plasmid is digested with Bal 31 nuclease for 5 min in order to digest nucleotides and create a mutation. After this digestion, the construct is ligated with T4 ligase. After transformation into JM109 *Escherichia coli*-competent bacteria, several clones are analyzed for the size of the mutated human HO-1 cDNA (mHHO-1) resulting from the digestion by Bal 31 and for the presence of *Eco*47 III restriction site (*see* **Note 1**).

One clone is selected with approx 50 bp truncated from the original cDNA of HHO-1. This clone is amplified and the insert excised from the vector with *Hind*III and purified. This insert is used as the internal standard in the PCR reactions.

3.2. RT-PCR

1. Total RNA from human liver are isolated using the method of Chomczynski and Sacchi ***(24)*** and quantified by spectrophotometry. The quality of RNA is checked by gel electrophoresis and ethidium-bromide staining.
2. RNA is reverse transcribed using the first-strand cDNA synthesis kit from Clontech (Palo Alto, CA). Briefly, RNA in 12.5 µL of oligo $(dT)_{18}$ (final concentration–0.2 µ*M*) is denatured at 70°C for 2 min. The denatured RNA is placed on ice and 6.5 µL of the reverse transcription mixture containing: 50 m*M* Tris-HCl, pH 8.3, 75 m*M* KCl, 3 m*M* $MgCl_2$, 0.5 m*M* of each dNTP, 1 U/µL of RNAse inhibitor, and 200 Unit of MMLV reverse transcriptase is added. The reaction tube is then placed at 42°C for 1 h followed by heating to 95°C to stop the reaction, and then placed on ice.
3. The PCR reaction is performed by adding to the reverse-transcription reaction tube the PCR mix to a final volume of 100 µL. The PCR mix contained: 10 m*M* Tris-HCl, pH 8.3, 50 m*M* Kcl, 0.001% gelatin, 1.5 m*M* $MgCl_2$; 250 µ*M* of each dNTP, 1 µ*M* of sense and anti sense primers, 2.5 Unit of Taq DNA polymerase, and 1 µCi of ^{32}P-dCTP.
4. The reaction mixture is overlaid with 2 drops of mineral oil and subjected to 40 cycles as followed: 95°C for 1 min, 55°C for 1 min, and 72°C for 2 min. After the last cycle, a final extension is performed at 72°C for 10 min.

3.3. Analysis of the Amplified Template

The amplification products are analyzed on a 6% acrylamide-bisacrylamide (37.5:1) gel in 1X Tris-glycine-ethylenediamine tetraacetic acid (EDTA) buffer (1X : 25 m*M* Tris, 200 m*M* glycine, and 1 m*M* EDTA). After drying the gel, it is exposed for autoradiography. The autoradiogram is then analyzed by either densitometry or the bands are cut from the gel and counted for radioactivity.

Two points have to be clarified before the quantitative nature of our RT-PCR system can be examined: the amplified templates have to be clearly distinguished, and the internal standard has to be amplified as efficiently as the sample templates.

3.4. Distinguishing Between the Amplified Templates

1. Several clones are analyzed for the size of the mutated cDNA (MHHO-1) resulting from the digestion by Bal 31 and for the presence of the *Eco*47 III restriction site.
2. After transformation into competent bacteria by the mutated plasmid, PCR products obtained with our primers from several clones are analyzed on acrylamide

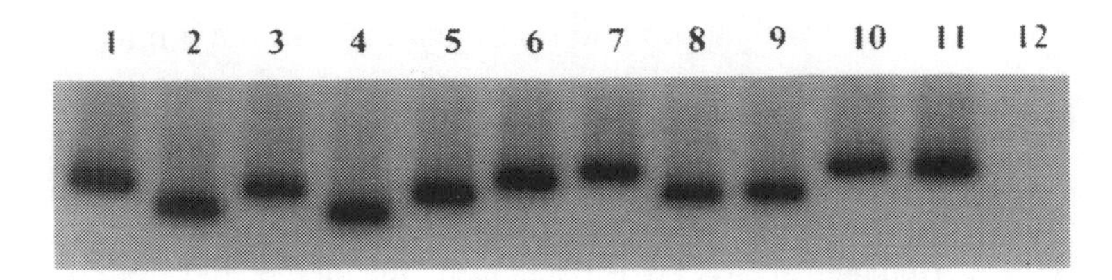

Fig. 1. PCR products obtained from the amplification of several clones selected after modification of the *Eco*47 III site (cf. methods). Lane 1–10, different clones tested after mutation; lane 11, PCR product from the pCMV-HHO1 as positive control for the reaction and as size marker for the original size; lane 12, PCR without DNA, negative control.

gel 6% (**Fig. 1**). As a result, clone number 2 is chosen for the following experiments and named pCMV-mHHO1. This clone is also examined for size and the presence or absence of the *Eco*47 III restriction site (**Fig. 2**). The difference in size between the two is approx 50 bp (lanes 5 and 6) and in the mutated insert the *Eco*47 III site is missing as predicted (lanes 8 and 9).

3. To better evaluate the size of the PCR products, the two inserts (HHO1 and mHHO1) are amplified and analyzed on 2% agarose gel. A difference in 50 bp is demonstrated, as expected, from the previous results (**Fig. 3**).

3.5. Competitive vs Noncompetitive Amplification

1. In order to achieve a reliable quantitation, it is important that the internal-standard and sample template(s) are amplified equally under our reaction conditions. We examined whether the 10% (50 bp) difference in the sequence between the two templates caused a discrepant amplification efficiency (**Fig. 4A**).
2. Various amounts of total RNA from human liver are subjected to RT-PCR alone or as a mixture with a fixed amount of the mutated insert (mHHO1). In the combined PCR, the mHHO1 is added at the PCR step, the RT is performed using oligo(dT)18 primers as described in the Methods section (*see* **Note 2**)..
3. Respective templates are identified by their difference in size, examined by electrophoresis on acrylamide gel, and are quantified by densitometry of the bands on the autoradiogram. As can be seen in **Fig. 4B**, when the internal standard (mHHO1) is included in the PCR mixture, the amount of amplified liver template decreased. The decrease diminished as the amount of liver RNA sample increased. These results indicate that the sample and the internal-standard templates are both amplified in the same manner. As the amount of one template increased, the chance of the other template being amplified declined. Eventually, the portion of the other template becomes too small to be significant.
4. To achieve quantitation of specific messages, it is important that a proper ratio be determined for each sample. It can be noted that, in the presence of the internal

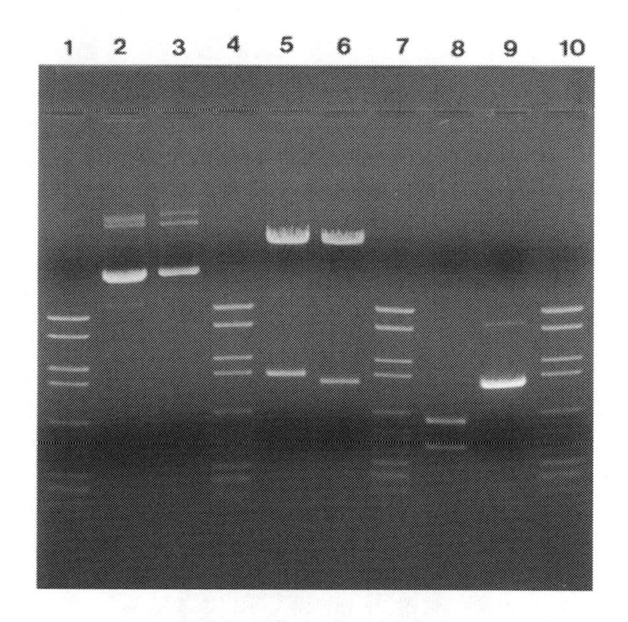

Fig. 2. Analysis on 1% agarose gel of clone 2, which is selected as internal standard. Lane 2, pCMV-HHO1; lane 3, pCMV-mHHO1; lane 5, pCMV-HHO1 digested by *Hind* III; lane 6, pCMV-mHHO1 digested by *Hind*III; lane 8, HHO1 insert digested by *Eco*47 III; lane 9, mHHO1 insert digested by *Eco*47 III; lane 1, 4, 7, and 10, DNA molecular-weight marker (pBR 328 DNA-Bg1I - pBR 328 DNA-Hinf I; marker VI, Boehringer Mannheim).

standard, the amplification of the liver template does not have a linear relationship with its input. This reflects the saturation of the system, as noted by other investigators ***(17,18)***. However, the amount of specific target present at input can be determined from the ratio of the amplified liver template to the internal standard.

3.6. Quantitation of HHO1 mRNA

1. Once our RT-PCR system had been shown to be specific for designated targets and the sample and internal-standard templates could be amplified with an equivalent efficiency, the quantitative nature of the system could be examined.
2. In the course of testing for the competitiveness of the sample and the internal-standard templates for amplification, we found that 10 fg of standard is the appropriate amount required in the reaction mixture from human liver RNA at the range of 1–200 ng to be amplified and unambiguously quantified (**Fig. 5**).
3. Because the internal standard and sample templates are shown to be amplified efficiently under the conditions described, it is therefore possible to determine the human HHO1 mRNA content. The amount of HHO1 cDNA is calculated by extrapolating from the intersection of the curve with a ratio of 1.0 down to the X-axis (**Fig. 6**).

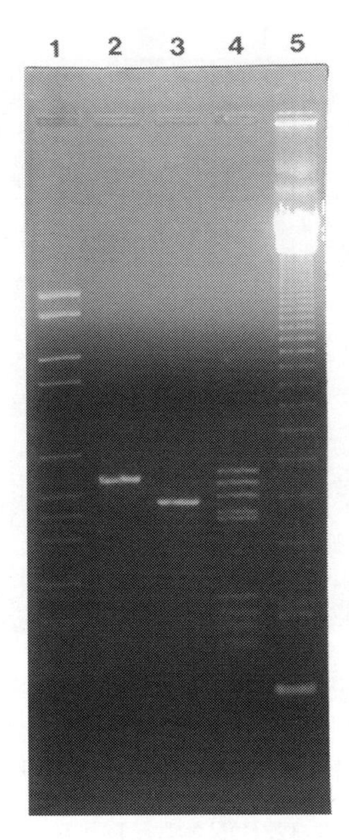

Fig. 3. Analysis on 2% agarose gel of the PCR products obtained from the original insert (HHO1) compared to the mutated one (mHHO1). Lane 2, PCR product from HHO1 insert (555 bp); lane 3, PCR product from mHHO1 insert (500 bp); lane 1, DNA molecular-weight marker (pBR 328 DNA-Bg1I + pBR 328 DNA-Hinf I; marker VI Boeringer Mannheim); lane 4, DNA molecular weight marker (pBR 322 DNA-Hae III; marker V Boehringer Mannheim); lane 5, DNA molecular-weight marker (123 bp DNA ladder, Gibco-BRL).

4. Because the number of molecules in the standard is known, the actual number of target DNA molecules added to the PCR reaction can be calculated. In turn, the number of mRNA molecules can be calculated in the RNA sample used for RT, assuming that the efficiency of cDNA synthesis in 100%. The actual efficiency is less than that value, thus such calculation gives the minimum number of mRNA molecules.
5. As described previously, there are 500 bases in the standard that account for a molecular weight of approx 3.3×10^5. Thus, 10 fg of standard has an equivalency of 1.82×10^4 molecules. The ratio is calculated taking into account the difference in base composition between the two templates. The number of molecules of HHO1 in human liver is estimated to be between 750 and 980 per nanogram of total RNA from normal human liver.

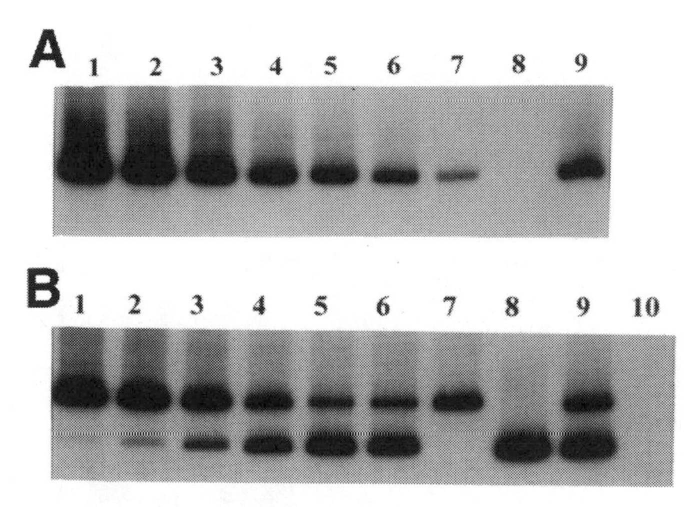

Fig. 4. (**A**) Analysis of RT-PCR product obtained from total RNA of human liver. Lane 1, 200 ng; lane 2, 100 ng; lane 3, 50 ng; lane 4, 25 ng; lane 5, 5 ng; lane 6, 1 ng; lane 7, negative control (no RNA is used in the reaction); lane 8, positive control of the PCR reaction, amplification of the HHO1 insert (555 bp). (**B**) Competitive amplification of human liver total RNA and mutated insert (mHHO1). Internal standard (mHHO1) at 10 fg is mixed with human liver total RNA at 200 ng, 100 ng, 50 ng, 25 ng, 5 ng, and 1 ng respectively (lanes 1 to 6); lane 7, positive control of the PCR reaction, amplification of the HHO1 insert (555 bp); lane 8, positive control of the PCR reaction, amplification of the mHHO1 insert (500 bp); lane 9, the PCR product from the two inserts (HHO1 and mHHO1) are mixed to serve as size reference; lane 10, negative control (no RNA is used in the reaction).

6. Owing to its powerful amplification capacity, PCR has been widely used to detect DNA and mRNA species, which are present in low levels. Various groups have employed this technique to trace residual malignant cells in leukemic patients after bone-marrow transplantation ***(25–27)***.
7. The great power of amplification of PCR presents a challenge, when the technique is used to quantify copy numbers of genes and mRNAs. Minute differences in any of the variables that affect the efficiency of amplification can dramatically alter product yield. The problem has been circumvented by co-amplifying a target sequence with a standard template st that the reaction is internally controlled ***(17–19)***. Accordingly, our system to evaluate the genetic expression of HHO1 is developed with the following considerations in mind. The internal standard used is identical to the sample template in the target region; the difference is 10% in length owing to the modification described in the Methods section. The primers are derived from regions where the standard and sample(s) templates have identical sequences. Consequently, differences in the melting temperatures of cDNA templates and primer/template duplexes, which can greatly influence the amplification are

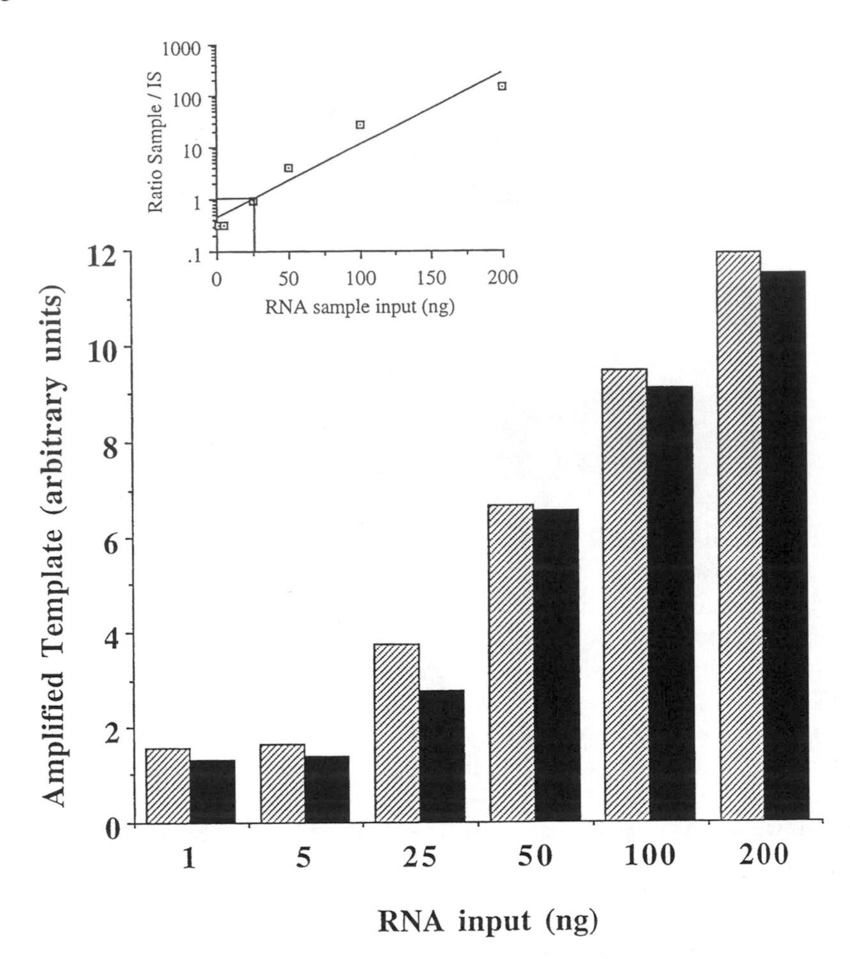

Fig. 5. Competitive vs noncompetitive amplification of human liver total RNA. Total RNA from liver at the amount indicated is subjected to RT-PCR alone (▧) or as a mixture with 10 fg of internal standard (mHHO1) (■). Reaction products are analyzed on a 6% acrylamide gel. The amounts of amplified human template are determined by densitometry on the autoradiograms.

minimized. However, the standard and the sample templates can be distinguished easily on an acrylamide gel, owing to their difference in size (50 bp).

8. As expected, the great sequence similarity between the standard and the sample templates rendered them good candidates for the amplification. The input ratio of internal-standard vs sample RNA is therefore critical for a clear-cut measurement of the amount of each template amplified. This required that each sample be individually determined. When the ratio is appropriate, there existed a linear relationship between the amount of input sample RNA and the amount of amplified

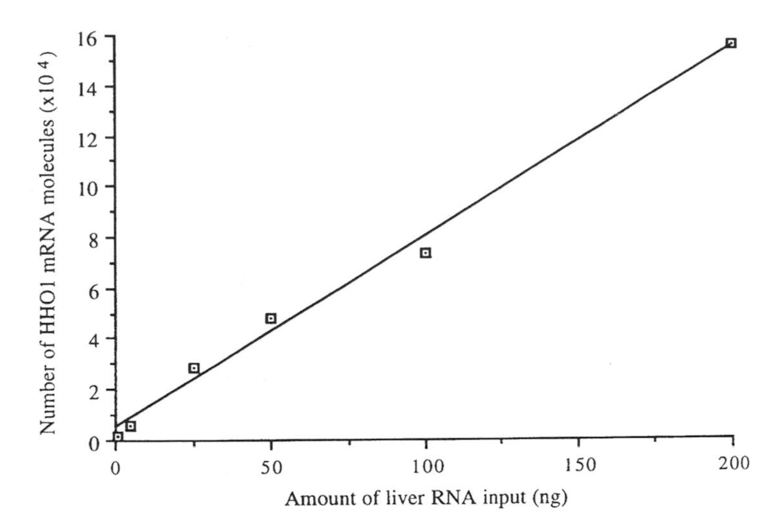

Fig. 6. Competitive amplification of human liver total RNA and mutated insert (mHHO1). Internal standard (mHHO1) at 10 fg is mixed with human liver total RNA at 200 ng, 100 ng, 50 ng, 25 ng, 5 ng and 1 ng respectively. Reaction products are analyzed on a 6% acrylamide gel. The amounts of amplified human template are determined by densitometry on the autoradiogram and by counting the radioactivity of the various bands after excision from the gel (the two methods gave the same results). The ratio of the human RNA vs internal standard is used to calculate the amount of HHO1 mRNA content. The human liver mRNA content thus derived is plotted against the amount of input sample RNA.

sample template after the latter is normalized against that of the amplified internal standard. We estimated that there is approx 1500 HHO-1 mRNA molecules in each nanogram of total RNA from human liver using this technique. The comparative levels of enzyme activity and the total number of PCR gives an accurate indication of the role of HO-1 in normal and pathological state.

4. Notes

1. The internal standard size described in **Subheading 3.1.** must be accurately determined since most calculations of the number of HO mRNA are dependent on this purity.
2. The use of much lower RNA from samples with high levels of HO-1 is recommended in order for the RT/PCR to achieve the ideal ratio of the target mRNA and the internal standard as previously suggested in **Subheading 3.5.** For example, the spleen and liver contain several-fold higher mRNA than that of the intestine, kidney, testes, and blood vessels. Performing six to eight different RNA concentrations for spleen or liver is, therefore, recommended.

Acknowledgment

This study was supported, in part, by NIH Grant #HL-54138.

References

1. Shibahara, S., Muller, R. M., and Taguchi, H. (1987) Transcriptional control of rat heme oxygenase by heat shock. *J. Biol. Chem.* **262,** 12,889–12,892.
2. Mitani, K., Fujita, H., Sassa, S., and Kappas, A. (1989) Heat shock induction of heme oxygenase mRNA in human Hep3B hepatoma cells. *Biochem. Biophys. Res. Commun.* **165,** 437–441.
3. Keyse, S. M. and Tyrrell, R. M. (1989) Heme oxygenase is the major 32–kDa stress protein induced in human skin fibroblasts by UVA radiation, hydrogen peroxide, and sodium arsenite. *Proc. Natl. Acad. Sci. USA* **86,** 99–103.
4. Taketani, S., Kohno, H., Yoshinaga, T., and Tokunaga, R. (1989) The human 32–kDa stress protein induced by exposure to arsenite and cadmium ions is heme oxygenase. *FEBS Lett.* **245,** 173–176.
5. Abraham, N. G., Lin, J. H-C., Schwartzman, M. L., Levere, R. D., and Shibahara, S. (1988) The physiological significance of heme oxygenase. *Int. J. Biochem.* **20,** 543–558.
6. Stocker, R., Yamamoto, Y., McDonagh, A. F., Glazer, A. N., and Ames, B. N. (1987) Bilirubin is an antioxidant of possible physiological importance. *Science* **235,** 1043–1046.
7. Nath, K. A., Balla, G., Vercellotti, G. M., Balla, J., Jacob, H. S., Levitt, M. D., and Rosenberg, M. E. (1992) Induction of heme oxygenase is a rapid, protective response in rhabdomyolysis in the rat. *J. Clin. Invest.* **90,** 267–270.
8. Paller, M. S. and Jacob, H. S. (1994) Cytochrome P-450 mediates tissue-damaging hydroxyl radical formation during reoxygenation of the kidney. *Proc. Natl. Acad. Sci. USA* **91,** 7002–7006.
9. Stocker, R. (1990) Induction of haem oxygenase as a defence against oxidative stress. *Free Rad. Res. Comm.* **9,** 101–112.
10. McCoubrey, W. K., Jr., Ewing, J. F., and Maines, M. D. (1992) Human heme oxygenase-2: characterization and expression of a full-length cDNA and evidence suggesting that the two HO-2 transcripts may differ by choice of a polyadenylation signal. *Arch. Biochem. Biophys.* **295,** 13–20.
11. Shibahara, S., Yoshizawa, M., Suzuki, H., Takeda, K., Meguro, K., and Endo, K. (1993) Functional analysis of cDNAs for two types of human heme oxygenase and evidence for their separate regulation. *J. Biochem.* **113,** 214–218.
12. Cantoni, L., Rossi, C., Rizzardini, M., Gadina, M., and Ghezzi, P. (1991) Interleukin-1 and tumour necrosis factor induce hepatic haem oxygenase. *Biochem. J.* **279,** 891–894.
13. Rizzardini, M., Terao, M., Falciani, F., and Cantoni, L. (1993) Cytokine induction of haem oxygenase mRNA in mouse liver. B*iochem. J.* **290,** 343–347.
14. Lutton, J. D., da Silva, J-L., Moquattash, S., Brown, A. C., Levere, R. D., and Abraham, N. G. (1992) Differential induction of heme oxygenase in the

hepatocarcinoma cell line (Hep3b) by environmental agents. *J. Cell Biol.* **49,** 259–265.
15. Alam, J. and Zhining, D. (1992) Distal AP-1 binding sites mediate basal level enhancement and TPA induction of the mouse heme oxygenase-1 gene. *J. Biol. Chem.* **267,** 21,894–21,900.
16. Lavrovsky, Y., Schwartzman, M. L., Levere, R. D., Kappas, A., and Abraham, N. G. (1994) Identification of NFkB and AP-2 binding sites in the promoter region of the human heme oxygenase-1 gene. *Proc. Natl. Acad. Sci. USA* **91,** 5987–5991.
17. Becker-Andre, M. and Hahlbrock, K. (1989) Absolute mRNA quantification using the polymerase chain reaction (PCR): a novel approach by a PCR aided transcript titration assay (PATTY). *Nucleic Acids Res.* **17,** 9437–9446.
18. Wang, A. M., Doyle, M. V., and Mark, D. F. (1989) Quantitation of mRNA by the polymerase chain reaction. *Proc. Natl. Acad. Sci. USA* **86,** 9717–9721.
19. Gilliland, G., Perrin, S., Blanchard, K., and Bunn, H. F. (1990) Analysis of cytokines mRNA and DNA: detection and quantitation by competitive polymerase chain reaction. *Proc. Natl. Acad. Sci. USA* **87,** 2725–2729.
20. Cross, N. C. (1995) Quantitative PCR techniques and applications. *Br. J. Haematol.* **89,** 639–647.
21. Murphy, L. D., Herzog, C. E., Rudick, J. B., Fojo, A. T., and Bates, S. E. (1990) *Biochemistry* **29,** 10,351–10,356.
22. Kutty, G., Hayden, B., Osawa, Y., Wiggert, B., Chader, G. J., and Kutty R. K. (1992) Heme oxygenase: expression in human retinal and modulation by stress agents in a human retinoblastoma cell model system. *Curr. Eye Res.* **11,** 153–160.
23. Yoshida, T., Biro, P., Cohen, T., Muller, R. M., and Shibahara, S. (1988) Human heme oxygenase cDNA and induction of its mRNA by hemin. *Eur. J. Biochem.* **171,** 457–461.
24. Chomczynski, P. and Sacchi, N. (1987) Single-step method of RNA isolation by acid guanidium. *Anal. Biochem.* **162,** 156–159.
25. Cross, N. C., Feng, L., Chase, A., Bungey, J., Hughes, T. P., and Goldman, J. M. (1993) Competitive polymerase chain reaction to estimate the number of BCR-ABL transcripts in chronic myeloid leukemia patients after bone marrow transplantation. *Blood* **82,** 1929–1936.
26. Frenoy, N., Chabli, A., Sol, O., Goldschmit, E., Lemonnier, M. P., Misset, J. L., and Debuire, B. (1994) Application of a new protocol for nested PCR to the detection of minimal residual bcr/abl transcripts. *Leukemia* **8,** 1411–1414.
27. Lion, T., Gaiger, A., Henn, T., Horth, E., Haas, O. A., Geissler K., and Gadner H. (1995) Use of quantitative polymerase chain reaction to monitor residual disease in chronic myelogenous leukemia during treatment with interferon. *Leukemia* **9,** 1353–1360.

21

A Fluorescent-Based Assay for Measurement of Phospholipase A_2 Activity

A Facile Assay for Cell Sonicates

Thomas M. Nicotera and Glenn Spehar

1. Introduction

Phospholipases A_2 (PLA_2) catalyze the hydrolysis of the ester bond at the sn-2 position of phospholipids to generate the lysophospholipid and free fatty acid *(1)*. Several classes of PLA_2s have been identified *(2)*. They include the low molecular weight-secreted enzymes, which demonstrate a requirement for millimolar concentrations of Ca^{2+} for full activity and are generally involved in the inflammatory response. The high molecular-weight, cytosolic PLA_2s are activated by submicromolar calcium, have specificity for unsaturated fatty acids and arachidonic acid in particular *(3,4)*. The arachidonic acid released can have many direct functions or can serve as a precursor in the formation of the numerous prostaglandin and leukotriene products *(5)*. Thus, it is becoming increasingly clear that this latter class of enzymes are the principle regulators of arachidonic-acid release and therefore play an important role in signal transduction *(6)*.

As such, PLA_2s have commanded considerable attention in terms of the number and variety of assays developed in recent years *(7)*. The most common assays require the use of radiolabeled arachidonic acid, in which the arachidonic acid is an intrinsic component of the phospholipid substrate, or the arachidonic acid is incorporated into the phospholipid following long incubations. Many reports simply measure the radioactivity released as an indication of phospholipase activity. These extracts contain not only the released arachidonic acid but also include all the eicosanoid products derived from arachidonic acid. Therefore, a more accurate means of assessing enzymatic activity is to extract the radiolabeled arachidonic acid *(8)*, separate it from other prostag-

From: *Methods in Molecular Biology, vol. 108: Free Radical and Antioxidant Protocols*
Edited by: D. Armstrong © Humana Press Inc., Totowa, NJ

landin and leukotriene products, usually by thin layer chromatography (TLC), and quantitate the radioactivity specifically associated with pure standard ***(9,10)***.

Recently, a number of fluorescent assays have been reported that provide alternatives to the use of radiolabeled compounds and that provide similar levels of sensitivity and specificity ***(7,11,12)***. We have adapted a PLA_2 assay used for intact cells for use with cellular sonicates ***(13)***. Our modification is an attempt to avoid problems inherent in a two-phase system commonly used to measure PLA_2 activity and have incorporated dimethyl sulfoxide (DMSO) in order to generate a true solution phase ***(14)***. The presence of DMSO yields considerably higher levels of activity than similar assay preparations using micelles ***(15)***. The assay makes use of the fluorescent phospholipid substrate C_6-NBD-PC, which yields the C6-NBD-labeled fatty acid that is not further metabolized, nor is it incorporated into cells ***(16)***. The C6-NBD-label is covalently bonded at the C-2 position of the phospholipid and will therefore provide unambiguous specificity for the PLA_2 activity. This latter consideration is significant, because under certain conditions, $cPLA_2$ also possesses lysophospholipid activity as well as simple fatty acid esterase activity ***(12)***.

2. Materials

2.1. Instruments

1. Perkin Elmer (Norwalk, CT) 650-10S Fluorescence Spectrometer.
2. Shimadzu (Columbia, MD) UV-1201 UV-Vis Spectrophotometer.
3. Fluorescence cuvets, Quartz (4-sided), 3 mL.
4. Branson (Danbury, CT) Sonicator.
5. Low speed, tabletop centrifuge.

2.2. Reagents

1. 1-acyl-2-(N-4-nitrobenzo-2-oxo-1,3-diazole) aminocaproyl phosphatidylcholine (C_6-NBD-PC) (Avanti Polar Lipids, Birmingham, AL. Also available from Molecular Probes, Eugene, OR).
2. Caproic acid-NBD, Used as product reference standard (Avanti Polar Lipids).
3. Isopropyl ether (Aldrich, Milwaukee, WI; Caution: extremely flammable!).
4. DMSO (Fisher Scientific, Pittsburgh, PA; ACS certified.)
5. 0.5 *M* Tris buffer, pH 7.4, prepared by the addition of HCl to Tris base.
6. Isopropyl alcohol (Fisher, ACS certified)
7. Heptane (Fisher, ACS certified)

2.3. Cell Lines

The lymphoblastoid cell lines GM3299 and GM3403 were obtained from the NIGMS Human Genetic Mutant Cell Repository (Camden, NJ). Cells were cultured in RPMI-1640 medium (GIBCO, Grand Island, NY), supplemented with 10% fetal bovine serum (FBS; Hyclone, Logan, UT), and containing 100 U/mL penicillin and 50 g/μL streptomycin and 1 m*M* fresh glutamine.

2.4. Extraction Solvents and Standard

1. Isopropyl alcohol/isopropyl ether/heptane (1.2/1.5/1.0, v/v/v) is used to stop the reaction as well as to remove unreacted substrate.
2. The remaining enzymatic product (C6-NBD) was, solubilized by the addition of chloroform/methanol/saline in 0.2*N* HCl, (1: 1:0.05, v/v/v).
3. A caproic acid standard (C6-NBD) was used to calibrate the spectrophotometer during each assay.

3. Methods

3.1. Preparation of Cell Sonicate

1. Cells grown in log phase are collected, counted, and their viability determined using the trypan-blue exclusion method.
2. Cells are centrifuged at 300*g* for 5 min at room temperature, washed twice with PBS, and the pelleted material resuspended in 10 m*M* Tris buffer at pH 7.4.
3. Approximately 10^6 cells are disrupted by sonication using a Branson sonifer for two 10 s bursts at 50 watts.

3.2. Authentication of C6-NBD Product

Incubation of substrate with commercially available porcine PLA_2 or cell sonicate yields NBD-labeled fatty acid, which is extracted and quantitated by fluorescence spectroscopy. The component extracted in the organic phase co-migrates with unreacted substrate when separated by TLC and detected by exposure to either iodine vapors or to ultraviolet (UV) light. Only one reaction product is observed in the aqueous layers, which co-migrated with authentic caproic acid-NBD standard and is in agreement with a previous report ***(10,13)***. The porcine PLA_2 reaction is completely inhibited using the monoalide-analog inhibitor BMY30204 as observed on TLC plates (*see* **Note 1**).

3.3. Assay Protocol

The sonicate or membrane suspension is aliquoted (620 μL) into tubes containing substrate (2 to 14 μ*M*) in 160 μL of DMSO and 20 μL of 0.24 *M* $CaCl_2$. The assay mixture is incubated for 30 min art 37°C with mixing. The reaction is stopped by the addition of 3.0 mL aliquot of isopropyl alcohol/isopropyl ether/heptane (1.2/1.6/1.0, v/v/v) with mixing. This mixture is centrifuged for 5 min at 300*g* and the top organic layer containing unreacted substrate extract carefully removed. The extraction procedure is repeated twice and the remaining enzymatic product (C6-NBD) solubilized by the addition of chloroform/methanol/saline in 0.2 *N* HCl, (1:1:0.05, v/v/v) with mixing and the mixtures

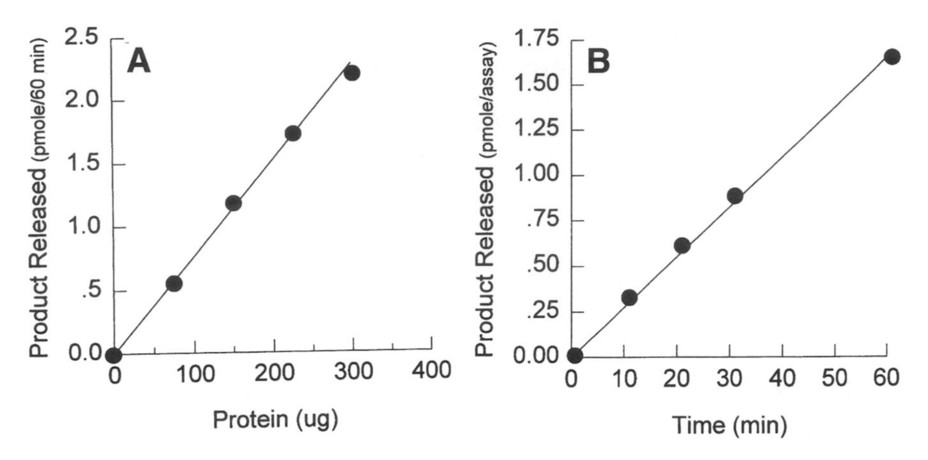

Fig. 1. The effect of protein concentration (**A**) and time (**B**) on phospholipase A_2 activity in cellular extracts derived from a human B-lymphoblastoid cell line GM3299. Phospholipase A_2 activity was in the linear range for both the time and protein concentrations used. The concentration of protein used in the time course (**Fig. 1B**) was 175 µg per assay point. This concentration of protein corresponds with approx 10^6 cells per experimental point. Each point represents the average of duplicate experiment.

centrifuged for 5 min at 300*g*. This last centrifugation step is necessary in order to pellet any remaining membrane lipid to the bottom of the tube. The lipid component is found to interfere with the fluorescence reading if not removed.

One milliliter aliquots of the bottom organic layer are pipetted from the bottom layer of the tubes, placed into a quartz cuvet and the fluorescence immediately measured at 540 nm with an excitation wavelength of 450 nm (see **Note 2**). A caproic acid standard is used to calibrate the spectrophotometer during each assay (*see* **Note 3**). The fluorescence measured is converted to picomoles of substrate released per microgram of protein, protein being determined by the Lowry method *(17)*.

3.4. Time and Protein Linearity

Figure 1A,B indicates that the assay is linear for both protein (cell sonicate) and time, respectively, under the conditions selected. We have noted considerable variability in phospholipase content in various cell types, particularly in tumor cells (unpublished data). It is therefore necessary to demonstrate linearity when first developing the assay and when changes are made.

3.5. Assay Example

Figure 2 compares the phospholipase A_2 activities in the genetic disorder Bloom's syndrome B-lymphoblastoid cell line (GM3403), control (GM3299),

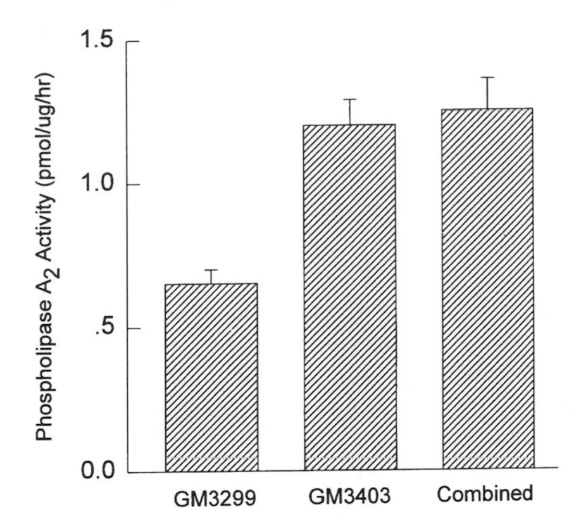

Fig. 2. Phospholipase A_2 activity in whole-cell sonicates derived from two B-lymphoblastoid cell lines and in a mixture containing equal amounts of protein from each cell line. These data not only distinguish between the enzymatic activity associated with each cell line but also implicates the presence of a PLA$_2$ activating factor present in the Bloom's syndrome cell line GM3403 but not present in the control cell line GM3299.

and an equal mixture of each (average of triplicate samples ± SE). These results clearly indicate that this assay can be useful in the quantitation of phospholipase A_2 without the incorporation of radioactive substrates.

4. Notes

1. When using inhibitors, it is of particular importance to include the inhibitor in the blank. In addition to its affect on fluorescence, one should consider differential extraction of inhibitor by the organic solvent and its potential effect on fluorescence levels.
2. The quartz cuvets used to measure fluorescence should not be washed with detergent. Any trace detergent can interfere wit]h this lipid-based assay. It is therefore preferable to acid-wash the cuvets.
3. A separate blank is required for each experimental point, which is then subtracted from the corresponding data point. This should be kept in mind when optimizing conditions such as pH or buffer concentration because fluorescence readings are sensitive even to minor changes in conditions.

References

1. Smith, W. L. (1992) Prostanoid biosynthesis and mechanisms of action. *Am. J. Physiol.* **263,** F 181–F 191.

2. Ackerman, E. J., Kempner, E. S., and Dennis, E. A. (1994) Ca^{2+}-independent cytosolic phospholipase A_2 from macrophage-like P388D_1 cells. *J. Biol. Chem.* ***269,*** 9227–9223.
3. Clark, J. D., Lin, L-L., Kriz, R. W., Ramesha, C. S., Sultzman, L. A. Lin, A. Y., Milona, N., and Knopf, J. L. (1991) A novel arachidonic acid-selective cytosolic PLA_2 contains a Ca^{++}-dependent translocation domain with homology to PKC and GAP. *Cell* **65,** 1043–1051.
4. Nemenoff, A., Winitz, S., Qian, N. X., Van Putten, V., Johnson, G. L., and Heasly, L. (1993) Phosphorylation and activation of a high molecular weight form of phospholipase A_2 by p42 microtubule-associated protein 2 kinase and protein kinase C. *J. Biol. Chem.* **268,** 1960–1994.
5. Smith, W. L. (1989) The eicosanoids and their biochemical mechanisms of action. *Biochem. J.* **259,** 315–324.
6. Shimidzu, T. and Wolfe, L. .S. (1990) Arachidonic acid and signal transduction. *J Neurol.* **55,** 1–15.21.
7. Reynolds, L. J., Washburn, W. N., Deems, R. A., and Dennis, E. A. (1991) Assay strategies and methods for phospholipases. *Methods Enzymol.* **197,** 3–23.
8. Thigh, E. G. and Dyer, W. J. (1959) A rapid method of total lipid extraction and purification. *Can. J. Biochem. Physiol.* **37,** 911–917.
9. Fischer, S. M., Patrick, K. E., Lee, M. L., and Cameron, G. S. (1991) 4β- and 4α-12-O-Tetradecanoylphorbol-13-acetate elicit: arachidonate release from epidermal cells through different mechanisms. *Cancer Res.* **51**, 850–856.
10. Wikiel, H., Zhao, L., Gessner, T., and Bloch, A. (1994) Differential effect of growth- and differentiation factors on the release of eicosanoids and phospholipids from ML-1 human myeloblastic leukemia cells. *Biochim. Biophys. Acta* **1211,** 161–170.
11. Dagan, A. and Yegdar, S. (1987) A facile method for direct determination of phospholipase A_2 in intact cells. *Biochem. Int.* **15,** 801–808.
12. Huang, Z., Laliberte, F., Tremblay, N. M., Weech, P. K., and Street, I. P. (1994) A continuous fluoresecene-based assay for the human high-molecular-weight cytosolic phospholipase A_2. *Anal. Biochem.* **222,** 110–114.
13. Nicotera, T. M. (1994) Free radical mechanisms for chromosomal instability in Bloom's syndrome, in *Free Radicals in Diagnostic Medicine: A Systems Approach to Laboratory Technology, Clinical Correlations and Antioxidant Therapy.* (Armstrong, D., ed.), Plenum, New York. *Adv. in Expt. Med. Biol.* **366,** 29–41.
14. Ballou, L. R. and Cheung, W. Y. (1983) Inhibition of human platelet phospholipase A_2 activity by unsaturated fatty acids. *Proc. Natl. Acad. Sci.USA* **80,** 5203–5207.
15. Bergers, M., Verhagen, A. R., Jongerius, M., and Mier, P. D. (1986) A new approach to the measurement of phospholipase A_2 in tissue homogenates and its application to human skin. *Biochim. Biophys. Acta* **876,** 327–332.
16. Yegdar, S., Reisfeld, N., Halle, D., and Yuli, I. (1987) Medium viscosity regulates the activity of membrane-bound and soluble phospholipase A_2. *Biochemistry* **26,** 3395–3401.
17. Lowry, O. H., Rosebrough, N. J., Farr, A. L., and Randall, R. J. (1951) Protein measurement with the folin phenol reagent. *J. Biol. Chem.* **193,** 265–275.

22

Autofluorescent Ceroid/Lipofuscin

Dazhong Yin and Ulf Brunk

1. Introduction

Autofluorescent ceroid/lipofuscin-type pigments are usually classified and defined as follows: Lipofuscin is an intracellular, age-related, fluorescent, cytoplasmic, granular pigment. It is mainly present in secondary lysosomes of post-mitotic cells, such as neurons, cardiac myocytes, and retinal pigment epithelial (RPE) cells. Ceroid is a group of biopigments, also with an intralysosomal location, which are rapidly produced as a result of various pathologies and experimental conditions, such as x-irradiation, E-vitamin deficiency, starvation, and intoxication. Ceroid is probably akin to lipofuscin and may share the same mechanisms of formation, although it does not accumulate in relation to aging. It may be present in a variety of cells, e.g., in the kidney, thymus, pancreas, testis, prostate, seminal vesicles, uterus, and adrenal gland of man and animal ***(1–4)***. Advanced glycation end-products (AGEs) are mainly found extracellularly in association with long-lived proteins, e.g., in the cataractic lens of the eye ***(5,6)***, and in cross-linked collagen of the skin, arteries, lungs, and kidneys ***(7–9)***. They are formed during complex non-enzymatic glycation-reactions involving Maillard- and Amadori-type chemistry. AGEs may also, however, be present intracellularly, and they probably constitute a substantial part of lipofuscin and ceroid. A variety of synthetic age pigment-like fluorophores (APFs) can be produced by oxidation/peroxidation of different biological materials. Most such fluorophores are badly characterized and need to be studied further before they can be compared to ceroid/lipofuscin and AGEs.

The age-related, progressive lysosomal accumulation in post-mitotic cells of the yellow-brown lipofuscin pigment, which emits yellowish-reddish fluorescence when excited with UV or blue light, is consistently recognized in

From: *Methods in Molecular Biology, vol. 108: Free Radical and Antioxidant Protocols*
Edited by: D. Armstrong © Humana Press Inc., Totowa, NJ

human and animals. This accumulation is regarded as an important hallmark of aging ***(10–13)***. There are good evidences that lipofuscin is formed within secondary lysosomes, being strongly accelerated by iron-catalyzed oxidative reactions of autophagocytosed cellular material. Critical parameters seem to be the amount of intralysosomal low-molecular-weight iron in redox-active form, degree of autophagocytosis, and influx of hydrogen peroxide to the lysosomal apparatus ***(14–16)***.

It must be pointed out that ceroid/lipofuscin greatly differs between various cell types with respect to its chemical composition, being a consequence of what type of intracellular material that mainly was autophagocytosed.

Much additional investigation is obviously needed within this complex area which is firmly related to only superficially understood processes such as cellular aging, handling of oxidative stress, and the regulation of intra-cellular turnover. Thus, there is a need not only of better understanding of ceroid/lipofuscin biochemistry, but also, and more importantly, of obtaining information about the influence, if any, on the function of the cellular acidic vacuolar compartment when loaded with ceroid/lipofuscin.

2. Methods

2.1. Experimental Formation of Ceroid/Lipofuscin Outside and Within Cells

A few techniques used to produce ceroid/lipofuscin, and some methods to quantitate and characterize such pigments, are described in this chapter, such as: The induction of ceroid/lipofuscin in cultured cells by growing them under enhanced oxidative stress, with ensuing fluorometric measurement of the formed autofluorescent pigments. Formation of artificial ceroid/lipofuscin by ultraviolet (UV)-irradiation using various cellular material as substrates, its characterization, and its use in cell culture experiments.

The study of lipofuscin, ceroid, and age pigment-like fluorophores involves different techniques. Four of those are of major importance:

1. Cytochemistry and histochemistry,
2. Transmission electron microscopy,
3. Microfluorometry, and
4. Spectrofluorometry.

The latter two methodologies are currently the most used ones, hence some comparisons between them are given.

2.1.1. Induction of Enhanced Formation of Ceroid/Lipofuscin in Cultured Cells

The culture of cells at high oxygen tension, results in enhanced accumulation of lipofuscin ***(16–19)***. The reason for this is that the increased oxygen concentration around the cells' mitochondria will allow an increased amount of superoxide anion radicals to form along the inner mitochondrial membranes. These radicals will subsequently dismutate, spontaneously or catalyzed by superoxide dismutases (SOD), and form hydrogen peroxide. The increased cytosolic concentration of hydrogen peroxide will, in turn, result in enhanced intralysosomal Fenton-chemistry, with increased lipofuscin formation as a consequence of enhanced diffusion of hydrogen peroxide into the acidic vacuolar apparatus. Thus, the growth of cells in culture at 40% ambient oxygen is an efficient way of obtaining lipofuscin-loaded cells much quicker than otherwise would be possible. It should be understood, however, that postmitotic cells, such as myocardial cells, will accumulate lipofuscin in culture much quicker even when grown under a gas phase of the 21% oxygen that is normal for the atmosphere, than under conditions in vivo where the oxygen partial pressure may be only 5–8%.

2.1.2. Preparation of Artificial Ceroid/Lipofuscin and Its Uptake by Cultured Cells

1. Subcellular fractions, and noncellular biomaterials, e.g., collagen and elastin, may be used to produce artificial ceroid/lipofuscin. Such material can be obtained using common procedures for cell and tissue separation.
2. The biomaterials are finally suspended in phosphate buffer and the protein concentration adjusted to a level of 3.0 mg protein/mL.
3. The suspensions are then transferred to open plastic Petri dishes, without lids.
4. The suspension is exposed overnight to UV light, preferable in a sterile laminar air hood.
5. The material under irradiation is usually completely dried up after a few hours, owing to the ventilation in the box, and converted to a yellowish gel.
6. This material is re-suspended by the addition of distilled water to regain the initial volume.
7. Intensive sonification is often needed to re-suspend the yellowish aggregate.
8. When the material is studied in the fluorescence microscope, normal autofluorescence of lipofuscin-type is observed, and in the transmission electron microscope, it has the ordinary lipofuscin-like character of an osmiophilic amorphous mass (**Fig. 1**).
9. Artificial, mature, and well-polymerized lipofuscin may be given to a variety of cells in culture, and then become endocytosed by them. Granules of artificial lipofuscin (GAL) may be fed to cells by the addition of a suitable amount of stock solution (*see* **Subheading 3.3.2.**) to their growth medium. Phagocytotically very active cells, such as macrophages and RPE cells, may be given relatively large conglomerates, whereas less endocytotic active cells, such as fibroblast and myocardial cells, require the granules to be small. By homogenization and/or sonification, a suitable GAL size may be obtained. As a result of the uptake process, GAL will be accumulated in the cells' secondary lysosomes, many of

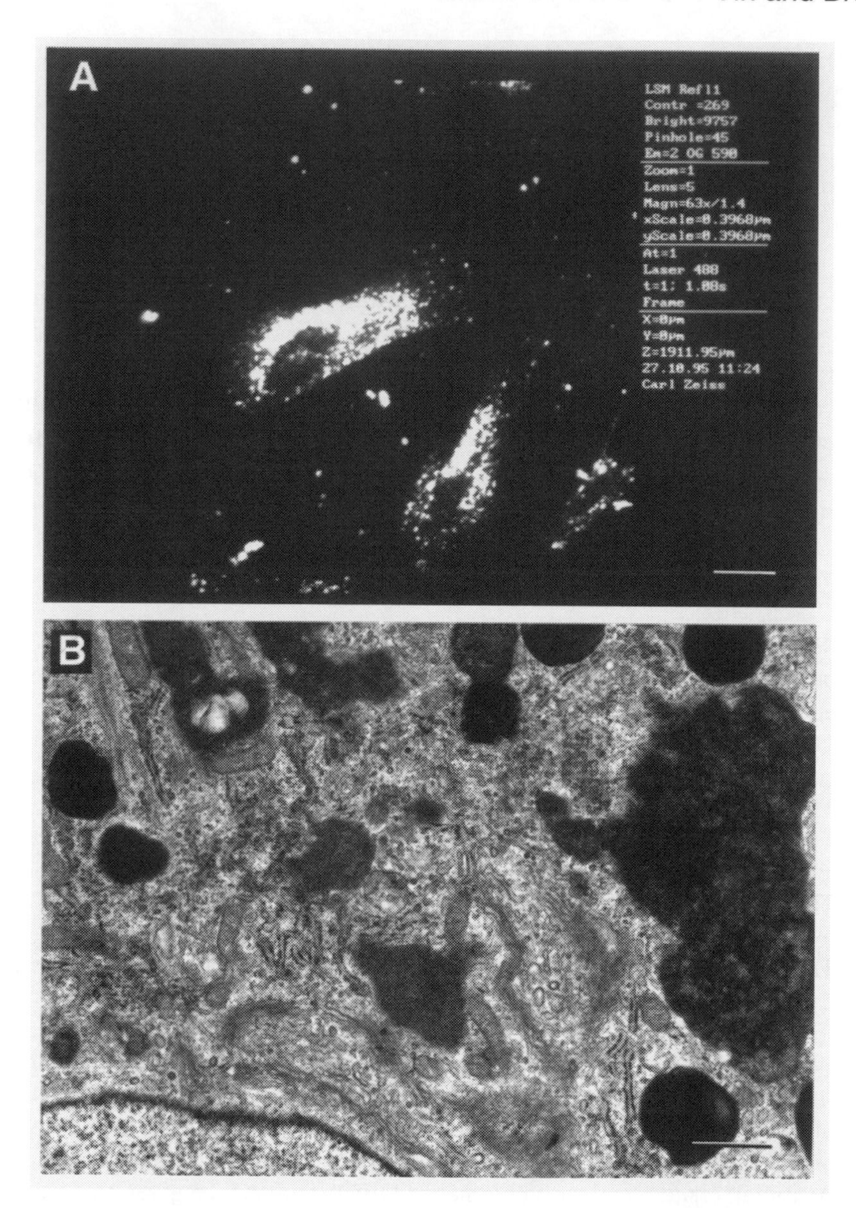

Fig. 1. Artificial lipofuscin is endocytosed by various cells in culture, converting them into an "old" phenotype. (**A**) Fibroblasts as seen in a Zeiss inverted confocal laser scanning microscope. (**B**) Part of a retinal pigment epithelial (RPE) cell as visualized in a Jeol 2000 TEM. Endocytosed, autofluorescent, osmiophilic lipofuscin-like material is located within distended lysosomes in both types of cells. Bars are 25 and 1 µm, respectively.

which, thus, will be transformed to lipofuscin-loaded "residual bodies," transforming the cells into a phenotype that is characteristic of old postmitotic cells *(20,21)*. Endocytotic active cells undergo this transformation rather quickly, or within a few days time, whereas cells that endocytose slowly may require more than a week to obtain a maximal accumulation, especially if the GAL are not very small.

10. Using artificial lipofuscin to quickly convert "young" cells in culture to "aged" lipofuscin-loaded ones will allow the study of whether, and how, a lipofuscin-loaded cellular acidic vacuolar apparatus influences cell metabolism. The only difference between such cells would then be their different contents of lysosomal lipofuscin (*see* **Note 1**).

2.2. Microfluorometric Studies of Ceroid/Lipofuscin

Early studies on lipofuscin/ceroid were mainly performed using morphological and histochemical techniques. In the light (bright field) microscope lipofuscin within postmitotic cells, such as neurones and cardiac myocytes, was observed as yellowish-brown, irregularly shaped pigment granules with a distribution and size corresponding to secondary lysosomes *(2,10–12,22)*. Later, the occurrence of lipofuscin within secondary lysosomes was confirmed by electron microscopy *(22)*. Ceroid/lipofuscin shows a characteristic autofluorescence that can be used for its characterization, and for measurement of its amount as well.

2.2.1. Selection of Filter Combinations for Microfluometry

1. In early lipofuscin studies, excitation with UV-light was often applied. However, by that practice a relatively weak lipofuscin autofluorescence is obtained and native protein fluorescence can sometimes cause disturbances.
2. At present, most frequently used conditions for excitation and emission during microfluorometry are excitation with visible blue light (390–490 nm) in combination with a 510–560 nm emission barrier filter. By the use of differently sized masks the amount of lipofuscin in whole cells, or their parts, may be measured.
3. By using barrier filters up to 560 nm is possible to avoid almost all unspecific autofluorescence, even in living cells, while obtaining lipofuscin-specific autofluorescence only.
4. The standardization is done against a uranyl standard, and the lipofuscin amounts are expressed in arbitrary units (AU).
5. By measuring 100–200 cells in a static, computer-based fluometric system, histograms and statistics can easily be obtained.
6. A comparison of fluorometric and microfluorometric spectral data is sometimes necessary. Significant differences between the outcome of these two techniques are given in **Note 2**.

2.3. Spectrofluorometric Studies of Ceroid/Lipofuscin

Spectrofluorometric studies of ceroid/lipofuscin and APFs often require isolated and dissolved pigment material. Extensive studies on the separation of lipofuscin were performed by Björkerud and Siakotos ***(23,24)***. Many difficulties were encountered following attempts to isolate and dissolve lipofuscin pigments ***(24,25)***. Resistance to extraction by different solvents was often found to be enhanced by increased "maturation" of the lipofuscin ***(26)***. A variety of extraction techniques using water, chloroform/methanol, ethanol, ethanol/diethylester, sodium dodecylsulfate (SDS), etc. have been documented in the literature ***(27–31)***. Chloroform/methanol (2:1) is so far the most popular solvent for lipofuscin extraction ***(27,29,32,33)***. However, the use of SDS, having the advantage of also dissolving protein-derived compounds, recently has drawn much attention. The solubility and extractability of lipofuscinand APFs in water and organic solvent has been comprehensively reviewed ***(34,35)***.

2.3.1. Influence of Inner Filter Effects and Other Conditions

1. Cuvet selection: Using plastic cuvets is convenient for the fluorometric assay of many samples and avoids contamination. However, such cuvet often absorbs light under 300 nm. Precise absorption information of a plastic cuvets itself can be easily obtained by scanning its absorption spectrum using a spectrophotometer. Many organic solvents, such as chloroform, tend to dissolve plastic cuvet and thus cause problems. Quartz cuvets are required when one intends to get an overview of the whole fluorescence spectrum, particularly when absorption and fluorescence of proteins are of interest.
2. Absorption and autofluorescence of proteins: Estimation of protein absorption (at 280 nm) and fluorescence (ex/em 280/335 nm) are both excellent checkingstones, but a major disturbance during fluorometric assays when fluorophores of cellular fractions are studied. Careful discrimination of the possible overlap of fluorescence between protein and lipofuscin-like fluorophores is critical in order to obtain a true characterization of the biopigments.
3. Quenching effects: Fluorophores at high concentrations cause inner-filter effects (*see* **Note 3**). The strongest quenching can be seen at the wavelength of the absorption peak. A relatively lower excitation maximum is often encountered mainly owing to such quenching. To obtain true fluorescence spectra, a preliminary control of the absorption of the solution, or the sample suspension, is required. For a clear solution without apparent turbidity (turbidity often causes absorption over the whole spectrum), its dilution enough to obtain absorption peak values <0.05 is suggested.
4. Turbidity and scattered light: Even slight turbidity disturbs absorption spectral studies. Turbidity is less dangerous in studies of fluorescence spectra as long as the scatter peaks are distinguishable from the fluorescent maxima. Main scatter peaks are Rayleigh scatters, which appear right at the excitation wavelength; and Raman scatters, which appear at about 40–45 nm higher wavelength in aqueous solutions. Molecular aggregates in suspension often enhance scatter lines (peaks).

Owing to weak fluorescence of biopigments, Raman lines can become a significant disturbance when further signal amplification is needed. When excitation and emission peaks are recorded during scanning of spectra, there are second-order Rayleigh scatters appearing in both excitation and emission scanning spectra at twice or half values of the wavelength of the real peaks. Biopigments also have secondary peaks. When protein samples are excited at 280 nm, one can get a first fluorescent, and real, peak at 335 nm and a secondary, artificial one, at 670 nm (the intensity of secondary peak is usually lower). Such false fluorescent peaks sometimes cause misunderstanding and formed owing to a grating system need to be discriminated during spectral studies.

2.3.2. Estimation of Excitation/Emission Maxima

1. Quinine sulfate standard: Fluorescence intensity is often reported in arbitrary units. In order to compare results of different experiments and from different laboratories, the fluorescence intensity is recommended to be related to a quinine sulfate standard. When a quinine solution is excited at 350 nm, a fluorescence spectrum with an emission maximum at 450 nm is obtained. When the emission monochromator then is set at 450 nm, an excitation spectrum with a fluorescent maximum at 350 nm is obtained. The intensity yield of the excitation/emission at a certain slit-width is used for calculation. Frequent checks of quinine-standard values are helpful to avoid instrumental defects.
2. Spectral correction: Spectral bias, owing to different photon sensitivities of the photomultipliers of presently available fluorometers, are significant at wavelengths >500 nm. When lipofuscin-like pigments are assayed, having fluorescent maxima in the blue region, spectral correction is not recommended. An unsophisticated spectral correction often induces artifactual values (*see* **Note 4**).
3. Location of excitation/emission fluorescence maxima: To obtain a general idea about the fluorescent regions of lipofuscin/ceroid, stepwise procedures are sometimes needed. Using the absorption maximum as the primary excitation wavelength is a theoretically correct way, but it may not always be practical. Setting the excitation at 350 nm, while scanning the emission spectrum, may be a useful way to start up the search of fluorescence maxima of ceroid/lipofuscin. When such an emission maximum is then observed, the emission monochromator can be adjusted to the obtained emission peak to allow the scanning of the excitation spectrum. When an excitation maximum is subsequently obtained, the process is repeated until a satisfactory spectrum has been registered. If several peaks (including shoulders) are obtained, they should be checked one by one using the same setting-scanning procedure. Fluorescent biopigments are often a mixture of many different fluorophores. Chromatographic separations, such as thin layer chromatography (TLC) or high-pressure liquid chromatography (HPLC), are helpful to gain insight into their different components.
4. Dealing with overlaps: When fluorescent peaks overlap, the use of a very narrow slit bandwidth can be helpful. This technique is also useful when the excitation

Table 1
Differences of the Microfluorometric and Spectrofluorometric Techniques

	Materials under study	Accuracy of spectral data	Light path-length	Special advantages
Microfluorometry	solid (concentrated)	distorted, or limited, by filters used	microns	visible fluorescence
Spectrofluorometry	solutions (very diluted)	accurate	mm – cm	quantitative and sensitive

and emission maxima are very close. Syncronic scanning may also be used to avoid the overlap of excitation- and emission spectra.

3. Notes

1. Artificial lipofuscin may be produced from a variety of biological materials, including lipids, proteins, carbohydrates, carotenoids, ascorbic acid, and possibly nucleic acids. Oxidative stress strongly accelerates their formation. The crosslinking between carbonyls and amino groups, as recently reviewed by us ***(13)***, probably represents a process common to the formation of most, if not all, ceroid-lipofuscin materials.
2. The technical differences between the microfluorometric and spectrofluorometric procedures are often overlooked, although the differences are important. As shown in **Table 1** lipofuscin within intact cells and tissues, observed in the fluorescence microscope and measured by microfluorometry, usually is in the form of tiny pigment granules, condensed within secondary lysosomes, whereas extracted, and much diluted, fluorophores are measured by spectrofluorometry.
3. Concentration-dependent fluorescence shift is usually not observable during microfluorometry using an epi-illumination system, where the effective light path-lengths are within the order of a few microns. Quenching, or inner filtering, effects only occurs at very high fluorophore concentrations. When lipofuscin/ceroid is accumulated to excessive concentrations within secondary lysosomes, the fluorescence red-shift, however, may become substantial even by epifluorometry. When accumulated in high concentration, and/or mixed with other light-absorbing materials, ceroid/lipofuscin fluorophores would not exhibit their original and individual fluorescence maxima that, typically, are found at low concentrations. The occurrence of quenching-related fluorescence shifts should be taken into consideration when comparing in vitro biochemical with *in situ* cytochemical findings.
4. Although lipofuscin generally shows yellowish autofluorescence when studied *in situ* by microfluorometric techniques, the fluorescence maxima of lipofuscin-

extracts when studied by spectrofluorometric methods, are consistently reported to be in the blue (400–500 nm) region reviewed in **ref. *13***. Such blue emissions are often also obtained from fluorophores derived from various biological model systems, as long as retinoids and carotenoids are not included. The practice of spectral correction, as suggested by Eldred and Katz *(36)*, does not resolve the aforementioned fluorescence discrepancies, as long as ceroid/lipofuscin from sources outside retinal pigment epithelial cells is concerned. An alternative explanation, based on concentration-dependent fluorescence shift, of the aforementioned discrepant fluorescent data for ceroid/lipofuscin was recently suggested by Yin and Brunk *(37)*.

References

1. Strehler, B. L. (ed.) (1977) *Time, Cells, and Aging*. 2nd ed. Academic Press, New York.
2. Ordy, J. M. and Brizzee, K. R. (eds.) (1975) *Neurobiology of Aging*. Plenum Press, New York.
3. Feeney, L. (1978) Lipofuscin and melanin of human retinal pigment epithelium. *Invest. Ophthal. Vis. Sci.* **7,** 583–600.
4. Miquel, J., Oro, J., Bensch, K. G., and Johnson, J. E. (1977) Lipofuscin: fine-structural and biochemical studies, in *Free Radicals in Biology Vol. III* (Pryor, W. A., ed.), Academic Press, New York, pp. 133–182
5. Duncan, G. (ed.) (1981) *Mechanisms of Cataract Formation in the Human Lens*. Academic Press, New York.
6. Monnier, V. M. and Cerami, A. (1983) Detection of nonenzymatic browning products in the human lens. *Biochim. Biophys. Acta* **760,** 97–103.
7. Verzar, F. (1957) The ageing of connective tissue. *Gerontologia* **1**, 363–378.
8. Monnier, V. M. and Cerami, A. (1983) Nonenzymatic glycosylation and browning of proteins in vivo. in *The Maillard Reaction in Foods and Nutrition* (Waller, G. R. and Feather, M. S., eds.), American Chemical Society, Washington DC, pp. 431–449.
9. Baynes, J. W. and Monnier, V. M. (eds.) (1989) *The Maillard Reaction in Aging, Diabetes, and Nutrition*. Alan R. Liss, Inc., NY.
10. Strehler, B. L. (1964) On the histochemistry and ultrastructure of age pigment, in *Advances in Gerontological Research* (Strehler, B. L., ed.), Academic Press, New York, pp. 343–384.
11. Porta, E. A. and Hartroft, W. S. (1969) Lipid pigments in relation to aging and dietary factors, in *Pigments in Pathology* (Wolman, W., ed.), Academic Press, NY, pp. 191–235.
12. Sohal, R. (ed.) (1981) *Age Pigments*. Elsevier, Amsterdam.
13. Yin, D. (1996) Biochemical basis of lipofuscin, ceroid, and age pigment-like fluorophores. *Free Rad. Biol. Med.* **21,** 871–888.
14. Brunk, U. T. and Sohal, R. S. (1991) Mechanisms of lipofuscin formation, in *Membrane Lipid Oxidation vol. II* (Vigo-Pelfrey, C., ed.), CRC Press, Boca Raton, FL, pp. 191–201.

15. Brunk, U. T., Marzabadi, M. R. and Jones, C. B. (1992) Lipofuscin, lysosomes and iron, in *Iron and Human Disease* (Lauffer, R. B., ed.), CRC Press, Boca Raton, FL, pp. 237–260.
16. Brunk, U. T., Jones, C. B., and Sohal, R. S. (1992) A novel hypothesis of lipofuscinogenesis and cellular aging based on interactions between oxidative stress and autophagocytosis. *Mut. Res.* **275,** 395–403.
17. Sohal, R. S. and Brunk, U. T. (1990) Lipofuscin as an indicator of oxidative stress of aging, in *Lipofuscin and Ceroid Pigments* (Porta, E. ed.), Plenum, New York. Adv. Exp. Med. Biol. Vol. 226,. pp. 17–26.
18. Sohal, R. S. and Brunk, U. T. (1990) Mitochondrial production of proxidants and cellular senescence. *Mut. Res.* **275,** 295–304.
19. Marzabadi, M. R. Yin, D., and Brunk, U. T. (1992) Lipofuscin in a model system of cultured cardiac myocytes, in *Free Radicals and Aging* (Emerit, I. And Chance, B., eds.), Birkhäuser Verlag, Basel Switzerland, pp. 78–88.
20. Brunk, U. T. Wihlmark, U., Wrigstad, A., Roberg, K., and Nilsson, S-E. (1995) Accumulation of lipofuscin within retinal pigment epithelial cells results in enhanced sensitivity to photo-oxidation. *Gerontology* **41 (suppl 2),** 201–212.
21. Yin, D. and Nilsson, E. (1997) The preparation of artificial ceroid/lipofuscin by UV-oxidation of subcellular organelles. *Mech. Ageing Dev.* **99,** 61–78.
22. Brunk, U. T. and Ericsson, J. L. E. (1972) Electron microscopical studies on rat brain neurons: localization of acid phosphatase and mode of formation of lipofuscin bodies. *J. Ultrastruct. Res.* **38,** 1–15.
23. Björkerud, S. (1964) Isolated lipofuscin granules– a survey of a new field. *Adv. Gerontol. Res.* **1,** 257–288.
24. Siakotos, A. N. and Koppang, N. (1973) Procedures for the isolation of lipopigments from brain, heart and liver, and their properties: a review. *Mech. Ageing Dev.* **2,** 177–200.
25. Elleder, M. (1981) Chemical characterization of age pigment, in *Age Pigments* (Sohal, R. S., ed.), Elsevier, Amsterdam, pp. 204–241.
26. Patro, N., Patro, I. K., and Mathur, R. (1993) Changes in the properties of cardiac lipofuscin with age and environmental manipulation. *Asian J. Exp. Sci.* **7,** 57–60.
27. Fletcher, B. L., Dillard, C. J., and Tappel, A. L. (1973) Measurement of fluorescent lipid peroxidation products in biological systems and tissues. *Anal. Biochem.* **52,** 1–9.
28. Desai, I. D., Fletcher, B. L., and Tappel, A. L. (1975) Fluorescent pigments from uterus of vitamin E deficient rats. *Lipids* **10,** 307–309.
29. Csallany, A. S. and Ayaz, K. L. (1976) Quantitative determination of organic solvent soluble lipofuscin pigments in tissues. *Lipids* **11,** 412–417.
30. Maeba, R., Shimasaki, H., Ueta, N., and Inoue, K. (1990) Accumulation of ceroid-like pigments in macrophages cultured with phosphatidylcholine liposomes in vitro. *Biochim. Biophys. Acta* **1042,** 287–293.
31. Kikugawa, K., Kato, T., Yamaki, S., and Kasai, H. (1994) Examination of the extraction methods and re-evaluation of blue fluorescence generated in rat tissues in situ. *Biol. Pharm. Bull.* **17,** 9–15.

32. Ettershank, G., Macdonnell, I., and Croft, R. (1983) The accumulation of age pigment by the fleshfly *sarcophage bullata* parker (diptera: Sarcophagidae). *Aust. J. Zoology* **31,** 131–138.
33. Ettershank, G. (1984) A new approach to the assessment of longevity in the Antarctic krill *euphausia superba. J. Crustac. Biol.* **4,** 295–305.
34. Tsuchida, M., Miura, T., and Aibara, K. (1987) Lipofuscin and lipofuscin-like substances. *Chem. Phys. Lipids* **44,** 297–325.
35. Porta, E. A., Mower, H. F., Moroye, M., Lee, C., and Palimbo, N. E. (1988) Differential features between lipofuscin (age pigment) and various experimentally produced ceroid pigments, in *Lipofuscin-1987: State of the Art* (Zs.-Nagy, I., ed.), Excerpta Medica, Amsterdam,. pp. 341–374.
36. Eldred, G. E. and Katz, M. L. (1989) The autofluorescent products of lipid peroxidation may not be lipofuscin-like. *Free Rad. Biol. Med.* **7,** 157–163.
37. Yin, D. and Brunk, U. T. (1991) Microfluorometric and fluorometric lipofuscin spectral discrepancies: A concentration dependent metachromatic effect? *Mech. Ageing Dev.* **59,** 95–109.

23

In Vivo Technique for Autofluorescent Lipopigments

François C. Delori and C. Kathleen Dorey

1. Introduction

Lipofuscin—a generic term applied to autofluorescent lipopigment—is a mixture of protein and lipid that accumulates in most aging cells, particularly those involved in high lipid turnover (e.g., the adrenal medula) or phagocytosis of other cell types (e.g., the retinal pigment epithelium or RPE; macrophage).

1.1. Lipofuscin and Oxidative Damage

The content of the lipofuscin granule is widely understood as the accumulated undigested consequence of prior oxidative damage to lipids and proteins *(1,2)*. Oxidative damage is of particular concern in the eye, because the photoreceptors function in an environment of high tissuc oxygcn, photosensitizers (rhodopsin), and sunlight. The photosensitive outer segments of the photoreceptors are particularly vulnerable because they contain both the highest concentration of photosensitizer in the body, and the highest concentration of the easily oxidized polyunsaturated fatty acids.

These highly fragile tissues are rapidly replaced; a new outer segment is generated approx every 10–12 d by addition of new material from the proximal side. The consequent constant elongation is balanced by the daily removal of the distal tips by the RPE cells). These tips are ingested, rapidly digested, and cleared from the system. However, prior oxidative damage slows the digestive process, and undigested outer segment byproducts build up in the lysosomes of the RPE cells. These accumulations, called lipofuscin, increase over time; in an aging cell they may account for up to 25% of the cytoplasmic volume not occupied by mitochondria or the nucleus *(3)*.

From: *Methods in Molecular Biology, vol. 108: Free Radical and Antioxidant Protocols*
Edited by: D. Armstrong © Humana Press Inc., Totowa, NJ

1.2. Lipofuscin and Macular Degenerations

The large quantities of lipofuscin seen in the aging macular RPE have led many to suggest that its consequence might be impaired RPE-cell function and that lipofuscin may be involved in the early events leading to age-related macular degeneration (AMD) ***(3–8)***, the most common cause of blindness in Caucasians over age 65. These concepts are supported by the finding that elevated levels of lipofuscin in donor eyes are associated with decreased numbers of photoreceptors ***(9)***, and by the striking parallels in the age-relation and topography of lipofuscin in donor eyes and AMD ***(7)***. Excessive accumulation of fluorescent lipopigments have also been implicated in the pathogenesis of several inherited retinal degenerations (e.g., Stargardt's disease), where pathology has been linked to the progressive accumulation of materials in the lysosomes ***(10–13)***.

1.3. Noninvasive Measurement of Lipofuscin

Because factors modulating oxidative damage (e.g., smoking, dietary antioxidants, and dietary lutein and zeaxanthin) are significant risk factors for AMD ***(14)***, it has become important to evaluate the relationship between lipofuscin (a cumulative marker for oxidative damage) and risk for, and progression of, AMD in the living human eye.

Ex-vivo studies in donor eyes have shown that lipofuscin fluoresces optimally under excitation at 365 nm (range: 300–500 nm) and emits over a broad range between 450 and 800 nm with a maximum at 570–610 nm ***(6,15,16)***. The ability to excite lipofuscin with visible light (>400 nm) makes it feasible to elicit this fluorescence in vivo ***(17)***. We have developed a retinal fluorophotometer, which is described in this chapter, for noninvasive measurement of the intrinsic retinal fluorescence ***(18)***.

With this noninvasive method in human eyes, we have shown ***(19)*** that fluorescence emitted from the retina originates behind the photoreceptors and in front of the choroid. This places the source of the fluorescence in the RPE or its underlying acellular matrix, the Bruch's membrane. We have further demonstrated that the dominant fluorophore exhibits the same spectral characteristics, topographic distribution, and age relationship as shown for RPE lipofuscin in donor eyes.

The following describes the instrumentation and procedures we developed to measure RPE lipofuscin in vivo, to accurately account for individual variation in absorbers and other fluorophores within the eye, and to distinguish the spectral signature of lipofuscin (*see* **Note 1**). These procedures are now being utilized to study factors that modulate accumulation of lipofuscin in vivo, to examine the relationship between lipofuscin accumulation and pathology in

AMD, and to determine the relationship between amount of lipopigment present in some inherited disorders and the rate of visual loss.

2. Materials

2.1. Instrumentation

1. The Fundus Spectrophotometer (FSP; **ref.** *18)* comprises several integrated light paths called optical channels (fluorescence excitation, fluorescence detection, retinal illumination, retinal observation, and pupil alignment).
2. These channels are relayed to the subject's eye by the coupling optics and the central optical system (**Fig. 1**).
3. Each channel is associated with a distinct area within the dilated pupil (**Fig. 2A**): light enters and exits the eye in separate areas, insuring accurate excitation and detection of the fluorescence, and minimizing the contributions of light reflected and scattered by the cornea and crystalline lens.

2.1.1. Excitation

A 150 W Xenon lamp, an IR filter, and one of eight excitation filters (*see* **Subheading 2.1.3.**) are used to provide excitation between 430 and 570 nm. The filters are mounted on a motorized filter wheel for rapid recording of excitation spectra.

The retinal excitation field is a circle 890 μ*M* in diameter defined by an aperture imaged on the retina via the central optical system and the optics of the eye. The duration of excitation (180 ms) is controlled by a shutter.

2.1.2. Detection

1. The excitation light enters the eye in one side of the subject's pupil, and retinal fluorescence and reflected light exits the eye through a slit-shaped area on the opposite side (**Fig. 2A**). The retinal-sampling field is defined an aperture located in a plane conjugate to the retina and thus confocal to excitation aperture. The sampling field is a circular area, 585 μ*M* in diameter, centered within to the excitation field (**Fig. 2B**).
2. Scattered and reflected excitation light is rejected by one of several blocking filters (*see* **Subheading 2.1.3.**) mounted on a second motorized filter wheel.
3. The fluorescence is focused on the entrance slit of an f/3.8 Czerny-Turner monochromator (Jarrell-Ash, MonoSpec 18, Franklin, MA). The entrance slit is conjugated to the subject's pupil.
4. The dispersed spectrum is measured by an optical multichannel analyzer (Princeton Applied Research, OMA Model II, Princeton, NJ) consisting of an image intensifier (type 1455R, multi-alkali photocathode), a fiber-optic coupling plate, and a 512-channel diode-array detector, cooled to 10°C (*see* **Note 2**). Integration by the diode array occurs simultaneously in all channels for a duration of 180 ms. The spectral range is360–900 nm (512 channels, 1.05 nm/channel) and the spectral resolution is 6 nm.

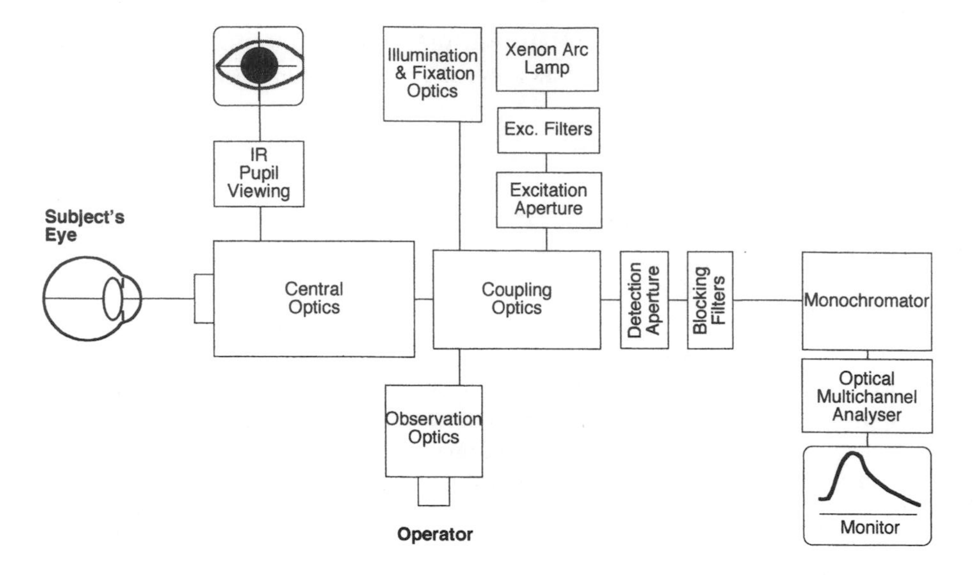

Fig. 1. Schematic diagram of the Fundus Spectrophotometer (FSP). The excitation, detection, illumination, and observation channels are combined by the coupling optics (prisms, mirrors).

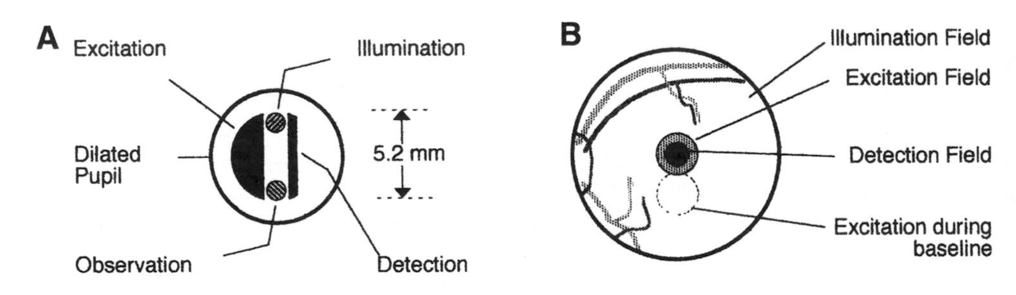

Fig. 2. (**A**) Pupil configuration: Configuration of the different channels in the subject's pupil. (**B**) Retinal configuration: Configuration of the excitation and detection fields on the retina.

2.1.3. Excitation and Blocking Filters

As in all epifluorescence measurements, the selection of the excitation and blocking filters is critical for adequate detection of the weak retinal fluorescence. Any leakage of reflected excitation light through the blocking filter will spuriously contribute to the fluorescence signal. Retinal fluorescence is about 100 times weaker than reflected blue light, and about 1000 times less that in red light. Excitation filters are narrow band interference filters (Omega Optical,

Brattleboro, VT) with center wavelengths 430, 450, 470, 490, 510, 530, 550, 570 nm (halfwidths: ≈20 nm).

The blocking filters are individually matched to each excitation filter using high pass absorbing filters (GG, KV, OG, and RG; Schott Duryea, PA). Transmissions of all filters are less than 10^{-5} at all wavelengths outside their main transmission band. Combined throughput for all filters pairs is less than 10^{-4} in the excitation bands and less than 5×10^{-6} in the spectral range of fluorescence emission.

2.1.4. Correction for Lens Fluorescence. Baseline

The crystalline lens exhibits fluorescence from tryptophan (excitation: 250–300 nm, peak emission: ≈300 nm) as well as from accumulating oxidized proteins (excitation: 400–600 nm, peak emission: ≈520 nm). This fluorescence can contribute to the measured retinal fluorescence by multiple scattering within the lens and/or by retinal reflection in the sampling field. To correct for these contributions, the excitation field on the retina is optically displaced outside of the detection field (**Fig. 2B**) without changing the volume of irradiated crystalline lens.

The detection field remains in its original position. The detected spectrum then represents lens fluorescence scattered within the lens and reflected by the retina in the detection field. This baseline signal also includes instrumental fluorescence and dark/leakage current in the detector. The baseline spectrum is subtracted from the fluorescence spectrum (excitation and detection optically superimposed) to yield the emission spectrum.

2.1.5. Retinal-Reflectance Measurements

In addition to fluorescence measurements, the FSP is also used to measure the retinal-reflectance spectrum. This spectrum give information on pigments, and is particularly important in estimating the absorption of the ocular media (*see* **Subheading 3.3.**).

One position of the excitation filter wheel is occupied by a 2.5-ND filter, providing incident white light, and the corresponding position on the blocking filter wheel is open. The measured spectrum is the spectral distribution of the reflected light, which, after accounting for the distribution of the incident light, yields the reflectance spectrum. A baseline spectrum is also acquired in this measurement.

2.1.6. Illumination and Observation, Fixation, and Pupil Alignment

1. Direct observation of the retina with the FSP makes it possible to obtain fluorescence spectra from particular retinal features. Illumination over a 4.5-mm diameter retinal field is provided by a separate light source (540–620 nm).

2. The observation optics incorporate a reticule for alignment of the fluorescence sampling field on the retinal field of interest. By directing the subject's gaze with a fixation target (668 nm, laser diode) the area of interest is guided into the sampling field.
3. Accurate focusing of the confocal optical system is achieved by projecting images of the excitation (attenuated) and detection apertures on the retina. Focus is achieved when the images of the excitation and detection fields are concentric. Alignment of subject's pupil to the FSP optical axis is achieved continuously under infrared (IR) illumination with IR video camera. A reticule on the video monitor allows for the measurement of the pupil diameter.

2.1.7. Electronics and Microcomputers

The FSP is operated under microcomputer control. Following initiation by the operator, a single spectrum is acquired by sequential activation of various filter wheels, shutters, digital logic circuits, and the multichannel analyzer. The system is monitored for proper and safe operation, and all relevant parameters (gain, fixation position, filter combination, and so on) are recorded.

The FSP operation is menu-driven, allowing selection of different protocols for retinal measurements, system calibration, calibration monitoring procedures (*see* **Subheading 2.3.2.**), and file management.

2.2. Light Safety

Retinal light exposures at each excitation wavelength are typically 6–8 mJ/cm^2 (exposure time: 0.18 s), which is 1.5% of the maximum permissible exposure recommended by the American National Standards Institute (ANSI) 136.1–1993 light safety standards. Observation and focusing exposures are also well below the ANSI standard.

Protocol for each retinal site involves exposures from 5 excitations, as well as from the other light sources. The total retinal energy density is then about 50 mJ/cm^2 (delivered within 5 min), which is 10% of the maximum permissible exposure.

2.3. Calibration and Data Reduction

2.3.1. System Calibration

General calibration of the FSP, performed every 3–4 mo, consists of:

1. Measurement of the excitation energies ($Q_{ex,\Lambda}$) for each excitation wavelength (Λ) using a calibrated photometer.
2. Calibration of the spectral sensitivity (S_λ) for all emission wavelengths (λ): A standard lamp calibrated for spectral irradiance (OL220M, Optronics Labs, Silver Spring, MD) illuminates a $BaSO_4$ standard reflector located on the FSP's optical axis. FSP recordings of the known spectral radiance allow derivation of the sensitivity. The relative sensitivity S_λ/S_{585} is derived (*see* **Subheading 2.3.2.**).

3. Determination of the transmission ($T_{b,\lambda}$) of the blocking filters by repeating the aforementioned standard lamp measurement with and without filters.
4. Determination of the spectral distribution of the radiant energy ($Q_{re,\lambda}$) of the incident white light used in reflectance measurements, using the $BaSO_4$ standard reflector.
5. Wavelength calibration using a HgNe standard spectral lamp.
6. Calibration of the intensifier gain (G) at different settings.
7. Determination of all conversion factors of the electro-optical sensors that are used in the calibration monitoring of the system, described in **Subheading 2.3.2.**

2.3.2. Calibration Monitoring

The above calibration procedure is time consuming (1 d) and does not account for progressive change that can occur in the system between calibrations, including oxidation of the filters, change in the output of the xenon lamp, changes in alignment, and so on. To monitor daily variations, a simplified version of this calibration was developed. This procedure, performed entirely under computer control after each measurement session, includes the measurements of:

1. The excitation energies ($Q_{ex,\Lambda}$) at all excitation wavelengths.
2. The spectral sensitivity at 585 nm (S_{585}).
3. The distribution of the incident energy distribution ($Q_{re,\lambda}$) used in the reflectance measurement.

2.3.3. Fluorescence and Data Reduction

Retinal fluorescence $F(\Lambda,\lambda)$ is defined as the spectral radiant energy emitted at an emission wavelength λ in a solid angle of 1 steradian (sr), for an excitation energy of 1 J at the excitation wavelength Λ (units for $F(\Lambda,\lambda)$: nJ.nm-1.sr–1 /J). The fluorescence $F(\Lambda,\lambda)$ is calculated from the optical multichannel analyzer signal (OMA_λ, in counts.pixel^{-1}) and its baseline ($OMA_{\lambda,bas}$, *see* **Subheading 2.1.4**) using:

$$F(\Lambda,\lambda) = K_{opt} \cdot n^2 \cdot f_e^2 \cdot \frac{(OMA_\lambda - OMA_{\lambda,bas})}{G \cdot Q_{ex,\Lambda} \cdot T_{b,\lambda} \cdot S_\lambda} \qquad (1)$$

with:

- K_{opt} instrumental constant (K_{opt} = 95.5 pixel.cm^{-2}.sr^{-1}.nm^{-1}).
- n refractive index of the vitreous (n = 1.336).
- f_e anterior focal length of the eye (f_e = 1.68 cm, emmetropic eye).
- G gain of the image intensifier.
- $Q_{ex,\Lambda}$ energy (J) entering the pupil for excitation wavelength Λ.
- $T_{b,\lambda}$ transmission of the blocking filter at λ.
- S_λ sensitivity of the detecting system at λ for $G = 1$ (OMA counts.J^{-1}).

The energies $Q_{ex,\Lambda}$ and the spectra $T_{b,\lambda}$ and S_{λ} are obtained from system calibration (*see* **Subheading 2.3.1.**). Note that $F(\Lambda,\lambda)$ is the fluorescence determined through the ocular media; correction for absorption by the lens is derived from the reflectance data (*see* **Subheading 3.3.**).

2.3.4. Spectral Reflectance and Data Reduction

Retinal reflectance $\rho(\lambda)$ is defined as the ratio of light reflected by the retina at a wavelength λ to the light reflected by a perfectly diffusing surface located at the retina ($\rho(\lambda)=1$ for all λ). This spectral reflectance $\rho(\lambda)$ is calculated from the optical multichannel analyzer signal (OMA_{λ} in counts.pixel^{-1}) and its corresponding baseline ($OMA_{\lambda,bas}$), using:

$$\rho(\lambda) = K_{opt} \cdot \pi \cdot n^2 \cdot f_e^2 \cdot \frac{(OMA_{\lambda} - OMA_{\lambda,bas})}{G \cdot Q_{re,\lambda} \cdot S_{\lambda}} \tag{2}$$

with all parameters defined as above (*see* **Subheading 2.3.2.**) except for $Q_{re,\lambda}$, which is now the spectral distribution of the incident energy $Q_{re,\lambda}$(J.nm^{-1})

3. Methods

3.1. Procedure for In Vivo Measurements

1. After obtaining informed consent, each subject is examined by an ophthalmologist to ensure that there are no ophthalmic contraindications to the performance of any aspect of the test. Systemic blood pressure and intraocular pressure are measured.
2. The pupil of the subject's test eye is dilated with 1% Tropicamide. The dilated pupil diameter must be larger than 6 mm to include the 5.2-mm diameter circle necessary for the instrumental pupils (**Fig. 2A**) and to allow for some movement.
3. The subject is seated in front of the FSP and his/her head is positioned on a chin-forehead holder.
4. By adjusting the holder, the operator aligns the subject's pupil to the optical axis of the FSP and brings the pupil into focus.
5. While observing the retina, the operator directs the subject's gaze using the fixation target, and modifies the fixation position until the retinal field of interest corresponds to the measuring field of the FSP.
6. Accurate focus is achieved by aligning concentric images of the excitation and detection fields on the subject's retina (*see* **Subheading 2.1.6.**).
7. **Steps 4–6** are repeated until optimal alignment is obtained.
8. The operator initiates data acquisition for one excitation wavelength. The fluorescence spectrum appears immediately on the monitor, allowing the operator to assess data quality.
9. A baseline spectrum (*see* **Subheading 2.1.4.**) is recorded at the same excitation wavelength without realignment.

10. The excitation wavelength is changed and **steps 8** and **9** are repeated while monitoring fixation and pupil position. The measurement at Λ=510 nm is repeated. The subject is asked to blink between exposures in order to keep the corneal tear film uniform and hence optimize the optical quality of the media.
11. **Steps 4–10** are repeated until completion of protocol (different sites).
12. The results are inspected and general data quality-assessed. Immediate remeasurement at a wavelength or test site is performed, if necessary, owing to subject movements, blinking, and so on.
13. After completion of the subject's measurements, calibration monitoring is performed (*see* **Subheading 2.3.2.**).

3.2. Data Analysis

All data from one experimental session are imported into a data-analysis program (Igor, WaveMetrics, Lake Oswego, OR) where the data are corrected for the excitation energies and spectral sensitivity of the system.

1. Data from the calibration monitoring (*see* **Subheading 2.3.2.**) yield the excitation energies ($Q_{ex,\Lambda}$), the distribution of the incident white light for reflectance measurements ($Q_{re,\lambda}$), and the sensitivity S_{585} at 585 nm. This sensitivity, together with the relative sensitivity at other wavelengths (S_λ/S_{585}, determined during system calibration), give the spectral sensitivity S_λ pertinent to the experimental session.
2. Fluorescence spectra $F(\Lambda,\lambda)$ for different excitation wavelengths and sites are calculated using **Eq. 1**.
3. Reflectance spectra $\rho(\lambda)$ for different sites are calculated using **Eq. 2**.
4. Characteristics of the emission spectra attained with different excitation wavelength and at different sites are derived; this includes the wavelength of maximal fluorescence (determined from a 50-nm wide parabolic fit around the maximum), the spectral width at half height, and various spectral ratios. These characteristics help define the spectral shape of the emission spectra and verify that the emission measured is owing to lipofuscin.
5. Fluorescence at fixed emission wavelengths are averaged in 10-nm wide bands centered on these wavelengths, and the results tabulated. For example, the fluorescence $F(510, 620)$ frequently used in our analyses, is derived by averaging the spectral data in the spectral band 610–630 nm.
6. Fluorescence for 2 measurements at Λ = 510 nm are averaged, and an intra-session coefficient of variation calculated.
7. Excitation spectra are obtained by plotting the fluorescence intensity at λ = 620 nm as a function of the excitation wavelength Λ.

3.3. Correction for Absorption of the Crystalline Lens

Fluorescence and reflectance are measured through the "transparent" ocular media and are therefore affected by light lost during the double pass through

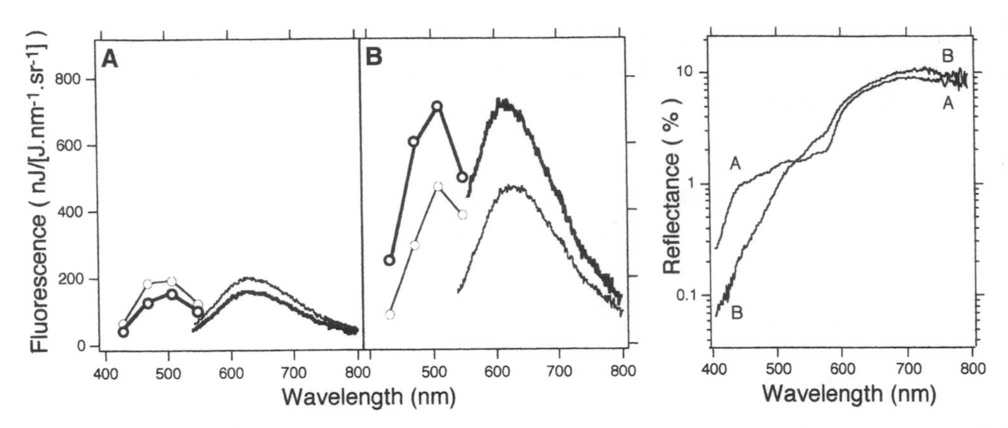

Fig. 3. Left: Subject A, age 23; Subject B, age 63. Fluorescence spectra measured at a site 2.1 mm temporal to the fovea in two normal subjects, ages 23 (**A**) and 63 (**B**). The spectra are shown before (thin lines) and after (thick lines) correction for lens absorption. The continuous curves are the emission spectra for an excitation at 510 nm, and filled squares connected by a line are the excitation spectra for an emission at 620 nm. Emission spectra for 430, 470, and 550 nm excitation were omitted for clarity, and have approx the same shape as the emission at 510 nm excitation. Right: Subjects A and B. Reflectance spectra (logarithmic scale) for the same two subjects (**A** and **B**).

these media, especially the crystalline lens. Accumulation throughout life of water-insoluble protein fractions in the lens causes a decrease in light transmission and an increase in scatter, especially at shorter wavelengths ***(20)***. The short wavelength end of the retinal-reflectance spectra is most affected by this absorption, because the light traverses the lens twice (this is illustrated in **Fig. 3**). Fluorescence is less affected because only the excitation light is absorbed strongly; emission occurs at longer wavelengths that are less absorbed. Light loss in the lens is derived from an analysis of the reflectance spectrum providing an individual correction for lens absorption in each subject studied (*see* **Note 3**).

This correction for lens absorption is based on using the retina (at a fixed location in all subjects, 2.1 mm temporal to the fovea) as a reflector for a double-pass measurement of lens optical density ***(21)***. If $\rho(485)$ and $\rho(520)$ are the spectral reflectances measured at the cornea at 485 and 520 nm, then the lens optical density D_λ at a wavelength λ is given by:

$$D_\lambda = \frac{K_\lambda}{2 \cdot (K_{485} - K_{520}} \cdot \log \frac{\rho(520)}{\rho(485)} - D_{\lambda,44} \quad (3)$$

where K_λ, K_{485}, and K_{520} are the known extinction coefficients of the lens at λ, 485, and 520 nm, respectively ***(20)***. The lens density is not found in absolute terms, but relative to the average density $D_{\lambda,44}$ at age 44 yr (derived empirically from the population average). The retinal fluorescence corrected for the influence of lens absorption, $F_{mc}(\Lambda,\lambda)$, is then:

$$F_{mc}(\Lambda,\lambda) = F(\Lambda,\lambda)\cdot 10^{(D_\Lambda + D_\lambda)} \quad (4)$$

where D_Λ and D_λ are the lens densities (**Eq. 3**) at the excitation and emission wavelength, respectively. All data are therefore corrected as if fluorescence in all subjects was measured through the media of a 44-yr-old average subject, thus providing a means to compare retinal fluorescence in individuals of different ages.

3.4. Results

3.4.1. Representative Spectra in Normal Subjects

Figure 3 shows fluorescence and reflectance spectra measured at 2.1 mm temporal to the fovea in two individuals (ages: 23 and 63). The spectra are shown before and after correction for lens absorption. Retinal fluorescence is emitted over a broad band (500–800 nm) with a maximum at 620–640 nm; optimal excitation occurs for $\Lambda \approx 510$ nm excitation. The emission spectra at different excitations (not shown) have approx the same shapes, indicating the predominance of lipofuscin fluorescence with peak emission at about 630 nm. The reflectance spectra measured at the same location in the same individuals reveal absorption bands of blood (540 and 575 nm). The reflectance spectrum of the older subject (**B**) is more attenuated at shorter wavelengths than that of the young subject (**A**), illustrating the age-related increase in absorption of the crystalline lens. This effect is also seen in the fluorescence data; the excitation spectrum for the older subject is more attenuated at shorter wavelengths.

The fluorescence $F(510,620)$ is 193 and 479 nJ/(J.nm^{-1}.sr^{-1}) for subjects A and B, respectively. The lens density at $\lambda = 510$ nm (**Eq. 3**), referenced to that of a 44-yr-old average subject, is –0.087 and +0.156 density units for the younger and older subjects, respectively. After correcting for the lens (**Eq. 4**), the fluorescence $F_{mc}(510,620)$ is 154 and 720 nJ/(J.nm^{-1}.sr^{-1}) for subjects A and B, respectively.

3.4.2. Reproducibility

Intra-session coefficients of variation associated with two measurements (without head repositioning) of $F(510,620)$ are typically less than 5%. Inter-session reproducibility was assessed in nine normal subjects by repeated measurements of the fluorescence and reflectance spectra in different sessions less than 40 d apart. The reproducibility is defined as the absolute difference in the measures expressed in percent of their mean. For the fluorescence $F(510,620)$, the reproducibility was 4% on average (range: 0–9%). For the reflectance ρ(660), the reproducibility was 10% on average (range: 1–21%). The average absolute difference in the D_{510} estimates for lens absorption was 0.04 density units (range: 0.01–0.06). For the final lens corrected fluorescence $F(510,620)_{mc}$, the reproducibility was 5% on average (range: 1–14%).

3.4.3. Applications

Retinal fluorescence $F(510,620)$, lens density D_{510}, and corrected retinal fluorescence $F(510,620)_{mc}$ in normal subjects ($n = 92$, ages: 15–65) all increase significantly with age ($p < 0.001$). The fluorescence $F(510,620)_{mc}$ at age 60 is on average ≈5 times higher than at age 20 (*see* **Note 4**). Fluorescence emission spectra have a relatively constant shape at all ages; the wavelength of maximal emission is 631 ± 4 nm (mean ± standard deviation [S.D]) and the spectrum width at half height is 168 ± 5 nm.

Fluorescence levels in early AMD are not significantly different than normal (*see* **Note 5**), but the fluorescence levels decrease with progression of the disease. Measurements in areas of photoreceptor atrophy demonstrate very low fluorescence (and spectral distortions) consistent with degeneration of the RPE. Deviations from the normal emission spectrum are observed in patients with subretinal deposits (drusen) that are pathonomic for AMD ***(22)***. These drusen cause disturbance in the emission spectrum that are consistent with their green fluorescence in chemically fixed specimens from eyes of subjects with AMD. The possibility of following progression of AMD by measuring the contribution of drusen is under evaluation.

Increased accumulation of fluorescent material are part of the pathology in Stargardt's disease or fundus flavimaculatus ***(10,11)***. Noninvasive measurements (*see* **Note 6**) in patients with Stargardt's demonstrated that the emission spectra matched the spectrum lipofuscin, and that young patients had significantly elevated levels of lipofuscin compared to normal subjects of the same age ***(23)***.

4. Notes

1. Determination of the emission spectrum is essential to verify that the fluorescence measured is derived from the source one wants to measure. Accumulation of certain drugs, pathological changes in the eye, and system errors can result in fluorescence spectra that are not representative of the lipopigments (lipofuscin) accumulated with increased age.
2. Cooling the diode array detector reduces thermal noise of the photoelectric process, but also stabilizes the sensitivity of the detection system. If an uncooled system is used, it will require repeated daily calibration to compensate for variation in room temperature.

3. Changes in ocular media will influence any noninvasive measurement of fluorescence or reflectance from the retina. Without individual correction for ocular media, the results will be highly influenced by the absorption by the lens. Moreover, because the accumulation of lipofuscin and the changes in the lens are both positively correlated with prior history of oxidative damage, the measurements of retinal fluorescence will be confounded in ways that simple corrections for age cannot undo.
4. The RPE cells in which the fluorescent lipopigments accumulate also contain significant amounts of melanin. The amount of RPE melanin varies across the retina (highest in the fovea), and also changes with age ***(5,16,24)***. RPE melanin in younger subjects is distributed apical to the lipofuscin, whereas in older subjects, the melanin tends to be incorporated into a central core of melanolipofuscin. It is thus possible that some of the age-related increase in lipofuscin observed by our technique is owing to the redistribution of the melanosomes within the RPE cell; however, the increase seen in vivo is consistent with that found by others in sections of donor eye tissue.
5. Measurements in patients with macular degeneration are consistently less reproducible than in normal subjects. Patients have often very poor fixation, and the measurement site can move from one measurement to the next.
6. Measurements in the human fovea are significantly reduced by absorption of the excitation light by the macular pigment, a local accumulation of dietary carotenoids lutein and zeaxanthin ***(14)***. The cones in the retina of a number of other species, including primates, frogs, turtles, and birds, also accumulate carotenoids. Non-invasive measurements of retinal reflectance or fluorescence in these areas require correction for the absorption by these carotenoids ***(25)***.

References

1. Blackett, A. D. and Hall, D. A. (1981) Tissue vitamin E levels and lipofuscin accumulation with age in the mouse. *J. Gerontol.* **36,** 529–533.
2. Monji, A., Morimoto, N., Okuyama, I., Yamashita, N., and Tashiro, N. (1994) Effect of dietary vitamin E on lipofuscin accumulation with age in the rat brain. *Brain Res.* **634,** 62–68.
3. Feeney-Burns, L., Berman, E. R., and Rothman, H. (1980) Lipofuscin of human retinal pigment epithelium. *Am. J. Ophthalmol.* **90,** 783–791.
4. Boulton, M. and Marshall, J. (1986) Effects of increasing numbers of phagocytic inclusions on human retinal pigment epithelial cells in culture: a model for aging. *Br. J. Ophthalmol.* **70,** 808–815.
5. Weiter, J. J., Delori, F. C., Wing, G., and Fitch, K. A. (1986) Retinal pigment epithelial lipofuscin and melanin and choroidal melanin in human eyes. *Invest. Ophthalmol. Vis. Sci.* **27,** 145–152.
6. Eldred, G. E. (1987) Questioning the nature of the fluorophores in age pigments. *Adv. Biosci.* **64,** 23–36.

7. Dorey, C. K., Staurenghi, G., and Delori, F. C. (1993) Lipofuscin in aged and AMD eyes, in *Retinal Regeneration* (Holyfield, J. G., ed.), Plenum Press, New York, pp. 3–14.
8. Taylor, A., Jacques, P. F., and Dorey, C. K. (1993) Oxidation and aging: impact on vision. *J. Toxicol. Indust. Health* **9,** 349–371.
9. Dorey, C. K., Wu, G., Ebenstein, D., Garsd, A., and Weiter, J. J. (1989) Cell loss in the aging retina: Relationship to lipofuscin accumulation and macular degeneration. *Invest. Ophthalmol. Vis. Sci.* **30,** 1691–1699.
10. Eagle, R. C., Lucier, A. C., Bernadino, V. B., and Janoff, M. (1980) Retinal pigment epithelial abnormalities in fundus flavimaculatus: a light and electron microscopic study. *Ophthalmology* **87(12),** 1189–1200.
11. Birnbach, C. D., Jarvelainen, M., Possin, D. E., and Milam, A. H. (1994) Histopathology and immunocytochemistry of the neurosensory retina in fundus flavimaculatus. *Ophthalmology***101(7),** 1211–1219.
12. Weingeist, T. A., Kobrin, J. L., and Watzke, R. C. (1982) Histopathology of Best's macular dystophy. *Arch. Ophthalmol.* **100,** 1108–1114.
13. Samuelson, D., Dawson, W. W., Webb, A. I., and Dowson, J. (1985) Retinal pigment epithelial dysfunction in early ovine ceroid lipofuscinosis: electrophysiologic and pathological correlates. *Ophthalmologica* **190,** 150–157.
14. Snodderly, D. M. (1995) Evidence for protection against age-related macular degeneration by carotenoids and antioxident vitamins. *Am J Clin Nutr.* **62,** 1448S–1461S.
15. Eldred, G. E. and Katz, M. L. (1988) Fluorophores of the human retinal pigment epithelium: separation and spectral characterization. *Exp. Eye Res.* **47,** 71–86.
16. Boulton, F., Dayhaw-Barker, P., Ramponi, R., and Cubeddu, R. (1990) Age-related changes in the morphology, absorption and fluoresence of melanosomes and lipofuscin granules of the retinal pigment epithelium. *Vision Res.* 3**0,** 1291–1303.
17. Kitagawa, K., Nishida, S., and Ogura, Y. (1989) In vivo quantification of autofluorescence in human retinal pigment epithelium. *Ophthalmologica* **199,** 116–121.
18. Delori, F. C. (1994) Spectrophotometer for noninvasive measurement of intrinsic fluorescence and reflectance of the ocular fundus. *Appl. Optics.* **33,** 7439–7452.
19. Delori, F. C., Dorey, C. K., Staurenghi, G., Arend, O., Goger, D. G., and Weiter, J. J. (1995) In vivo fluorescence of the ocular fundus exhibits retinal pigment epithelium lipofuscin characteristics. *Invest. Ophthalmol. Vis. Sci.* **36,** 718–729.
20. Pokorny, J., Smith, V. C., and Lutze, M. (1987) Aging of the human lens. *Appl. Optics.* **26,** 1437–1440.
21. Delori, F. C. and Burns, S. A. (1996) Fundus reflectance and the measurement of crystalline lens density. *J. Opt. Soc. Am.* **13(2),** 215–226.
22. Arend, O. A., Weiter, J. J., Goger, D. G., and Delori, F. C. (1995) In-vivo fundus-fluoreszenz-messungen bei patienten mit alterabhangiger makulardegeneration. *Ophthalmologie* **92,** 647–653.
23. Delori, F. C., Staurenghi, G., Arend, O., Dorey, C. K., Goger, D. G., and Weiter, J. J. (1995) In-vivo measurement of lipofuscin in Stargardt's disease/Fundus flavimaculatus. *Invest. Ophthalmol. Vis. Sci.* **36,** 2337–2331.

24. Docchio, F., Boulton, M., Cubeddu, R., Ramponi, R., and Dayhaw-Barker, P. (1991) Age-related changes in the fluorescence of melanin and lipofuscin granules of the retinal pigment epithelium: a time-resolved fluorescence spectroscopy study. *J. Photochem. Photobiol.* **54,** 247–253.
25. Delori, F. C. (1993) Macular pigment density measured by reflectometry and fluorophotometry: noninvasive assessment of the visual system. *OSA Tech. Digest* **3,** 240–243.

24

An Ex Vivo Erythrocyte Model for Investigating Oxidative Metabolism

Luigia Rossi and Mauro Magnani

1. Introduction

The technology of opening and resealing the erythrocytes can be used successfully to investigate several basic aspects of cellular metabolism, including oxidative metabolism. Human and animal erythrocytes can be loaded with different kinds of molecules (enzymes, metabolites, immunoglobulins, exogenous substances, etc.) to modify cellular metabolism, through procedures easy to perform, reproducible, and highly conservative *(1,2)*. Use of these procedures, notably those based on hypotonic hemolysis, isotonic resealing, and reannealing leads to intact and fully viable erythrocytes.

Mammalian erythrocytes are specialized cells whose main function is oxygen transport. To maximize oxygen delivery, the majority of the cell's organelles and metabolic pathways, typical of the undifferentiate cells, are lost during the erythrocyte maturation. Thus, circulating red blood cells (RBC) produce adenosine tryphosphate (ATP) only through the anaerobic glycolytic pathway and generate reducing equivalents through the hexose monophosphate shunt (HMS) pathway. Glutathione is the predominant reducing agent in these cells (**Fig. 1**) *(3)*. This limited metabolic ability of erythrocytes allows a survival of 120 d in circulation in humans. Alterations in the cell abilities to produce ATP or to maintain glutathione/nicotinamide adenine dinucleotide phosphate ($NADPH^+$) in reduced state result in shortening of in vivo life-span. RBC are also regularly subjected to high oxygen tension and are among the first body cells exposed to exogenous oxidative substances that are ingested, injected, or inhaled.

Encapsulation in RBC of one or more enzymes or enzyme-inactivating antibodies permits the remodeling of erythrocyte metabolic pathways, and thus

From: *Methods in Molecular Biology, vol. 108: Free Radical and Antioxidant Protocols*
Edited by: D. Armstrong © Humana Press Inc., Totowa, NJ

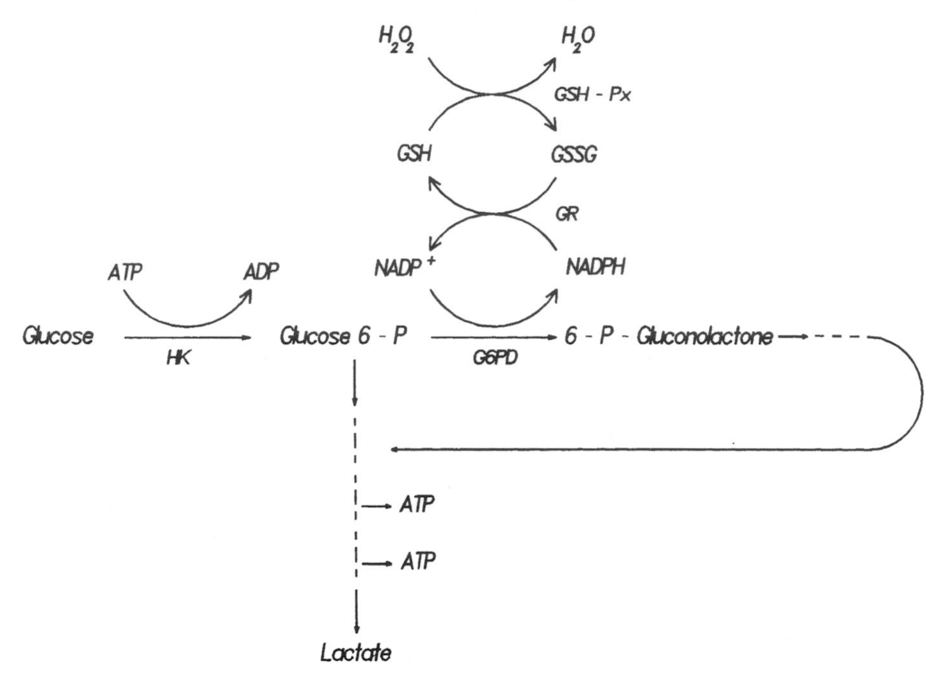

Fig. 1. Schematic reactions of glucose and glutathione metabolism of erythrocytes. The following abbreviations are used: HK, hexokinase; G6PD, glucose 6-phosphate dehydrogenase; GSH, reduced glutathione; GSSG, oxidized glutathione; GSH-Px, glutathione peroxidase; GR, glutathione reductase.

investigation of the role that each step plays in its energy and/or oxidative metabolism. Hexokinase (HK) catalyzes the first step of glucose metabolism in several cell types, including erythrocyte, in which it is currently believed to be a crucial rate-limiting enzyme of the glycolytic pathway. Moreover, we have showed that it limits the HMS flux under stress conditions ***(4)***. Here, we report the method of HK and anti-HK antibodies encapsulation into RBC as an ex vivo model to investigate the regulation of the erythrocyte HMS under oxidative stress or unstressed conditions.

When control mechanisms of erythrocyte-encapsulated enzyme proteins are under investigation, the ideal source of these proteins should be the erythrocyte from the same species. Use of heterologous proteins may prove misleading because of different kinetic properties ***(5)***, incorrect compartimentalization ***(5)*** or protein instability in the intraerythrocytic environment ***(6)***. However, the established identity of the erythrocyte enzyme with the species present in human placenta and purification to homogeneity of this enzyme allowed encapsulation of homologous HK to be made. More recently the same isozyme was obtained as a recombinant protein in *Escherichia coli* and purified to homogeneity. Thus, recombinant human HK is now available in large quantities.

Furthermore, enzymatic systems generating intracellular peroxides (i.e., glucose oxidase, lactate oxidase, etc.) can also be investigated by this approach *(7,8)*. Finally, it is worth noting that the levels of glutathione can also be easily modified by encapsulation of this tripeptide to obtain up to 5–10 times the basal concentrations *(9)*. In this chapter we report the manipulations performed on human erythrocytes to investigate the role of glucose 6-phosphate production in energy and redox metabolism. This manipulation is shown only as an example of the many potential applications of this technology.

2. Materials

2.1. Instruments

2.1.1. HK Assay

1. Uvikon 860 Kontron (Zurich, Switzerland) spectrophotometer with thermostated cell heater or similar equipment.
2. Precision cuvets of optic special glass 10 mm.

2.1.2. Anti-Hexokinase IgG Preparation

1. Precision cuvets of quartz glass 10 mm.
2. 1.5 × 5 cm Chromatographic column.
3. Peristaltic pump.

2.1.3 HK and Anti-HK IgG Loading into RBC

1. Coulter Counter ZM (Luton, Beds, UK), or similar.
2. Rotating plate.
3. Automatic osmometer (Roebling, Berlin, Germany).
4. Dialysis tube (Spectrapor, Spectrum Medical Ind., Los Angeles, CA) molecular size cut off, 12–14 kDa.

2.1.4. Hexose Monophosphate Pathway Determination

1. Glass fiber filters 2.5 cm Whatman.
2. Polyvinyl chloride (PVC) vials.
3. Liquid Scintillation Analyzer Tri-Carb 2100 TR Packard (Downers Grove, IL), or similar.

2.1.5. Lactate Production and Glucose Concentration Determination

Analox GM7 Analyzer (Analox Instruments, LTD, London, UK). Tubes supplied for Analox analyzers containing fluoride, heparin, and nitrite.

2.2. Reagents

All reagents must be of the highest quality to ensure accurate and reproducible results. We usually obtain biochemicals from Sigma Chemical Co. (St. Louis, MO) or Boehringer Mannheim (Indianapolis, IN).

2.2.1. HK Assay

1. 0.25 *M* Glycilglycine, pH 8.1.
2. 0.05 *M* Glucose.
3. 0.05 *M* ATP Mg^{2+}.
4. 0.005 *M* $NADP^+$.
5. 0.05 *M* $MgCl_2$.
6. Glucose-6-phosphate dehydrogenase grade II 140 U/mg (1:40 in distilled water).
   Glycylglycine, glucose, and $MgCl_2$ solutions are stable at 4°C for up to 3 wk; ATP-Mg^{2+} and $NADP^+$ are stable at 4°C for up to 1 wk, whereas glucose-6-phosphate dehydrogenase is diluted fresh every day.

2.2.2. Anti-HK IgG Preparation

1. Complete Freunds's adjuvant.
2. Immobilized r Protein A™ RepliGen (Cambridge, MA).
3. 0.1 *M* Phosphate buffer, pH 8.0.
4. 0.1 *M* Sodium cytrate, pH 3.5.
5. 1.0 *M* Tris-HCl.
   Phosphate buffer and sodium-cytrate solution are stored at 4°C for 2 wk. Anti-HK IgG in 0.1 *M* sodium cytrate, neutralized with 1 *M* Tris-HCl, are stored at −80°C, and are stable for years.

2.2.3. Hexokinase and Anti-HK IgG Loading into RBC

1. Heparin 5000 UI/mL.
2. Washing buffer: 5 m*M* sodium phosphate; 0.9% (w/v) NaCl; 5 m*M* glucose pH 7.4.
3. Dialysis buffer: 10 m*M* sodium phosphate; 10 m*M* sodium bicarbonate, 20 m*M* glucose, pH 7.4.
4. Resealing solution: 5 m*M* adenine, 100 m*M* inosine, 100 m*M* sodium pyruvate, 100 m*M* sodium phosphate, 100 m*M* glucose, 12% (w/v) NaCl, pH 7.4.
   Washing and dialysis buffers are made fresh every time, whereas resealing solution is stored at −20°C and it is stable for months.

2.2.4. Hexose Monophosphate Pathway Determination

1. Emulsifier Scintillator Plus from Packard.
2. [1-^{14}C] glucose (58 mCi/mmol).
3. 1.0 N KOH.
4. $HClO_4$ 10% (v/v).

[1-^{14}C] glucose is stored at –20°C. KOH may cause severe burns. $HClO_4$ is oxidizing and corrosive.

2.2.5. Lactate Production and Glucose-Concentration Determination

1. Lactate Reagent Buffer solution, lactate oxidase enzyme, and lactate standard, 8.0 mmol/L, provided with the Lactate II Reagent Kit from Analox Instruments, LTD.
2. Glucose Reagent buffer and glucose oxidase as a single reagent provided with the Glucose Reagent Kit from Analox Instruments, LTD.
3. Glucose standard, 8.0 mmol/L.

2.2.6. Hemoglobin Estimation

Drabkin's solution: NaCN 100 mg/L, $K_3[Fe(CN)_6]$ 300 mg/L; reagent is toxic, causing serious risk of poisoning by inhalation, swallowing, or skin contact.

2.2.7. Protein Estimation

BIO-RAD Protein assay dye reagent: Caution: the reagent contains phosphoric acid and methanol and is harmful if inhaled or swallowed.

3. Methods

3.1. HK Assay

1. HK (EC 2.7.1.1.) type I is purified from human placenta by a combination of ion-exchange chromatography, affinity chromatography, and dye-legand chromatography (as described in **ref. *10***). The specific activity is 190 ± 5 IU/mg protein and found to be homogeneous as shown by sodium dodecyl sulfate gel electrophoresis.
2. HK activity is measured at 37°C spectrophotometrically in a system coupled with glucose-6-phosphate dehydrogenase (EC 1.1.1.49). The assay mixture contains, in a total volume of 1 mL: 0.125 *M* Glycylglycine, pH 8.1, 5 m*M* glucose, 5 m*M* ATP-Mg^{2+}, 0.25 m*M* $NADP^+$, 5 m*M* $MgCl_2$, 0.5 IU of glucose-6-phosphate dehydrogenase (G6PD).
3. Sample (5–50 μL according to dilution) is added before G6PD. Final volume is brought to 1 mL with distilled water.
4. Initial rate measurements are performed by following the reduction of $NADP^+$ at 340 nm in a Uvikon 860 spectrophotometer in the kinetic version. For each molecule of glucose utilized, 1 molecule of $NADP^+$ is reduced. One unit of hexokinase activity is defined as the amount of enzyme which catalyzes the formation of 1 μmol of glucose-6-phosphate/min at 37°C.

3.2. Anti-HK IgG Preparation

1. Antiserum against HK type I from human placenta is raised in rabbits. The first injection is with complete Freund's adjuvant, followed by two further injections of the enzyme at 10-d intervals. Each injection consists of 100 μg of protein.
2. IgG are prepared from this serum and normal control rabbit serum by chromatography on immobilized Protein A as follows: Protein A is equilibrated with 50 mL

0.1 *M* phosphate buffer, pH 8.0, then it is put under agitation with plasma for 4 h at 4°C to allow binding of IgG to protein A.

3. After centrifugation at 220*g* for 10 min at 4°C, supernatant is removed while pellet is resuspended in the aforementioned phosphate buffer and loaded onto a chromatographic column.
4. The column is washed with phosphate buffer until no protein absorbance at 280 nm is present. IgG are eluted by 0.1 *M* sodium citrate pH 3.5 and immediately neutralized by 1 *M* Tris HCl. 50 µg of anti-HK IgG are found to be able to inactivate 0.25 IU of HK activity. No HK inactivation occures with IgG from normal rabbits.

3.3. Encapsulation of HK or Anti-HK IgG in Erythrocytes

1. Encapsulation of HK in human erythrocytes is obtained according to Ropars et al. ***(11)*** as described ***(12,13)***. This procedure involves three sequential steps (**Fig. 2**): hypotonic hemolysis, isotonic resealing, and reannealing of erythrocytes as follows: blood is collected in heparin immediately before use and centrifuged at 3000 rpm at 4°C to separate the plasma, which is then maintained at 0°C until use. Erythrocytes are washed twice in washing buffer and finally resuspended in the same buffer containing HK (10 IU/mL packed erythrocytes) at a hematocrit of 70% in the dialysis tube (*see* **Note 1**).
2. Hypotonic lysis of erythrocytes is obtained by dialysis of 2 mL of cell suspension in a Falcon 50-mL sterile tube containing the dialysis buffer, in which 2 m*M* ATP and 3 m*M* GSH are added before use and rotated at 15 rpm for 1 h at 4°C. The hemolysate is then collected and 1 volume of resealing solution is added to every 10 vols of hemolysate.
3. Reannealing of the cells is then performed at 4°C with a physiological saline solution. Finally, erythrocytes are resuspended in their native plasma and utilized for metabolic studies.
4. To encapsulate anti-HK IgG in human erythrocytes, the same procedure is used, except that anti-HK IgG (3.0 mg/mL packed erythrocytes) resuspended in washing buffer are added to washed erythrocyte during the dialysis step instead of HK (*see* **Note 2**). Two additional erythrocyte suspensions are generally used, the first one (referred to as "controls") corresponding to the native, untreated cells and the second one (designed "unloaded") corresponding to erythrocytes that have been processed as in the entrapment technique, but without any HK or anti-HK IgG addition.
5. During the loading procedure, the following HK activities are evaluated:
   a. HK activity of HK type I purified from human placenta.
   b. HK activity of blood donor measured on a hemolysate 1:20 in distilled water of RBC suspension 1:2 in the physiological saline solution.
   c. Total HK activity evaluated on a hemolysate 1:40 of RBC after dialysis step.
   d. Final HK activity determined on a hemolysate 1:20 of RBC at 20% Ht in plasma or physiological saline solution.
6. Percent of encapsulation is estimated as the ratio between the final HK activity in the erythrocytes submitted to the loading procedure and the HK activity determined after the dialysis step.

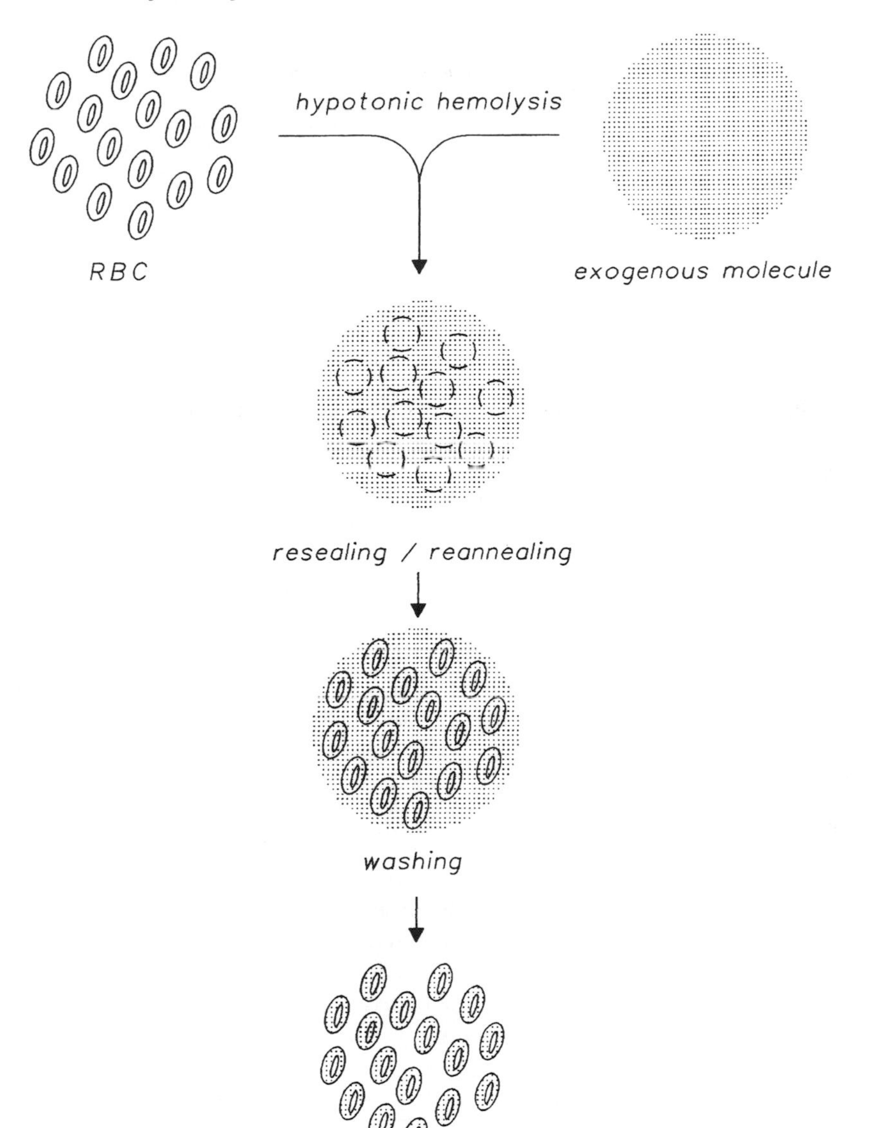

Fig. 2. Scheme of the procedure used to encapsulate exogenous molecules into erythrocytes.

7. With the loading procedure previously described a 19.7 ± 6.2% of entrapment of HK and a cell recovery of 70–80% is obtained (*see* **Notes 3** and **4**). In **Table 1**, the HK activities of human erythrocytes after encapsulation of HK or anti-HK IgG are reported. The opened/resealed erythrocytes have almost normal hematological parameters, with the exception of a reduced cellular volume, and are slightly hypochromic (*see* **Note 5**).

Table 1
Hexokinase Activity in Human Erythrocytes after Loading Procedure

Red Blood Cells	Hexokinase activity (IU/mL RBC)
Controls	0.33 ± 0.1
Unloaded	0.32 ± 0.1
Loaded with hexokinase	4.77 ± 0.75
Loaded with antihexokinase IgG	0.04 ± 0.002

3.4. Determination of the Amount of Glucose Metabolized Through the Pentose Shunt (HMP)

1. Because the pentose shunt is the only source of CO_2 in the erythrocyte, and the carbon atom of CO_2 is derived from the first carbon of glucose, it is possible to evaluate the amount of glucose metabolized through the pentose-phosphate pathway from the measurement of $^{14}CO_2$ production derived from [1-^{14}C]-glucose.
2. Erythrocyte suspensions at about 20% Ht are preincubated 30 min at 37°C. Then, at 1.1 mL of sample, 1.1 μL of [1-^{14}C] glucose containing 50 μCi/200 μL are added and incubation is continued for other 30 min to equilibrate the mixture. 100 μL of RBC are utilized to determine glucose concentration (*see* **Subheading 3.5.**), while the remaining 1 mL is placed in a vial closed with a glass-fiber filter paper wet with 100 μL 1*N* KOH (**Fig. 3**). After 1 h of incubation at 37°C, the reaction is stopped by adding 500 μL of 10% (v/v) $HClO_4$ and incubation continues for other 30 min. The glass-fiber filter paper used to collect $^{14}CO_2$ is counted in 20 mL scintillation fluid. Results are expressed in nmol CO_2/min/mL RBC given the specific activity of glucose in the suspension (cpm/nmol glucose).
3. The HMP rate in the HK-overloaded RBC is not significantly different from controls (*see* **Note 6**) in the absence of the oxidant agent methylene blue (resting conditions), but shows values 1.5 times higher in its presence. Under this condition, RBC with HK activity of 0.04 ± 0.002 IU/mL RBC (a value that represents 12% of the activity found in control RBC) shows a dramatic decrease as shown in **Fig. 4**. Therefore, it is possible to say that a different rate of HMP under oxidative stress is observed changing the HK activity level and that RBC with HK levels comparable to those found in HK-deficiency are not able to respond to oxidizing agents with increased HMP rates. These cells can now be used as an ex vivo cellular model to the oxidative metabolism.

3.5. Determination of Lactate Production and Glucose Concentration

1. By evaluating lactate production, it is possible to estimate the amount of glucose metabolized by RBC in the glycolytic pathway.

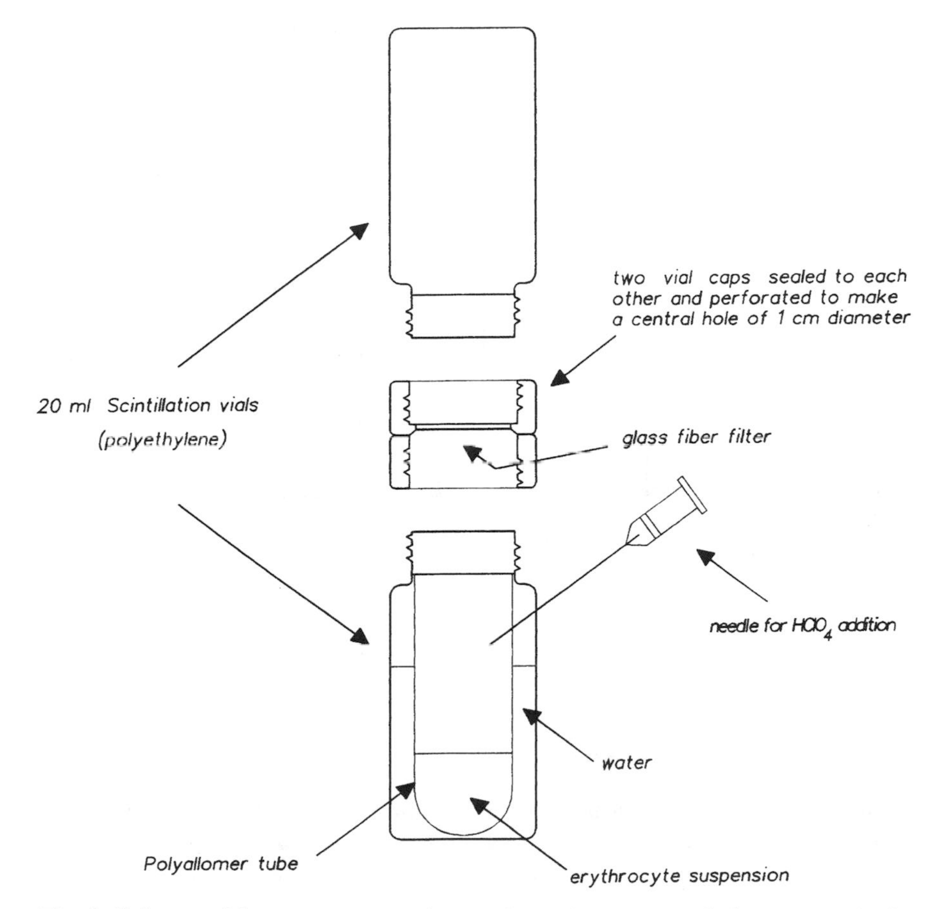

Fig. 3. Scheme of the apparatus used to evaluate the amount of glucose metabolized through the pentose shunt.

2. Lactate production is measured on red-cell suspensions at 20% hematocrit in a GM7 Analyzer. Briefly, the erythrocyte suspensions are incubated at 37°C and samples taken every 15 min and placed in the tubes supplied for Analox analyzer. 7 µL samples are then taken and lactate content determined by measuring O_2 consumption by L-lactate oxygen reductase. Lactate production is usually tested over 1 h of incubation at 37°C and is expressed in µmol/mL RBC.
3. Loading procedure allows one to modify the cellular ability to metabolize glucose not only through the HMS pathway, but also in the Embden-Meyerhof-Parnas pathway. RBC with increased HK activity have in fact, a greater glycolytic ability than controls, whereas RBC with a decreased HK activity metabolize a lower amount of glucose.
4. To evaluate glucose concentration, the same procedure is used. 10 µL (instead of 7) are taken and glucose content determined by measuring O_2 consumption by glucose oxidase.

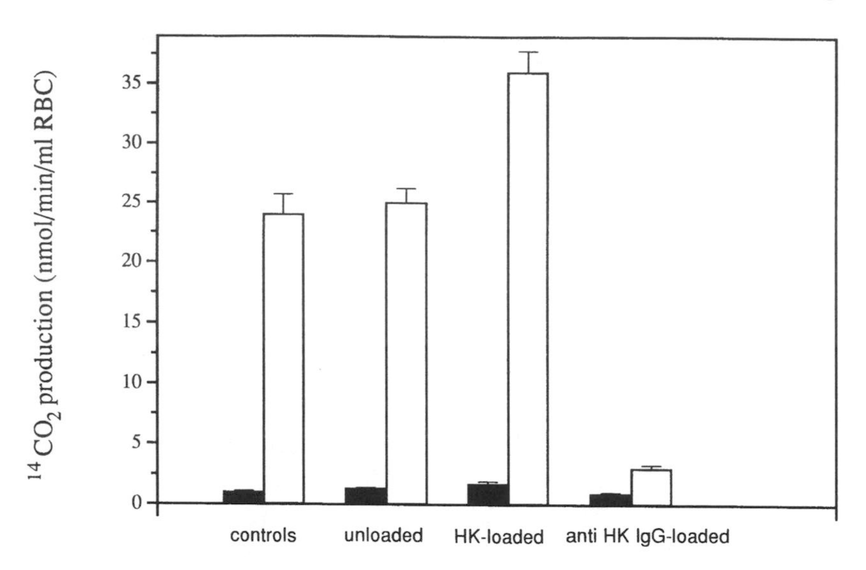

Fig. 4. HMP in human erythrocytes with different levels of hexokinase activity in absence and in presence of the oxidant agent Met Blue. HK activities are as reported in **Table 1**. All values are mean ± SD of 4 experiments. (■), without Met Blue; (□), 100 μ*M* Met Blue.

3.6. Hemoglobin Estimation

Hemoglobin concentrations are spectrophotometrically determined at 540 nm with Drabkins's solution, as described by Beutler *(14)*. Commercially available cyanmethemoglobin standard has been used to provide a standard curve.

3.7. Protein Estimation

Protein concentrations are spectrophotometrically determined at 595 nm with the dye reagent Bio-Rad Protein Assay. Commercially available albumin standard solutions have been used to provide a standard curve.

4. Notes

By the technology of opening and resealing the erythrocytes, researchers will have an extraordinary opportunity to specifically modify the erythrocytes by the introduction of enzymes that generate new metabolic abilities, antibodies that inactive single metabolic steps, or metabolites that can influcence other cell properties. Furthermore, engineered erythrocytes can behave as circulating bioreactors for the degradation of toxic metabolites or inactivation of xenobiotics.

The loading procedure is very simple to perform, but some peculiarities must be kept in mind to successfully operate:

1. It is important to work with a high hematocrit level (70%) during the dialysis step.
2. The substance to be encapsulated has to be added before dialysis if its molecular weight is greater than the cut-off of the dialysis tube; otherwise, the substance has to be added after the dialysis step, incubating dialyzed RBC with the substance directly in a tube at room temperature for 15 min before adding the resealing solution.
3. To evaluate the osmolarity of dialysis buffer (it has to be about 60–70 mOsm) it must be kept in mind that a correct RBC osmolarity (~120 mOsm) after the dialysis step is crucial to obtain a good entrapment and a good cell recovery.
4. To evaluate the time of dialysis and modify it, or the mOsm of dialysis buffer, or both, to optimize the results, it must be remembered that a longer time of dialysis or a decrease mOsm of dialysis buffer leads to a decrease in RBC recovery.
5. The loaded RBC must be washed in physiological saline solution at 2000 rpm instead of 3000 because of their major fragility.
6. "Unloaded" erythrocytes should be used as control, because addition of the resealing solution causes some metabolic perturbations. Usually, steady-state metabolite concentrations are better measured after the first hour of incubation.

Acknowledgment

The work described in this chapter was supported by C.N.R. and M.U.R.S.T. Rome, Italy.

References

1. De Loach, J. R. and Sprandel, U. (eds.) (1985) *Red Blood Cells as Carriers for Drugs.* Karger Press, Basel, Switzerland.
2. Ropars, C., Chassaigne, M., and Nicolau, C. (eds.) (1987) *Red Blood Cells as Carriers for Drugs: Potential Therapeutic Applications.* Pergamon Press, Oxford, UK.
3. Grimes, A. J. (1980) *Human Red Cell Metabolism*, Blackwell Scientific Publications, Oxford, UK, pp. 192–258.
4. Magnani, M., Rossi, L., Bianchi, M., Serafini, G., and Stocchi, V. (1988) Role of hexokinase in the regulation of erythrocyte hexose monophosphate pathway under oxidative stress. *Biochem. Biophys. Res. Commun.* **155(1),** 423–428.
5. Morelli, A., Benatti, U., Salamino, F., Sparatore, B., Michetti, M., Melloni, E., Pontremoli, S., and De Flora, A. (1979) In vitro correction of erythrocyte glucose 6-phosphate dehydrogenase (G6PD) deficiency. *Arch. Biochem. Biophys.* **197,** 543–550.
6. De Flora, A., Morelli, A., and Grasso, M. (1986) Intraerythrocyte stability of normal and mutant G6 PD, in *Glucose 6 phosphate Dehydrogenase* (Yoshida, A. and Beutler, E., eds.), Academic Press, NY, pp. 133–152.
7. Rossi, L., Bianchi, M., and Magnani, M. (1992) Increased glucose metabolism by enzyme-loaded erythrocytes in vitro and in vivo normalization of hyperglycemia in diabetic mice. *Biotechnol. Appl. Biochem.* **15,** 207–216.

8. Garin, M., Rossi, L., Luque, J., and Magnani, M. (1995) Lactate catabolism by enzyme-loaded red blood cells. *Biotechnol. Appl. Biochem.* **22,** 295–303.
9. Fazi, A., Mancini, U., Piatti, E., Accorsi, A., and Magnani, M. (1992) Xenobiotic detoxification by GSH-loaded erythrocyte, in *The Use of Resealed Erythrocytes as Carriers and Bioreactors* (Magnani, M. and DeLoach, J. R., eds,), Plenum Press, NY, pp. 195–201.
10. Magnani, M., Stocchi, V., Serafini, G., Chiarantini, L., and Fornaini, G. (1988) Purification, properties and evidence for two subtypes of human placenta hexokinase type I. *Arch. Biochem. Biophys.* **260,** 388–399.
11. Ropars, C., Chassaigne, M., Villereal, H. C., Avenard, G., Hurel, C., and Nicolau, C. (1985) Resealed red blood cells as a new blood transfusion product, in *Red Blood Cells as Carriers for Drugs* (DeLoach, J. R. and Sprandel, U., eds.), Karger Press, Basel, pp. 82–91.
12. De Flora, A., Benatti, U., Guida, L., and Zocchi, E. (1986) Encapsulation of adriamycin in human erythrocytes. *Proc. Natl. Acad. Sci. USA.* **83,** 7029–7033.
13. De Flora, A., Guida, L., Zocchi, E., Tonetti, M., and Benatti, U. (1986) Construction of glucose oxidase-loaded human erythrocytes: a model of oxidative cytotoxicity. *Ital. J. Biochem.* **35,** 361–367.
14. Beutler, E. (1984) *Red Cell Metabolism: A Manual of Biochemical Methods.* 3rd Ed., Grune & Stratton, Orlando, FL.

25

Purification of Vesicular Carriers from Rat Hepatocytes by Magnetic Immunoadsorbtion

Lucian Saucan

1. Introduction

Vesicular carriers represent a cellular entity involved in the transfer of proteins between different intracellular compartments. They were first identified and described by Jamieson and Palade *(1)* in their studies on the process of protein secretion in guinea-pig pancreas. Most of the evidence obtained since then is in favor of the vesicular nature of the intracellular transport *(2,3)*.

The secretory pathway originates in the cytosol, where proteins are first synthesized. Proteins destined for secretion are co-translationally translocated into the ER, where they undergo structural and biochemical modifications (oligomerization, core glycosylation, thiol bridge formation, etc.). From the ER, proteins are translocated via vesicular carriers to the Golgi complex, where proteins undergo terminal glycosylation as well as other modifications like sulfation and phosphorylation. Proteins move vectorially across the Golgi stacks, starting at the cis side; once they reach the trans-Golgi network, they are sorted in different batches, loaded in different classes of vesicular carriers, and shipped to many destinations throughout the cell *(4)*. So far, several different classes of vesicular carriers have been described, e.g., ER-Golgi, intra-Golgi, Golgi-plasma membrane, Golgi-lysosomes, clathrin-coated vesicles, endocytic and transcytotic vesicles, etc.

Traditionally, intracellular organelles and vesicles are isolated through a combination of physical methods based on differences in their size and density. These methods are based on cell fractionation followed by differential density-gradient centrifugation *(5)*. Most of the methods employed are long and tedious with low yields and heavy intercontamination, owing to similarities in physical properties between different organelles.

From: *Methods in Molecular Biology, vol. 108: Free Radical and Antioxidant Protocols*
Edited by: D. Armstrong © Humana Press Inc., Totowa, NJ

Immuno-isolation was introduced as an alternative method and is based on an antibody-antigen reaction in which the antibodies are coupled to a solid support and the antigen represents an epitope present on the surface of the organelle to be isolated *(6)*. The magnetic beads were introduced in 1980 by J. Ugelstad and coworkers *(7)*, and, owing to their properties, are widely used today as a solid support for immuno-isolation *(8)*.

In this chapter, I describe the immuno-isolation protocol that is used in our laboratory (using encapsulated magnetic beads) for purification of carrier vesicles that shuttle between trans-Golgi network and the sinusoidal plasmalemma in rat hepatocytes *(9)*. This method can be easily adapted for purification of other types of vesicles or organelles if a specific antibody is available.

2. Materials

1. Dry paramagnetic beads were a kind gift of Dr. J. Ugelstad (Trondheim, Norway).
2. Activated beads, M-500 (Dynal Inc., Great Neck, NY).
3. Sprague-Dawley rats (Bantin & Kingman, Fremont, CA).
4. Protease inhibitors; Leupeptin, Pepstatin, and Aprotinin (Boehringer-Mannheim, Indianapolis, IN).
5. [^{35}S] Trans-label met-cys (ICN, Costa Mesa, CA), ^{125}I (Amersham, Arlington Heights, IL), Iodobeads (Pierce, Rockford, IL).
6. Protein A-Sepharose (Pharmacia, Piscataway, NJ).
7. XAR-50 film (Kodak, Rochester, NY).
8. Anti-rabbit IgG(Fc) affinity purified antibodies (Biodesign Int, Kennebunkport, ME), antibodies directed to the endodomain of the polymeric IgA receptor were raised in our laboratory using as antigen a synthetic peptide representing the last 11 a.a. of its carboxy-terminus. All other reagents used were of analytical grade.
9. 0.05 *M* Borate buffer, pH 9.5.
10. Binding buffer: 5% fetal calf serum (FCS).
11. 2 m*M* ethylenediamine tetraacetic acid (EDTA) in phosphate buffered saline (PBS), pH 7.5.
12. Washing buffer: 2.5% FCS, 2 m*M* EDTA, in PBS, pH 7.5.
13. Magnetic tube holder (Dynal).
14. Rotating wheel with adjustable speed (Scientific Instruments Inc., Bohemia, NY).
15. Beckman L65 ultracentrifuge and rotors 70Ti and SW28.

3. Methods

3.1. Preparation of Beads

The magnetic beads that have free -OH groups on their surface are activated with para-toluene-sulfonyl-chloride (tosyl-chloride) in the presence of pyridine as a catalyst for 20 h at room temperature with gentle mixing *(10)*. Because the beads are now available already activated this step can be omitted.

The activated beads are initially coated with a secondary antibody (*see* **Notes 1** and **2**) directed against the Fc portion of the primary IgG; in our case, goat anti-rabbit IgG(Fc). The beads are incubated with the secondary antibodies (affinity-purified) in 0.05 *M* borate buffer, pH 9.5, for 20 h with mixing. The recommended ratio of antibody to beads is ~10 µg antibody/mg of beads. After the incubation, the beads are retrieved from the mixture using the magnetic tube holder, the supernatant saved, and the beads washed several times with a binding buffer that contains extraneous protein for blocking any remaining binding sites (see **Subheading 2.** and **Note 3**). The saved supernatant can be checked by ultraviolet (UV) or protein assay in order to determine how much antibody has been coated on the beads (by making the difference between the input and output). Also, at this point a small amount of beads is solubilized and run in parallel with a standard IgG preparation on a sodium dodecyl sulfate (SDS) gel to quantitate the amount of IgG on the beads. In our laboratory, we usually coat ~5 µg IgG/mg of beads. The primary antibodies are diluted in binding buffer and added to the coated beads at a ratio of 2:1 (primary vs secondary antibodies) and should also be affinity-purified (if they are not monoclonals). The incubation is usually overnight (between 2 and 12 h) at 4°C with mixing. At the end, the beads are retrieved again on the magnetic stand and washed 3 × 10 min in binding buffer. At this point, the beads are ready to be incubated with the sample.

3.2. Labeling

For in vivo labeling, the [^{35}S] trans-label is injected through the portal vein of anesthetized rats, and chased for 30 min at which time the liver is removed and processed through homogenization and density-gradient flotation.

3.3. Subcellular Fractionation

Rats are anesthetized and their livers flushed with ice-cold isotonic sucrose containing a cocktail of protease inhibitors. Livers are homogenized in the same buffer and processed through a fractionation protocol *(9)* in which the total microsomal fraction is layered at the bottom of, and floated through, a sucrose density gradient consisting of 1.22, 1.15, 0.86, and 0.25 *M* layers. The Golgi light fraction (GLF) is collected from the 0.86–0.25 *M* interface and used as the starting material for the purification of vesicles (*see* **Notes 4** and **5**).

3.4. Immuno-isolation

GLF (300 µL or 75 µg total protein) is diluted in binding buffer at a 3:7 ratio, added to 10 mg of beads coated with the primary antibody in a 1.5 mL Eppendorf tube and incubated overnight at 4°C with gentle mixing. After incubation, the beads are retrieved with the magnetic stand and the supernatant, which represents the nonbound subfraction (NB), saved for further analysis. The beads are then washed in washing buffer for up to 6 × 15 min.

The ratio of beads to starting material (*see* **Note 6**) is determined empirically by using the same volume of sample (GLF) and increasing amounts of coated magnetic beads starting from 1–15 mg. As a marker for the efficiency of the immuno-isolation we have used the newly synthesized polymeric IgA receptor labeled with [^{35}S] trans-label, in an in vivo pulse-chase. Ten milligrams of beads coated with anti-pIgA-R antibody retrieved over 90% of the receptor carrying vesicles from the GLF starting material.

3.5. Endocytic Uptake

[^{125}I] asialofetuin (3×10^8 cpm/100 μg) was injected in the portal vein and chased for 6 min. The liver is then removed and processed as above.

Dimeric IgA produced in a hybridoma cell line, is labeled in culture with 100 μCi [^{35}S] trans-label/mL and the tissue culture supernatant concentrated 10 times using a Centriprep device with 30K MW cut off. 5×10^4 cpm in dIgA as determined by immunoprecipitation are injected in the portal vein and chased for 12 min, followed by liver removal and fractionation.

3.6. Assays

Through immuno-isolation, the starting material (SM) is resolved in two subfractions NB and bound (B). These fractions are analyzed through a variety of assays.

1. Immunoprecipitation, or immunoblottingμ to follow the presence and/or distribution of specific markers. Quantitation is obtained either by using radiolabeled samples for immuno-isolation, followed by immunoprecipitation and scintillation counting or fluorography, or by immunoblotting followed by radioiodinated secondary antibodies (or protein-A) and autoradiography.
2. One or two dimensional SDS-polyacrylamide gel electrophoresis (PAGE) followed by silver or Coomassie-Blue staining.
3. Electron microscopy: SM, NB, and B are fixed in 4% PFA, 1% glutaraldehyde in 0.1 *M* cacodylate-HCl buffer, pH 7.2, for 45 min followed by 1% osmium tetroxide in the same buffer for 1 h. The samples are stained in block with 2% uranyl acetate (1 h), dehydrated in ethanol and propylene oxide and embedded in EPON.
4. Immunocytochemistry is performed on the B subfraction if different species antibodies are available. After the washes, the vesicles on the beads are reacted with the primary antibody (1 h), washed, and reacted with a secondary IgG conjugated to colloidal gold (1 h). After washing, the beads are fixed and processed as noted for TEM.
5. Functional assays in which the vesicles still on the beads (*see* **Note 7**) are incubated with cytosol using an intracellular buffer (20 m*M* HEPES, pH 7.2/140 m*M*

K+ (129 m*M* as potassium acetate and 11 m*M* as KOH)/10 m*M* Na^+/0.5m*M* Mg^{2+} 10.5 m*M* Cl−], adenosine triphosphorphate (ATP) and an ATP-regenerating system.

Following incubation at 37°C for 15 min the beads are retrieved from solution, washed in buffer without cytosol and any associated proteins can then be eluted with a high-molarity Tris solution (0.5 *M* Tris, pH 7.5). In this way, cytosolic proteins that associate with this class of vesicular carriers are identified and their role in trafficking investigated.

3.7. Results

1. Immuno-isolation is a method with high specificity that makes use of an antigen-antibody interaction. Its efficiency depends on the affinity of the antibodies for the antigen of interest and the abundance (surface density) of the antigen, which ideally, should be restricted to one compartment.
2. **Figure 1A,B** shows the morphology of the GLF (SM) and the NB subfraction; **Fig. 2A,B** shows the B subfraction as well as the control magnetic beads which are coated only with secondary antibodies. Out of a heterogeneous mixture of elements in SM, the B subfraction is highly enriched in one type of vesicles. Morphologically over 90% of the vesicles on the beads are small in size and have a clear content; a few large secretory vacuoles are seen as well as scattered Golgi cisternae. Owing to the preliminary centrifugation step, elements like mithocondria, lysosomes, or rough ER are practically excluded from the SM. In order to check for the content of these vesicles, we have surveyed several secretory and membrane proteins ***(9)***. Except for albumin and apolipoprotein B, which represent the newly synthesized proteins labeled in vivo with [^{35}S] met-cys and detected by immunoprecipitation and fluorography, in all other cases quantitation was done by SDS-PAGE and immunoblotting, followed by iodinated secondary antibodies and autoradiography. We found up to 90% segregation in the distribution of secretory and membrane proteins between the NB and B subfractions.
3. In order to exclude endocytic and transcytotic vesicles, we first labeled these vesicles by uptake in vivo using a marker for receptor-mediated endocytosis (asialofetuin) and a marker for transcytosis (dimeric IgA). The distribution of these markers in the SM as well as NB and B subfractions showed that there was practically no contamination with these classes of vesicles ***(9)***.
4. By combining a fractionation protocol with an immuno-isolation procedure we have obtained a highly pure population of vesicles that shuttle between trans Golgi network and the plasma membrane.
5. The vesicles on the beads are stable; they can be further manipulated and we have used them in cell-free assays like reconstituted fusion with the plasma membrane as well as incubations with cytosol under conditions mimicking the intracellular milieu.

Immuno-isolation on magnetic beads has been used mostly for cell purging but also for isolation of DNA, RNA, and, more recently, for isolation of subcel-

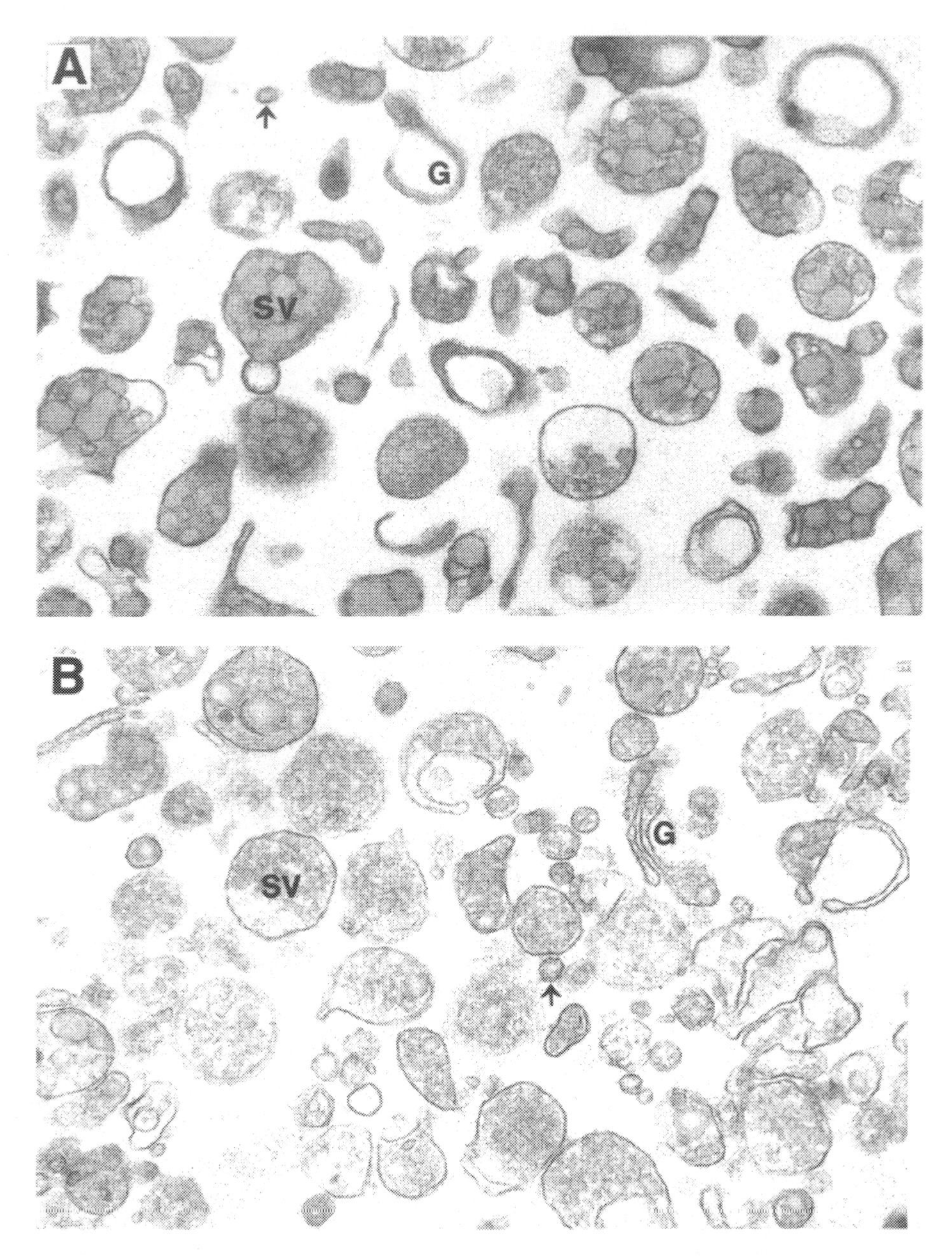

Fig. 1. (**A**) The GLF (SM) contains large secretory vacuoles (SV) filled with lipoproteins particles, unstacked Golgi cisternae (G) and small vesicles free or apparently tethered to the larger elements; (**B**) The NB subfraction (after overnight incubation of the beads with the SM), shows an abundance in secretory vacuoles, but Golgi cisternae can be seen as well as a few small vesicles. Their contents are of lower density suggesting partial extraction. Bar = 500 nm.

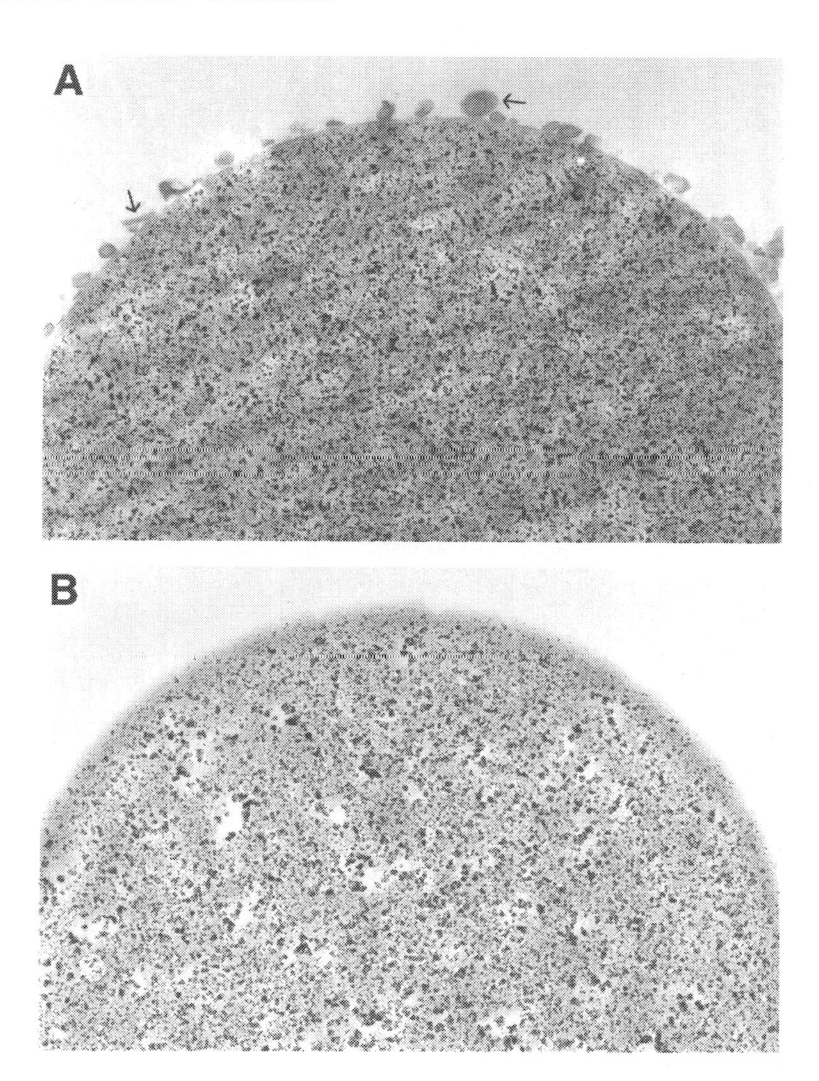

Fig. 2. (**A**) A section of a magnetic bead is shown: out of the SM only small vesicles with a clear content are isolated; occasionally unstacked Golgi cisternae or larger vesicles carrying few lipoprotein particles are seen (arrows); (**B**) a control bead, which has been coated with an unrelated antibody, shows no vesicles or other elements attached (note the smooth, nonporous surface). Bar = 500 nm.

lular organelles *(9,11)*. As more organelle specific antigens are being described and more antibodies against cytoplasmic epitopes available, we expect immuno-isolation on magnetic beads to be used for purification of other classes of vesicular carriers as well as organelles like peroxisomes *(12)*, mithocondria,

ER, nuclei, or plasma membranes. In this way, pure populations of specific organelles as well as specific proteins and their interacting partners could be studied in their natural organellar environment (*see* **Note 8**).

4. Notes

1. The magnetic beads are first coated with secondary antibodies for three reasons: to serve as an extension arm, to amplify the binding and give the right orientation to the primary antibodies. When IgM's are used as primary antibodies, they can be bound directly to the beads ***(13)***.
2. Polyclonal antibodies (PAb) have to be affinity-purified: secondary antibodies, because any other protein present would bind to the activated beads and compete for the binding; primary antibodies, because of the yield (any immune serum contains the most 10% of the specific antibody). Once coated with antibodies, the beads can be stored at 4°C in binding buffer containing 0.02% sodium azide for several months.
3. The buffers used for binding and washing of the beads should, in principle, contain extraneous protein for the reasons stated in the Methods section. Other investigators ***(6)*** have recommended 1% albumin in PBS for binding and 0.1% for washing. We found FCS to be more efficient in lowering nonspecific binding; we also added 2 m*M* EDTA as an inhibitor of lipid peroxidation, which occurs in the membranes during the incubation period ***(14)***.
4. Immuno-isolation is a powerful technique, but is limited by the availability of a specific antibody as well as the availability of an antigen restricted to a specific organelle. In this respect, the pIgA-R is not an ideal antigen because it is present at different concentrations throughout the secretory pathway. This is why we have used a preliminary-gradient centrifugation step.
5. The concentrated sucrose solutions should be diluted; isotonicity of the buffers might be important for some organelles.
6. The ratio of magnetic beads: sample should be determined by each investigator because it depends on many factors such as: affinity of the antibodies, efficiency of coupling to the beads, presence and density of the antigen in the sample, and incubation times.
7. Elution of the vesicles or organelles off the beads is possible either by using a competing peptide or excess antigen when it is available. Dynal Inc. offers a special buffer for elution of cells of magnetic beads, other investigators have recommended mechanical elution ***(8,15)***.
8. Magnetic beads represent one of the best solid supports available and this is owing to several important qualities they posses: they are monosized (equal in size), they have smooth surfaces, are nonporous, and have low hydrophobicity (less nonspecific binding), they can be retrieved from solution with the help of a magnet (avoiding repeated centrifugation steps and contamination), and are easy to resuspend (they are paramagnetic—lose their magnetism once removed from the magnet). The magnetic beads are far superior compared to the classical types of solid supports ***(6)***.

Acknowledgments

I wish to thank Drs. John Ugelstad and Ruth Schmid from Sintef-Polymer Group in Trondheim, Norway and Dynal Co. (Great Neck, NY) for providing us with the magnetic beads, and Dr. George Palade for his advice and critical reading of this manuscript. This work was supported by a National Cancer Institute program project (CA58689).

References

1. Palade, G. E. (1975) Intracellular aspects of the process of protein (synthesis) secretion. *Science* **89,** 347–358.
2. Rothman, J. E. (1994) Mechanisms of intracellular protein transport. *Nature* **372,** 55–63.
3. Pryer, K. N., Wuestehube, L. J., and Schekman, R. (1992) Vesicle-mediated protein sorting. *Annu. Rev. Biochem.* **61,** 471–516.
4. Pfeffer, S. R. and Rothman, J. E. (1987) Biosynthetic protein transport and sorting by the endoplasmic reticulum and Golgi. *Annu. Rev. Biochem.* **56,** 829–852.
5. Beaufay, H. and Amar-Costesec, A. (1976) Cell fractionation techniques. *Methods Membr. Biol.* **6,** 1–99.
6. Howell, K. E., Schmid, R., Ugelstad, J., and Gruenberg, J. (1989) Immunoisolation using magnetic solid supports: subcellular fractionation for cell-free functional studies, in *Laboratory Methods in Vesicular and Vectorial Transport* (Tartakoff, A., ed.), Academic Press, Inc., New York, pp. 171–198.
7. Ugelstad, J., Mork, P. C., Kaggerud, K. H., Ellingsen, T., and Berge, A. (1980) Swelling of oligomer-polymer perticles. new methods of preparation of emulsions and polymer dispersions. *Adv. Colloid Interface Sci.* **13,** 1201–1243.
8. Ugelstad, J., Kilaas, L., Aune, O., Biorgum, J., Herje, R., Schmid, R., Stenstad, P., and Berge, A. (1994) Monodisperse polymer particles: preparation and new biochemical and biomedical applications, in *Advances in Biomagnetic Separation* (Uhlen, M., Hornes, E., and Olsvik, O., eds.), Eaton Publishing Co., Natick, MA, pp. 1–19.
9. Saucan, L. and Palade, G. E. (1994) Membrane and secretory proteins are transported from the Golgi complex to the sinusoidal plasmalemma of hepatocytes by distinct vesicular carriers. *J. Cell Biol.* **125(4),** 733–741.
10. Nustad, K., Danielsen, H., Reith, A., Funderud, S., Lea, T., Vartdal, F., and Ugelstad, J. (1988) Monodisperse particles in immunoassays and cell separation, in *Microspheres: Medical and Biological Applications* (Rembaum, A. and Tokes, Z. A. eds.), CRC Press, Boca Raton, FL, pp. 53–75.
11. Howell, K. E., Crosby, J. R., Ladinsky, M. S., Jones, S. M., Schmid, R., and Ugelstad, J. (1994) Magnetic solid supports for cell-free analysis of vesicular transport, in *Advances in Biomagnetic Separation* (Uhlen, M., Hornes, E., and Olsvik, O., eds.), Eaton Publishing Co., Natick, MA, pp. 195–204.
12. Luers, G., Hartig, R., Fahimi, H. D., Cremer, C., and Volkl, A. (1996) Immunomagnetic isolation of peroxisomes. *Ann. NY Acad. Sci.* **804,** 698–700.

13. Kvalheim, G., Fodstad, G., Phil, A., Nustad, K., Pharo, A., Ugelstad, J., and Funderud, S. (1987) Elimination of B-Lymphoma cells from human bone marrow: model experiments using monodisperse magnetic particles coated with primary monoclonal antibodies. *Cancer Res.* **47,** 846–851.
14. Howell, K. E. and Palade, G. E. (1982) Heterogeneity of lipoprotein particles in hepatic Golgi fractions. *J. Cell Biol.* **92,** 833–845.
15. Ghetie, V., Mota, G., and Sjoquist, J. (1978) Separation of cells by affinity chromatography on SpA-Sepharose 6MB. *J. Immunol. Methods* **21,** 133–141.

II

Techniques for the Measurement of Antioxidant Activity

26

Simultaneous Determination of Serum Retinol, Tocopherols, and Carotenoids by HPLC

Richard W. Browne and Donald Armstrong

1. Introduction.

Epidemiological studies have confirmed that dietary supplementation with vitamin E significantly inhibits oxidation of low-density lipoproteins *(1)* and reduces the risk of atherosclerosis *(2)* and coronary heart disease *(3)*. Retinol and β-carotene (β-caro) are important micronutrient vitamins in reducing the risk of age-related macular degeneration *(4)* and cancer *(5)*. The simultaneous determination of retinol, tocopherols, and carotenoids by high-pressure liquid chromatography (HPLC) using ultraviolet-visible absorbance (UV-VIS) detection has become the method of choice for investigators interested in determining fat-soluble vitamin levels in health and disease *(6)*. This method achieves the desired separation and provides quantitation of numerous fat-soluble vitamins in biological specimens *(7)*. Furthermore, the performance characteristics of the method have been extensively studied by the National Institute of Standards and Technology (NIST/NCI Micronutrients Measurement Quality Assurance Program, Chemistry B208, Gaithersburg, MD), which now supplies both quality-control material and offers proficiency testing for multiple analytes, as well as workshops on HPLC analysis of these analytes.

We describe here a variation of this methodology that allows for the isocratic, automated determination of 11 different fat-soluble vitamins in a minimum amount of time.

2. Materials

2.1. Equipment

1. Vortex/mixer (*see* **Note 1**).
2. Organic solvent evaporator.

From: *Methods in Molecular Biology, vol. 108: Free Radical and Antioxidant Protocols*
Edited by: D. Armstrong © Humana Press Inc., Totowa, NJ

3. Centrifuge.
4. UV-VIS spectrophotometer.
5. Analytical HPLC instrumentation.
   a. Shimadzu (Kyoto, Japan) LC-6A HPLC pump.
   b. Shimadzu SPD-M6A UV-VIS photodiode array (*see* **Note 2**).
   c. Shimadzu SIL-7A autosampler/injector.
   d. Shimadzu LPM-600 low-pressure mixer.
   e. 486Dx PC.
   f. Shimadzu Class-VP chromatography software.
   g. Kontes (Vineland, NJ) HPLC solvent reservoir and degassing system.
   h. Supelco (Bellemonte, PA) LC-18 HPLC column (4.6 × 150 mm, 5 µL particle size,100Å pore).
   i. Supelcoguard C18 guard column(4.6 × 20 mm, 5µL Particle size, 100Å pore).
6. Conical glass HPLC autosampler vial inserts (300 µL) Supelco.
7. Vacutainer Tubes (Baxter Inc., McGraw Park, IL).

2.2. Reagents

1. The following solvents are HPLC grade obtained from Fisher Scientific (Pittsburgh, PA) or J. T. Baker (Phillipsburg, NJ): acetonitrile (ACN), methanol (MeOH), dichloromethane (Cl_2CH_3), triethylamine (TEA), absolute ethanol (EtOH), hexane (Hex), and ammonium acetate (NH_4Ac).
2. Nitrogen and helium (prepurified).
3. The Internal standard (ISTD) solution consisted of 10 mg of tocopherol acetate and 30 mg of butylated hydroxytoluene (BHT) in 500 mL ethanol. Mix and store at 0–4°C. Stable at least 6 mo.

3. Methods

3.1. Sample Processing

1. Collect whole blood into a Vacutainer-type tube containing no anticoagulant. Wrap immediately in aluminum foil to protect the sample from exposure to light (*see* **Note 3**).
2. Let stand at room temperature for 30 min for clotting, then centrifuge at 3000*g* for 10 min, decant serum, and analyze immediately, or store at –70°C in a light protected container.

3.2. Sample Extraction

1. Vortex samples and combine 200 µL serum, 200 µL ethanol, and 200 µL ISTD. Vortex again and inspect to see that proteins have precipitated.
2. Add 2 mL Hex, cap, and vortex each tube vigorously for 45 s. Centrifuge tubes for 1 min to separate phases and remove upper layer. Repeat extraction of remaining lower layer with an additional 2 mL of Hex; pool the extracts.

3. Evaporate extracts under nitrogen in the dark. Reconstitute in 200 μL mobile phase and sonicate for 30 s. Transfer to appropriate autosampler rack and analyze by HPLC.

3.3. HPLC Procedure

3.3.1. Mobile Phase

A multiple-reservoir solvent system consisting of three solvent reservoirs is used. The premixed mobile phase is placed in reservoir A and pure ACN and water are placed in reservoir B and C, respectively. This allows for rapid solvent change-over when washing and rinsing the analytical system. Reservoir A contains 600 mL ACN, 200 mL MeOH, 0.77 g of NH_4Ac, 200 mL Cl_2CH_3 and 1.0 mL TEA in 1 L total volume. All mobile-phase solvents should be filtered through 0.22-μm nylon membranes using a suitable filtration unit.

3.3.2. Standards/Calibrators

Pure standards (Sigma Chemical Co., St Louis MO; Indofine Chem. Co., Somerville, NJ; Fluka Chemical Co., Ronkonkoma NY) are dissolved in ETOH or HEX and prepared using the extinction coefficients (Σ) and absorbance maxima shown in **Table 1**.

1. Dilute raw calibrator in appropriate solvent and dissolve and filter.
2. Read absorbance against a solvent blank at the appropriate wavelength. Dilute as necessary to achieve a solution with absorbance between 0.5 and 0.75 absorbance units.
3. Determine concentration of standard by applying appropriate extinction coefficient (Σ) and light path in cm (b) to Beer's law:

$$[\text{Conc. Std. g/dl}] = \text{Abs}_{(\lambda\text{max})}/\Sigma b$$

4. Run pure standard solution on HPLC and determine the purity of the standard (*see* **Note 4**). Correct standard concentration by multiplying the concentration by the percent of the total area generated by the standard under optimal HPLC conditions.
5. Dilute this final solution to make standards that encompass the range of sample concentrations that you expect to encounter.

3.3.3. HPLC Analysis

Samples or standards are placed into 300 μL conical, glass inserts, which are then loaded into spring-loaded, amber-glass autosampler vials with Teflon septa screw caps. Fifty μL is injected and samples are eluted isocratically using the mobile phase described in **Subheading 3.3.1.**, at a flow rate of 1.5 mL/min The typical pressure of the system is approx 70 kg/cm^3. Absorbance data is collected at 0.64-s intervals on five separate channels at 284 nm, 292 nm, 326 nm, 452 nm,

Table 1
Spectrophotometric Characteristics of Standards

Compound Name	Solvent	λ-Max	Σ (dl/g cm)	Vendor
α-tocopherol	Ethanol	292	75.8	Sigma
γ-tocopherol	Ethanol	298	91.4	Sigma
δ-tocopherol	Ethanol	297	91.2	Sigma
tocopherol acetate	Ethanol	284	ISTD	Sigma
α-carotene	Hexane	444	2800	Sigma
β-carotene	Hexane	452	2592	Sigma
lycopene	Hexane	472	3450	Indofine
lutein	Ethanol	445	2765	Indofine
zeaxanthin	Ethanol	452	2416	Indofine
β-cryptoxanthin	Ethanol	452	2486	Indofine
retinol	Ethanol	325	1850	Fluka
retinol palmitate	Ethanol	325	975	Sigma

and 478 nm, all with 4-nm slit widths using a photodiode array detector. Three-dimensional raw data is processed and stored by the on-line computer (PC) and resident chromatography software. This allows for post-analysis processing of data as well as 2-D, 3-D, or contour display of the chromatograms.

Automated analysis is accomplished under the control of the SIL-7A unit, which injects a sample and initiates data collection every 10.5 min. The PC and resident software begin data collection at the prompt of the autosampler and automatically stops and resets at the end of the 9.0 min separation period while awaiting the next prompt.

3.4. Results

3.4.1. Chromatographic Separation

A representative chromatogram acquired from human serum is shown in **Fig. 1**. The acquisition parameters of the software call for a 1.4-min acquisition delay, which removes troublesome solvent front peaks from peak identification and integration. Other integration events such as width and threshold should be determined by individual laboratories.

1. An advantage to this procedure is the short chromatographic time. In **Fig. 1A**, the three isomers of vitamin E elute in under 5 min and in **Fig. 1B**, retinol elutes under 1 min. The carotenoids are eluted within 7.5 min, as shown in **Fig. 1C**.
2. The unidentified shoulders associated with xanthophyll and β-cryptoxanthin (β-crypt) probably represent isomers. Chromatograph software allows tangent skimming and removal of these shoulders so that the major peak, identified by identical retention time as those of the standards, can be quantified.

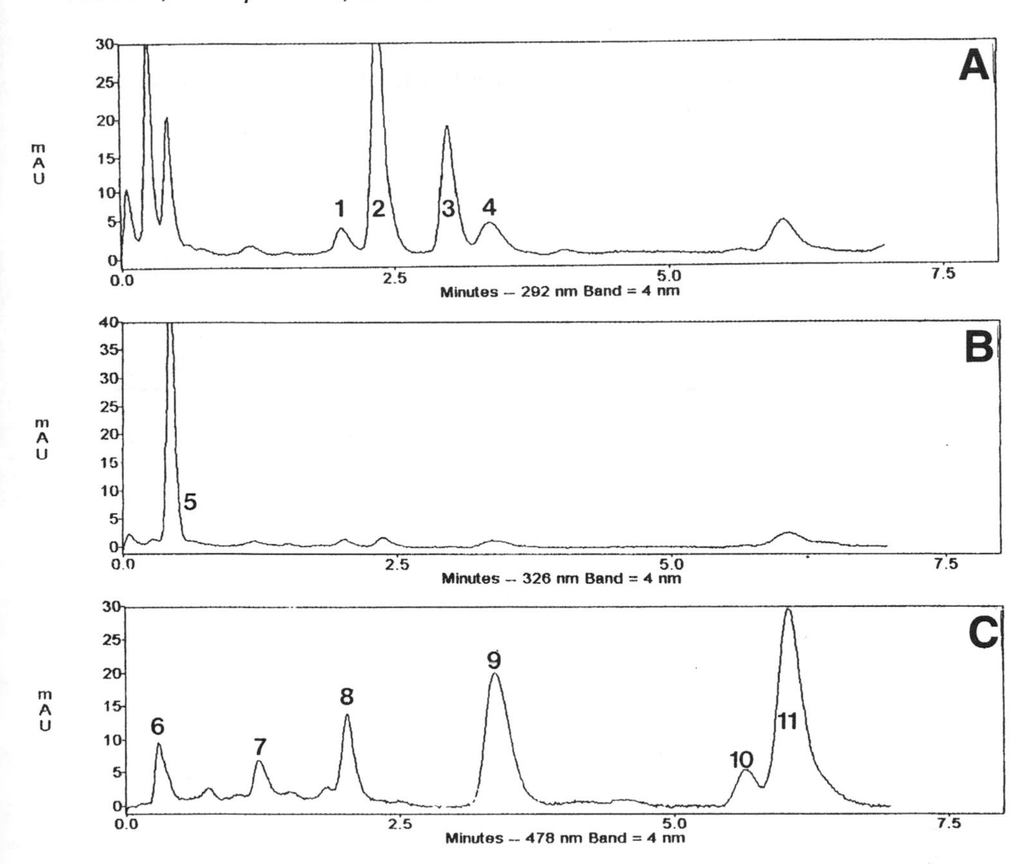

Fig. 1. Representative chromatogram of human serum analyzed by photodiode array detector. The abscissa is milli absorbance units and the ordinate is retention time in min at 292 nm (**A**), 326 nm (**B**), and 452 nm (**C**). 1, γ-tocopherol, 2, α-tocopherol, 3, tocopherol acetate (internal standard), 4, Δ-tocopherol, 5, all-trans retinol, 6, lutein and zeaxanthine, 7, canthaxanthine, 8, β-cryptoxanthine, 9, lycopene, 10, α-carotene, 11, β-carotene.

3. Spectrum Max Plot capability of the Class-VP software allows all peaks in **Fig. 1A–C** to be printed as one scan, which provides a summary of the maximum absorbance at each point along the chromatogram.
4. For automated batch analysis, 5 samples can be analyzed per h. The method is efficient and uses less than 1 L of mobile phase to analyze 40 samples.

3.4.2. Observed Range of Vitamins in Human Sera

Table 2 depicts the range of values our laboratory has seen in normal human-serum samples for some of the more common analytes.

Table 2
Vitamin Levels in a Sample Population of Presumably Normal Healthy Individuals[a]

Vitamin	Mean	SD	Minimum	Maximum
α-toc	9.80	3.11	5.140	19.120
Vit A	0.57	0.17	0.065	0.860
β-Caro	0.22	0.19	0.034	0.690
Lyco	0.61	0.42	0.111	1.537
Lu/Zx	0.18	0.08	0.042	0.336
β-Crypt	0.11	0.08	0.032	0.393

[a]All values are in μg/mL serum.

1. In normal human serum α-tocopherol (α-toc) is the major vitamin E form. γ-tocopherol and δ-tocopherol are normally present in small amounts. β-tocopherol (β-toc) is only present in trace quantities and is not normally detected by this method.
2. Retinol (Vit A) elutes within the first 2 min. Retinol palmitate elutes after β-caro by this method and is not shown. Extending the run time allows for measurement of this retinoid, which elutes at approx 8.5 min.
3. For the quantitation of carotenoid peaks, leutein and zeaxanthin (Lu/Zx) are measured together because they are difficult to resolve. The rank order of carotenoids in serum is lycopene (Lyco) > β-caro > Lu/Zx and β-crypt > α-caro. The remaining compounds are generally low in concentration, but can vary substantially, probably related to diet.

4. Notes

1. There are a multitude of instruments available that satisfy the requirements for this analysis. We use a simple "home-made" nitrogen-evaporating manifold consisting of peristaltic pump tubing connected to a series of "Y" type joints and titanium probes, allowing the simultaneous evaporation of 16 samples. However, it is important to clean degassing probes thoroughly to prevent carry-over or contamination from sample to sample. The spectrophotometer used should be able to resolve particular wavelengths with the narrowest possible bandwidth in order to prevent overestimation of the standard absorbance. The vortexer and centrifuge are only necessary to provide for the proper mixing and separation of the hexane extraction phase.
2. Refer to **Fig. 1** for the relative retention times of the analytes of interest. The photodiode array is employed because it allows for the analysis of each individual peak at its maximum wavelength of absorbance (i.e., one channel for each wavelength). The minimum capability of the detector for this assay requires a two-channel programmable UV-VIS absorbance detector or a combination of UV-Vis/Fluorescence detection. In the first case one channel of the UV-Vis detector can be programmed to detect peaks with wavelength maxima at 325 nm for the first 2 min

of analysis (for the determination of retinol) and then at 284–298 nm for the determination of tocopherols. The second channel can be programmed to switch between 450 and 478 nm for the determination of carotenoids. The combination of UV-Vis/Fluorescence detection requires that the UV detector collect data at 326 nm for the first 2 min of the chromatogram and then switch the carotenoid wavelengths of 450–478 nm. The fluorescence detector continuously monitors the tocopherols at and excitation wavelength of 295 nm with emission at 335 nm.

3. Perform extraction procedure so as to protect samples from light, i.e., keep tubes covered with foil or use opaque (amber) tubes. Use poly-styrene plastic or glass tubes for the extraction.
4. This step is done in order to correct for any contaminants present in the standard mixture. When determining the percent purity, it is not necessary to consider peaks that are associated with the solvent front. Simply integrate the entire chromatogram (excluding the solvent front) and determine the percent of the total area which is attributed to the peak of interest. This is the percent purity and, considering the high quality of the standards now available from many manufactures, is usually between 95 and 100% pure.

References

1. Reaven, P. D., Herold, D. A., Barnett, J., and Edelman, S. (1995) Effects of vitamin E on susceptibility of low-density lipoprotein and low-density lipoprotein subfraction to oxidation and or protein glycation in NIDDM. *Diabetes Care* **18,** 807–816.
2. Nyyssoner, K., Pokkala, E., Saloner, R., Korpela, H., and Saloner, S.F. (1994) Increase in oxidation resistance of atherogenic serum lipoprotein following antioxidant supplementation: a randomized double-blind placebo-controlled clinical trial. *Eur. J. Clin. Nutr.* **48,** 633–642.
3. Stampfer, M. S., Hennekens, C. H., Manson, J. E., Colditz, G. A., Rosner, B., and Willett, W. C. (1993) Vitamin consumption and the risk of coronary disease in women, *N. Engl. J. Med.* **328,** 1444–1449.
4. Seddon, J. M., Ajani, V. A., Sperdutto, R. D., Hiller, R., Blair, N., Burton, T. C., Farher, M. D., Gragondas, E. S., Haller, J., Miller, D. T., Yannuzzi, L. A., and Willet, W. (1994) Dietary carotinoids, vitamin A, C and E and advanced age-related macular degeneration. *JAMA* **272,** 1413–1420.
5. Poppel, G. (1993) Carotinoids and cancer: an update with emphasis on human intervention studies. *Eur. J. Cancer* **29A,** 1335–1344.
6. Mitton, K. P. and Trevittrick, J. R. (1994) High-performance liquid chromatography-electrochemical detection of antioxidants in vertebrate lens: glutathione, tocopherol, and ascorbate, in *Oxygen Radicals in Biological Systems, Methods in Enzymology (vol.233)* (Abelson, J. N. and Simon, M. I., eds.), Academic Press, Inc. San Diego, CA, pp. 523–539.
7. Epler, K. S., Ziegler, R. G., and Craft, N. E. (1993) Liquid chromatographic method for the determination of carotenoids, retinoids and tocopherols in human serum and in food. *J. Chromatog.* **619,** 37–48.

27

EPR Measurements of Nitric Oxide-Induced Chromanoxyl Radicals of Vitamin E

Interactions with Vitamin C

Valerian E. Kagan and Nikolai V. Gorbunov

1. Introduction

Tocopherols (vitamin E) are the major lipid-soluble chain-breaking antioxidants of membranes and blood plasma ***(1–3)***. Antioxidant function of tocopherols is confined to their chromanol nucleus with 6-hydroxy group, and is based on the ability of the latter to reduce different radicals (e.g., lipid peroxyl radicals [LOO*], oxygen radicals) yielding chromanoxyl radicals ***(2–4)***:

$$\text{LOO*} + \text{Toc-OH} \rightarrow \text{LOOH} + \text{Toc-O*} \quad (1)$$

Nitric oxide (NO) is among the radicals that can be directly reduced by 6-hydroxy-chromanes (e.g., upon exposure of tocopherol solutions in organic solvents to NO gas ***(5)***:

$$\text{NO*} + \text{Toc-OH} \rightarrow \text{NO}^- + \text{Toc-O*} + \text{H}^+ \quad (2)$$

Thus overproduction of NO in vivo may cause antioxidant depletion ***(5,6)***.

High efficiency of tocopherols in protecting polyunsaturated lipids in membranes and lipoproteins against free-radical damage is, to a large extent, supported by the recycling mechanisms ***(3,7,8)***. These include redox reactions in which physiological reductants with appropriate redox potentials (e.g., ascorbate, thiols) regenerate vitamin E at the expense of their own oxidation to the respective radicals (e.g., semidehydroascorbyl radicals, thiyl radicals) ***(3,7,8)***:

$$\text{Toc-O*} + \text{AH, (GSH)} \rightarrow \text{Toc-OH} + \text{A}\cdot\text{, (GS}\cdot\text{)} \quad (3)$$

Interplay between tocopherols, NO*, and intracellular reductants, such as ascorbic acid, may thus be important for the overall antioxidant balance of

From: *Methods in Molecular Biology, vol. 108: Free Radical and Antioxidant Protocols*
Edited by: D. Armstrong © Humana Press Inc., Totowa, NJ

cells, especially those exposed to oxidant stress, e.g., during the inflammatory response *(9)*.

This chapter describes several different electron paramagnit spectroscopy resonance (EPR) procedures that may be useful for studies of the reactions of α-tocopherol, ascorbate, and combination thereof with NO in aqueous environments. Additionally, we have included examples of artifacts resulting from the presence of NxOy in aerobic NO gas compositions.

2. Materials

2.1. Instruments and Devices

1. JEOL JES-RE1X spectrometer with 100 kHz magnetic field modulation.
2. Xenon lamp, model 6251, lamp housing model 66002 (Oriel Corp., Stratford, CT).
3. Interference filters (Oriel Corp.).
4. Gas-permeable Teflon tubing, 0.8 mm i.d., 13 μ*M* wall thickness (Alpha Wire Co., Elizabeth, NJ).
5. A standard quartz flat cell for EPR measurements.

2.2. Reagents

All reagents should be of the highest analytical quality to ensure accurate and reproducible results.

1. D,L-α-Tocopherol (Sigma Chemical Co., St. Louis, MO).
2. L-Ascorbic acid (Fisher Scientific, Fair Lawn, NJ).
3. Sodium dodecyl sulfate (SDS) (Sigma).
4. NO gas: (Valley Welding Supply, Pittsburgh, PA) 99.5% (poison gas, corrosive). An additional purification was achieved by passing NO gas through a glass tube filled with sodium hydroxide.
5. Oxygen gas (O_2): (Valley Welding) high-purity.
6. Nitrogen gas (N_2): (Valley Welding) 99.999%. High-purity grade is essential to avoid the presence of oxygen in anaerobic atmosphere. Trace amounts of oxygen in the NO/N_2-composition leads to the formation of a mixture of strong oxidants NxOy (*see* **Note 1**).
7. Chelating resin Chelex-100: (Bio-Rad, Hercules, CA) analytical grade.

2.3. Solutions

Aqueous buffers and solutions are treated with Chelex-100 prior to all experiments. Chromanoxyl (α-tocopheroxyl) radicals with an extended lifespan can be generated in charged micellar suspensions (100 m*M* SDS in 100 m*M* phosphate buffer, pH 7.4) to avoid radical-radical quenching of chromanoxyl radicals.

3. Methods

3.1. Sample Preparation

Samples of 0.1 m*M* α-tocopherol (solubilized in 100 m*M* solutions of SDS at a 1000:1 molar ratio of SDS to α-tocopherol) and/or 0.5 m*M* ascorbate in phosphate buffer containing 100 m*M* SDS are used to produce chromanoxyl radicals and to activate chromanoxyl radical-dependent formation of semidehydroascorbyl radicals. Samples of 60 µL total volume are placed into a standard quartz flat cell for ultraviolet (UV)-irradiation experiments, or into gas-permeable Teflon tubing for chemical-reaction experiments. To react aqueous compositions with gases, a gas-permeable Teflon tubing filled with sample is placed into a standard quartz tube adjusted for EPR measurements.

3.2. UV Irradiation

Samples in quartz tubes (or in a standard flat quartz cuvet) inserted into cavity of the EPR spectrometer are exposed to filtered UV light through a window in the front of the resonane cavity. UV light is generated by a xenon lamp (operating at 120 W power and placed at a distance of 20 cm from the sample). Interference filters are selected to allow for passage of photons at or near the UV absorption maximum of the analytc of interest (260 ± 4.5 nm for ascorbate, 290 ± 5.0 nm for α-tocopherol).

3.3. Gas Composition Replacements

Reactions of antioxidants with NO gas can be carried out under aerobic or anaerobic conditions. The latter are achieved by placing and preincubating gas-permeable Teflon tubing with a sample in a standard quartz tube that is purged with N_2 gas for 15 min before addition of NO. Samples are exposed to NO by purging the N_2 atmosphere with NO gas for additional 15 min. Anaerobic conditions in NO-exposed samples are confirmed by the absence of semidehydroascorbyl radicals that immediately become detectable upon exposure to trace amounts of O_2 reacting with NO to yield NxOy.

3.4. EPR Measurements

EPR measurements are performed at room temperature by using a JEOL JES-RE1X spectrometer (with a 100-kHz magnetic-field modulation). The spectrometer settings are: center field 335.0 mT; modulation amplitude 0.25 or 0.05 mT; scan range 10 mT; scan time 8 min; time constant 10 ms; microwave power 10 mW. Hyperfine coupling constants can be evaluated relative to external standards of Mn^{2+} in MgO.

3.5. Results

3.5.1. EPR Detection of Chromanoxyl Radicals Produced Upon Exposure of α-Tocopherol to NO

Addition of NO to SDS micelles containing α-tocopherol (1000:1 molar ratio of SDS to α-tocopherol) under either aerobic or anaerobic conditions in the dark produces an EPR spectrum of chromanoxyl radicals with seven resolved components (**Fig. 1**, Line 2, Line 4, respectively). Using filtered UV light (290 ± 5.0 nm, which is overlapping with the absorbance maximum for α-tocopherol at pH 7.4), chromanoxyl radicals with identical EPR features are generated from α-tocopherol under anaerobic conditions (**Fig. 1**, Line 5). NO enhances EPR signal of the chromanoxyl radical, when added to UV-exposed α-tocopherol (**Fig. 1**, Line 6).

Similarly, NO generates chromanoxyl radicals from a water-soluble analog of α-tocopherol–Trolox (1 m*M* in 100 m*M* phosphate buffer, pH 7.4, under anaerobic conditions) (**Fig. 2**, Line 2).

3.5.2. Reduction of Chromanoxyl Radicals by Ascorbate in the Presence of NO: EPR Detection of Redox Cycling of α-Tocopherol

1. Upon exposure of combinations of α-tocopherol and ascorbate to UV light (290 ± 5.0 nm) under anaerobic conditions, an EPR spectrum a two-line signal characteristic of semidehydroascorbyl radical is observed (**Fig. 3**, Line 2). No EPR signals are detected when the solutions are exposed to UV light in the range 260 ± 4.5 nm (maximum absorbance for ascorbate) (**Fig. 3**, Line 3). These results suggest that UV-induced oxidation of α-tocopherol to chromanoxyl radical triggered the recycling of chromanoxyl radicals back to α-tocopherol by ascorbate (*see* **Note 2**).
2. Similar experiments can be performed to demonstrate ascorbate-dependent NO-induced recycling of chromanoxyl radicals generated in anaerobic conditions. Incubation of α-tocopherol/ascorbate mixtures with NO in anaerobic conditions yields a two-line semidehydroascorbyl radical EPR signal (**Fig. 3**, Line 4). NO does not produce semidehydroascorbyl radicals (under anaerobic conditions) in the absence of α-tocopherol (**Fig. 3**, Line 5). Aerobic incubation of NO and ascorbate, however, immediately results in the appearance of semidehydroascorbyl radical EPR signal (**Fig. 3**, Line 6):

$$NO^* + O_2 \rightarrow OONO^* \quad (4)$$

$$A^- + OONO^* \rightarrow A^* + NO_3^- \quad (5)$$

3. NO is capable of directly oxidizing α-tocopherol to chromanoxyl radical that, in the presence of ascorbate, is reduced back to tocopherol at the expense of ascor-

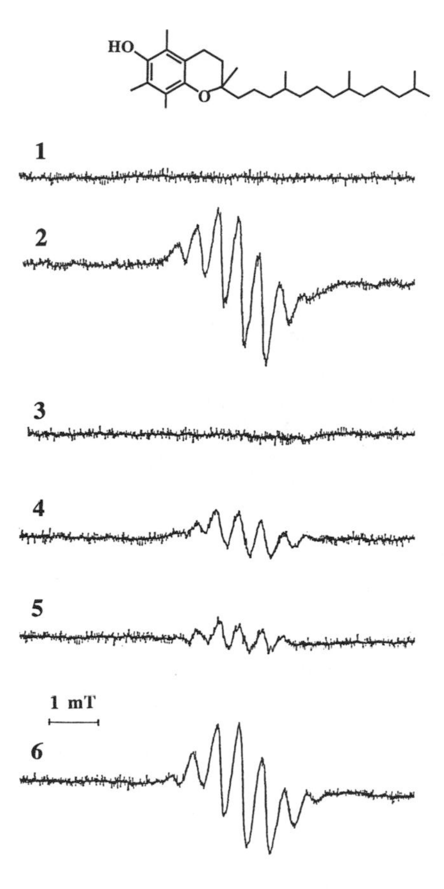

Fig. 1. EPR spectra of chromanoxyl radical generated from α-tocopherol by NO and/or UV light under anaerobic or aerobic conditions. 1, A suspension of 0.1 m*M* α-tocopherol in 100 m*M* SDS in aerobic atmosphere. 2, A suspension of 0.1 m*M* α-tocopherol in 100 m*M* SDS after purging NO gas (50 mL/min, 2 min) under aerobic conditions. 3, Same as 1 except that the air was completely substituted for N_2 by (purging for 15 min). 4, Same as 3, but after purging NO gas in N_2 atmosphere (50 mL/min, 2 min). 5, Same as 3, but after 2 min of UV- irradiation (290 ± 5.0 nm). 6, Same as 4, but after 2 min of UV irradiation (290 ± 5.0 nm).

bate oxidation to semidehydroascorbyl radical. In accord with this, is the time-course of NO-induced chromanoxyl semidehydroascorbyl radical EPR signals (in SDS micelles under anaerobic conditions, **Fig. 4**). The seven-line EPR spectrum of the chromanoxyl radical appears immediately upon exposure of α-tocopherol to NO when ascorbate is not a part of the reaction mixture (**Fig. 4**). The presence of ascorbate in the reaction mixture delays the appearance of

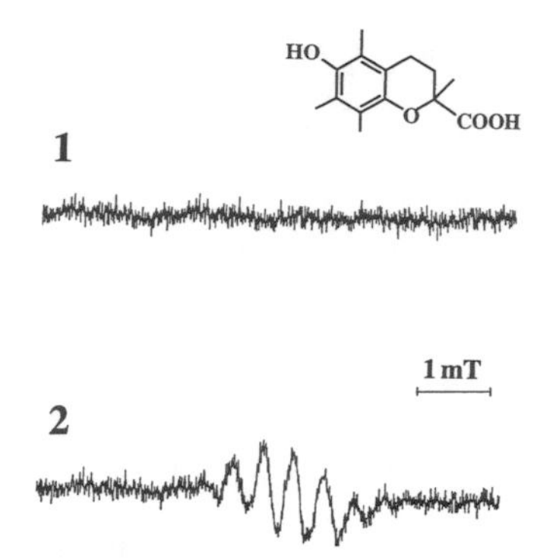

Fig. 2. EPR spectra of chromanoxyl radical obtained from Trolox anaerobically exposed to NO. 1, Solution of Trolox (1 m*M* in 100 m*M* phosphate buffer, pH 7.4) in N_2 atmosphere. 2, Same as 1, but after purging NO gas in N_2 atmosphere (50 mL/min, 2 min).

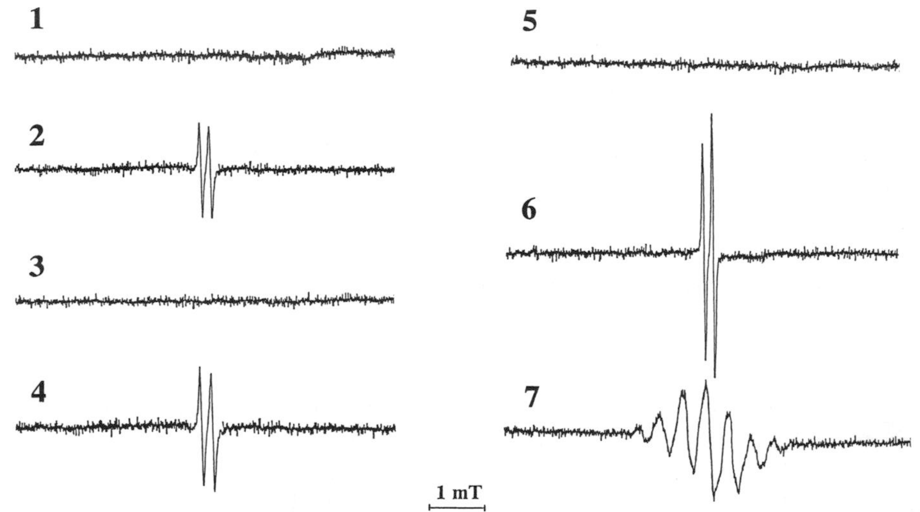

Fig. 3. EPR spectra obtained from α-tocopherol/ascorbate composition anaerobically exposed to NO plus UV light. 1: 0.1 m*M* α-tocopherol in 100 m*M* SDS and 0.5 m*M* ascorbate in N_2 atmosphere (dark). 2, Same as 1, but after 2 min of UV- irradiation (290 ± 5.0 nm). 3, Same as 2, except that α-tocopherol was omitted from the system. 4, Same as 1, but after purging NO gas in N_2 atmosphere (50 mL/min, 2 min). 5, Same as 4, except that α-tocopherol was omitted from the system. 6, Same as 5, except that trace amounts of oxygen were present. 7, Same as 4, except that ascorbate was omitted from the system.

chromanoxyl radical; instead, the EPR signal of semidehydroascorbyl radical is observed. The chromanoxyl radical could be only detected when ascorbate is no longer present in the composition that as evidenced by the disappearance of semidehydroascorbyl radical signal (**Fig. 4**).

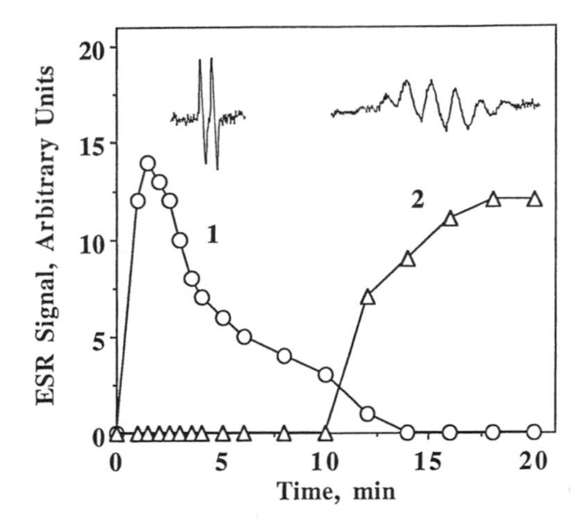

Fig. 4. Time-course of EPR signals of semidehydroascorbyl radicals and chromanoxyl radicals produced by exposure of α-tocopherol to NO in the presence of ascorbate. 1, Semidehydroascorbyl radicals (open circles). 2, Chromanoxyl radicals (open triangles). Conditions: 0.1 m*M* α-tocopherol in 100 m*M* SDS and 0.5 m*M* ascorbate were incubated in the presence NO under anaerobic conditions.

4. Notes

1. Prooxidant effects of NO are potentiated in the presence of trace amounts of oxygen. To avoid contamination of anaerobic atmosphere with oxygen, high-purity nitrogen gas (99.999%) should be used.
2. Only in the presence of oxygen (even in trace amounts) is nitric oxide able to directly oxidize ascorbate to semidehydroascorbyl radical in aqueous solutions. This can be used for monitoring completeness of oxygen removal from solutions containing NO. The semidehydroascorbyl radical EPR signal disappears upon complete substitution of oxygen for nitrogen. Thus conditions can be determined to provide for anaerobic reactions of NO with reagents of interest.
3. The efficacy of the free radical reactions induced by NO anaerobically depends on experimental conditions.

Acknowledgments

N.V.G. is a recipient of NRC Fellowship Award.

References

1. Tappel, A. L. (1962) Vitamin E as the biological antioxidant. *Vitam. Horm.* **20,** 493–510.
2. Frei, B. B. and Ames, B. N. (1993) Relative importance of vitamin E in antiperoxidative defenses in human blood and low-density lipoprotein (LDL), in

Vitamin E in Health and Diseases (Packer, L. and Fuchs, J., eds.), Marcel Dekker, New York, pp. 131–141.

3. Packer, L. and Kagan, V. E. (1993) Vitamin E: the antioxidant harvesting center of membranes and lipoproteins, in *Vitamin E in Health and Diseases* (Packer, L. and Fuchs, J., eds.), Marcel Dekker, New York, pp. 179–193.
4. Liebler, D. C. (1993) Peroxyl radical trapping reactions of α-tocopherol in biomimetic system, in *Vitamin E in Health and Diseases* (Packer, L. and Fuchs, J., eds.), Marcel Dekker, New York, pp. 85–95
5. Janzen, E. G., Wilcox, A. L., and Manoharan, V. (1993) Reactions of nitric oxide with phenolic antioxidants and phenoxyl radicals. *J. Organ. Chemistry.* **58,** 3597–3599.
6. Burkart, V., Gross-Eick, A., Bellman, K., Radons, J., and Kolb, H. (1995) Suppression of nitric oxide toxicity in islet cells by α-tocopherol. *FEBS Lett.* **364,** 259–263.
7. Packer, J. E., Slater, T. F., and Willson, R. L. (1979) Direct observation of a free radical interaction between vitamin E and vitamin C. *Nature* **278,** 737, 738.
8. Reed, D. J. (1993) Interaction of vitamin E, ascorbic acid, and glutathione in protection against oxidative damage, in *Vitamin E in Health and Diseases* (Packer, L. and Fuchs, J., eds.), Marcel Dekker, New York, pp. 269–281.
9. Huie, R. E. (1994) The reaction kinetics of NO_2. *Toxicology* **89,** 193–216.
10. Hogg, N., Singh, R. J., Goss, S. P., and Kalganaraman, B. (1996) The reaction between nitric oxide and alpha-tocopherol: a reappraisal. *Biochem. Biophys. Res. Commun.* **224(3),** 696–702.

28

Nonvitamin Plasma Antioxidants

Nicholas J. Miller

1. Introduction

The chemical composition of human blood plasma is widely studied and well-known, however, the exact nature of the antioxidants of plasma is very much open to dispute. The study of plasma antioxidants has been given a fresh stimulus in recent years by the development of new methods for the measurement of hydrogen-donating antioxidant activity, the majority using fluorescent or chemiluminescent endpoints. The feature that these systems have in common with the older techniques *(1)* is that they are all inhibition assays (**Fig. 1**). In these assays, a free radical is generated and its generation is then linked to an endpoint that can be observed and quantified. Addition of an antioxidant inhibits the development of this endpoint and in each case from this measured inhibition, the antioxidant activity is quantitated *(2)*.

The most commonly used assay of hydrogen-donating antioxidant activity has been the total peroxyl radical-trapping antioxidant potential (TRAP) assay of Wayner et al. *(3)* and its subsequent developments *(4,5,6)*. In the original system *(3)*, pyrolytic decomposition of the synthetic azo compound 2,2′-azobis(2-amidinopropane) (ABAP) in the aqueous phase was linked to oxygen consumption as an endpoint. The introduction of a luminol-enhanced chemiluminescence endpoint greatly improved the precision of the assay *(4)*. Another approach, which remains to be thoroughly investigated, is to generate peroxyl radicals in the nonpolar lipid phase, instead of the aqueous phase, which can be achieved using 2,2′-azobis(2,4-dimethylvaleronitrile) (AMVN) *(5)*. This might prove useful in measuring the activity of nonpolar antioxidants.

A number of technical problems and important safety considerations have somewhat restricted the use of these radical-generating azo compounds. How-

From: *Methods in Molecular Biology, vol. 108: Free Radical and Antioxidant Protocols*
Edited by: D. Armstrong © Humana Press Inc., Totowa, NJ

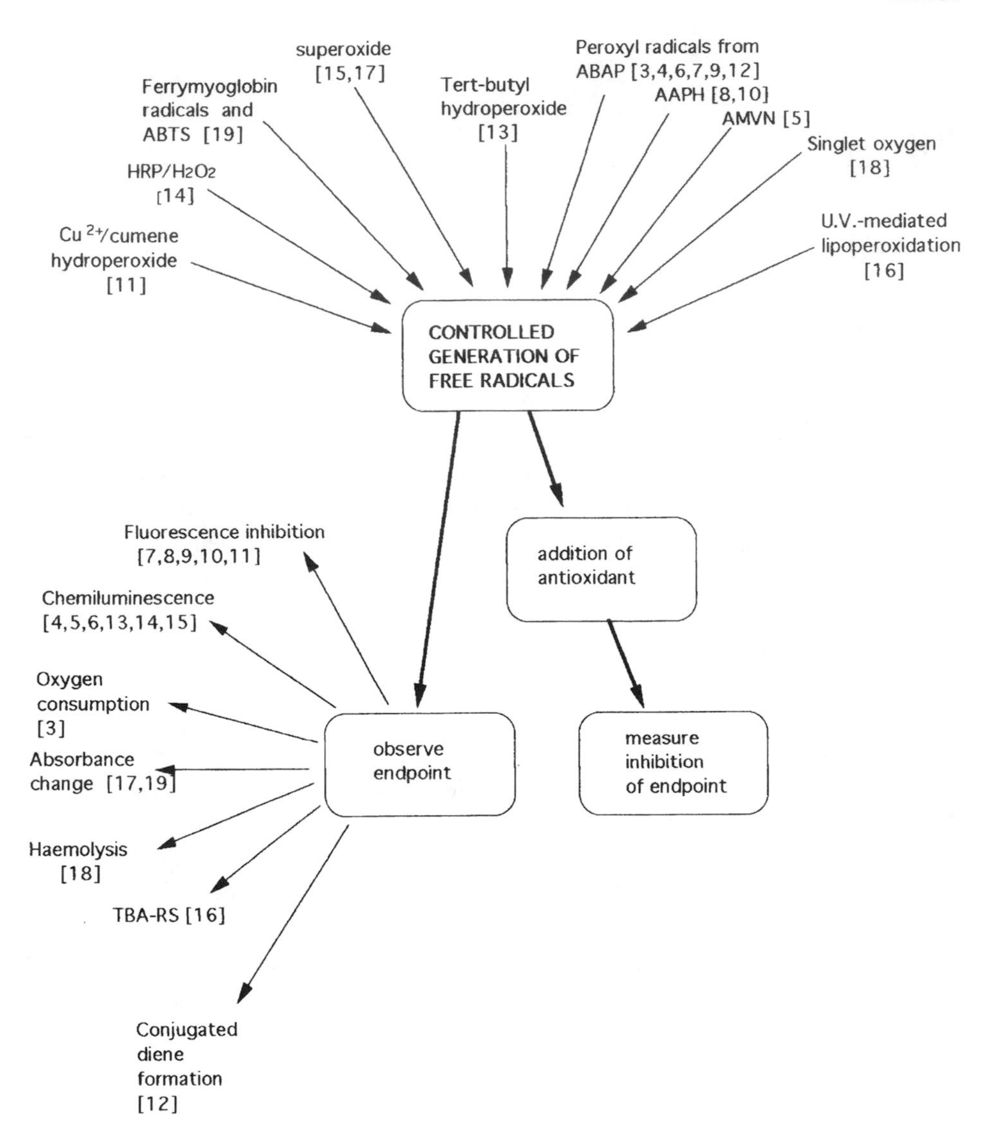

Fig. 1. Current methods for the measurement of antioxidant activity. (Refs. as per text; adapted **ref. *1*.**)

ever, further results with the β-phycoerythrin fluoresence inhibition assay of Delange and Glazer ***(7)*** have been reported ***(8,9)*** using thermal decomposition of 2,2′-azobis(2-amidinopropane) dihydrochloride (AAPH) the dihydrochloride of ABAP, as a source of peroxyl radicals, with loss of inhibition of β-phycoerythrin fluorescence as the antioxidant-related endpoint. The response parameter for quantitation of antioxidant activity in the method of Cao et al.

(8) is the integrated fluorescence output per unit time, expressed with reference to the extent of fluorescence quenching by Trolox. In the original method of Delange and Glazer *(7)*, the length of the lag phase to the initiation of phycoerythrin peroxidation was used as the endpoint and the same approach has been followed by Ghiselli et al. *(9)*. However, there are three problems relating to the use of lag-phase measurements in such an assay:

1. Use of lag-phase measurements assumes that the reaction rate between the peroxyl radical generator and β-phycoerythrin is linear;
2. Samples of body fluids, particular those with a high protein content, as well as many pure substances, do not necessarily produce a clearly defined lag phase, especially at low concentrations;
3. The contribution to the total antioxidant activity made by plasma proteins will not affect the final result; this is regarded as an advantage by those researchers who do not wish to include albumin among their list of hydrogen-donating antioxidants.

There are several other current methods in use. Laranjinha et al. *(10)* use AAPH, but detect peroxyl radicals by means of fluorescence quenching of *cis*-parinaric acid previously incorporated into low-density lipoprotein. McKenna et al. *(11)* use *cis*-parinaric acid incorporated into red-cell ghost membranes as a site of lipid peroxidation. Pryor et al. *(12)* use ABAP as a source of peroxyl radicals, followed by conjugated diene formation in added linoleic acid as an endpoint. Flecha et al. *(13)* use tert-butyl hydroperoxide as a source of peroxyl radicals and endogenous tissue chemiluminescence as an endpoint. Whitehead et al. *(14)* use a procedure in which radicals are generated from the interaction between horseradish peroxidase and hydrogen peroxide (H_2O_2), enhanced by para-iodophenol, and coupled to luminol chemiluminescence. Lewin and Popov *(15)* have developed a system using the generation of superoxide from the ultraviolet (UV) irradiation of water linked to chemiluminesence from luminol oxidation. Negre-Salvayre et al. *(16)* who use UV irradiation of isolated LDL to induce lipid peroxidation, measuring TBA reactivity as an endpoint. Yuting et al. *(17)* have described a method in which superoxide is generated from a phenazin methosulphate-NADH system and the radical generated measured by reduction of nitroblue tetrazolium blue. Jan et al. *(18)* use singlet oxygen as the primary radical, followed by oxidation of α-tocopherol and then, as an endpoint, red-cell hemolysis with hematoporphyrin as a photosensitiser.

It has not been demonstrated that these methods *(10–18)* can be equally applied to the measurement of:

1. The antioxidant activity of pure compounds,
2. Simple mixtures of compounds in solution, and

3. Mixtures of such compounds in a biological matrix, such as is found in human plasma *(2)*.

Because both the 2,2′-azinobis(3-ethylbenzothiazoline-6-sulfonate) (ABTS) method ***(19)*** and the TRAP-based methods ***(3–9)*** meet these three criteria, they are the methods of choice for studying plasma antioxidants. In addition, the methods of Lewin and Popov ***(15)*** and Yuting et al. ***(17)*** will only measure scavengers of superoxide, whereas that of Jan et al. ***(18)*** will only be applicable to scavengers of singlet oxygen. These last features might, of course, be desirable under certain circumstances.

The method of Miller et al. ***(19)*** uses a nitrogen-centered radical, the ABTS radical cation. The endpoint is the measurement of its absorbance, taken at a predetermined time (the point at which depletion occurs of the antioxidant activity of the highest concentration calibration standard). Trolox, introduced by Wayner et al. ***(3)*** remains the antioxidant standard of choice, although not all researchers agree with the use of the stoichiometric factors introduced by these authors for the derivation of the results ***(8,19)***.

2. Materials

2.1. Total Antioxidant Activity (TAA)

1. 5 mmol/L Phosphate-buffered saline (PBS), pH 7.4: prepared from mixing 1.7907 g/L of $Na_2HPO_4.12H_2O$ and 0.655 g/L of $NaH_2PO_4.H_2O$ to a pH of 7.4 (approx 8:1) with the addition of 9.0 g/L sodium chloride. This buffer should not contain sodium azide or any other additional substances. Stable at 4°C for 6 mo.
2. 2.5 mmol/L Trolox (M_r 250.29) (Aldrich Chemical Co., Milwaukee, WI, Cat. no. 23,881-3): prepared in PBS for use as an antioxidant standard (0.1564 gm of Trolox in 250 mL of buffer). At this pH and concentration, the upper limit of Trolox's solubility is approached, and gentle ultrasonication is required to dissolve the crystals. Frozen (–20°C) Trolox dissolved in PBS at this concentration is stable for more than 12 mo. Fresh working standards (0.5, 1.0, 1.5, 2.0 mmol/L) are prepared immediately before use by mixing 2.5 mmol/L Trolox with PBS.
3. H_2O_2 (British Drug Houses "Aristar" or equivalent): working solutions are prepared from stock Aristar H_2O_2 after an initial dilution in PBS to a concentration of 500 mmol/L. Aristar H_2O_2 is supplied as a 30% solution, with a specific gravity of 1.10 (1.099–1.103). Because the M_r is 34, a 500 mmol/L solution is prepared by diluting 515 µL of stock H_2O_2 to 10 mL in PBS (solution A). The required initial working concentration of H_2O_2 is then prepared by diluting solution A with PBS (solution B: working H_2O_2). Working H_2O_2 solutions are freshly prepared prior to use (or every 6 h).
4. ABTS diammonium salt (Aldrich, Cat. no. 27,172-1): A colorless powder when freshly re-crystallized; with storage it takes on a pale green color. 5 mmol/L ABTS (M_r 548.68) is prepared in 18 MΩ water (0.1372 gm in 50 mL or 0.02743 gm in 10 mL). This solution should be kept at 4°C in the dark and is stable for more than 1 wk.

5. Metmyoglobin (MetMb) (Sigma Chemical Co., St. Louis, MO, myoglobin M-1882 from horse heart): The purification of MetMb and the determination of its concentration are fully described in Chapter 31. Purified MetMb at 4°C is stable for at least 8 wk, whereas frozen at –20°C it is stable for more than 6 mo.

3. Methods

3.1. Manual Method for TAA by Inhibition of $ABTS^{\bullet+}$

1. The formation of the ferrylmyoglobin radical from MetMb and H_2O_2 in the presence of the peroxidase substrate ABTS (λ_{max} 342 nm) produces the ABTS radical cation ($ABTS^{\bullet+}$), and an intense blue/green chromogen with characteristic absorption maxima at 417, 645, 734, and 815 nm. The formation of this colored radical cation is suppressed by the presence of hydrogen-donating antioxidants, and the extent of this suppression is directly related to the antioxidant capacity (activity) of the sample being investigated (*see* **Note 1**).
2. The method can be used for human serum or plasma, as well as saliva and other body fluids. In the case of blood samples, no special precautions are required for the separation of the serum from the red cells. The antioxidant activity of serum is stable for at least 6 mo at –20°C. The method has also been shown to work with serum samples from a variety of nonhuman mammalian species (mouse, rat, rabbit, primate).
3. A 1.0 mmol/L working solution of H_2O_2 (solution B): Prepared by diluting 100 µL of solution A to 50 mL with PBS (solution B).
4. Assay protocol:
   a. Add 8.4 µL of sample to a glass culture tube.
   b. Add PBS to give a final volume of 1000 µL.
   c. Add 18 µL of 140 µmol/L MetMb.
   d. Add 30 µL of 5 mmol/L ABTS (*see* **Note 2**).
   e. Bring to 30°C.
   f. Vortex-mix.
   g. Start the reaction with 75 µL of 1.0 mmol/L H_2O_2, vortex-mix and maintain at 30°C.
   h. Read the absorbance at 734 nm after a time determined to be equivalent to the lag phase observed for a 2.5 mmol/L Trolox standard (6 min under the conditions described).
   i. In the absence of hydrogen-donating antioxidants, an immediate and continuous absorbance change is observed after the addition of H_2O_2. No change in absorbance at 734 nm is recorded until after 6 min, when using an aliquot of a 2.5 mmol/L Trolox standard (final concentration 21 µmol/L), at which time the absorbance will start to develop (*see* **Note 3**). The exact timing of the end of the lag phase depends on temperature as well as reagent concentrations.
   j. Because absorbance changes continue after the measuring point, exact timing and reproducibility of conditions is essential. Trolox standards of 2.5, 2.0, 1.5, 1.0, and 0.5 mmol/L initial concentration (21.0, 16.8, 12.6, 8.4, 4.2 µmol/L final concentration) are used to construct the dose-response curve. The highest

absorbance value will be observed in the buffer blank. The division among the reagents of the total assay volume of 1000 µL is shown in **Table 1**.

5. A quantitative relationship exists between the percentage inhibition of absorbance at 734 nm and the antioxidant activity of the added sample or standard:

Antioxidant activity ∝ [Buffer blank (A734 nm) – Test (A734 nm)]/Buffer blank (A734 nm)

3.2. Automated Method for TAA by Inhibition of $ABTS^{•+}$

1. The method is straightforward to automate with a spectrophotometric kinetic analyzer system. Sample or standard is pipetted with mixed ABTS/myoglobin reagent, and hydrogen peroxide dispensed to start the reaction. With the enhanced precision obtained by using an automated analyzer, a higher concentration of H_2O_2 may be used to overcome positive interference from endogenous peroxidase activity in certain samples (for example, hemolysed plasma) *(2)*. Increasing the H_2O_2 concentration drives the reaction faster and hence makes the exact timing of absorbance readings more critical.
2. A 5.4 mmol/L working solution of H_2O_2 (solution B): Prepared by diluting 540 µL of solution A to 50 mL with PBS (solution C).
3. In the protocol devised for the Cobas Bio centrifugal analyzer (Roche Diagnostics Welwyn, Garden City, UK), 300 µL of ABTS/myoglobin reagent is mixed with 3 µL of sample, the probe flushed with 30 µL of diluent, and then H_2O_2 (25 µL of solution C) added as a starter. The total incubation volume is thus 358 µL and reagents are prepared so that on dilution into this incubation volume they are at the desired concentration of 2.5 µ*M* MetMb, 150 µ*M* ABTS, 375 µ*M* H_2O_2, and 0.84% sample fraction.
4. At a temperature of 30°C and a final H_2O_2 concentration of 375 µ*M* H_2O_2, the end of the lag phase for the 2.5 mmol/L standard is at 3.25 min. Minor fluctuations in assay conditions caused by differences between instruments can be compensated for by appropriate adjustment of this measuring time (*see* **Note 4**).
5. Instrument settings used for the ABTS antioxidant assay on the Cobas Bio centrifugal analyzer, using a final H_2O_2 concentration of 375 µ*M*, are shown in **Table 2**.

3.3. Trolox Equivalent Antioxidant Capacity (TEAC) Determination

The TEAC is equal to the millimolar concentration of a Trolox solution having the antioxidant capacity equivalent to a 1.0 mmol/L solution of the substance under investigation. The TEAC can be used to compare the antioxidant activity of different compounds and to explore the antioxidant content of complex mixtures.

The protocol is as follows:

1. A 10 mmol/L solution of the pure substance is prepared in ultra-pure water. Ethanol or methyl sulfoxide (DMSO) may be used if the compound is not sufficiently water-soluble.

Table 1
Manual Protocol for TAA Determination

Addition (initial concentration shown)	Blank (μL)	Sample (μL)	Final concentration
Sample or standard	0	8.4	0.84% of initial concentration
PBS	877	869	
MetMb (140 μ*M*)	18	18	2.5 μ*M*
ABTS (5.0 m*M*)	30	30	150 μ*M*
H_2O_2 (1.0 m*M*)	75	75	75 μ*M*

Table 2
Instrument Settings on the Cobas Bio for the Determination of TAA by Inhibition of $ABTS^{\cdot+}$ Formation

Function	Setting
Standard concentrations	0.0, 0.5, 1.0, 1.5, 2.0, 2.5 m*M* (initial concentrations)
Temperature	30°C
Wavelength (nm)	734
Sample volume	3 μL
Diluent volume (water)	30 μL
Reagent volume	300 μL
Incubation time (s)	20
Start reagent volume	25 μL
Reading time (s)	210

2. This solution is analyzed for TAA.
3. If it proves to have antioxidant activity, serial dilutions of the compound are analyzed until an estimate is obtained of the TEAC value.
4. Three different dilutions of the compound are then selected that produce absorbance values in the most linear region of the Trolox dose-response curve (40–80% inhibition of the blank value, equivalent to an initial concentration of 1.0–2.0 mmol/L Trolox).
5. At a minimum, these three different dilutions of each stock solution are analyzed in triplicate on three separate days (n = 3, i.e., 27 determinations *in toto*).
6. The TEAC is calculated for each dilution and the mean value of all the results derived.

3.4. Results

1. Intra-assay imprecision is in the range of 0.5–1.5%. Inter-assay imprecision is c. 5.0%. Results are most reproducible in the range of the Trolox dose-response

curve (**Fig. 2**) giving 40–80% inhibition of the blank value, equivalent to an final concentration of 8.4–16.8 μmol/L Trolox, or 1.0–2.0 mmol/L initial concentration (*see* **Note 5**).

2. The reference interval for the antioxidant activity of human serum and plasma by this method has been determined to be 1.32–1.58 mmol/L (n = 312) ***(19)***.
3. TEAC values for a number of endogenous plasma components are shown in **Table 3**. 1.0 mmol/L solutions of α-tocopherol, ascorbic acid, and uric acid all give activity values of 1.0 m*M* (*see* **Note 6**). Substances in **Table 3** with TEAC values of 0.00 do not act as antioxidants by these criteria. It should be noted, for example, that whereas phenylalanine does not have antioxidant activity, its hydroxylated metabolite tyrosine is an antioxidant. Bilirubin has a TEAC value of 1.5 m*M*, i.e., it is a more active antioxidant than α-tocopherol (*see* **Note 7**).
4. The principal antioxidants (by mass and activity) of human plasma are albumin and uric acid, which account for more than 50% of the total antioxidant activity of most samples. The residual activity can be referred to as the "antioxidant gap," which is calculated from the TAA (μmol/L), the concentrations of albumin and uric acid (μmol/L) in the plasma sample, and the TEAC values for albumin and uric acid (*see* **Note 8**).

Antioxidant gap = TAA – ([albumin concentration × TEAC] + [uric acid concentration × TEAC])

The antioxidant gap therefore reflects the combined activity of plasma antioxidants other than albumin and uric acid, i.e. ascorbic acid, α-tocopherol, bilirubin, transferrin, haptoglobulin, β-carotene, and various other substances. **Table 4** shows the TAA, albumin, uric-acid, and antioxidant-gap values (95% confidence interval) obtained for human plasma.

5. These relationships only become self-evident if albumin concentrations are expressed in S.I. units, which can be achieved using the known molecular weight for human albumin of 65,000 ***(20)***. Albumin concentrations far exceed those of any other plasma protein (transferrin is the next most abundant plasma protein) and, even allowing for the low activity of albumin (TEAC = 0.69), is the most significant antioxidant in terms of mass and activity in this complex mixture. The source of its hydrogen-donating antioxidant activity is its tyrosine (TEAC = 0.38) and sulphydryl (TEAC = 0.28) residues. In this respect, the action of albumin is directly comparable to that of the lower molecular-weight antioxidants. The significantly greater molecular mass of albumin may account for its slower rate of reaction as an antioxidant in comparison to the other substances listed in **Table 3**.
6. The physiological significance of the TEAC values of the nonvitamin plasma constituents depends on the circulating concentrations of these substances as well as their partition coefficients between the aqueous and lipid phases of plasma. Protection of lipid membranes and lipoproteins requires the presence of lipid-soluble antioxidants (such as α-tocopherol). Of particular interest, therefore, is the finding that bilirubin, which is relatively nonpolar, is also a good antioxidant. Normally, bilirubin circulates in the plasma at concentrations below 20 μmol/L, but in jaundiced patients it can be much more abundant. For example, in prema-

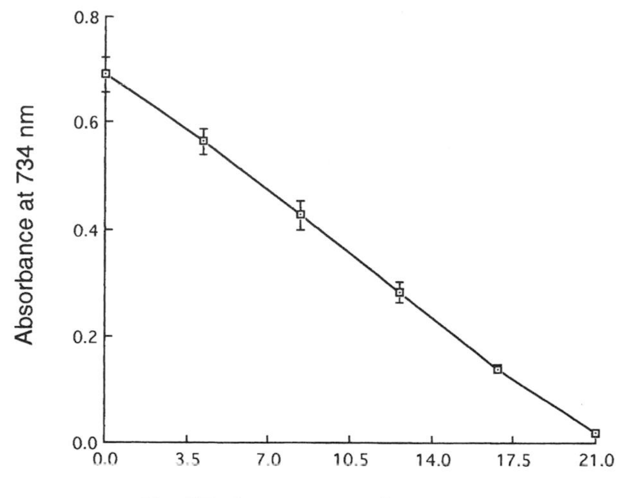

Fig. 2. Dose-response curve of final Trolox concentration against absorbance at 734 nm. Conditions: 2.5 μ*M* MetMb, 150 μ*M* ABTS, 75 μ*M* H_2O_2, temperature 30°C, incubation time 6 min. Mean ± 1 SD is plotted, n = 16.

ture infants, who are also α-tocopherol deficient, plasma bilirubin can exceed 600 μmol/L. In the Special Care Baby Unit, strenuous efforts are made to promote the breakdown and removal of bilirubin from the circulation by phototherapy. The knowledge that bilirubin, which can have a toxic effect at high concentrations, also has a beneficial free radical-scavenging action that adds a different dimension to the biochemical monitoring of these patients.

7. **Figure 3** depicts the relative contributions to the TAA of albumin, uric acid, and the "antioxidant gap" substances. Most antioxidant-deficient individuals with a low-plasma TAA also have a low value for the antioxidant gap. However, premature babies with a low TAA have a raised antioxidant gap, suggesting the presence of significant undefined plasma antioxidants. This raised antioxidant gap may be found in infants who are not significantly jaundiced and cannot, therefore, be accounted for by high-plasma bilirubin concentrations.

4. Notes

1. Because "plateau conditions" are not established in this assay, the incubation temperature and reagent concentrations as well as the timing of the final reading are critical in obtaining reproducible results.
2. The concentration of ABTS is the limiting factor in increasing the sensitivity of the assay: sensitivity increases as ABTS decreases. However the stabilization of the $ABTS^{•+}$ radical cation (λ_{max} 417 nm, peaks at 645, 734, and 815 nm) depends partially on the presence of a mass of unreacted ABTS. A resonance-stabilized monoradical cation is formed if an excess of ABTS substrate is used. Under dif-

Table 3
TEAC (Trolox Equivalent Antioxidant Capacity) Values for Selected Plasma Analytes

Substance	TEAC
alanine	0.00
albumin	0.69
ascorbic acid	1.00
bilirubin	1.50
citric acid	0.00
creatinine	0.00
cysteine	0.28
fructose	0.00
glucose	0.00
glutathione	0.90
lactic acid	0.00
methionine	0.00
phenylalanine	0.00
proline	0.00
α-tocopherol	1.00
tyrosine	0.38
uric acid	1.00
urea	0.00
xanthine	0.01

ferent conditions (e.g., lower pH, insufficient ABTS, excess H_2O_2), the reaction may also proceed further to form a radical dication (λ_{max} 513 nm), whose appearance will invalidate the reaction conditions.

3. Although $ABTS^{\bullet+}$ has a broad absorption spectrum, there are substantial advantages in using the 734 nm peak. At this wavelength, in the near-infra red region, myoglobin, together with many other potential interferents, will not absorb light. Nonspecific interference, for example from lipaemia or mild sample turbidity, also does not affect the absorbance at this wavelength, and hence makes the assay adaptable to a wide range of different specimens.
4. Albumin reacts with ABTS to form a purple complex that absorbs at 580 nm. Under the reaction conditions previously described, the measurement of antioxidant activity will be finished before this complex develops; however, a characteristic purple color may be observed in the reaction cuvets after measurements have been completed.
5. A nonlinear curve fit is required for automated plotting of final absorbance at 734 nm against standard concentration. A logit/log 4 plot has been found satisfactory in this respect and is readily available as commercial software.

Table 4
Total Antioxidant Activity, Albumin, Uric Acid, and Antioxidant Gap Values for Normal Human Serum (95% Confidence Interval)[a]

Component	Concentration (μmol/L) (95% confidence interval)
TAA	1320–1580
Albumin	538–770
Uric acid	210–420
Antioxidant gap	450–800

[a]Antioxidant gap, TAA – ([albumin concentration × TEAC] + [uric acid concentration × TEAC)]).

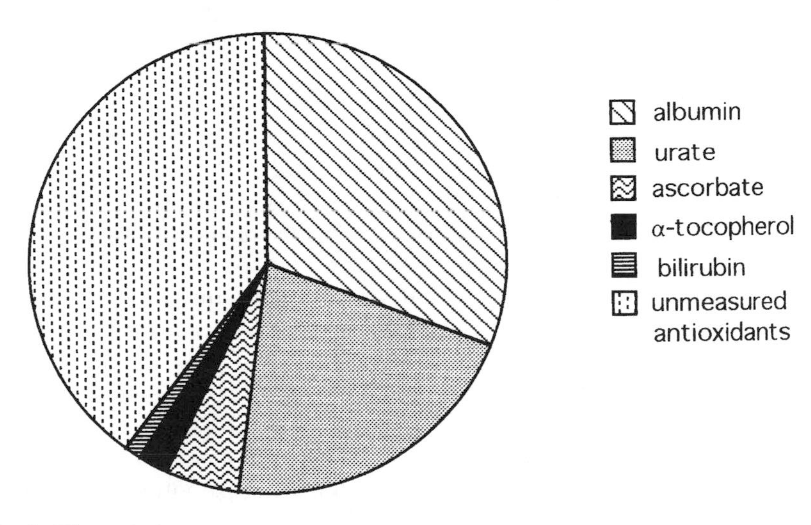

Fig. 3. The relative contributions to the TAA of albumin, uric acid, and the "antioxidant gap," TAA μ*M* – ([albumin μ*M* × TEAC] + [uric acid μ*M* × TEAC]). In the diagram, the antioxidant gap is the activity of the ascorbate + α-tocopherol + bilirubin + unmeasured antioxidants.

6. Analyticals can be added to the incubation mixture as an aliquot of an aqueous or an ethanolic solution. Ethanol at this dilution does not react with the $ABTS^{\bullet+}$ radical cation. DMSO can also be used (e.g., 70% aqueous DMSO) and does not react with the radical.
7. For TEAC determination, bilirubin can be mixed with PBS by adding 0.5 mL of 0.1*M* KOH to 0.00585 gm of bilirubin, followed by 4.5 mL of PBS containing 2% Nonidet P-40 (detergent), giving a 2.0 m*M* solution. If a KOH/PBS/Nonidet blank is analyzed, it will be found to have no antioxidant activity. Cysteine can be dissolved in 10% aqueous ethanol and uric acid in 0.5 g/L aqueous lithium carbonate (which again has no antioxidant activity).

8. Because albumin and uric acid constitute the major part of the antioxidant activity of plasma, these substances form a background against which fluctuations, for example in ascorbic acid, will be relatively insignificant, even on supplementation.

References

1. Rice-Evans, C. and Miller, N. J. (1994) Total antioxidant status in plasma and body fluids. *Methods Enzymol.* **234,** 279–293.
2. Miller, N. J. and Rice-Evans, C. A. (1996) Spectrophotometric determination of antioxidant activity. *Redox. Rep.* **2,** 161–171.
3. Wayner, D. D. M., Burton, G. W., Ingold, K. U., and Locke, S. (1985) Quantitative measurement of the total peroxyl radical-trapping antioxidant capability of human blood plasma by controlled peroxidation: the important contribution made by human plasma proteins. *FEBS Lett.* **187,** 33–37.
4. Metsa-Ketela, T. (1991) Luminescent assay for total peroxyl radical-trapping antioxidant activity of human blood plasma, in *Bioluminescence and Chemiluminescence: Current Status.* (Stanley, P. and Kricka, L., eds.), John Wiley and Sons, Chichester, UK, pp. 389–392.
5. Metsa-Ketela, T. and Kirkkola, A.-L. (1992) Total peroxyl radical-trapping capability of human LDL. *Free Rad. Res. Commun.* **16S,** 215.
6. Lissi, E., Salim-Hanna, M., Pascual, C., and Del Castillo, M. D. (1995) Evaluation of total antioxidant potential (TRAP) and total antioxidant reactivity from luminol-enhanced chemiluminescence measurements. *Free Rad. Biol. Med.* **18,** 153–158.
7. Delange, R. J. and Glazer, A. N. (1989) Phycoerythrin fluorescence-based assay for peroxyl radicals: a screen for biologically relevant protective agents. *Anal. Biochem.* **177,** 300–306.
8. Cao, G., Alessio, H. M., and Cutler, R. G. (1993) Oxygen-radical absorbance capacity assay for antioxidants. *Free Rad. Biol. Med.* **14,** 303–311.
9. Ghiselli, A., Serafini, M., Maiani, G., Azzini, E., and Ferro-Luzzi, A. (1995) A fluorescence-based method for measuring total plasma antioxidant capability. *Free Rad. Biol. Med.* **18,** 29–36.
10. Laranjinha, J. A. N., Almeida, L. M., and Madeira, V. M. C. (1994) Reactivity of dietary phenolic acids with peroxyl radicals: antioxidant activity upon low density lipoprotein oxidation. *Biochem Pharmacol.* **48,** 487–494.
11. McKenna, R., Kezdy, F. J., and Epps, D. E. (1991) Kinetic analysis of the free-radical-induced lipid peroxidation in human erythrocyte membranes: evaluation of potential antioxidants using cis-parinaric acid to monitor peroxidation. *Anal. Biochem.* **196,** 443–450.
12. Pryor, W. A., Cornicelli, J. A., Devall, L. J., Tait, B., Trivedi, B. K., Witiak, D. T., and Wu, M. (1993)A rapid screening test to determine the antioxidant potencies of natural and synthetic antioxidants. *J. Org. Chem.* **58,** 3521–3532.
13. Flecha, B. G., Llesuy, S., and Boveris, A. (1991) Hydroperoxide-initiated chemiluminescence: an assay for oxidative stress in biopsies of heart, liver and muscle. *Free Rad. Biol. Med.* **10,** 93–100.

14. Whitehead, T. P., Thorpe, G. H. G., and Maxwell, S. R. J. (1992) Enhanced chemiluminescent assay for antioxidant capacity in biological fluids. *Anal. Chim. Acta.* **266,** 265–277.
15. Lewin, G. and Popov, I. (1994) Photochemiluminescent detection of antiradical activity; III: a simple assay of ascorbate in blood plasma. *J. Biochem. Biophys. Methods* **28,** 277–282.
16. Negre-Salvayre, A., Alomar, Y., Troly, M., and Salvayre, R. (1991) Ultraviolet-treated lipoproteins as a model system for the study of the biological effects of lipid peroxides on cultured cells. The protective effect of antioxidants (probucol, catechin, vitamin E) against the cytotoxicity of oxidized LDL occurs in two different ways. *Biochim. Biophys. Acta.* **1096,** 291–300.
17. Yuting, C., Rongliang, Z., Zhongjian, J., and Yong, J. (1990) Flavonoids as superoxide scavengers and antioxidants. *Free Rad. Biol. Med.* **9,** 19–21.
18. Jan, C. Y., Takahama, U., and Kimura, M. (1991) Inhibition of photooxidation of a-tocopherol by quercetin in human blood cell membranes in the presence of haematoporphyrin as a sensitiser. *Biochim. Biophys. Acta.* **1086,** 7–14.
19. Miller, N. J., Rice-Evans, C., Davies, M. J., Gopinathan, V., and Milner, A. (1993) A novel method for measuring antioxidant capacity and its application to monitoring the antioxidant status in premature neonates. *Clin. Sci.* **84,** 407–412.
20. He, X. M. and Carter, D. C. (1992) Atomic structure and chemistry of human serum albumin. *Nature* **358,** 209–215.

29

Regulatory Antioxidant Enzymes

C. E. Pippenger, Richard W. Browne, and Donald Armstrong

1. Introduction

Enzymes play an important role in the production of radicals and their metabolism. Techniques to measure pro-oxidant conditions that generate radicals and end-products are described in various chapters throughout this book. The major defense enzymes are superoxide dismutase (SOD), which converts the superoxide radical to hydrogen peroxide: (H_2O_2), catalase (CAT), selenium (SE)-dependent glutathione peroxidase (GSHPx), and leukocytic myeloperoxidase, which degrade inorganic and lipid hydroperoxides formed by interaction with reactive oxygen species and glutathione-S-transferase (GST; which also has peroxidase activity, but is selenium-independent) *(1)*. The activity of GSHPx is coupled to glutathione reductase (GSSG-R), which maintains reduced glutathione (GSH) levels *(2)*. Enzyme activity can be decreased by negative feedback from excess substrate or from damage by oxidative modification *(3)*.

1.1. SOD

The superoxide radical ($O_2^{\bullet -}$)—also referred to by its systematic name dioxide (1-) according to the Commission on the Nomenclature of Inorganic Chemistry *(4)*—is produced primarily from reactions involving xanthine oxidase, nicotinamide adenine dinucleotide phosphate (NADPH) oxidase, cytochrome P450, and protaglandin synthase, from mitochondrial respiration and from autoxidation *(5)*. It is regulated by SOD. A cytoplasmic isoform requires copper (Cu) and zinc (Zn) as cofactors, whereas the mitochondrial isoform requires manganese (Mn) *(6)*. An immunodistinct extracellular SOD also requires Cu/Zu *(7)*.

These metalloenzymes catalyze the dismutation of two molecules of $O_2^{\bullet -}$ into H_2O_2 and oxygen at a rate 10,000 times faster than the spontaneous reac-

From: *Methods in Molecular Biology, vol. 108: Free Radical and Antioxidant Protocols*
Edited by: D. Armstrong © Humana Press Inc., Totowa, NJ

tion *(8)*. Iron (Fe) can react with $O_2^{\bullet -}$, H_2O_2 *(9)* and lipid hydroperoxides *(10)* to liberate hydroxyl radicals (OH). $O_2^{\bullet -}$ can react with nitric oxide (NO) and also generate $OH^{\bullet}$ *(11)*. SOD therefore, serves as the first line of enzymatic defense against radical toxicity and thus protects metabolizing cells from the harmful effects of $O_2^{\bullet -}$ formed during the univalent reduction of oxygen. Endogenous activity can be inhibited by diethyldithiocarbamate.

1.2. CAT

CAT is a ubiquitous heme protein that reduces H_2O_2 to water *(6)*. It is located in the cytosol and in peroxisomes. CAT is considered one of the antioxidant enzymes because it regulates H_2O_2 levels, which, if elevated, can lead to $OH^{\bullet}$ excess through metal catalzed Fenton (Fe/Cu) and Haber-Weiss ($O_2^{\bullet -}$/H_2O_2) reactions. In addition to this catalytic activity, CAT can also act in peroxidatic reactions involving substrates that readily donate hydrogen ions. Because there are no enzymes that regulate the potent $OH^{\bullet}$, CAT becomes an important enzyme in oxidative stress. Endogenous activity can be inhibited by 3-amino-1,2,4-triazole.

1.3. GSHPx

GSHPx is a selenoenzyme found in cytoplasmic and mitochondrial fractions; a deficiency of selenium dramatically decreases its activity *(1,12)*. Like CAT, GSHPx hydrolyzes H_2O_2 but does so at low concentration *(5)*. It primarily acts on lipid hydroperoxide (LHP) substrates that are released from membrane phospholipids by phospholipase A2 *(13)* and can utilize cholesterol hydroperoxide substrates as well *(14)*. Other GSHPxs include a selenoglycoprotein termed plasma GSHPx *(15)* and one insoluble enzyme associated with membranes called phospholipid hydroperoxide GSHPx. *(16)*. Besides the selenocysteine active site, GSH serves as an electron donor cofactor along with riboflavin, which maintains intracellular GSH levels. *(17)* Some tissues such as erythrocytes contain high levels of SOD, CAT, and GSHPx, whereas others have low levels and are more vulnerable to radical damage. GSHPx is inhibited by sodium azide.

1.4. GST

GST is a multifunctional cytosolic, microsomal and membrane-bound Phase II detoxifying enzyme *(18)*, serving both in detoxication of exogenous electrophile and alkylating agents, i.e., carcinogens, as well as 4-hydroxy-2-3-nonenal, a cytotoxic byproduct of LHP metabolism *(19,20)*. Conjugation of the SH group from GSH produces a water soluble, less toxic, more easily destroyed and excretable product *(2,21)*. Seven isoforms of GST including (cationic), μ (neutral), and (antonic) class genotypes have been described, which are selectively expressed in tissue with specific detoxification needs *(22)*. Owing to the peroxidase activity of the form towards LHP and phospholipid hydroperoxides *(23)*, GST can be considered a back-up enzyme to CAT and GSHPx. GST are inhibited by S-alkyl GSH *(24)* and inactivated by monobromobimane *(25)*. Specific inhibitors for the three class forms are described in **ref.** *22*.

1.5. GSSG-R

GSSG-R is a flavoprotein that catalyzes the NADPH-dependent reduction of glutathione disulfide (GSSG). This reaction is essential for the availability of GSH levels in vivo. (GSSG-R, therefore, plays a major role in GSHPx and GST reactions as an adjunct in the control of peroxides and radicals. For example, when CAT levels are decreased, the GSH dependent enzymes become activated *(26)*. In Se deficiency, not only does GSHPx activity decrease, but GSSG-R activity does so as well *(27)*. A deficiency of GSSG-R is characterized by hemolysis owing to increased sensitivity of the erythrocite (RBC) membrane to H_2O_2 which leads to increased osmotic fragility *(28)*. The enzyme is inhibited by 1,3 bis 2-chloroethyl-1-nitrosourca and paraquat.

1.6. Rationale

Oxidative stress has been implicated in aging *(29–32)* and in the pathogenesis of a number of disorders. The extent of injury is generally related to an increase, or decrease of one or more free-radical scavenging enzymes previously mentioned. These enzymes have been identified in blood cells, extracellular fluids (plasma, serum, spinal fluid, lymph), tissues, and cultured cells. To date, antioxidant enzymes have been evaluated in patients with atherosclerosis; ischemia/reperfusion injury related to heart, brain, retina, and gastrointestinal mucosa; hemolytic diseases; emphysema/respiratory distress; diabetes; arthritis; renal failure; hepatitis/hepatotoxicity; parasitic infection; Alzheimer's disease; Down's syndrome; Amyotrophic lateral sclerosis; Parkinson's disease; retinopathy of prematurity; alcoholism; pancreatitis; sepsis/shock; radiation damage; cancer; autoimmune disease; endometriosis; idiopathic infertility; and trauma *(33)*.

The number of new disorders where antioxidant-enzyme analysis is proving useful is growing rapidly and we can expect these analyses to aid greatly as biomarkers in future diagnosis and management. The objective of this chapter is to describe procedures for SOD, CAT, GSHPx, GST, and GSSGR, illustrating optimal conditions with results demonstrating sensitivity, specificity accuracy/ reproducibility, and precision, along with their adaptability to automation.

2. Materials

2.1. SOD

2.1.1. Reagents

1. Reagent A: 50 m*M* sodium carbonate buffer: Prepared by mixing 220 mL of 0.1 *M* Na_2CO_3, 180 mL of 0.1 *M* $NaHCO_3$ and 400 mL Type 1 water. 23.4 mg ethylenediamine tetraacetic acid (EDTA) is added and either 1*N* NaOH or 2*N* HCl to obtain a pH of 10. Stable for 3 mo at room temperature.
2. Reagent B: Three mmol/L xanthine: Prepare by dissolving 26.1 mg xanthine disodium in 50 mL Type 1 water; add two drops of 1*N* NaOH. This solution must be prepared every wk and can be stored at room temperature.
3. Reagent C: 10.5 μ*M* cytochrome C, horse heart, Type III (Sigma Chem. Co., St. Louis, MO): Add 13 mg cytochrome C to 100 mL volumetric flask and take to the mark with reagent A. Mix with reagent B at a ratio of 5.7:1 (28.5 mL xanthine and 0.5 mL cytochrome C). Light-sensitive and stable for only 24 h at room temperature, or 2 wk if frozen at –70°C.
4. Reagent D: Xanthine oxidase grade III, buttermilk, (Sigma): Diluted with reagent A to achieve a change in absorbance of 0.025/min at 550 nm for the reduction of cytochrome C in the absence of SOD. It is necessary to check the ferricytochrome C concentration in each manufacturers lot and adjust to 10 μmoles.
5, Reagent E: CuZu and Mn SOD standards (Sigma) are prepared at 10 U/mL concentration. Stable for 1 yr at –70°C.

2.2. CAT

2.2.1. Reagents

1. Reagent A: 0.2 *M* potassium phosphate, dibasic: Add 17.4 g K_2HPO_4 to 400 mL of Type 1 water, mix until completely dissolved, adjust to 500 mL.
2. Reagent B: 0.2 *M* potassium phosphate, monobasic: Add 13.6 g KH_2PO_4 to 400 mL of Type 1 water, mix until completely dissolved, adjust to 500 mL.
3. Reagent C: 0.1 *M* potassium phosphate buffer: Mix 122 mL of reagent A and 78 mL of reagent B with 400 mL of Type 1 water. Adjust to pH 7.0 with either 2*N* HCl (20 mL/100 mL water) or 1*N* NaOH (4g/100 mL water). Stable for 6 mo at room temperature.
4. Reagent D: 20 m*M* H_2O_2 stock solution: Add 1.1 mL of 30% H_2O_2 ACS grade to 500 mL volumetric flask and bring to the mark with Type 1 water. Mix well. Stable for 3 mo at –4°C.
5. Reagent E: 10 m*M* H_2O_2 in 50 m*M* potassium phosphate buffer: Mix 200 mL of reagent C and 200 mL of reagent D. Stable for 2 mo at –4°C.

2.3. GSHPx

2.3.1. Reagents

1. Reagent A: 50 m*M* potassium phosphate buffer, pH 7.0: Add 91.5 mL of 0.2 *M* K_2HPO_4 and 58.5 mL of 0.2 *M* KH_2PO_4 to 450 mL of Type 1 water. Adjust pH with 2 *N*. HCl and 1 *N* NaOH. Stable for 6 mo at room temperature.

2. Reagent B: 50 m*M* potassium phosphate with EDTA and sodium azide pH. 7.0: Add 658 mg of EDTA and 109 mg of NaN_3 to 450 mL of reagent A. Adjust pH. Stable for 6 mo at room temperature.
3. Reagent C: 0.1% sodium bicarbonate: Add 2 mg $NaHCO_3$ to 200 mL of Type 1 water. Stable for 1 yr at room temperature.
4. Reagent D: 8.4 m*M* nicotinamide adenine dinucleotide phosphate: Add 18 mg NADPH to 2.5 mL of reagent C. Stable for 24 h at 4°C or 3 d if frozen.
5. Reagent E: 5.8 m*M* GSH: Add 927 mg of GSH to 52 mL of reagent A. Stable for 24 h at 4°C.
6. Reagent F: 2.2 m*M* H_2O_2: Dilute 10 mL of reagent D with 10 mL of Type 1 water and take three readings at 240 nm on a spectrophotometer. Average the readings and calculate according to the following equation:

   $$21.81 \times \text{O.D.} - 0.36 = \text{actual concentration in n}M\text{/L}$$

   Actual concentration × V1 = 2.2 × V2. Subtract 2.5 from V2 to dctcrminc amount of water needed to make a final concentration of 2.2.
7. Reagent G: Start reagent. Mix 2 parts of reagent F and 1 part of reagent D just before each run.
8. Reagent H: GSSG-R (Sigma): Mix 0.1 mL of 10 U/mL GSSG-R and 2.4 mL of reagent B. Stable for 8 h at room temperature.
9. Reagent I: Working GSSG-R and GSH. Mix 1 mL of 10 U/mL GR and 26 mL of reagent E.

2.4. GST

2.4.1. Reagents

1. Reagent A: 0.1 *M* potassium phosphate buffer, monobasic, pH 6.25: Mix 46.5 mL of 0.2 *M* K_2HPO_4 and 203.5 mL of 0.2 *M* KH_2PO_4 with 250 mL of Type 1 water. Adjust pH with 2*N* HCl or 1*N* NaOH. Stable for 6 mo at room temperature.
2. Reagent B: 45 mmol 1-chloro-2,4-dinitrobenzene (CDNB) substrate solution (Sigma): Add 911.7 mg CDNB to 100 mL of 95% ethanol. Dissolve slowly and let stand overnight. Seal bottle with parafilm and store at room temperature. Stable for 2 mo.
3. Reagent C: Working CDNB buffer: Add 0.5 mL of reagent B slowly to 27 mL of reagent A immediately before use. Stable for 1 d at room temperature.
4. Reagent D: 15 mmol GSH: Dissolve 23 mg GSH in 5 mL of Type 1 water immediately before use. Stable for 1 d at room temperature.

2.5 GSSG-R

2.5.1. Reagents

1. Reagent A: 0.12 *M* potassium phosphate buffer, pH 7.0, with EDTA: Mix 183 mL of 2.1 reagent A, 117 mL of reagent B, 175.4 mg of EDTA, and 200 mL of Type 1 water. Adjust pH. Stable for 6 mo at room temperature.
2. Reagent B: 0.1 *M* TRIS: Add 1.2 g of TRIS to 100 mL of Type 1 water.

3. Reagent C: 0.1 *M* TRIS-HCl stock solution: Mix 50 mL of reagent B, 46.6 mL 0.1*M* HCl and 3.4 mL of Type 1 water. Stable for 6 mo at room temperature.
4. Reagent D: 1.5 m*M* NADPH in 0.01 *M* Tris-HCl: This preparation is based on the manufacturer's labeled purity, which varies according to lot. The amount of NADPH in a 6-mL vial of 97% purity is determined by substituting a higher or lower % in the following formula:

$$833.4 \text{ (M.W. of NADPH)} \times 1.5 \times 0.006 \div 0.97 = 7.7 \text{ mg NADPH/vial}$$

5. Reagent E: 15 m*M* GSH, oxidized form: Mix 36.8 mg of GSSG with 4 mL of Type 1 water. Stable for 8 h at 4°C.
6. Reagent F: Substrate solution: Mix 25 mL of reagent A, 2 mL of reagent D, and 2 mL of reagent E. Stable for 8 h at 4°C.

2.6. Special Supplies and Equipment

Drabkin's cyanmethemoglobin: Dissolve 2.0 gm $NaHCO_3$, 100 mg potassium cyanide, and 400 mg potassium ferricyanide in 500 mL of Type 1 water. Store in amber bottle. Stable for 1 yr.

For all assays, a Shimadzu (Kyoto, Japan) UV160 or similar double-beam spectrophotometer and a Cobas Mira auto analyzer (Roche Diagnostic Systems, Inc., Nutley, NJ) was used. Disposable supplies i.e., sample cups (#10-0678-9), reagent tips (#10-2114-1) reagent boats with covers (#10-0676-2; #10-0677-0) and cuvet rotors (#10-0679-7) should be purchased from the manufacturer. Substitutions are not recommended.

3. Methods

3.1. Instructions for Washing Erythrocytes

1. The assays for analyzing antioxidant enzymes utilize a red blood cell (RBC) hemolyzate sample. Therefore, separation and washing are required for preparation of the specimen.
2. Collect one 7-mL tube (Vacutainer or Monovet) containing sodium EDTA (purple top) and refrigerate immediately. Enzymes are stable for 6 h at room temperature; however, specimens should be processed as quickly as possible. Centrifuge at 3000*g* for 10 min, decant plasma and freeze, discard buffy coat, transfer approx 2 mL of packed erythrocytes into a 13 × 100 mm polypropylene tube, add approx 2 volumes of physiological saline at 4°C, cap and mix gently by inversion, centrifuge at 3000*g* for 10 min, discard saline and residual buffy coat. After the third saline wash, add an equal volume of Type 1 water at 4°C, vortex for 1 min, and centrifuge at 3000*g* for 10 min. The hemolyzate can be stored at –70°C until assay and is stable for 6 mo. Hemoglobin (Hb) concentration in the hemolyzate is measured by the cyanmethemoglobin method ***(34)***. Appropriate dilutions of the original hemolyzate are made for each enzyme assay.
3. Quality control: Collect a donated unit of fresh whole blood with sodium EDTA as anticoagulant, or pool multiple samples to achieve desired volume. Divide

into 15 mL centrifuge tubes and centrifuge at 3000*g* for 10 min, discard plasma, and wash RBC according to **Subheading 3.1.2., step 2.**

Aliquot hemolyzate into 13 × 100 polypropylene tubes and freeze at –70°C. Assay endogenous activity for each enzyme a minimum of 20 times (no more than 2 samples/run) and calculate ± 2 and ± 3 standard deviation (SD) *(35)*. If the coefficient of variation (CV) for each triplicate sample is >10%, re-run and determine new mean ± SD.

3.2. Instrument Settings

3.2.1. Preliminary Preparations.

Thaw reference and samples to be analyzed. Turn on instrument and perform function checks according to manual. Make reagents that need to be prepared fresh. Pipet reagents into appropriate well of autoanalyzer reagent rack. Pipet reagent blanks into analyzer rotor at position 1, 2, and 24 and unknown samples in triplicate into positions 3–23.

3.2.2. Parameters

The parameters shown in **Table 1** are for the Cobas Mira autoanalyzer. Parameters are adaptable to any other analyzer with the ability to pipet at least two different reagents (i.e., primary reagent/Diluent and starting reagent). Manual determinations can be achieved with a standard spectrophotometer by scaling up the reaction volumes proportionately and using an appropriate blank and manual reaction timing.

3.3. SOD

1. Sample preparation: Prior to analysis, an appropriate volume of hemolyzate is diluted with Type 1 water to approx 4–5 mg/mL of Hb. Add 0.25 μL of absolute ethanol to 1 mL of diluted hemolyzate, mix by vortex, add 0.15 mL of chloroform, vortex for 2 min, centrifuge at 3000*g* for 15 min at 4°C, and save clear red upper layer.
2. Assay conditions: Triplicate samples and controls are transferred to sample cups. Cytochrome C/xanthine oxidase and xanthine solutions are transferred to the reagent well and starting well of the reagent boat, respectively. Mount cuvets in rotor and begin run.
3. Calculation: The percent inhibition of the reduction of cytochrome C by SOD is determined from the following equation:

$$\frac{A_{blank} - A_{sample}}{A_{blank}} \times 100\%$$

1 unit of SOD activity is defined as the amount of enzyme that inhibits the reaction by 50%. Specific activity is expressed in units/g Hb. Samples with a percent inhibition above 50% should be diluted with 25% ethanol.

Table 1
Analytical Settings for Antioxidant Enzymes[a]

General	SOD	Catalase	GSH-Px	GST	GSSG-R
Reaction Mode	R-S-SR1	R-S	R-S-SR1	R-S-SR1	R-S
Wavelength	550 nm	240 nm	340 nm	340 nm	340 nm
Decimal Position	2	2	0	0	0
Unit	A/min	A/min	U/L	U/L	U/L
Analysis					
Sample					
Volume	10 µL	5 µL	10 µL	10 µL	2 µL
Dil	0 µL	0 µL	20 µL	20 µL	10 µL
Reagent					
Volume	290 µL	330 µL	270 µL	275 µL	290 µL
Start Reagent 1					
Cycle	2	-	2	2	-
Volume	5 µL	-	20 µL	20µL	-
Calculation					
Reac. Direction	Increase	Decrease	Decrease	Decrease	Decrease
Readings (s)					
First	2	3	5	3	3
Last	9	8	13	11	8
Calibration					
Factor	1	1	8576	6250	1608

[a]The following settings are common to all enzyme assays: measurement mode is set for Absorb; calibration mode is set for Factor; reagent blank is set for Reag/Dil; sample dilution name is set for Water; sample and reagent cycles are set at 1; reagent diluting and start reagent 1 are set at zero; reaction direction check is set at On; conversion factor is set at 1 and offset at zero; the number of steps is set on 1 and calculation step A is set on Kinetic. No settings are required for Cleaner mode, Post Dilution Factor and Concentration Factor, Sample Limit, Antigen Excess, High and Low Normal Range, and Reaction Limit.

3.4. CAT

Assay conditions: Prior to analysis, an appropriate volume of hemolyzate is diluted with Type 1 water to approx 1.0 to 1.2 mg/mL of Hb.

Spectrophotomatically match three empty cuvets. Place Type 1 water in one cuvet and adjust spectrophotometer to zero. Place reagent E (see **Subheading 2.2.1., item 5**) in the other two cuvets and read absorbance at 240 nm. Calculate H_2O_2 concentration, which should be 10 + 0.5 m*M* using the following regression equation:

$$H_2O_2 \text{ conc. } (\text{m}M/\text{L}) = 21.81 \times (\text{A240}) - 0.36$$

If out of range, adjust accordingly with reagent D (**Subheading 2.2.1., item 4**), or reagent A (**Subheading 2.3.1., item 1**). Add 3 mL of reagent E (**Sub-**

heading 2.2.1, item 5) to a cuvet and pre-warm at 25°C for 5 min. Add 20 µL of hemolyzate. Rinse pipetter tip at least 3 times in cuvet volume. Place in spectrophotometer and record change in absorbance between 30 and 210 s.

Calculation: One unit of activity is defined arbitrarily as the amount of enzyme which induces a change in A_{240} of 0.43 during the 3-min incubation. Activity is expressed in KU/g Hb calculated from the following equation:

$$CAT = A_{240} \times 0.43 \times 0.02 \text{ (mg/mL Hb)}$$

3.5. GSHPx

Assay conditions: Prior to analysis, dilute sample to 6–7 mg/mL of Hb. Place reagent G and reagent I (**Subheading 2.3.1., items 7** and **9**) in appropriate cups in reagent rack.

Calculation: Subtract the mean of the three reagent blanks from the mean of the triplicate samples and divide by the Hb concentration. The changes in absorbance at 340 nm in samples containing enzyme will be lower that the blank owing to consumption of NADPH in the reaction.

3.6. GST

Assay conditions: Prior to analysis, dilute sample to 18–22 mg/mL of Hb. Place GSH reagent D (**Subheading 2.3.1., item 4**) and the working CDNB buffer solution C (**Subheading 2.4.1, item 3**) into the appropriate well of the reagent rack. Run the analysis in the multi-run mode.

Calculation: Subtract the mean of the three reagent blanks from the mean of the triplicate samples and divide by the Hb concentration. Absorbance at 340 nm in samples containing enzyme will be lower than the blank owing to consumption of NADPH in the reaction.

3.7. GSSG-R

Assay conditions: Prior to analysis, dilute sample to 10–12 mg/mL of Hb. Place reagent F (**Subheading 2.5.1., item 6**) in appropriate reagent boat.

Calculation: Subtract the mean of the three reagent blanks from the mean of the triplicate samples and divide by the Hb concentration. Absorbance at 340 nm in samples containing enzyme will be lower that the blank owing to consumption of NADPH in the reaction.

3.8. Results

3.8.1. Linearity

Examples of the expected relationship between activity and % inhibition for the SOD assay, as well as Hb concentration is shown in **Figs. 1–3**.

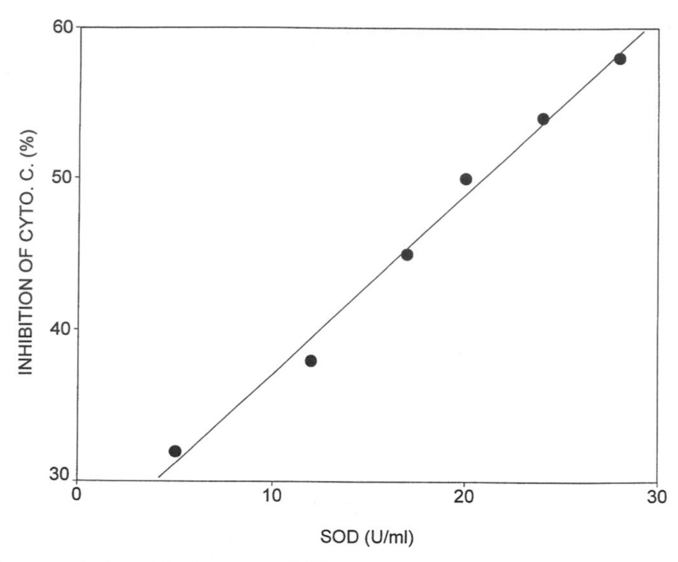

Fig. 1. Linear relationship between SOD activity and the inhibition of cytochrome C.

3.8.2. Reproducibility

Table 2 shows the precision statistics for the five assays in terms of coefficient of variation (CV) for different periods of time.

3.8.3. Expected Values

Table 3 shows the expected activities of the 5 enzymes in human erythrocyte samples, which are assayed at 25°C.

4. Notes

1. New lot numbers of chemicals and enzymes must be checked by assaying in parallel with current stock reagents.
2. Normal values must be determined on the local population for RBC, serum, and plasma. *(35)*.
3. Enzyme activity can also be calculated on the basis of volume, g protein, net weight, and number of cells (time culture).
4. If the CV for each triplicate sample set is >10%, re-run to rule out technical error.
5. The CV for clinical samples should be less than 5%.
6. Specimens should be kept cold, processed as quickly as possible and frozen at –70°C if they cannot be analyzed within the same d of collection.
7. Caution should be used in handling hazardous chemicals i.e., KCN, ethanol (flammable), and CDNB.
8. For temperature corrections, the conversion factor from 25°C to 37°C is 1.22 × recorded activity + 4.29.
9. Disposable cuvets should be used whenever possible.

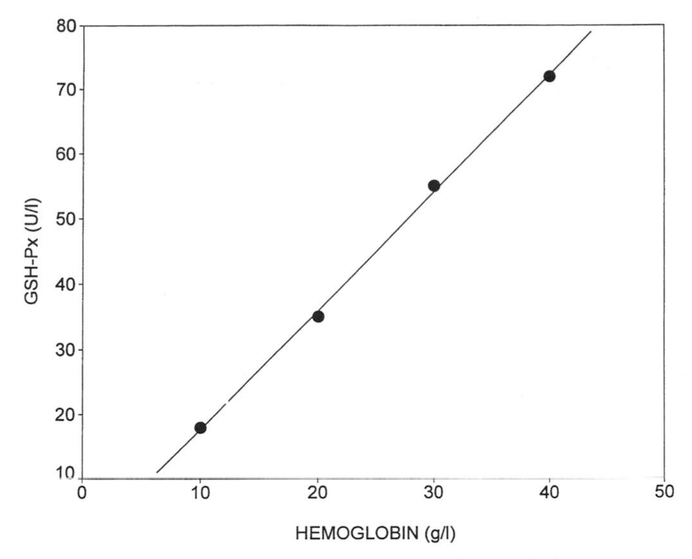

Fig. 2. Linear relationship between sample hemoglobin concentration and GSH-Px activity.

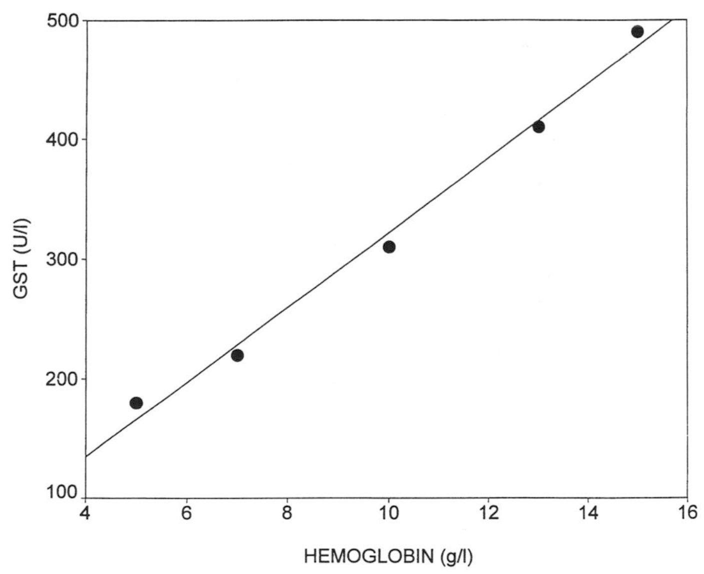

Fig. 3. Linear relationship between sample hemoglobin concentration and GST activity.

10. It is important to dilute high CAT assays because oxygen bubbles produced during the reaction alter the absorbance reading.
11. Performing the SOD assay at pH 10.0 gives six times higher activity than at pH 7.8. The micro-method described here requires only 1/10 the reagent and sample volumes of the original method ***(8)*** and requires less incubation time.

Table 2
Precision and Reproducibility of Methods

TEST		Within-run	Between-run	D-to-d
SOD				
	mean ± SD	4933 ± 50	4863 ± 52	4881 ± 145
	CV	1.0	1.1	3.0
CAT				
	mean ± SD	19.03 ± 0.45	19.25 ± .51	19.15 ±.62
	CV	2.36	2.64	3.24
GSHPx				
	mean ± SD	35.7 ± 1.4	36.2 ± 0.9	35.8 ± 1.6
	CV	3.8	2.5	4.6
GST				
	mean ± SD	1.92 ± 0.051	1.958 ± .081	1.89 ± 0.082
	CV	2.65	4.13	4.34
GSSG-R				
	mean ± SD			6.5 ± .54
	CV			8.31

Table 3
Representative Erythrocyte Enzyme Activities in Normal Human Population

Test	Activity (U/gHgb)
GSHPx	30–44
GST	0.8–24
GSSG-R	5.4–8.8
SOD	6500–14,500
CAT	16,500–26,500

12. GSSG-R can be measured in a similar automated system (*see* **ref.** ***2***).
13. Expressing enzyme activity on the basis of Hb will correct in anemia, but cannot be used when hemoglobinopathy is present.
14. GST peroxidase activity can be determined using cumene hydroperoxide or substrate.

References

1. Michiels, C., Raes, M., Toussant, O., and Remacle, J. (1994) Importance of Se glutathione peroxidase, catalase and Cu/Zu - SOD for cell survival against oxidative stress. *Free Rad. Biol. Med.* **17,** 235–248.

2. Bompart, G. J., Prevot, D. S., and Bascands, J-L. (1990) Rapid automated analysis of glutathione reductase, peroxidase, and S-transferase activity. *Clin. Biochem.* **23,** 501–504.
3. Tabatabaie, T. and Floyd, R. A. (1994) Susceptibility of glutathione peroxidase and glutathione reductase to oxidative damage and the protective effect of spin trapping agents. *Arch. Biochem. Biophys.* **314,** 112–119.
4. Koppenol, W. H. (1990) What is in a name? Rules for radicals. *Free Rad. Biol. Med.* **9,** 225–227.
5. Grisham, M. B. (Ed.) (1992) *Reactive Metabolites of Oxygen and Nitrogen in Biology and Medicine.* R. G. Landes Co., Austin, TX, pp. 104.
6. Diplock, A. (1994) Antioxidants and free radical scavengers, in *Free Radical Damage and its Control* (Rice-Evans, C. A. and Burton, R. H., eds.), Elsevier Science, Amsterdam, pp. 113–130
7. Marklund, S. L. (1990) Expression of extracellular superoxide dismutase by human cell lines. *Biochem. J.* **266,** 213–219.
8. McCord, J. M. and Fridovich, I. (1969). Superoxide dismutase: an enzymic function for erythrocuprein (hemocuprein). *J. Biol. Chem.* **244,** 6049–6055.
9. Liochev, S. and Fridovich, I. (1994). The role of O_2^- in the production of HO in vitro and in vivo. *Free Rad. Biol. Med.* **16,** 29–33.
10. Yagi, K., Ishida, N., Komura, S., Ohishi, N., Kusai, M., and Kohno, M. (1992) Generation of hydroxyl radical from linoleic acid hydroperoxide in the presence of epinephrine and iron. *Biochem. Biophys. Res. Comm.* **183,** 945–951.
11. Rubbo, H., Radi, R., Trujillo, M., Telleri, R., Kalyanaraman, B., Barnes, S., Kirk, M., and Freeman, B. (1994) Nitric oxide regulation of superoxide and peroxynitrite-dependent lipid peroxidation: formation of novel nitrogen-containing oxidized lipid derivatives. *J. Biol. Chem.* **269,** 26,066–26,075.
12. Spallholz, J. E. and Boylan, L. M. (1991). Glutathione peroxidase: The two selenium enzymes, in *Peroxidases in Chemistry and Biology, (vol. 1)* (Everse, J., Everse K. E., and Grisham. M. B., eds.), CRC Press, Boca Raton, FL, pp. 259–291.
13. van Kuijk, F. J. G. M., Sevanian, A., Handelman, G. J., and Dratz, E. A. (1987) A new role for phospholipase A2: protection of membranes from lipid peroxidation damage. *TIBS* **12,** 31–34.
14. Thomas, J. P., Maiorino, M., Ursini, F., and Girotti, A. W. (1990) Protective action of phospholipid hydroperoxide glutathione peroxidase against membrane-damaging lipid peroxidation; in situ reduction of phospholopid and cholesterol hydroperoxides. *J. Biol. Chem.* **265,** 454–461.
15. Takahashi, K., Arissar, N., Whitin, J., and Cohen H. (1987) Purification and characterization of human plasma glutathione peroxidase: a selenoglycoprotein distinct from the known cellular one. *Arch. Biochem. Biophys.* **256,** 677-686.
16. Ursini, F., Maiorino, M., and Gregolin, C. (1986). Phospholipid hydroperoxide glutathione peroxidase. *Int. J. Tiss. Reac.* **8,** 99–103.
17. Thurnham, D. I. (1990) Antioxidants and prooxidants in malnourished populations. *Proc. Nutr. Soc.* **49,** 247–259.

18. Mantle, T. J. (1995) The glutathione S-transferase multigene family: a paradigm for xenobiotic interactions. *Biochem. Soc. Trans.* **23,** 423–425.
19. Leonarduzzi, G., Parola, M., Muzio, G., Garramone, A., Maggiora, M., Robino, B., Poli, G., Dianzani, M. V., and Canuto, R. A. (1995) Hepatocellular metabolism of 4- hydroxy-2,3-nonenal is impaired in conditions of chronic cholestasis. *Biochem. Biophys. Res. Comm.* **214,** 669–675.
20. Singhal, S. S., Awasthi, S., Srivastava, S. K., Zimniak, P., Ansari, H. H., and Awasthi, Y. C. (1995) Novel human ocular glutathione S-transferases with high affinity toward 4- hydroxynonenal. *Invest. Ophthalmol. Vis. Sci.* **36,** 142–150.
21. Philbert, M. A., Beiswanger, C. M., Manson, M. M., Green, J. A., Novak, R. F., Primiano, T., Reuhl, K. R., and Lowndes, H. E. (1995) Glutathione S-transferases and gamma-glutamyl transpeptidase in the rat nervous systems: a basis for differential susceptibility to neurotoxicants. *Neurotoxicology* **16,** 349–362.
22. Nijhoff, W. A., Grubben, M. J., Nagengast, F. M., Jansen, J. B., Verhagen, H., van Poppel, G., and Peters, W. H. (1995) Effects of consumption of Brussel sprouts on intestinal and lymphocytic glutathione S-transferase in humans. *Carcinogenesis* **16,** 2125–2128.
23. Tsuchida, S. and Sate. K. (1992) Glutathione transferases and cancer. *Crit. Rev. Biochem. Mol. Biol.* **27,** 337–384.
24. Ogawa, N., Hirose, T., Tsukamoto, M., Fukushima, K., Suwa, T., and Satoh, T. (1995) GSH-dependent denitration of organic nitrate esters in rabbit hepatic and vascular cytosol. *Res. Comm. Mol. Pathol. Pharmacol.* **88,** 153–161.
25. Hu, L. and Coleman, R. F. (1995) Monobromobimane as an affinity label of the xenobiotic binding site of rat glutathione S-trasnerase 3-3. *J. Biol. Chem.* **270,** 21,875–21,883.
26. Gastani, G. F., Kirkman, H. N., Mangerini, R., and Ferraris, A. M. (1994) Importance of catalase in the disposal of hydrogen peroxide within human erythrocytes. *Blood* **84,** 325–330.
27. Castano, A., Ayala, A, Rodriguez-Gomez, J. A., de la Cruz, C. P. Revilla, E., Cano, J., and Machado, A. (1995) Increase of dopmine turnover and tyrosine hydroxylase enzyme in hippocampus of rats fed on low selenium diet. *Neurosci. Res.* **42,** 684–691.
28. Harmening, D. (ed.) (1992) *Clinical Hematology and Fundamentals of Hemostesis* (2nd ed.), F. A. Davis, Co., Philadelphia, PA, pp. 251, 540.
29. Emerit, I. and Chance, B. (eds.) (1992) *Free Radicals and Aging.* Birkhauser Verlag, Basel and Boston, p. 437.
30. Yu, B. P. (ed.) (1993) *Free Radicals in Aging.* CRC Press, Boca Raton, FL, p. 303.
31. Ames, B., Shigenaga, M. K., and Hagen, T. M. (1993). Oxidants, antioxidants, and the degenerative diseases of aging. *Proc. Nat'l. Acad. Sci. USA* **90,** 7915–7922.
32. American Health Research Institute (1994) Antioxidants and Effects on Longevity. ABBE Publication Association., Washington, DC.
33. Armstrong, D. (ed.) (1994) Free Radicals in Diagnostic Medicine. Plenum Press, New York, pp. 1–464.

34. Fairbanks, V. F. and Klee, G. G. (1987). Biochemical aspects of hematology, in, *Fundamentals of Clinical Chemistry, (3rd ed.)* (Tietz, N. W., ed.), W. R. Saunders Co., Philadelphia, PA, pp. 803–804.
35. Solberg, H. E. (1987) Establishment and use of reference values in, *Fundamentals of Clinical Chemistry (3rd ed.)* (Tietz, N. W., ed.), W. R. Saunders Co., Philadelphia, PA, pp. 197–212.

30

In Vitro Screening for Antioxidant Activity

Donald Armstrong, Tadahisa Hiramitsu, and Takako Ueda

1. Introduction

The protective "tone" of a tissue against oxidative stress depends on a balance between endogenous antioxidants (enzymatic and non-enzymatic) and the amount of pro-oxidant reactive oxygen lipid peroxides *(1–3)*. Lipid and cholesterol peroxides are transported in lipoproteins (primarily low-density lipoproteins [LDL]) and these pre-existing peroxides make LDL highly susceptible for propagation reactions *(4,5)*. Lipid peroxidation is thought to be an important signalling event in triggering pathophysiological change *(6)*. Increased levels of these compounds may respond to α-tocopherol, glutathione (GSH), phytonutrients like carotenoids and flavonoids, ascorbate, coenzyme Q10 (ubiquinone), N-acetylcysteine, trans-resveratol in red wine, and lipoic acid, which have been studied extensively as dietary supplements to improve antioxidant reserve and thereby help regulate oxidative stress *(7–17)*. Interest in synthetic compounds with radical scavenging activity and those obtained from plant/herbal sources is growing rapidly. Some examples are the nonhormonal, nonglucocorticoid 21-aminosteroids *(18)*, sulfonylurea *(19)*, enzyme replacement with high molecular weight (MW) polyethylene-superoxide dismutase (SOD) conjugates *(20)*, enzyme mimics such as Ebselen that have glutathione peroxidase (GSHPx) and scavenging activity *(2,21)*, as well as Manganese-based porphyrin and cyclopentane SOD mimetics *(22,23)* and GSH mimics *(24,25)*. Another novel approach is to induce gene expression of antioxidant enzymes by cytokine activation of nuclear factors *(26)*.

Antioxidants have been evaluated in a variety of animal models, including in vivo *(27–31)*, in vitro *(32–34)* and nonvertebrate species *(35)*. We have previously reported an in vitro assay that measures the inhibition of hydroxy radical ($OH^\bullet$) and superoxide radical ($O2^{\bullet -}$) induced lipid peroxidation as an index of drug efficacy *(36–37)*.

From: *Methods in Molecular Biology, vol. 108: Free Radical and Antioxidant Protocols*
Edited by: D. Armstrong

The purpose of this chapter is to describe an in vitro model that is suitable for testing antioxidant activity of one or more drugs in a variety of animal tissues, cultured cells, and body fluids. Retinal tissues and the underlying retinal pigment epithelium (RPE) are used in this model because they are rich in polyunsaturated fatty acids and are easily oxidized to lipid hydroperoxides (LHP-), which constitute the substrate for testing antioxidants' capacity. The model also provides simultaneous information on cytotoxicity; therefore, it is a useful tool to screen compounds with proposed antioxidant action as a preliminary step in targeting appropriate compounds for therapeutic intervention.

2. Materials

2.1. Thiobarbituric Acid Reaction (TBA)

1. 8.1% Sodium dodecyl sulfate (SDS).
2. 20% Glacial acetic acid.
3. 0.67% Thiobarbituric acid.
4. n-Butanol.
5. 1,1,3,3 – tetramethoxypropane.
6. Total protein kit (Sigma Chem. Co. St. Louis, MO, procedure no. 541).
7. Total cholesterol kit (Sigma, procedure no. 352).
8. Shimadzu (Kyoto, Japan) RF 5000 spectrofluorophotometer.

2.2. Preparation of Ocular Tissues

1. Sucrose.
2. Ringer's solution.
3. Camel-hair brush of a size appropriate for the eye being used.
4. Sonicator.

2.3. Peroxidation

5 m*M* Ferric chloride. GL-15 lamp emitting 254 nm light at a radiation intensity of 31 uW/cm^2 (Toshiba Co., Tokyo).

2.4. Antioxidants

1. α–tocopherol (Eisai Pharm. Co., Ltd., Japan).
2. Palm Oil tocopherol (PORIM, Malaysia).
3. Pycnogenol (Kaire Int'l., Longmont, CO).
4. EPC-K1: A combination of α-tocopherol and ascorbate (Senju Pharm. Co., Japan).
5. 2-methyl-6-(p-methoxyphenyl)-3,7-dihydroimidazo[1,2-a]pyrazin-3-one hydrochloride (MCLA) (Tokyo Kasei Corp., Japan).
6. S-17224 (Riker Labs, St. Paul, MN).
7. U-74500A Upjohn Pharm. Co., Kalamazoo, MI). All other compounds were purchased from Sigma Chem., Co., St. Louis, MO.

3. Methods

3.1. TBA Assay

The procedure for TBA analysis is detailed in Chapter 9. A standard curve of malondialdehyde over the range of 0.156–2.5 μmol/L is used.

3.2. Preparation of Ocular Tissues

1. External connective tissue is removed from enucleated bovine, ovine, porcine, or canine eyes, the anterior segment removed by making a circumferential incision just behind the lens, and the vitreous expressed.
2. Retina are gently peeled away from the eye cup with fine forceps, moved around the eye cup, severed at the optic nerve, and transferred to a beaker containing Ringer's solution, pH 7.4. Multiple samples are pooled so the yield is sufficient for multiple assays. The eye cups are saved for further processing of the RPE.
3. The photoreceptor rod outer segments (ROS) are harvested by shaking each retina for 1 min in Ringer's solution. The cloudy supernate is layered over a discontinuous 24–41% sucrose gradient in a clear 35-mL plastic polycarbonate tube, centrifuged at 100,000*g* for 4 h, the band that forms at approx 6 mm from the top is collected with a needle and syringe, transferred to a clean tube, diluted with 5 volumes of Ringer's solution, and recentrifuged at 15,000*g* for 5 min to pellet the ROS. The supernatant is discarded and the white pellet resuspended in 1 mL of Type 1 water (*see* **Note 1**), sonicated in an ice bath with two 10-s bursts at 40 W of power, and retained as a whole homogenate.
4. The pooled retinas minus ROS are disrupted with a sonicator in a ratio of 1 g retina/4 mL of Type 1 water and retained as a whole homogenate. Alternatively, the entire retina can be used.
5. RPE preparation: 1–1.5 mL of cold 0.32 *M* sucrose is added to the eye cup minus retina and the surface gently brushed in a circular direction for 30 s with a soft camel's-hair brush *(38)*. The volume used is dependant on eye size. Detached cells are removed with a pipet, pooled, centrifuged at 800*g* for 10 min, and the crude ROS pellet resuspended and washed twice with 0.2M sucrose. The final pellet is resuspended in 5 mL of Type 1 water, sonicated as previously described, and retained as a whole homogenate (*see* **Note 2**).

3.3. Oxidation of Samples

Ten microliters of $FeCl_3$ is added to 10 μL of each homogenate to generate OH·, or 10 μL of normal saline is used as the control. Samples (200 μL) are tested for TBA reactivity at zero time and at 30 min of incubation at 37°C. For ultraviolet (UV)-induced lipid peroxidation, which generates $O2^{\bullet -}$, the GL-15 lamp is positioned 10 cm above homogenate samples and exposed for 60 min (*see* **Notes 3** and **4**).

3.4. Effect of Antioxidant

Water-soluble antioxidants are dissolved in normal saline to the desired molarity and lipid-soluble compounds dissolved in absolute ethanol or another suitable solvent. Control tubes receive only saline or solvent (*see* **Note 5**).

In a separate tube, antioxidants are added simultaneously with $FeCl_3$ over a range from 10^{-7} *M* to 10^{-3} *M* concentration, shaken periodically during the incubation and 200 µL assayed for TBA reactivity at the two time points.

3.5. Calculation

The amount of TBA formed is normalized and expressed as mg/mL of cholesterol(chol) or total protein (TP) where Vo is $FeCl_3$ only at zero time, V1 is $FeCl_3$ and antioxidant at 30 min incubation, and V2 is $FeCl_3$ only at 30 min incubation.

$$\%\ \text{inhibition} - \frac{V1}{V2} - \frac{Vo}{Vo} \times 100 - \text{chol or/TP}$$

3.6. Results

TBA levels expressed in nmol/mg TP in control tissues at zero time and at 30 min are as follows: bovine retina (11.2 + 1.5 without and 50.6 + 3.2 with $FeC1_3$); porcine retina (14.9 + 2.3 without and 53.2 + 3.5 with $FeC1_3$); bovine ROS (9.0 + 3.1 without and 70.2 + 13.6 with $FeC1_3$); porcine ROS (8.5 + 1.5 without and 65.3 + 5.0 with $FeC1_3$); bovine RPE (9.9 + 1.7 without and 38.3 + 9.7 with $FeC1_3$); and porcine RPE (3.7 + 2.4 without and 16.0 + 1.6 with $FeC1_3$).

The representative effects of several antioxidants on OH· induced lipid peroxidation in bovine and porcine samples is shown in **Table 1**. Comparison reactions with UV-light-induced lipid peroxidation are shown in **Table 2**.

These data indicate that the in vitro assay is a sensitive measure for accessing the antioxidant capacity of compounds thought to be protective against OH- and O2-induced lipid peroxidation. Baseline and treatment differences can be observed between tissues and between species. The synergistic effect of drug combinations can also be determined ***(37)***. From these results presented here, a number of compounds can be rapidly screened and prioritized for in vitro ***(38)***, in vivo, physiological ***(39)*** and morphological ***(40)*** studies or tissue culture ***(41)***. Subfractions of retina can also be obtained by gradient centrifugation ***(38)***, which expands use of the posterior segment of the eye for comparing antioxidant activity relative to cell type within a single organ.

4. Notes

1. Water of high purity must be used to decrease the possibility of iron-complex formation.

Table 1
Iron-Induced Lipid Peroxidation

	Inhibition (%) of TBA Reaction					
	Bovine			Porcine		
Antioxidant	Retina	ROS	RPE	Retina	ROS	RPE
α-Tocopherol						
10^{-5} *M*				15		
10^{-2} *M*				100		
Palm-oil tocopherols and tocotrieneols						
10^{-7} *M*	29	55	0			
10^{-6} *M*	36	82	0	0	92	85
10^{-5} *M*	100	100	84	51	100	97
10^{-4} *M*	0	0	43	0	0	0
10^{-3} *M*	57	0 100	85	0	42	
EPC–KI						
10^{-6} *M*	10	0	0	0	0	0
10^{-4} *M*	91	100	100	95	89	99
S-17224						
10^{-7} *M*				18		
10^{-5} *M*				99		
U74500A						
10^{-6} *M*	10	9	0	0	39	0
10^{-4} *M*	70	78	82	88	89	95
10^{-3} *M*	98	100	100	98	94	100
Glutathione						
10^{-2} *M*				0		
Mannitol						
10^{-2} *M*				0		
MCLA						
10^{-6} *M*	26	28	30	0	0	89
10^{-5} *M*	39	48	94	45	38	
10^{-4} *M*	98	91	100			
Quercetin						
10^{-5} *M*	97					
EGCg						
10^{-4} *M*	88					
Pycnogenol						
10^{-5} *M*	91					

Table 2
UV- vs Iron-Induced Lipid Peroxidation In Porcine Retina

	Inhibition (%) of TBA Reaction	
Antioxidant	UV	Iron
α-Tocopherol		
10^{-5} *M*	32	15
10^{-4} *M*	77	18
10^{-3} *M*	93	55
S-17224		
10^{-5} *M*	30	99
10^{-4} *M*	40	100
10^{-3} *M*	63	100
10^{-2} *M*	91	100

2. Only fresh eyes can be used for tissue preparations. However, once isolated, the frozen fractions may be used up to a maximum of 6 mo.
3. Changing the position of the lamp during UV oxidation will decrease the effect if >10 cm, or increase the effect if <10 cm.
4. Because UV light does not penetrate samples well, and to ensure adequate exposure, the volume depth should not exceed 0.5 cm and tubes should be shaken periodically.
5. Antioxidants that are not readily soluble should be sonicated until no visible particles are observed visually in the solution.
6. TBA values of samples from other species will vary.

References

1. Krinsky, N. I. (1992) Mechanism of action of biological antioxidants. *Proc. Soc. Exp. Biol. Med.* **200,** 248–254.
2. Sies, H. (ed.) (1991) *Oxidative Stress, Oxidants and Antioxidants.* Academic Press, NY, pp. 1–650.
3. Flaherty, J. T. (1991) Myocardial injury mediated by oxygen free radicals. *Am. J. Med.* **91(Suppl. 3C),** 79–85S.
4. Thomas, J. P., Kalyanaraman, B., and Girotti, A. W. (1994) Involvement of preexisting lipid hydroperoxides in Cu (2+)-stimulated oxidation of low-density lipoprotein. *Arch. Biochem. Biophys.* **315,** 244–254.
5. Noorooz-Zadeh, J., Tajaddini-Sarmadi, J., Ling, K. L., and Wolff, S. P. (1996) Low-density lipoprotein is the major carrier of lipid hydroperoxides in plasma: relevance to determination of total plasma lipid hydroperoxide concentration. *Biochem. J.* **313,** 781–786.
6. Ramakrishnan, N., Kalinich, J. F., and Mc Clain, D. E. (1996) Ebselen inhibition of apoptosis by reduction of peroxides. *Biochem. Pharmacol.* **51,** 1443–1451.

7. Lialal, I. and Grundy, S. M. (1993) Effect of combined supplementation with alpha tocopherol, ascorbate and beta carotene on low-density lipoprotein oxidation. *Circulation* **88,** 2780–2786.
8. Kahler, W., Kuklinski, B., and Ruhlmann, C., (1993) Diabetes mellitus—a free radical associated disease: adjuvant antioxidant supplementation. *Z. Gesamte* **48,** 223–232.
9. Butcher, G. P., Phodes, J. M., and Walker, R. (1993) The effect of antioxidant supplementation on a serum marker of free radical activity and abnormal serum biochemistry in alcoholic patients admitted for detoxification. *J. Hepatol.* **19,** 105–109.
10. Schalch, W. and Weber, S. (1994) Vitamins and carotenoids—a promising approach to reducing the risk of coronary heart disease and eye diseases, in *Free Radicals in Diagnostic Medicine* (Armstrong, D., ed.), Plenum Press, NY, pp. 335–350.
11. Kandaswami, C. and Middelton, E. (1994) Free radical scavenging and antioxidant activity of plant flavonoids, in *Free Radicals in Diagnostic Medicine*, Adv. Exp. Med. Biol., vol. 366, (Armstrong, D., ed.), Plenum Press, NY, pp. 351–376.
12. Weber, C., Jakobsen, T. S., Mortensen, S. A., Paulsen, G., and Holmer, B. (1994) Effect of dietary coenzyme Q10 as an antioxidant in human plasma. *Mol. Aspects Med.* **15(Suppl.),** S97–102.
13. Packer, L., Witt, E. H., and Tritschler, H. J. (1995) Alpha-lipoic acid as a biological antioxidant. *Free Rad. Biol., Med.* **19,** 227–250.
14. van Zandwijk, N. (1995) N-acetyleysteine (NAC) and glutathione (GSH): antioxidant and chemopreventive properties, with special reference to lung cancer. *J. Cell. Biochem.* **22(Suppl.),** 24–32.
15. Goldberg, D. M., Hahn, S. E., and Parkes, J. C. (1995) Beyond alcohol: beverage consumption and cardiovascular mortality. *Clin. Chim. Acta* **237,** 155–187.
16. Fuhrman, B., Lavy, A., and Aviram, M. (1995) Consumption of red wine with meals reduces the susceptibility of human plasma and low density lipoprotein to lipid peroxidation. *Am. J. Clin. Nutr.* **61,** 549–554.
17. Malterud, K. E., Farbrot, T. L., Huse, A. E., Sund, R. B. (1993) Antioxidant and radical scavenging effects of anthroquinones and anthrones. *Pharmacology* **1(Suppl.),** 77–85.
18. Means, E. D. (1994) 21-aminosteroids (Lazaroids), in *Free Radicals in Diagnostic Medicine* (Armstrong, D., ed.), Plenum Press, NY, pp. 307–312.
19. Kilo, C., Dudley., J., and Kalb, B. (1991) Evaluation of the efficacy and safety of Diamicron in non-insulin-dependent diabetic patients. *Diabetes Res. Clin. Pract.* **14(Suppl),** S79–82.
20. Saifer, M. G. P., Somack, R., and Williams, L. D. (1994) Plasma clearance and immunological properties of long-acting superoxide dismutase prepared using 35,000 to 120,000 dalton poly-ethylene glycol, in *Free Radicals in Diagnostic Medicine* (Armstrong, D., ed.), Plenum Press, NY, pp. 377–387.
21. Sattler, W., Maiorino, M., and Stocker, R. (1994) Reduction of HDL- and LDL-associated cholesterylester and phospholipid hydroperoxides by phospholipid

hydroperoxide glutathione peroxidase and Ebselen (PZ51). *Arch. Biochem. Biophys.* **309,** 214–221.
22. Day, B. J., Shawen, S., Liochev, S. I., and Crapo, J. D. (1995) A metallo porphyrin superoxide dismutase mimetic protects against paraquat-induced endothelial cell injury, in vitro. *J. Pharmacol. Exp. Ther.* **275,** 1227–1232.
23. Hardy, M. M., Flickenger, A. G., Riley, D. P., Weiss, R. H., and Ryan, U. S. (1994) Superoxide dismutase mimetics inhibit neutrophil-mediated human aortic endothelial cell injury in vitro. *J. Biol. Chem.* **269,** 18,535–18,540.
24. Bhanumathi, P. and Devi, P. U. (1994). Modulation of glutathione depletion and lipid peroxidation by WR-77913, an 2-mercaptopropionylglycine in cyclophosphamide chemotherapy. *Indian J. Exp. Biol.* **32,** 562–564.
25. Fact sheet on the pharmacology, safety considerations and clinical application of Thiola (tiopronoin) a metabolic detoxicant and Rimatil (bucillamine) an antirhumatic, 1986. Santen Pharmaceutical Co., Ltd., Osaka, Japan.
26. Das, K. C., Lewis-Molock, Y., and White, C. W. (1995) Thiol modulation of TNF alpha and IL-1 induced Mn SOD gene expression and activation of NF-kappa B. *Mol. Cell. Biochem.* **148,** 45–57.
27. Purpura, P., Westman, L., Will, P., Eidelman, A., Dagan, V. E., Usipov, A. N., and Schor, N. F. (1996) Adjunctive treatment of murine neuroblastoma with 6-hydroxydopamine and Tempol. *Cancer Res.* **56,** 2336–2342.
28. Serafim, M., Ghiselli, A., and Ferro-Luzzi, A. (1996) In vivo antioxidant effect of green and black tea in man. *Eur. J. Clin. Nutr.* **50,** 28–32.
29. Jourdan, A., Agnejouf, O., Imbault, P., Doutremepuich, F., Inamo, J., and Doutremepuich, C. (1995) Experimental thrombosis model induced by free radicals. Application to aspirin and other different substrates. *Thromb. Res.* **79,** 109–123.
30. Steele, V. E., Moon, R. C., Lubet, R. A., Grubbs, D. J., Reddy, B. S., Wargovich, M., Mc Cormick, D. L., Pereira, M. A., Crowell, J. A., and Bagheri, D. (1994) Preclinical efficacy evaluation of potential chemoprotective agents in animal carcinogenesis models: methods and results from the NCI Chemoprevention Drug Development Program. *J. Cell. Biochem.* **20(Suppl.),** 32–54.
31. Rahl, H., Khoschsorur, G., Colombo, T., Jatzber, F. and Esterbauer, H. (1992) Human plasma lipid peroxide levels show strong transient increase after successful revascularization operations. *Free Rad. Biol. Med.* **13,** 281–288.
32. Isom, G. E., and Borowitz, J. L. (1995) Modification of cyanide toxicodynamics: mechanistic based antidote development. *Toxicol. Lett.* **82/83,** 795–799.
33. Zhang, J. R., Scherch, H. M., and Hall, E. D. (1996) Direct measurement of lipid hydroperoxides in iron-dependent spinal neuronal injury. *J. Neurochem.* **66,** 355–361.
34. Li, L. and Lau, B.H. (1993) A simplified in vitro model of oxidant injury using vascular endothelial cells. *In Vitro Cell Dev. Biol. Anim.* **29A,** 531–536.
35. Batcabe, J. P., Mac Gill, R. S., Zaman, K., Ahmad, S., and Pardini, R. S. (1994) Mitomycin C induced alterations in antioxidant enzyme levels in a model insect species, Spodoptera eridania. *Gen. Pharmacol.* **25,** 569–574.
36. Hiramitsu, T. and Armstrong, D. (1991) Preventive effect of antioxidants on lipid peroxidation in the retina. *Ophthalmic Res.* **23,** 196–203.

37. Ueda, T., Ueda, T., and Armstrong, D. (1996) Preventive effect of natural and synthetic antioxidants on lipid peroxidation in the mammalian eye. *Ophthalmic Res.* **28,** 184–192.
38. Armstrong, D., Santangelo, G., and Connole, E. (1981) The distribution of peroxide regulating enzymes in the canine eye. *Curr. Eye Res.* **1,** 225–242.
39. Armstrong, D., Hiramitsu, T., Gutteridge, J., and Nilsson, S. E. (1982) Studies on experimentally induced retinal degeneration. 1: effect of lipid peroxides on electroretinographic activity in the albino rabbit. *Exp. Eye Res.* **35,** 157–171.
40. Armstrong, D. and Hiramitsu, T. (1990) Studies on experimentally induced retinal degenerative: 2: early morphological changes produced by lipid peroxides in the albino rabbit. *Jap. J. Ophthalmol.* **34,** 158–173.
41. Gadoth, N. and Armstrong, D. (1984) The concentration of four antioxidant enzymes in cells isolated from the canine retina: implications for aging. *AGE* **7,** 107–110.

31

Antioxidant Activity of Low-Density Lipoprotein

Nicholas J. Miller and George Paganga

1. Introduction

Oxidation of low-density lipoprotein (LDL) is of great interest for epidemiological and clinical diagnostic reasons, as well as for basic scientific research, because it is thought to precede the development of atherosclerotic lesions *(1)*. Measurement of individual antioxidants such as α-tocopherol, is, of course, a primary route to the study of LDL oxidation. However, such measurements are not always possible and do not take into account the potential contribution to the antioxidant activity of LDL from a wide variety of different antioxidants whose exact nature may not be known. Oxidized LDL can itself be quantitatively measured *(2)* as well as LDL oxidizability *(3–5)*. However, the measurement of LDL oxidizability involves a prolonged and variable incubation period and the results obtained from the studies previously quoted are, in a number of respects, contradictory. LDL oxidizability measurements clearly reflect a number of factors other than the antioxidant content of LDL (for example, the presence of pre-formed lipid hydroperoxides and the ratio of saturated to unsaturated fatty acids) *(6)*, which may, of course, be a desirable feature because oxidized LDL itself is the end product of a multi-factorial process.

Direct measurement of LDL total antioxidant activity (LDL-TAA) is the obvious diagnostic approach to this situation, although the correlation of LDL-TAA with clinical endpoints is yet to be ascertained. A rapid direct method is described here for the measurement of LDL-TAA, using the ferrylmyoglobin/2,2′-azinobis(3-ethylbenzothiazoline-6-sulfonic acid) diammonium salt) (ABTS) assay *(7)*. The principle behind this technique utilizes the peroxidatic interaction between metmyoglobin (MetMb) and hydrogen peroxide (H_2O_2) to form ferrylmyoglobin radicals, and in turn the radical mono cation of ABTS ($ABTS^{\bullet+}$) an intense blue/green chromogen. In the presence of antioxidant reductants, the

From: *Methods in Molecular Biology, vol. 108: Free Radical and Antioxidant Protocols*
Edited by: D. Armstrong © Humana Press Inc., Totowa, NJ

formation of this radical cation is suppressed *(8)*. The extent of the suppression can be calculated and directly related to the antioxidant capacity (activity) of the LDL sample under investigation. Measurement at 734 nm, in the near-I.R. region of the spectrum, reduces analytical interference from lipaemia and sample turbidity and this wavelength is also well beyond the absorption maxima of a wide range of potentially interfering substances that might be present in lipoprotein particles (*see* **Notes 1** and **2**).

A simple method is also described for the measurement of LDL oxidizability using ferryl myoglobin and H_2O_2 as the oxidizing agents, followed by spectroscopic determination of conjugated dienes as an endpoint. Myoglobin, used in the way described, provides a milder oxidizing stimulus for LDL than, for example, cupric ions, and is not influenced by the presence in the buffer of chelating agents such as ethylenediamine tetraacetic acid (EDTA). Although this measurement is frequently carried out by recording absorbance changes at 234 nm, the exact absorption maximum that will be observed is influenced by the stereochemistry of the conjugated dienes formed, which will, in turn, vary according to the fatty-acid composition of the sample. In general, *cis-trans* hydroperoxy dienes have an absorbance maximum at 242 nm, whereas the *trans-trans* dienes have an absorbance maximum at 233 nm. Thus, it is recommended to monitor the changes in the ultraviolet (UV) spectrum between 220 and 300 nm and determine the actual wavelength at which the diene peak develops. The lag time to diene formation or the time to 50% oxidation of LDL is the preferred endpoint.

2. Materials

2.1. LDL-TAA

1. 5 mmol/L Phosphate buffered saline, (PBS), pH 7.4: Prepared from mixing 1.7907 g/L of $Na_2HPO_4.12H_2O$ and 0.655 g/L of $NaH_2PO_4.H_2O$ to a pH of 7.40 (approximately 8:1) with the addition of 9.0 g/L sodium chloride. This buffer should not contain sodium azide or any other additional substances. Stable at 4°C for 6 mo.
2. 2.5 mmol/L Trolox (M_r 250.29) (Aldrich Chemical Co., Milwaukee, WI, Cat. no. 23,881-3): Supplied in crystalline form; prepare in PBS for use as an antioxidant standard (0.1564 gm of Trolox in 250 mL of buffer). Care must be taken when weighing Trolox, which is electrostatically attracted to charged surfaces. Precoating of plastic weighing boats with Trolox may be necessary before zeroing the balance. At this pH and concentration, the upper limit of Trolox's solubility is approached, and gentle ultrasonication is required to dissolve the crystals. Frozen (–20°C) Trolox dissolved in PBS at this concentration is stable for more than 12 mo. Fresh working standards of 20, 40, 60, 80, and 100 µmol/L are prepared immediately before use by mixing 2.5 mmol/L Trolox with PBS.

3. H_2O_2 (British Drug Houses "Aristar" or equivalent): Working solutions are prepared from stock Aristar H_2O_2 after an initial dilution in PBS to a concentration of 500 mmol/L. Aristar H_2O_2 is supplied as a 30% solution, with a specific gravity of 1.10 (1.099–1.103) Because the M_r is 34, a 500 mmol/L solution is prepared by diluting 515 µL of stock H_2O_2 to 10 mL in PBS (solution A). The required initial working concentration of H_2O_2 (3.13 mmol/L) is then prepared by diluting 313 µL of solution A to 50 mL with PBS (solution B). Working H_2O_2 solutions are freshly prepared prior to use (or every 6 h). H_2O_2 (in water) has an ε_{mM} of 72.4 (± 0.2) $cm^{-1}mol^{-1}$ at 230 nm. A 1.0 m*M* solution in water will therefore have an absorbance of 0.0724 at this wavelength ***(9)***.
4. ABTS (Aldrich, Cat. no.27,172-1): A colorless powder when freshly re-crystallised; with storage it takes on a pale green color. 5 mmol/L ABTS (M_r 548.68) is freshly prepared for use by dissolving 0.1372 gm in 50 mL of 18 MΩ water (or 0.02743 gm in 10 mL). This solution is stable for several wk in the dark, as judged by the absorbance maximum of unreacted ABTS at 342 nm, but it has not been firmly established to what extent "old" ABTS solutions are suitable for this particular application.
5. MetMb (Sigma Chemical Co., St. Louis, MO, myoglobin M-1882 from horse heart): Myoglobin with iron in the ferric state and should be purified prior to use. A 35 × 2.5 cm column with Sephadex G15-120 (Pharmacia, Piscataway, NJ) in PBS is suitable (use approx 80 gm of Sephadex), but any simple de-salting technique may be used (for example, dialysis). The concentration of myoglobin in the column eluate is calculated from the extinction coefficients, and aliquots of MetMb in PBS are stored frozen until required for use. The procedure is as follows:
   a. Prepare a 400 µmol/L solution of MetMb (0.0752 gm in 10 mL of PBS).
   b. Prepare a 740 µmol/L solution of potassium ferricyanide (potassium hexacyanoferrate[III) by dissolving 0.0244 gm in 100 mL of PBS.
   c. Add 10 mL of the freshly prepared ferricyanide solution to 10 mL of metmyoglobin solution.
   d. Mix the two solutions and allow the mixture to stand at room temperature for 30 min. (It is important to give the ferricyanide and myoglobin time to react.)
   e. Apply the mixture to the Sephadex column and elute with PBS.
   f. The first brown fraction is collected and the rest discarded.
   g. The collected eluate is mixed well, and its absorbance read at 490, 560, 580, and 700 nm.
   h. The absorbance reading at 700 nm is subtracted from the readings at 490, 560, and 580 nm to correct for background absorbance.
   i. Calculation of the relative proportions of the different forms of myoglobin present is made from the formulae ***(8)*** based on the algorithm of Whitburn ***(10)***:
      $[Met\ Mb] = 146\ A_{490} - 108\ A_{560} + 2.1\ A_{580}$
      $[Ferryl\ Mb] = -62A_{490} + 242\ A_{560} - 123\ A_{580}$
      $[MbO_2] = 2.8\ A_{490} - 127\ A_{560} + 153\ A_{580}$
   j. MetMb prepared in this way is not suitable for use unless it is more than 94% of the total haem species present.

k. The purified myoglobin is diluted to a concentration of approx 140 μ*M*, aliquoted, and stored until use.

l. Purified MetMb stored at 4°C is stable for at least 8 wk and at –20°C, it is stable for more than 6 mo.

2.2. LDL Oxidizability

1. 10 mmol/L PBS, pH 7.4: Prepared from mixing 3.58 g/L of $Na_2HPO_4.12H_2O$ and 1.31 g/L of $NaH_2PO_4.H_2O$ to a pH of 7.4 (approx 8:1) with the addition of 9.0 g/L sodium chloride. For incubation during LDL oxidizability measurements, 10 μmol/L (0.00372 g/L) EDTA is added (disodium EDTA, M_r 372.2). Store at 4°C. Stable for six mo.
2. H_2O_2: A 100 μmol/L working solution of H_2O_2 is prepared by diluting 10 μL of 500 mmol/L H_2O_2 (solution A - *see* **Subheading 2.1., item 3**) to 50 mL with PBS.
3. MetMb: The preparation and purification of metmyoglobin is described in **Subheading 2.1., item 5.**

3. Methods

3.1. LDL-TAA

1. LDL is isolated from plasma samples by density gradient ultracentrifugation *(11)* and the protein content of the LDL preparations measured by the method of Markwell *(12)*. Values for LDL antioxidant activity must be corrected for protein content to take into account variations in the concentration of these preparations (*see* **Note 3**). Typically, an LDL isolate containing 2 mg/mL of protein should be prepared because this will have a TAA value in the linear range of the dose-response curve using the protocol described in **Table 1.**
2. In the protocol devised for the LDL-TAA measurement, 250 μL of ABTS/myoglobin reagent is mixed with 50 μL of sample, the probe flushed with 20 μL of water, and then 20 μL of 5.40 mmol/L H_2O_2 is added, followed by 10 μL of water. The incubation volume is thus 350 μL and reagents are prepared so that on dilution into this incubation volume, they are at the desired concentrations, which are 4.36 μmol/L MetMb, 435.7 μmol/L ABTS, 178.6 μmol/L H_2O_2, and 14.3% sample fraction.
3. Using these reagent concentrations, the end of the lag phase for the 100 μmol/L Trolox standard (i.e., the time at which the absorbance will start to increase in the cuvet) is 248 s after the addition of the H_2O_2, which starts the reaction. It is this parameter that is used as the measuring time for the assay. Minor fluctuations in assay conditions, for example caused by differences between instruments, can be compensated for by appropriate adjustment of this measuring time.
4. The assay can be carried out using a Cobas Fara centrifugal analyzer (Roche Diagnostics Welwyn, Garden City, UK). The instrument settings used are shown in **Table 1**. A dose-response curve is derived using a logit/log 4 plot of absorbance at 734 nm against the concentration of the Trolox standards at 0, 2.86, 5.72, 8.58, 11.44, and 14.30 μmol/L final concentration.

Table 1
Instrument Settings used for LDL-TAA Measurement on the Cobas Fara Centrifugal Analyzer

Function	Setting
Measurement mode	Absorbance
Reaction mode	P-T-I-SR1-A
Calibration mode	logit/log 4
Reagent blank	none
Wavelength	734 nm
Temperature	30°C
Decimal position	2
Units	μmol/l
Sample volume	50 μL
Diluent (water)	20 μL
Reagent volume	250 μL
Incubation	20 s
M1 reading	15 s
Start reagent (hydrogen peroxide)	20 μL
Diluent (water)	10 μL
Firts reading	0.5 s
Number of readings	50
Time interval	8 s
Reaction direction	increase
Factor	1.0
Number of steps	1
First reading	M1
Last reading	31 (248 s)
Standard 1 concentration	0.0
Standard 2 concentration	20 μ*M* Trolox (initial concentration)
Standard 3 concentration	40 μ*M* Trolox
Standard 4 concentration	60 μ*M* Trolox
Standard 5 concentration	80 μ*M* Trolox
Standard 6 concentration	100 μ*M* Trolox

3.2. LDL Oxidizability

1. LDL samples, freshly isolated from subjects, with the concentration adjusted by dilution in 10 mmol/L PBS (EDTA-free) to 62.5 μg/mL of LDL protein (typically a 1/20 to a 1/40 dilution after initial isolation of the LDL), are incubated with 2.5 μmol/L MetMb and 3.13 μmol/L H_2O_2 (final concentration) in 10 mmol/L PBS with EDTA (*see* **Note 5**).

2. The reaction is carried out in quartz cuvets in a temperature-controlled Beckman DU-65 spectrophotometer, at 37°C. The reaction is scanned from 220 to 300 nm every 10 min for a 5-h period and the difference spectra recorded using Beckman Data Leader software on a linked computer (for example an IBM PS/2 model 70, with math co-processor).
3. The kinetics of the formation of conjugated dienes at the absorbance maximum (nm), which develop on sequential scanning, are used to estimate the lag phase of lipid peroxidation in the LDL isolates. This wavelength maximum may vary slightly from sample to sample (*see* **Note 6**).
4. The protocol is as follows:
   a. Calculate the volume of LDL isolate required to give a final concentration of 62.5 µg/mL in a 1.0 mL mixture (*see* **Note 7**).
   b. Calculate the volume of MetMb required to give a final concentration of 2.5 µmol/L in a 1.0 mL mixture.
   c. Calculate the volume of buffer required to give a final volume in the cuvet of 1.0 mL (= 1000 µL – [volume of LDL + volume of H_2O_2 + volume of MetMb]).
   d. Add 31.3 µL of 2 mg/mL LDL to a 1.0-mL quartz cuvet (final concentration 62.5 µg/mL).
   e. Add 920 µL of PBS with EDTA to the cuvet, if the other volumes are as shown (*see* **Note 8**).
   f. Add 31.3 µL of 100 µmol/L H_2O_2 to the cuvet (final concentration 3.125 µmol/L).
   g. Start the reaction with 17.9 µL of 140 µmol/L MetMb at a final concentration of 2.5 µmol/L.
   h. Mix the contents immediately, blank the cuvet, and incubate for 5 h at 37°C.
   i. Record spectra between 220 and 300 nm every 10 min.

3.3. Results

3.3.1. LDL-TAA

In blood samples obtained from patients attending a clinic for the treatment of lipid disorders, the LDL-TAA (range 13.9–43.6 n*M*/mg protein) correlated well with LDL α-tocopherol, measured by normal phase high-pressure liquid chromatography (HPLC) (range 10.2–47.9 n*M*/mg protein, **Table 2**). These results support the view that α-tocopherol is the major antioxidant substance in LDL (*see* **Note 3**). There is also a good correlation between LDL-TAA and the plasma α-tocopherol concentration and also between the LDL-TAA and the α-tocopherol:cholesterol ratio.

Figure 1 shows a plot of matched LDL-TAA values in n*M*/mg protein against LDL α-tocopherol concentrations in n*M*/mg protein in these samples (*see* **Note 4**).

Table 2
Antioxidant Activity of LDL ($N = 30$)

	Mean ± SD[b]	Range
LDL TAA (n*M*/mg protein)[a]	24.7 ± 3.7	17.7–31.4
LDL α-tocopherol (nmol/mg protein)	21.8 ± 8.2**	11.6–47.9
LDL cholesterol (mg/mg protein)	1.40 ± 0.22	0.94–1.92
LDL α-tocopherol:cholesterol ratio	15.5 ± 5.3**	7.7–31.3
plasma α-tocopherol (μmol/L)	55.4 ± 23.0***	35.8–151.6
plasma cholesterol (mmol/L)	6.00 ± 0.87	4.40–7.80
plasma α-tocopherol:cholesterol ratio	9.3 ± 3.9***	5.7–25.4

[a]All LDL results are expressed as "per mg of LDL protein." (Adapted with permission from **ref.** *7*.)

[b]Correlation with LDL-TAA (nmol/mg protein) *$p < 0.05$, **$p < 0.01$, ***$p < 0.005$ (Spearman rank-order correlation coefficient).

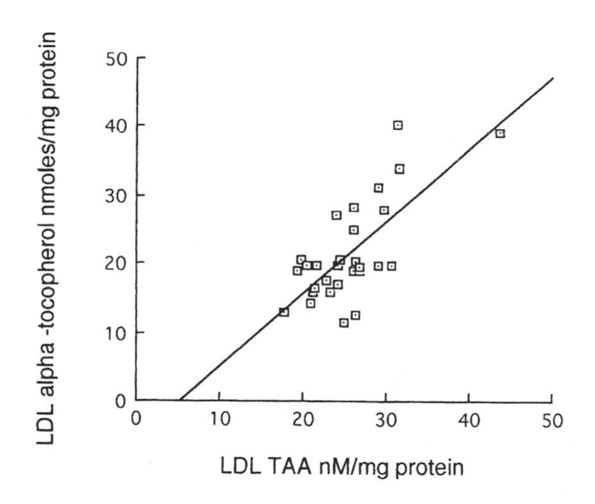

Fig. 1. Plot of matched LDL-TAA values (nmol/mg protein) against LDL α-tocopherol concentrations (nmol/mg protein), N = 30, $p < 0.05$. (Adapted with permission, from **ref.** *7*). y, –5.4789 + 1.0606x. R^2, 0.521.

3.3.2. LDL Oxidizability

Typical difference spectra of the formation of conjugated dienes in an LDL isolate are shown in **Fig. 2**. **Figure 3** shows the kinetics of oxidation of an LDL sample at 238 nm and demonstrates the derivation of the time to the end of the lag phase to lipid peroxidation. The time to 50% oxidation (T $50\%_{ox}$) is

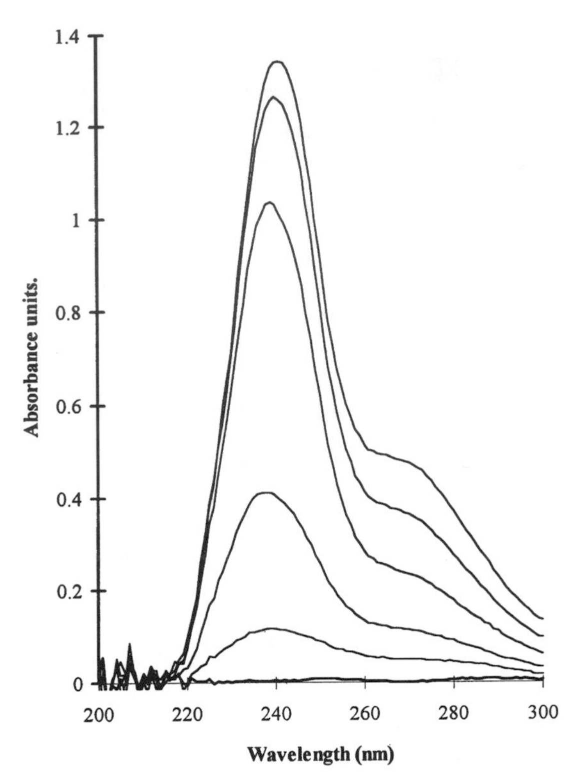

Fig. 2. Typical difference spectra showing the formation of conjugated dienes in an LDL isolate (62.5 mg/mL LDL protein) oxidized by 2.5 μ*M* MetMb and 3.125 μ*M* H_2O_2.

calculated by taking the time at which maximum absorbance is attained and dividing by two. A lengthening of the lag phase or the time to T $50\%_{ox}$ indicates an increase in antioxidant activity. Typical samples from healthy young subjects have a lag phase of between 80–180 min.

Lag phases estimated on LDL samples from patients attending a lipid clinic (*see* **Subheading 3.3., step 1**) ranged from 0 to 900 min (with a mean of 144 min) whereas the mean T $50\%_{ox}$ was 350 min. This demonstrates that some of these patients had LDL that was already partially oxidized ***(13)***. The presence of partially oxidized LDL in the circulation was found to correlate with the progression of arterial stenosis over a 6-mo study period. There is a significant negative correlation between the LDL-TAA and the T $50\%_{ox}$ ($p < 0.05$, Spearman rank-order correlation coefficient).

3.3.3. Estimation of Lag Phase (EndPoint)

It can be seen that this method does not require the quantitative estimation of conjugated dienes. **Figure 3** shows the reaction has:

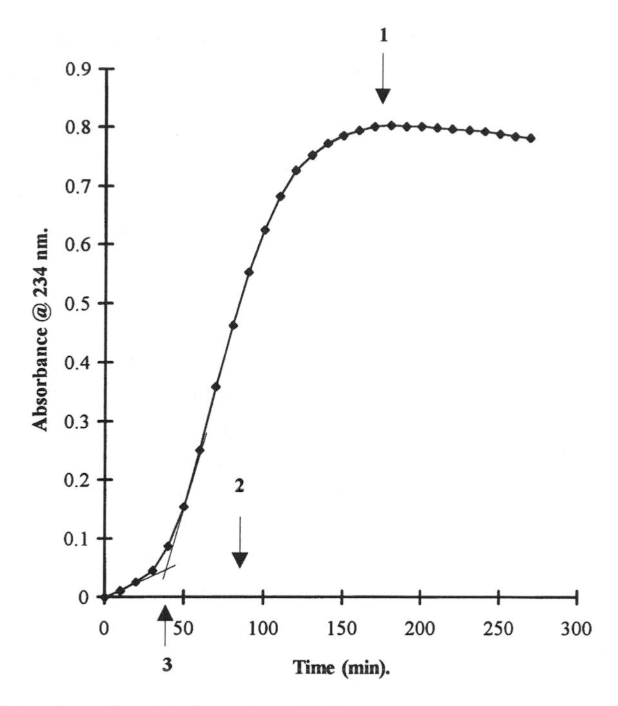

Fig. 3. The kinetics of oxidation of an LDL sample at 234 nm, showing the time to complete oxidation (**1**), and demonstrating the derivation of the time to 50% oxidation (**2**), and also the lag phase to lipid peroxidation (**3**).

1. An initial, slower lag phase, in which the rate of reaction is slowed by consumption of endogenous antioxidants, followed by
2. A faster propagation phase during which lipid peroxidation proceeds more rapidly—the endogenous antioxidants having previously been consumed, and
3. A termination phase, during which the absorbance at 234 nm finally stabilizes, indicating that all the lipid present is oxidized and that conjugated diene formation has ceased (point 1 in **Fig. 3**). The time to 50% oxidation is taken as half this time (point 2 in **Fig. 3**). The end of the lag phase (point 3 in **Fig. 3**) is determined by the time to the intercept of the slopes of the lag phase and the propagation phase.

4. Notes

1. In the $ABTS^{\bullet +}$ spectrophotometric method for the direct measurement of LDL total antioxidant activity, the antioxidants in lipoprotein particles suppress the formation of a colored free radical in the aqueous phase. This suppression is used to quantify their antioxidant activity. Even though the antioxidants in LDL particles are in the nonpolar phase, adequate mixing ensures that radical scavenging takes place.

2. Although there are certain advantages in using a centrifugal analyzer such as the Cobas Fara (excellent precision, enhanced mixing of reagents and samples), this protocol is readily adaptable to any spectrophotometric enzyme analyzer system. Not all instruments are capable of measuring at 734 nm; whereas this is the preferred wavelength, the wide absorption spectrum of $ABTS^{\bullet+}$ does makes it detectable at shorter wavelengths. However, careful evaluation of potential interference is required if measurements are to be made away from the 734 nm maximum.
3. The results obtained on LDL fractions with known α-tocopherol content confirm the view that α-tocopherol is the major antioxidant substance in LDL.
4. α-Tocopherol is not, however, the only antioxidant present. The "non-α-tocopherol antioxidant activity" in **Fig. 1** is represented by the activity at the x-intercept and amounts to approx 20% of the LDL-TAA. This finding is in general agreement with those of other authors ***(14)***. The exact nature of the non-α-tocopherol LDL antioxidant activity remains to be defined.
5. Pipetting the volumes indicated requires care and attention, but is easily achievable if appropriate positive displacement pipets are used e.g., SMI digital adjust micropettors (Baxter Diagnostics Inc., Muskegon, MI).
6. Scanning multiple samples for conjugated diene peaks between 220 and 300 nm permits the true absorbance maximum to be selected to measure the kinetics of lipid peroxidation. Quantitation of conjugated dienes solely at the widely quoted absorbance maximum of 234 nm may result in erroneous values.
7. 20 mL of fasted blood will yield approx 2.5 mL of an LDL isolate of 2 mg/mL LDL protein. However, the LDL concentration obtained will vary greatly from individual to individual.
8. Venipuncture often produces mild hemolysis, with the consequent release of iron into the plasma. Although this should be kept to a minimum, the presence of a chelator such as EDTA prevents variable amounts of iron and other transition metals in the incubation mixture from contributing to the oxidation of the LDL. Metmyoglobin-induced oxidation is not affected by the presence of EDTA.

Acknowledgments

The support of Roche Diagnostic Systems (NJM) and of the British Heart Foundation (GPP) is gratefully acknowledged.

References

1. Parthasarathy, S., Printz, D. J., Boyd, D., Joy L., and Steinberg, D. (1986) Macrophage oxidation of low density lipoprotein generates a modified form recognized by the scavenger receptor. *Arteriosclerosis* **6,** 505–510.
2. Wang, H., Chen, S., Kong, X., Wang, X., Chang, G., Xu, S., Luo, Z., and Xie, Y. (1993) Quantitation of plasma oxidatively modified low density lipoprotein by sandwich enzyme linked immunosorbent assay. *Clin. Chim. Acta* **218,** 97–103.

3. Esterbauer, H., Striegl, G., Puhl, H., and Rotheneder, M. (1989) Continuous monitoring of in vitro oxidation of human low density lipoprotein. *Free Rad. Res. Commun.* **6,** 67–75.
4. Cominacini, L., Garbin, U., Davoli, A., Micciolo, R., Bosello, O., Gaviraghi, G., Scuro, L. A., and Pastorino, A. M. (1991) A simple test for predisposition to LDL oxidation based on the fluorescence development during copper-catalyzed oxidative modification. *J. Lipid Res.* **32,** 349–358.
5. Regnstrom, J., Nilsson, J., Tornvall, P., Landon, C., and Hamsten, A. (1992) Susceptibility to LDL oxidation and coronary atherosclerosis in man. *Lancet* **339,** 1183–1186.
6. Esterbauer, H., Gebicki, J., Puhl, H., and Jurgens, G. (1992) The role of lipid peroxidation and antioxidants in oxidative modification of LDL. *Free Rad. Biol. Med.* **13,** 341–390.
7. Miller, N. J., Paganga, G., Wiseman, S., Van Nielen, W., Tijburg, L., Chowienczyk, P., and Rice-Evans, C. A. (1995) Total antioxidant activity of low density lipoproteins and the relationship with α-tocopherol status. *FEBS Letts.* **365,** 164–166.
8. Miller, N. J., Rice-Evans, C. A., Davies, M. J., Gopinathan, V., and Milner, A. (1993) A novel method for measuring antioxidant capacity and its application to monitoring the antioxidant status in premature neonates. *Clin. Sci.* **84,** 407–412.
9. Aebi, H. E. (1983) Catalase, in *Methods of Enzymatic Analysis* (vol. 3) (3rd ed.) (Bergmeyer, H. U., ed.), Verlag Chemie, Weinheim, Germany, p. 277.
10. Whitburn, K. D., Shieh, J. J., Sellers, R. M., Hoffman, M. Z., and Taub, I. A. (1982) Redox transformations in ferrimyoglobin induced by radiation-generated free radicals in aqueous solution. *J. Biol. Chem.* **257,** 1860–1869.
11. Chung, B. H., Wilkinson, T., Geer, J. C., and Segrest, P. J. (1980) Preparative and quantitative isolation of plasma lipoproteins: rapid single discontinuous density ultracentrifugation in a vertical rotor. *J. Lipid. Res.* **21,** 284–291.
12. Markwell, M. A. K., Haas, S. M., Bieber, L. L., and Tolbert, N. E. (1978) Modification of the Lowry procedure to simplify protein determination in membranes and lipoprotein samples. *Anal. Biochem.* **87,** 206–210.
13. Andrews, B., Burnand, K., Paganga, G., Browse, N., Rice-Evans, C. A., Sommerville, K., Leake, D., and Taub, N. (1995) Oxidisability of low density lipoproteins in patients with carotid or femoral artery atherosclerosis. *Atherosclerosis* **112,** 77–84.
14. Abuja, P. M. and Esterbauer, H. (1995) Simulation of lipid peroxidation in low density lipoprotein by a basic "skeleton" of reactions. *Chem. Res. Toxicology* **8,** 753–763.

32

Lipoic Acid as an Antioxidant

The Role of Dihydrolipoamide Dehydrogenase

Mulchand S. Patel and Young Soo Hong

1. Introduction

Free lipoic acid is present in biological systems at extremely low levels. The biological function of lipoic acid requires its covalent linkage to the N^6-amino group of lysine residues of several proteins *(1)*. The covalent attachment of lipoic acid to specific proteins is catalyzed by lipoate protein ligases in prokaryotes and eukaryotes *(1)*. Only the naturally occurring R form, not the S form, is covalently linked to the proteins by ligases. Recent evidence indicates that free lipoic acid, when provided exogenously in pharmacological doses, may serve as an antioxidant *(2)*. Lipoic acid has recently been used as a therapeutic agent against hepatic and neurological disorders *(2)*. Although the mechanism of action of lipoic acid is not known, the therapeutic effects might be the result of the free-radical scavenging role of reduced lipoic acid in preventing oxidative damage in the cell. For this purpose, oxidized lipoic acid needs to be reduced to dihydrolipoic acid. The enzyme, dihydrolipoamide dehydrogenase (E3), which reversibly oxidizes dihydrolipoyl moieties linked to protein, also catalyzes reversible reduction of free lipoic acid to dihydrolipoic acid (**Fig. 1**) *(3)*.

To a much lesser extent, glutathione reductase also catalyzes reversible interconversion of the oxidized and reduced forms of lipoic acid *(2)*. In this chapter, we will briefly discuss the properties of E3 and describe in detail the measurement of the forward and reverse reactions catalyzed by this enzyme *(1,4,5)*.

E3 (EC 1.8.1.4), a flavoprotein disulfide oxidoreductase, catalyzes the redoxidation of covalently attached dihydrolipoyl moieties as per the following reaction.

$$\text{Dihydrolipoyl-protein} + NAD^+ \leftrightarrow \text{Lipoyl-protein} + NADH + H^+$$

From: *Methods in Molecular Biology, vol. 108: Free Radical and Antioxidant Protocols*
Edited by: D. Armstrong © Humana Press Inc., Totowa, NJ

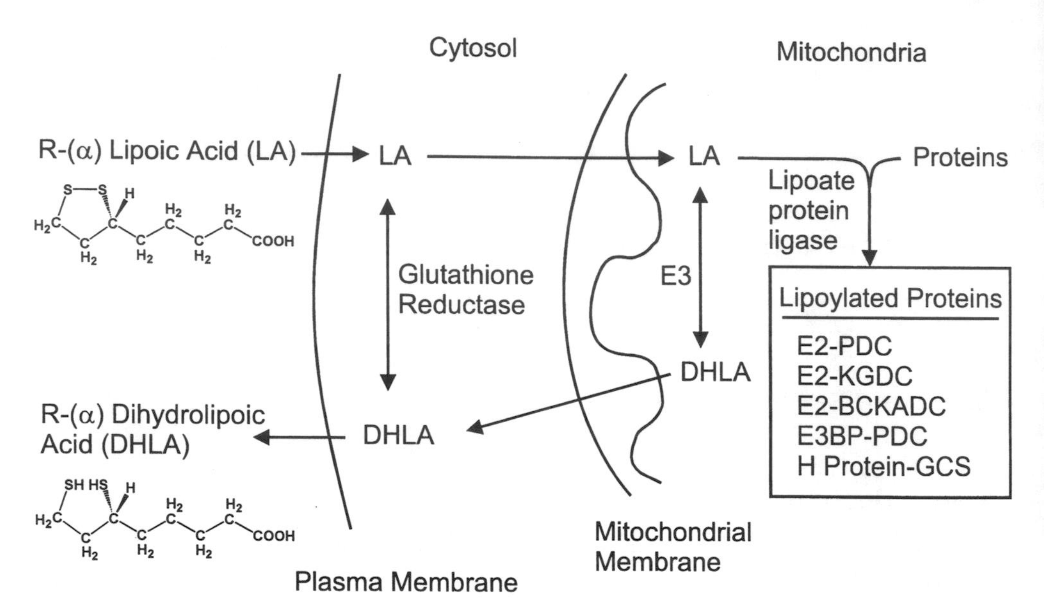

Fig. 1. Role of lipoic acid in cellular metabolism. E2, dihydrolipoamide acyltransferase; PDC, pyruvate dehydrogenase complex; KGDC, α-ketoglutarate dehydrogenase complex; BCKADC, branched-chain α-keto acid dehydrogenase complex; E3, dihydrolipoamide dehydrogenase; E3BP, E3-binding protein; GCS, glycine cleavage system.

Five lipoyl-containing proteins have been described in higher eukaryotes (**Fig. 1**) *(1)*. In these proteins, the lipoyl moiety is covalently linked to specific lysine residue(s). These proteins are components of four multienzyme complexes in eukaryotic mitochondria, namely the pyruvate dehydrogenase complex, the α-ketoglutarate dehydrogenase complex, the branched-chain α-keto acid dehydrogenase complex, and the glycine cleavage system *(1)*. In the mammalian cell, the dihydrolipoyl acyltransferase components of the three α-keto acid dehydrogenase complexes, the E3-binding protein (also known as protein X) of the pyruvate dehydrogenase complex only and the hydrogen carrier protein (H protein) component of the glycine cleavage system are the physiological substrates of E3.

E3 is a homodimer with a subunit size of approx 50 kDa. Each monomer contains a tightly but noncovalently bound FAD molecule, a redox disulfide, and a NAD^+-binding site *(3)*. Only the dimer and not the monomer is catalytically active because the active site is composed of specific amino acid residues from both subunits *(3)*. Owing to this arrangement, a dimer has two identical active sites *(3)*. During catalysis, electrons are sequentially transferred from the dihydrolipoyl moiety attached to the substrate protein to the redox disulfide, the FAD cofactor, and the NAD^+ cofactor *(3)*. E3 catalyzes a Ping-Pong bi-bi mechanism, and this reaction series can be divided into two half-reactions: the reduction of the oxidized enzyme to the two-electron reduced form and the transfer of electrons from the reduced form to NAD^+ forming reduced nicotinamide adenine dinucleotide (NADH) *(3)*. Some kinetic parameters of mammalian E3 from several species are shown in **Table 1** *(6,7)*.

Although a protein-bound lipoyl moiety serves as one of the substrates for E3 under normal physiological conditions, free lipoic acids, both R- and S-enantiomers, are also used as a substrate, albeit at reduced rates *(8)*. D,L-Dihydrolipoamide is used readily as a substrate, and this property is used to monitor the NAD^+-dependent oxidation of free D,L-dihydrolipoamide as a measure of E3 activity in the forward direction. E3 catalyzes the following reaction using free lipoamide compounds.

$$\text{Dihydrolipoamide} + NAD^+ \leftrightarrow \text{Lipoamide} + NADH + H^+$$

2. Materials

2.1. Chemicals

1. NAD^+ (FW = 663.4): Store at 4°C.
2. Ethylenediamine tetraacetic acid (EDTA) (disodium salt, dihydrate, FW = 372.24): Store at room temperature.
3. Sodium borohydride ($NaBH_4$, highly toxic, FW = 37.83): Store in protective container at room temperature.
4. D,L-Lipoamide (D,L-6,8-thioctic acid amide, FW = 205.3).
5. (5,5′-Dithio-*bis*(2-nitrobenzoic acid); 3-Carboxy-4-nitrophenyl disulfide [DTNB] Ellman's reagent; FW = 396.3): Store at room temperature.
6. Sodium acetate (CH_3COONa, anhydrous, FW = 82.0).
7. 95% Ethanol.
8. D,L-Dihydroipoamide (FW = 207.3): Store at –20°C. For the preparation of D,L-dihydrolipoamide from D,L-lipoamid (*see* **Subheading 3.2.**)
9. NADH (disodium salt, FW = 709.4): Store at 4°C.

2.2. Reagents

1. 200 m*M* potassium phosphate buffer, pH 8.0: Mix 94 mL of 1 *M* K_2HPO_4 and 6 mL of 1 *M* KH_2PO_4, check pH and adjust to pH 8.0 if necessary, and dilute to 500 mL with water. Store at 4°C.
2. 100 m*M* potassium phosphate buffer, pH 8.0: Dilute 200 m*M* potassium phosphate buffer, pH 8.0, with same volume of water. Store at 4°C.
3. 120 m*M* NAD^+: 79.6 mg/mL in 100 m*M* potassium phosphate buffer, pH 8.0. Stable up to 6 mo at –20°C.

Table 1
Kinetic Parameters of Mammalian E3 in Forward and Reverse Reactions

	Forward reaction			Reverse reaction		
Source	k_{cat}[a] (s^{-1})	K_mDHLA[b] (mM)	K_mNAD$^+$ (mM)	k_{cat} (s^{-1})	K_mLA[b] (mM)	K_mNADH (mM)
Human liver[c]	97	0.26	N/D[f]	719	1.39	N/D
Rat liver[d]	345	0.49	0.52	124	0.84	0.062
Pig heart[e]	555	0.3	0.2	N/D	N/D	N/D

[a]k_{cat} was calculated based on V_{max}.
[b]DHLA, Dihydrolipoamide; LA, Lipoamide.
[c]Forward reaction data are from **ref. *5***, and reverse-reaction parameters are based on unpublished data measured with highly purified over-expressed E3 in *E. coli.*
[d]**Ref. *6***.
[e]**Ref. *7***.
[f]N/D: Not determined.

4. 100 mM EDTA: 37.2 mg/mL in water. Store at 4°C.
5. 120 mM D,L-dihydrolipoamide: 24.9 mg/mL in 95% ethanol, prepare weekly, prevent exposure to light, and store at –20°C.
6. 0.1 M Sodium acetate (82 mg/10 mL of water). Store at room temperature.
7. 50 mM Potassium phosphate buffer, pH 7.6: Mix 86.6 mL of 0.1 M K_2HPO_4 and 13.4 mL of 0.1 M KH_2PO_4, check pH and adjust to pH 7.6 if necessary, and dilute to 200 mL with water. Store at 4°C.
8. 200 mM Potassium phosphate buffer, pH 6.3: Mix 23.5 mL of 1 M K_2HPO_4 and 76.5 mL of 1 M KH_2PO_4, check pH and adjust to pH 6.3 if necessary, and dilute to 500 mL with water. Store at 4°C.
9. 100 mM Potassium phosphate buffer, pH 6.3: Dilute 200 mM potassium phosphate buffer, pH 6.3 with same volume of water. Store at 4°C.
10. 10 mM NAD$^+$: 6.63 mg/mL in 100 mM potassium phosphate buffer, pH 6.3. Stable up to 6 mo at –20°C.
11. 10 mM NADH: 7.09 mg/mL in 100 mM potassium phosphate buffer, pH 6.3. Stable up to 6 mo at –20°C.
12. 60 mM D,L-lipoamide: 12.32 mg/mL in 95% ethanol, prepare wkly, avoid exposure to light. Store at –20°C.

2.3. Equipment

1. Spectrophotometer with a thermo-control unit.
2. Rotary evaporator.
3. Speed-Vac.
4. Water bath.

3. Methods

3.1. Assay of E3 in Forward Reaction

This reaction catalyzes the oxidation of dihydrolipoamide to lipoamide and concomitant reduction of NAD^+ to NADH.

$$\text{Dihydrolipoamide} + NAD^+ \rightarrow \text{Lipoamide} + NADH + H^+$$

E3 activity is measured by monitoring the rate of NADH production at 340 nm for the initial 2–3 min (*see* **Note 1**). The following protocol is recommended.

1. Set up the temperature control of both spectrophotometer and water bath to 37°C (30°C or room temperature can be also used).
2. Add components to a cuvet in the following order:
   a. 0.5 mL of 200 m*M* potassium phosphate buffer, pH 8.0.
   b. 25 µL of 120 m*M* NAD^+.
   c. 15 µL of 100 m*M* EDTA.
   d. 25 µL of 120 m*M* D,L-dihydrolipoamide.
   e. Add water to bring the final total reaction volume to 1 mL including the volume of the enzyme preparation to be added later on (10–50 µL as needed).

   The final reaction condition is 3 m*M* NAD^+, 3 m*M* D,L-dihydrolipoamide, and 1.5 m*M* EDTA in 100 m*M* potassium phosphate buffer, pH 8.0. E3 is diluted appropriately (0.1–2 mg/mL of crude E3 or 0.01–0.03 mg/mL of pure E3) in 100 m*M* potassium phosphate buffer, pH 8.0, and kept on ice.
3. Place the reaction cuvet in 37°C (or other appropriate temperatures) water bath for about 3 min to equilibrate the solution to the desired temperature. Transfer the cuvet to a spectrophotometer and take an initial reading of absorbance at 340 nm.
4. The reaction is started by addition of enzyme preparation followed by immediate mixing. As an experimental control, the same volume of 100 m*M* potassium phosphate buffer, pH 8.0, is used in place of enzyme (*see* **Note 2**). The order of the addition of D,L-dihydrolipoamide and enzyme can be reversed so that the reaction is started by the addition of the substrate.
5. Record the change in absorbance at 340 nm for 2–3 min by monitoring the linear increment of absorbance.
6. The specific activity of E3 is calculated as follows ***(9)***:

$$\text{Units} = \mu\text{mol of NADH produced per min} = (\Delta A_{340}/\text{min})/6.22$$

$$\text{Specific activity (units/mg protein)} = \text{units/mg E3 used in the reaction}$$

3.2. Preparation of D,L-Dihydrolipoamide from D,L-Lipoamide

Because D,L-dihydrolipoamide, one of the substrates of the E3 in the forward (physiological) reaction, is not available commercially at the present time, it is necessary to prepare D,L-dihydrolipoamide using D,L-lipoamide (D,L-6,8-thioctic acid amide; MW=205.3) with $NaBH_4$ as a reducing reagent ***(10)***.

The following protocol is based on the scale of 200 mg of D,L-lipoamide (D,L-6,8-thioctic acid amide) as a starting material, according to the method of Reed et al. ***(10)***

1. Dissolve 200 mg of D,L-6,8-thioctic acid amide in 5 mL of methanol-water mixture (4:1, v/v). Warming in a water bath at about 50°C is needed for fast and complete solubilization.
2. Chill suspension on ice. After complete solubilization, no precipitates should be formed.
3. Add 1 mL of ice-cold sodium borohydride (200 mg/ml in water) and stir on ice for 1–2 h (*see* **Note 3**). It will be colorless by reduction of D,L-6,8-thioctic acid amide.
4. Adjust the solution to pH 1.0–2.0 with 1 *M* HCl (to check, use pH paper).
5. Extract the reduced compound with an equal volume of chloroform by shaking, let it settle, decant chloroform phase (lower layer), and transfer to a new container. Repeat this extraction procedure twice.
6. Pool extracts and evaporate chloroform using a rotary evaporator or Speed-Vac with heat.
7. Resuspend the resulting white residue in 10 mL of benzene, immediately add 4 mL of *n*-hexane, mix well, and set on ice until white precipitate is formed (about 30 min).
8. Collect the precipitate by filtering through 3MM paper.
9. Dry the precipitate under vacuum, store as a powder at –20°C with protection from light. Dissolve appropriate amount in 95% ethanol just before use.

3.3. Evaluation of D,L-Dihydrolipoamide

The concentration of prepared D,L-dihydrolipoamide is estimated as follows:

1. Prepare 20 m*M* stock solution of DTNB in 0.1 *M* sodium acetate (7.93 mg DTNB/mL in 0.1 *M* sodium acetate) and store at –20°C.
2. Prepare 3 m*M* D,L-dihydrolipoamide solution in 95% ethanol (dilute 120 m*M* stock solution, 24.9 mg/mL of 95% ethanol, with 95% ethanol).
3. For assay, prepare fresh 1 m*M* DTNB by dilution of 20 m*M* stock solution with 50 m*M* potassium phosphate buffer, pH 7.6, and add 1 mL to a cuvet. The starting absorbance at 412 nm should be below 0.12.
4. To the same cuvet, add precisely 2 μL of 3 m*M* dihydrolipoamide, mix well, and record absorbance at 412 nm. Repeat this step several times to obtain a linear increment in absorbance after the addition of 3 m*M* D,L-dihydrolipoamide solution.
5. The following is used for the % reduction efficiency calculation (*see* **Note 4**).

$$\text{Slope} = \Delta A_{412}/\mu\text{L added D,L-dihydrolipoamide}$$

$$\text{Concentration (m}M\text{) of D,L-dihydrolipoamide} = \text{slope} \times 1000/27.2$$

$$\%\ \text{Reduction efficiency} = (\text{measured D,L-dihydrolipoamide in m}M/3\ \text{m}M) \times 100$$

3.4. Assay of E3 in Reverse Reaction

In the reverse reaction, E3 has an optimal pH at 6.3, unlike in the forward reaction, which has an optimal pH at 8.0 ***(5)***. In this direction, E3 catalyzes the oxidation of NADH in the presence of lipoamide.

$$\text{Lipoamide} + \text{NADH} + \text{H}^+ \rightarrow \text{Dihydrolipoamide} + \text{NAD}^+$$

The decrease of NADH in the reaction mixture is used for the measurement of the reverse reaction, by monitoring ΔA_{340}/min (*see* **Notes 5** and **6**). Either D,L-lipoamide or D,L-lipoic acid is used as one of the substrates. However, reaction with D,L-lipoamide gives a significantly higher (28-fold) turnover number than with D,L-lipoic acid ***(11)***.

The following protocol is suitable for the measurement of E3 activity in reverse reaction ***(5)***.

1. Set the temperature control of both spectrophotometer and water bath (37°C, 30°C, or room temperature).
2. Add components to a cuvet in the following order.
   a. 0.5 mL of 200 m*M* potassium phosphate buffer, pH 6.3.
   b. 10 µL of 10 m*M* NAD^+.
   c. 10 µL of 10 m*M* NADH.
   d. 15 µL of 100 m*M* EDTA.
   e. 10 µL of 60 m*M* D,L-lipoamide.
   f. Add water to a final volume of 1 mL including the volume of the enzyme preparation to be added later on (10–50 µL as needed).

   The final reaction conditions are 0.1 m*M* NAD^+, 0.1 m*M* NADH, 0.6 m*M* D,L-lipoamide, and 1.5 m*M* EDTA in 100 m*M* potassium phosphate buffer, pH 6.3. Diluted E3 samples (0.5–10 mg/mL of crude E3 or 0.05–0.15 mg/mL of pure E3) in 100 m*M* potassium phosphate buffer, pH 6.3, should be kept on ice.
3. Place the cuvet in a water bath for 3 min and equilibrate to the desired reaction temperature. Transfer the cuvet to a spectrophotometer and take an initial reading of absorbance at 340 nm.
4. The reaction is initiated by adding E3 to the cuvet followed by immediate mixing, and the ΔA_{340}/min is recorded for 2–3 min. An experimental control is made with same volume of 100 m*M* potassium phosphate buffer, pH 6.3 in place of enzyme in the reaction mixture.
5. The specific activity of E3 (reverse reaction) is calculated as follows ***(9)***:

$$\text{Units} = \mu\text{mol of NADH oxidized per min} = (\Delta A_{340}/\text{min})/6.22$$

Specific activity (units/mg protein) = units/mg E3 protein used in the reaction

3.5. Preparation of E3 from Various Sources

3.5.1. Bacterial Crude Cell Extracts

Samples containing E3 from bacterial sources are prepared by any method for the breakage of cells, e.g., osmotic shock (suspending frozen cells in appropriate buffer) *(12)*, sonication *(13)*, or using a French press. The clear E3-containing supernatant is obtained by centrifugation at 6000*g* for 10 min (*see* **Note 7**).

3.5.2. Eukaryotic-Cell Preparations

The following protocols are suitable for the preparation of crude E3 from eukaryotic cells *(4)*. Unlike bacterial cells, eukaryotic cells lack a cell wall, so relatively mild conditions are used for the lysis of cells.

Organelle or membrane-bound E3 is prepared by resuspending cells in hypotonic extraction buffer (20 m*M* potassium phosphate, pH 7.4, containing 1% Triton X-100) with 2 m*M* EDTA, 2 m*M* EGTA, and protease inhibitors (0.2 m*M* phenylmethylsulfonyl fluoride [PMSF], 0.5 mg/mL leupeptin, and 0.7 mg/mL pepstatin), followed by performing freeze–thaw steps several times. After 30 min of incubation on ice, E3 activity is determined.

For the preparation of mitochondrial E3, tissues or cells are minced and homogenized in buffer (for liver cells, 70 m*M* sucrose, 220 m*M* mannitol, 2 m*M* HEPES, and 1 m*M* EDTA, pH 7.4; for kidney or heart cells, 250 m*M* sucrose, 10 m*M* Tris-HCl, and 1 m*M* 2-mercaptoethanol, pH 7.4), and centrifuged at 650*g* for 10 min to remove unbroken cells. The supernatant is centrifuged at 12,000*g* for 10 min at 4°C, and the mitochondrial pellet is resuspended in a half volume of original homogenization buffer, and the homogenization-centrifugation procedure is repeated once more. The final mitochondrial pellet is resuspended in ice-cold phosphate buffered saline (PBS; 130 m*M* NaCl, 10 m*M* sodium phosphate buffer, pH 7.4) with protease inhibitors, and stored at –80°C in small aliquots until used. When E3 activity is assayed, the frozen mitochondrial E3 aliquot is thawed and E3 is extracted by hypotonic extraction buffer with protease inhibitors. The clear E3-containing supernatant, obtained by centrifugation at 20,000*g* for 30 min at 4°C, is used for the mitochondrial E3 assay.

4. Notes

1. Because NADH and lipoamide are generated by the forward reaction, the accumulation of NADH will exert a product inhibitory effect. Therefore, it is important to measure an initial rate for 2–3 min with a linear slope of less than 0.2 ΔA_{340}/min. An extinction coefficient of $6.22 \times 10^3\ M^{-1}\text{cm}^{-1}$ for NADH is used to calculate E3 activity. Also, the total amount of E3 will play the major role on the consumption of substrates. Preliminary measurements of activity with varying concentrations of E3 is advised to determine a desirable rate, avoiding either rapid depletion of substrates or too little absorbance change. It should be pointed out that E3 uses only L- or the (R)-enantiomer of dihydrolipoamide as a substrate, so when precise kinetic data are needed, further isolation of the natural substrate from D,L-mixture is necessary.
2. Usually the experimental control gives a negligible change of absorbance, e.g., slight negative ΔA_{340}/min when dihydrolipoamide is preincubated or slight posi-

tive ΔA_{340}/min when enzyme is pre-incubated. However, if ΔA_{340}/min is higher than 0.02, use a freshly prepared solution of NAD^+ or D,L-dihydrolipoamide.

3. When preparing the $NaBH_4$ solution, add $NaBH_4$ into water gradually in a small quantity to prevent any loss of chemicals by vigorous reaction. Also, since a flammable gas will be produced and $NaBH_4$ is highly toxic, prepare this solution in a hood with gloves.
4. An extinction coefficient of 27.2 mM^{-1}cm^{-1} for dihydrolipoyl-DTNB is used for the calculation of concentration of D,L-dihydrolipoamide.
5. A lag period at the start of the reverse reaction has been reported ***(5)***. The lag, which is enhanced by changing the pH to lower than 6.0 or increasing NADH concentration, can be abolished by including 0.1 m*M* of NAD^+ in the reaction mixture ***(5)***.
6. Higher NADH concentrations can inactivate E3 by over-reducing enzyme to the EH_4 state ***(11,14)***, so only 0.1 m*M* of NADH is used in the reaction. Also, because NAD^+, the product of the reverse reaction, is already included in the cuvet, the range of linear slope in absorbance change is shorter than that of the forward reaction.
7. E3 is relatively stable to temperature-induced inactivation. Over 90% of the activity of a highly purified E3 sample is retained after 4 d at room temperature (unpublished data), and 71% of the remaining activity is reported after treatment at 75°C for 5 min ***(15)***.

References

1. Patel, M. S. and Vettakkorumakankav, N. N. (1995) Lipoic acid-requiring proteins: recent advances, in *Biothiols in Health and Disease* (Packer, L. and Cadenas, E., eds.), Marcel Dekker, NY, pp. 373–388.
2. Packer, L., Witt, E. H., and Tritschler, H. J. (1995) Alpha-lipoic acid as a biological antioxidant. *Free Rad. Biol. Med.* **19,** 227–250.
3. Williams, Jr., C. H. (1992) Lipoamide dehydrogenase, glutathione reductase, thioredoxin reductase, and mercuric ion reductase: a family of flavoenzyme transhydrogenases, in *Chemistry and Biochemistry of Flavoenzymes vol. 3* (Muller, F., ed.), CRC, Boca Raton, FL, pp. 121–211.
4. Patel, M. S., Vettakkorumakankav, N. N., and Liu, T.-C. (1995) Dihydrolipoamide dehydrogenase: Activity assays. *Methods Enzymol.* **252,** 186–195.
5. Ide, S., Hayakawa, T., Okabe, K., and Koike, M. (1967) Lipoamide dehydrogenase from human liver. *J. Biol. Chem.* **242,** 54–60.
6. Reed, J. K. (1973) Studies on the kinetic mechanism of lipoamide dehydrogenase from rat liver mitochindria. *J. Biol. Chem.* **248,** 4834–4839.
7. Massey, V., Gibson, Q. H., and Veeger, C. (1960) Intermediates in the catalytic action of lipoyl dehydrogenase. *Biochem. J.* **77,** 341–351.
8. Yang, Y. and Frey, P. A. (1989) 2-Ketoacid dehydrogenase complexes of Escherichia coli: Stereospecificities of the three components for (R)-lipoate. *Arch. Biochem. Biophys.* **268,** 465–474.
9. Bergmeyer, H.-U. (1965) Experimental Techniques, in *Methods of Enzymatic Analysis* (Bergmeyer, H.-U., ed.) (2nd ed.), Academic, NY, pp. 32–42.

10. Reed, L. J., Koike, M., Levitch, M. E., and Leach, F. R. (1958) Studies on the nature and reactions of protein-bound lipoic acid. *J. Biol. Chem.* **232,** 143–158.
11. Kalse, J. F. and Veeger, C. (1968) Relation between conformations and activities of lipoamide dehydrogenase. I: Relation between diaphorase and lipoamide dehydrogenase activities upon binding of FAD by the apoenzyme. *Biochim. Biophys. Acta* **159,** 244–257.
12. Vettakkorumakankav, N. N., Danson, M. J., Hough, D. W., Stevenson, K. J., Davison, M., and Young, J. A. (1992) Dihydrolipoamide dehydrogenase from the halophilic archeobacterium *Haloferax volcanii*: characterization and N-terminal sequence. *Biochem. Cell. Biol.* **70,** 70–75.
13. Adamson, S. R. and Stevenson, K. J. (1981) Inhibition of pyruvate dehydrogenase complex from *Escherichia coli* with a bifunctional arseoxide: Selective inactivation of lipoamide dehydrogenase. *Biochemistry* **20,** 3418–3424.
14. Visser, J. and Veeger, C. (1968) Relation between conformations and activities of lipoamide dehydrogenase. III: Protein association-dissociation and the influence on catalytic properties. *Biochim. Biophys. Acta* **159,** 265–275.
15. Kim, H., Liu, T.-C., and Patel, M. S. (1991) Expression of cDNA sequences encoding mature and precursor forms of human dihydrolipoamide dehydrogenase in *Escherichia coli*. *J. Biol. Chem.* **266,** 9367–9373.

33

Reduced Glutathione and Glutathione Disulfide

Richard W. Browne and Donald Armstrong

1. Introduction

Glutathione (GSH; L- -glutamyl-L-cysteinl-glycine) plays an important role in the prevention of radical mediated injury to the body. It does so as a radical scavenger and by supplying GSH to the antioxidant enzymes described in Chapter 29. In conjunction with superoxide dismutase (SOD), which converts superoxide anions into hydrogen peroxide (H_2O_2), glutathione peroxidase (GSHPx) converts H_2O_2, into water *(1)*. As a result of the second conversion, GSH is oxidized to glutathione disulfide (GSSG). In this way, GSH acts as a cofactor in the removal of toxic radicals from the body. During oxidative stress, GSH levels decline and GSSG increases, which can influence signal transduction by stimulating NF-kB activation *(2)*. GSH is also thought to be a donor of glutamyl groups in amino-acid transport *(3)*.

This chapter discusses a simple and straightforward way to determine both GSH and GSSG concentrations in whole blood and other biological tissues. Because GSH is present in the body in relatively high concentrations, i.e., as high as 3.1 mg/g of tissue *(3,4)*, it can be determined with generally available chemical and enzymatic methods. GSSG, on the other hand, is present in much lower concentrations in the body and thus is not so easily determined. One method measures the disappearance of the nicotinamide adenine dinucleotide phosphate (NADPH) spectrophotometrically as glutathione reductase (GSSG-R) which converts GSSG back to GSH *(5)*. However, the incidental oxidation of GSH to GSSG was not taken into account. N-ethylmaleimide (NEM) has been used in some procedures to prevent oxidation of GSH but NEM is a potent inhibitor of glutathione reductase. Extensive measures must be taken in order to remove any NEM if GSSG-R is utilized.

We describe here a fluorometric method *(6)* for the determination of both GSH and GSSG using o-phthalaldehyde (OPT) as the fluorescent reagent

From: *Methods in Molecular Biology, vol. 108: Free Radical and Antioxidant Protocols*
Edited by: D. Armstrong © Humana Press Inc., Totowa, NJ

modified for whole blood. The specificity for GSH and the pH-dependent differentiation between GSH and GSSG make OPT an ideal fluorophore *(7)*. NEM is also utilized in the determination of GSSG and, because we do not use GSSG-R in our procedure, its removal is not incumbent upon the final measurement.

Finally, the similarities between the GSH and GSSG procedures allows for minimal alteration of sample preparation, assay conditions, and instrumentation. The sample extractions are identical for whole blood and other tissues requiring only the adjustment of standards and use of appropriate controls for correct determinations. Assay conditions vary only in the buffer used, pH, and removal of endogenous GSH in the GSSG procedure. Both methods require minimal sample manipulation resulting in recoveries near 100%.

2. Materials

2.1. Reagents

1. Meta-Phosphoric Acid Precipitating Reagent: Add 1.67 grams meta-phosphoric acid (Fisher Scientific Co., Springfield, NJ) to a 100 mL volumetric flask. Add approx 90 mL H_2O. Add 200 mg disodium ethylenediamine tetraacetic acid (Sigma Chemical Co., St. Louis, MO). Add 30 grams sodium chloride (Fisher). Mix on stir plate for 30 min. Filter through Whatman No. 1 filter paper (*see* **Note 1**).
2. O-Phthaldialdehyde Reagent: Add 25 mg of OPT (Sigma) to a 25 mL volumetric flask and complete to 25 mL with methanol (*see* **Note 2**).
3. GSH Buffer: Add 13.8 grams (0.1 mole) sodium phosphate, monobasic (Fisher) to a 1 L volumetric flask. Add 1.461 grams (0.005 mole) EDTA. Add approx 900 mL H_2O, adjust to pH 8.0 with 1*N* sodium hydroxide. Complete to 1 L with H_2O and store at room temperature.
4. Meta-Phosphoric Acid Extraction Solution: Add 1 mL of the precipitating reagent (*see* **Subheading 2.1., item 1**) to a 5 mL plastic test tube. Add 900 μL of H_2O. Store at 4°C.
5. N-Ethylmaleimide: Add 125.1 mg N-ethylmaleimide (Sigma) to a 25-mL volumetric flask and complete to 25 mL with H_2O. Prepare fresh for each analysis (*see* **Note 3**).
6. 1 *N* Sodium Hydroxide: Add 6 g of NaOH to a 150-mL volumetric flask and complete to 150 mL with H_2O.

2.2. Instruments

1. TJ-6 Centrifuge (Beckman Instrument Co., Palo Alto, CA).
2. Analytical Balance (Sartorius Co., Guttingen, Germany).
3. PC 351 Stir Plate (Corning Glass Works, Corning, NY).
4. Vortex Junior Mixer (Scientific Products Co., McGaw Park, IL).
5. Zeromatic SS-3 pH meter (Beckman).
6. RF-5000U Spectrofluorophotometer and FDU-3 recorder (Shimadzu Corp., Kyoto, Japan) programmed at the following configuration:

a. Operation set on local.
b. Scan modeset on quantitation.
c. Wavelength mode set on fixed.
d. The excitation wavelength is 350 nm.
e. The emission wavelength is 420 nm.
f. The excitation slit width is 5 nm.
g. The emission slit width is 1.5 nm.
h. Sensitivity is set on high.

3. Methods

3.1. Preparation of GSH Standards

Standards consist of a stock solution of 100 mg/dL reduced GSH (Sigma). This is prepared by adding 100 mg of GSH to 100 mL of GSH buffer and then diluted to yield 0, 10, 20, 30, 40, and 50 mg/dL working standards. Dilute 100 mg/dL GSH stock to make the 50 mg/dL standard by adding 2 mL H_2O to 2 mL of stock solution. Follow **Table 1** to prepare the other standards.

3.2. Preparation of Biological Samples

Prepare an extract of each sample by pipetting 100 µL of EDTA whole blood into 1 tube of the MPA extraction solution (**Subheading 2.1., item 4**) and vortex. Samples are centrifuged at 1000*g* for 10 min. The supernatant is decanted and frozen at –70°C.

3.3. Extraction of Standards

Prepare an extract of each standard by pipetting 100 µL of standard into 1 tube of the MPA extraction solution (**Subheading 2.1., item 4**) and vortex. Centrifugation is not required with standards (*see* **Note 5**).

3.4. GSH Assay Conditions

1. Add 2.0 mL of the GSH buffer to a 5-mL glass test tube.
2. Add 100 µL of standard, control, or sample to each tube.
3. Add 100 µL OPT (**Subheading 2.1., item 2**) reagent to each tube; mix well.
4. Incubate at room temperature for exactly 15 min.
5. Read fluorescence emission at 420 nm with excitation set at 350 nm.
6. Results are reported in mg GSH/dL packed RBC.
7. Results are normalized by whole blood GSH/hematocrit value.

3.5. Preparation of GSSG Standards

Standards consist of a stock solution of 24 mg/dL GSSG (Sigma). This is prepared by adding 24 mg of GSSG to 100 mL of 1*N* NaOH. This stock must

Table 1
Dilution Scheme for GSH Calibrators

Std. (mg/dL)	H_2O (mL)	50 mg/dL GSH (mL)
50	0.0	1.0
40	0.2	0.8
30	0.4	0.6
20	0.6	0.4
10	0.8	0.2
0	1.0	0.0

Table 2
Dilution Scheme for GSSG Calibrators[a]

Std. (mg/dL)	H_2O (mL)	4 mg/dL GSSG (mL)
4.0	0.0	1.0
2.0	0.5	0.5
1.0	0.75	0.25
0.5	0.875	0.125
0.0	1.0	0.0

[a]Preparation and extraction of samples and standards are the same as described in **Subheadings 3.2.** and **3.3.**; **Notes 1–4** are also applicable.

be diluted to yield 0, 0.5, 1.0, 2.0, 4.0 mg/dL working standards. Dilute the 24 mg/dL GSSG stock to make the 4 mg/dL standard by adding 5 mL H_2O to 1 mL of the stock solution. Follow **Table 2** to prepare the other standards.

3.6. GSSG Assay Conditions

1. Add 250 µL of the standard, control, or sample to a 5 mL glass test tube.
2. Add 250 µL of the 0.04 *M* NEM to each tube, vortex, mix.
3. Incubate at room temperature for 20 min.
4. Add 500 µL of the 1 *N* NaOH to each tube, vortex, mix.
5. Take 100 µL of the solution from **step 4** above and put it into another 5 mL test tube.
6. Add 2.0 mL of the 1 *N* NaOH.
7. Add 100 µL OPT (**Subheading 2.1., item 2**) to each tube. Mix well.
8. Incubate at room temperature for exactly 15 min.

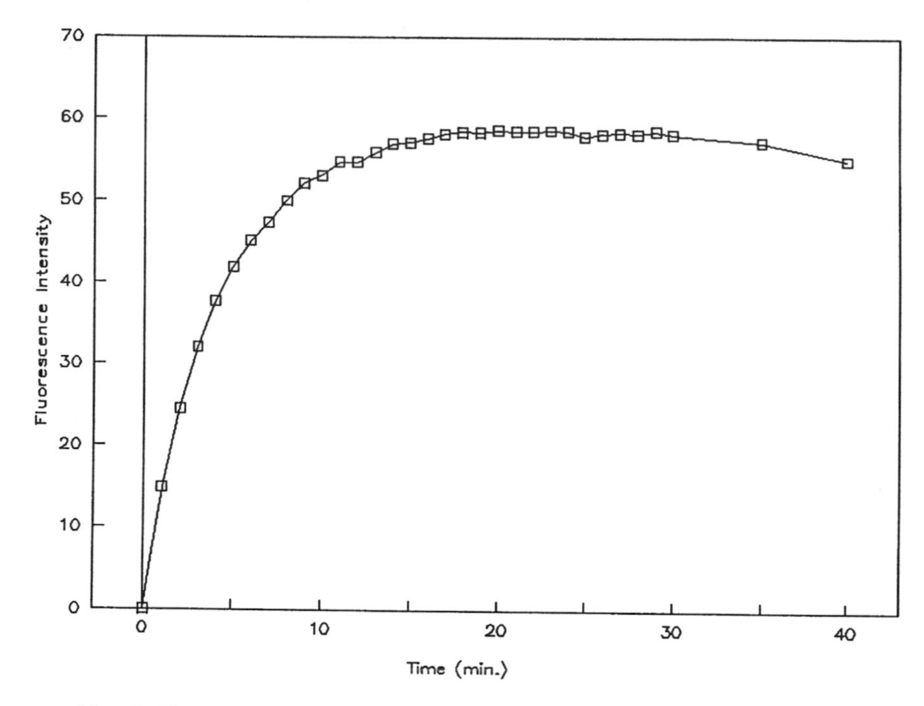

Fig. 1. Fluorescence intensity of GSH-OPT adduct as a function of time.

9. Read fluorescent emission at 420 nm with excitation at 350 nm.
10. Results are reported as described in **Subheading 3.4.**

3.7. Results

Figure 1 shows the relationship between the fluorescent intensity and the time of incubation for OPT and GSH. A minimum of 15 min should be allowed for maximum intensity to be achieved.

Figure 2 shows the optimal excitation peak at 350 nm for the GSH-OPT adduct on the left of the graph and the optimal emission peak at 420 nm for the same adduct on the right of the graph.

4. Notes

1. Prepare MPA precipitating reagent every 2 wk and store at 4°C.
2. Keep the OPT reagent protected from the light.
3. N-ethylmaleimide is toxic. Do not expose to the skin.
4. Make sure samples and standards are analyzed at the same time and are treated with the same precipitating reagent.

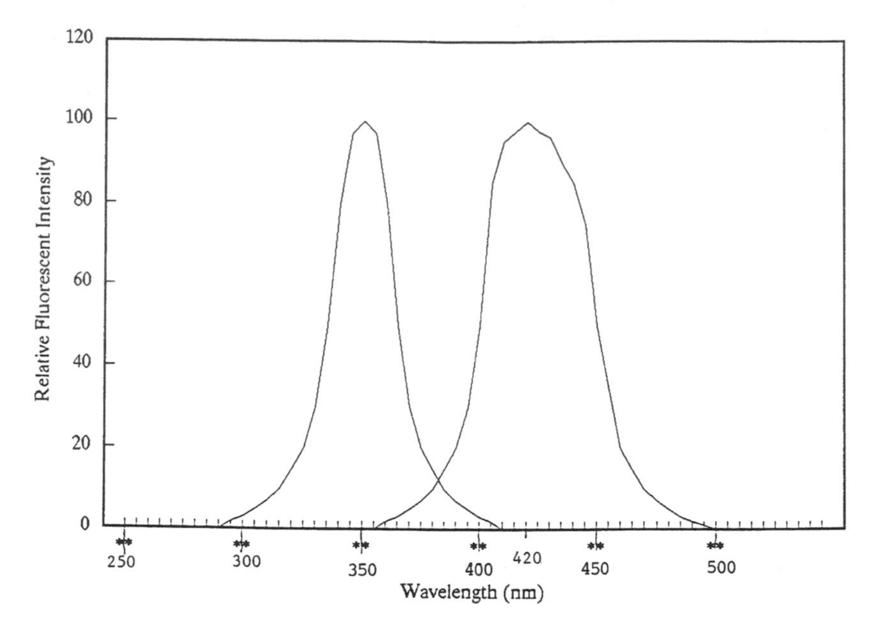

Fig. 2. Excitation/emission spectrum of GSH-OPT adduct.

References

1. Southorn, P. A. and Powis, G. (1988) *Mayo Clin. Proc.* **63,** 381–389.
2. Droge, W., Schulze-Orthoff, K., Mihm, S., Galter, D., Schenk, H., Eck, H. P., Roth, S., and Gmunder, H. (1994) Function of glutathione and glutathione disulfide in immunology and immunopathology. *FASEB-J.* **8,** 1131–1138.
3. Orlowski, M. and Meister, A. (1970) *Proc. Nat. Acad. Sci. USA*, **67,** 1248–1255.
4. Bhattacharya, S. K., Robson, J. S., and Steward, C. P. (1955) The determination of glutathione in blood and tissue. *Biochem J.* **60,** 696–702.
5. Boyd, S. C., Sasame, H. A., and Boyd, M. R. (1979) High concentrations of glutathione in glandular stomach: possible implications for carcinogenesis. *Science* **205,** 1010–1012.
6. Hissin, P. J. and Hilf, R. (1976) *Anal. Biochem.* **74,** 214–226.
7. Coutelle, C., Iron, A., Higueret, D., and Cassaigne, A. (1992) Optimization of a spectrophotometry assay of total and oxidized blood glutathione: comparison with a fluorometric method. *Ann. Biol. Clin. Paris.* **50,** 71–76.

34

Analysis of Coenzyme Q_{10} Content in Human Plasma and Other Biological Samples

Scott Graves, Marianna Sikorska, Henryk Borowy-Borowski, Rodney J. H. Ho, Tot Bui, and Clive Woodhouse

1. Introduction

Since its discovery in 1957, coenzyme Q_{10} (CoQ_{10}) has been extensively studied for its important role as a redox carrier in the mitochondrial electron transport chain and for its role as a potent antioxidant in animal tissues ***(1,2)***. In addition, several reports indicate that deficiencies in CoQ_{10} are implicated in heart disease, periodontal infections, and AIDS ***(3–5)***. These observations have generated ever-expanding needs for simple and sensitive analytical methods to assess CoQ_{10} content in biological materials.

The quinone group of CoQ_{10} is responsible for its redox properties, whereas the isoprenyl chain allows for insertion and mobility in lipid bilayer membranes. This latter property also contributes to the molecule's virtual insolubility in aqueous media, necessitating an organic solvent extraction step prior to assay. Several procedures involving extraction of CoQ_{10} in lower alcohols, such as methanol and ethanol, followed by a chromatographic separation of an identifiable CoQ_{10} peak; have been described ***(6–15)***. However, no sufficiently versatile method that could be applied with the same sensitivity and reliability to different biological samples, such as cultured cells, animal tissues and biological fluids, has been described so far. We present two methods in this chapter that have been developed with different objectives in mind. In the first example, we describe an alcohol and hexane extraction method that has been optimized for evaluating the CoQ_{10} content in human plasma samples. It is designed to be adaptable to semi-automated procedures, to minimize sample handling, and to allow for incorporation of an internal standard. In the second

From: *Methods in Molecular Biology, vol. 108: Free Radical and Antioxidant Protocols*
Edited by: D. Armstrong © Humana Press Inc., Totowa, NJ

example, we have selected 1-propanol for extraction of CoQ_{10} from a variety of different tissues, as it is a potent denaturant of CoQ_{10}-protein complexes. This method allows for detection of very low concentrations of CoQ_{10} and dispenses of the need for an internal standard.

1.1. Basic Principles

The standard method of evaluating CoQ_{10} in biological tissues involves the use of alcohol and hexane to extract contaminating proteins and solubilize CoQ_{10}. We have modified one reported method *(8)* in such a way that CoQ_{10} extraction can be accomplished in a minimal volume of solvent and with a limited volume of human serum. This is possible in part because the sensitivity of this method of extraction is well within the concentrations of CoQ_{10}, 0.79 ± 0.22 µg/mL, found in human blood of normal individuals *(16)*. The procedure has been simplified to a single extraction and evaporation process and therefore allows for efficient sample handling. Furthermore, our procedure takes advantage of the use of CoQ_9 as an internal standard. Because the levels of CoQ_9 in human blood are negligible, it is feasible to spike control plasma samples with this ubiquinone to validate the extraction process.

Alternatively, extraction of CoQ_{10} from a variety of biological tissues, in which the levels of CoQ_{10} are less optimal, is made facile with 1-propanol, because it can be mixed with water in any proportion. Additionally, CoQ_{10} has the highest solubility in 1-propanol relative to other alcohols. Quantitative recovery of CoQ_{10} using this solvent has been described previously *(17)*. When added in a 3:1 ratio to tissue homogenate, cell lysate, or plasma samples, the resulting 77% 1-propanol solution completely disrupts lipid-protein complexes and efficiently solubilizes CoQ_{10}. This eliminates the necessity of repeated extraction procedures that are often required with mixtures of either methanol or ethanol and n-hexane, mostly owing to formation of stable emulsions when tissues with high content of lipids are analyzed *(7)*. The 1-propanol extracts can be directly analyzed by high-pressure liquid chromatography (HPLC) if obtained from sources with a high CoQ_{10} content. However, when a sample contains a low level of CoQ_{10} its quantitative determination can be obscured by high ultraviolet (UV)-absorbing materials, which are always co-extracted during this procedure. In order to avoid such problems, the 1-propanol extracts can be further processed to remove most of the interfering material and remaining water. To achieve this level of purity, the samples are mixed with a double volume of n-hexane. After a simple phase separation by centrifugation, the volume of the lower phase will be equal to the volume of water present in the sample. The upper phase, which is a mixture of 1-propanol and n-hexane, contains all solubilized CoQ_{10}. After solvent evaporation, the dry residue can be reconstituted in a desired volume of ethanol for further analysis. The sample can also be oxidized with H_2O_2 and analyzed for total CoQ_{10} content using an HPLC apparatus equipped with reverse-phase column and UV detector. This simplified 1-propanol/n-hexane extraction procedure can be applied to different biological materials, is highly reproducible, allows for an excellent recovery of CoQ_{10}, eliminates the necessity of an internal standard, and allows for the assessment of the CoQ_{10} content from a standard concentration curve.

2. Materials

2.1. Reagents

Methanol, ethanol, 1-propanol and n-hexane are of analytical grade and are used without further purification. CoQ_{10} and CoQ_9 (Sigma, St. Louis, MO) is used for the standard curve preparation.

2.2. Instruments

For the rat-tissue samples, CoQ_{10} measurements are performed on a Beckman Gold System HPLC apparatus consisting of a 126 Solvent Module equipped with a Rheodyne 7725i loop injector and a UV 168 Detector. In the case of the human serum samples, CoQ_{10} measurements are done using a Waters Model 501 HPLC equipped with a WISP 712 loop injector and a model 440 UV detector set at 275 nm.

2.3. Tissue Sources

Rat tissues (plasma, liver, heart, and brain) are obtained from 300–400 g male Sprague-Dawley rats. Animals are fed with a standard certified pellet diet and with water available ad libitum. Rat adrenal pheochromocytoma PC12 (ATCC CRL 1721) cells are grown in Roswell Park Memorial Institute (RPMI) medium supplemented with 5% fetal bovine serum (FBS) and 10% horse serum. Human plasma is obtained from ethylenediamine tetraacetic acid (EDTA)-treated blood from the University of Washington Blood Bank or the cubital vein of normal donors with informed consent.

3. Methods

3.1. Methods for Ethanol/n-Hexane Extraction of CoQ_{10} from Human Plasma

After collection of human blood in sterile EDTA-containing 10-mL tubes, the plasma is separated from the blood cells by centrifugation at 300*g* for 15 min. Plasma aliquots are frozen at –20°C until tested.

CoQ_{10} is extracted from human plasma using the technique of Takada et al. *(8)*. In our studies, CoQ_9 is used as an internal standard for analysis of CoQ_{10} in human serum. Typically, 0.5 μg of CoQ_9 in 10 μL of ethanol/n-hexane (extraction solvent) is spiked in 300 μL of human plasma in glass tubes. The mixture is denatured by adding 2 mL of freshly prepared ethanol/n-hexane (2:5 v/v), vortexed for 2 min, and centrifuged at 300*g* for 10 min at 4°C to separate the organic n-hexane phase from the aqueous phase. A volume of 1.3 mL of the organic n-hexane is recovered and evaporated under nitrogen gas. The residue is reconstituted with 100 μL of ethanol/n-hexane (6:4 v/v). The extract is immediately transferred into auto injector vials (Waters Associated, Milford, MA) containing small glass inserts. A 50 μL sample is injected into the column for CoQ_{10} detection.

3.2. HPLC Analysis of CoQ_{10} from Human Plasma

Human serum plasma CoQ_{10} is analyzed on a C-8 column (4.6X250, 5-μ*M* pore size, Zorbax) with a methanol:hexane (6:4 v/v) as a mobile phase and 1 mL/min running time to detect CoQ_9 and CoQ_{10}. The UV detector is set at 275 nm for all assays. Peak height (PH) and peak height of CoQ_{10} to CoQ_9 ratios (PR) values are determined using the Maxima Program (Waters Associated).

3.3. Preparation of Biological Samples

1. 1–2 × 10^7 rat cells are harvested and pelleted by centrifugation at 300*g* for 5 min. The cell pellets are washed twice with phosphate-buffered saline (PBS) and lysed by osmotic shock in 100 μL of water. The lysate is subsequently frozen at –70°C.
2. Samples of heparinized rat blood (venous) are centrifugated at 3000*g* for 10 min. Plasma is collected, aliquoted into 300 μL samples, and frozen at –70°C immediately after sampling.
3. Tissues (brain, liver) are removed and placed in liquid nitrogen. Frozen samples (approx 100 mg) are placed in a Potter-Elvehjem apparatus and homogenized with 0.5 mL of water, carefully removed, and adjusted to a known volume. The homogenate is kept frozen at –70°C.

3.4. Extraction of CoQ_{10} from Rat Tissues

1. All rat tissue samples, prior to further extraction of CoQ_{10}, are subjected to repeated freezing/thawing steps. This is found to be essential for a full disruption of cell organelles. Sample aliquots, such as cell lysate, plasma samples or tissue homogenate, are brought to ambient temperature, mixed with 3 parts of 1-propanol, vortexed for 30 s and left to stand at room temperature for 1 min.
2. The samples (containing approx 77% 1-propanol), are mixed with 2 volumes of n-hexane, again vigorously vortexed for 30 s, and centrifugated for 1–5 min at maximum speed in a bench-top centrifuge to achieve a phase separation and to pellet the denatured proteins.
3. The upper phase, consisting of n-hexane and 1-propanol and containing the CoQ_{10} is collected and evaporated to dryness under argon.

4. The remaining dry residue is redissolved in 60 µL of ethanol and 2 µL of H_2O_2 in order to convert all CoQ_{10} to its oxidized form.
5. Sample aliquots are analyzed by HPLC. The amount of CoQ_{10} is calculated according to the measured peak-area of the analyzed sample.

3.5. HPLC Analysis of CoQ_{10} from 1-Propanol Extracted Tissues

Rat tissue and plasma samples are analyzed by a reverse-phase chromatography (RPC) on a Supelcosil LC-18-DB column (5-µ*M* particle size, 30 cm × 4.0 mm I.D.,Supelco) with the mobile phase of ethanol:methanol 80:20 (v/v) at the flow rate of 1 mL/min. Absorbance at 275 and 290 nm is monitored. A standard calibration curve of CoQ_{10} and Beckman System Gold Software is utilized for data quantification.

A standard curve is prepared from a CoQ_{10} ethanol solution of a concentration determined spectrophotometrically from its extinction coefficient of ε = 14,200 at 275 nm ***(18–20)***. The ethanol solution of CoQ_{10} is stored in sealed amber vials at –20°C.

3.6. Results

3.6.1. CoQ_{10} Content in Human Serum

1. Preliminary studies focused on assessing the ability of lower alcohols to separate CoQ_9 and CoQ_{10} chromatographically. Our studies revealed that with our HPLC system, methanol:hexane (6:4 v/v) is preferred over other alcohols (data not shown). Additional analyses are performed using ethanol:hexane extraction of CoQ_{10} from spiked PBS and human plasma samples.
2. For these studies, 300 µL of plasma or PBS is spiked with CoQ_{10} (1 µg final concentration), CoQ_{10} is extracted with 2 mL of ethanol:hexane (2:5 v/v), vortexed, and centrifuged to separate n-hexane from the aqueous phase. Next, 1.3 mL of hexane is dried under N2 gas and the residue reconstituted with 100 µL of ethanol:hexane (6:4 v/v).
3. A sample of 50 µL is injected into the HPLC column. Using this extraction procedure resulted in CoQ_{10} recovery of 78% from PBS and 98% from human serum (**Table 1**).
4. Chromatographic resolution of CoQ_{10} from human serum spiked with CoQ_9 is shown in **Fig. 1**. With more than sufficient resolution, endogenous and spiked CoQ_{10} as well as the CoQ_9 are detected under the defined extraction and chromatographic conditions. The resolution times for CoQ_9 is 5.92 ± 0.02 min whereas that of CoQ_{10} is 6.91 ± 0.08 min (**Fig. 1**).

3.6.2. Reproducibility of CoQ_{10} Evaluation in Human Serum

Having established our ability to chromatographically resolve CoQ_{10} and CoQ_9 in human plasma using the methanol:hexane procedure, we chose to use 0.5 µg/mL of CoQ_9 as an internal standard to determine the CoQ_{10} concentrations in the plasma of four human subjects (**Table 2**). We found that they are in

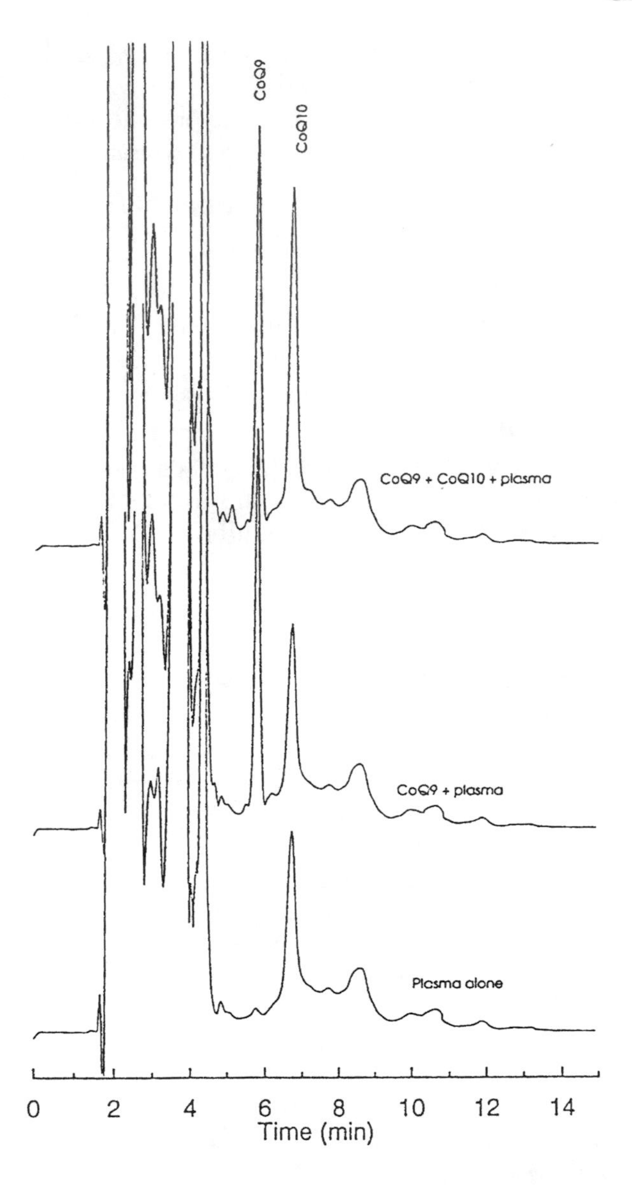

Fig. 1. Representative chromatographs of CoQ_{10} extracted from plasma. Normal human plasma, spiked with 0.5 μg/mL of internal standard CoQ_9 or plasma spiked with both CoQ_9 and CoQ_{10} (1 μg/mL) are extracted with ethanol:hexane (2:5, v/v) as described in **Table 3**. The carrier solvent (ethanol:hexane, 6:4, v/v), CoQ_9 alone or CoQ_9 and CoQ_{10} is separated on a Zorbax C-8 column and methanol:hexane (6:4, v/v) solvent running at 1 mL/min. The CoQ_9 and CoQ_{10} eluted are detected with a UV detector ($\lambda = 275$ nm).

Table 1
Efficiency of CoQ_{10} Extraction from Human Serum and Plasma

Extraction solvent	CoQ_{10} (1µg/mL) in	% Recovery[a]
Ethanol:Hexane[b]	PBS	78.3 ± 3.7
	human serum	97.7 + 9.2

[a]% recovery is calculated based on the PH-concentration standard curve after correcting for the endogenous CoQ_{10} in the serum.

Table 2
Reproducibility Tests Using the Methanol:Hexane Method for Human Plasma

Plasma from subject	Replicates n	Based on PH (µg/mL)	Based on PR (µg/mL)
BF	4	0.71 ± 0.03	0.80 ± 0.04
BS	4	0.68 ± 0.01	0.70 ± 0.01
CB	6	0.86 ± 0.04	1.04 ± 0.09
RH	6	0.96 ± 0.09	1.10 ± 0.06

the range of 0.7–1.1 µg/mL. Owing to the less than 100% recovery, the estimated CoQ_{10} concentration based on PH is always less than the PR (a recovery-independent) estimate. The reproducibility of the assay is tested using the plasma from two subjects. With six repeated determinations of CoQ_{10} for the two subjects (spiked with 0.5 µg/mL of CoQ_9, internal standard), we found the standard deviation of the sample replicates is less than 9% (**Table 2**).

3.6.3. Sensitivity of Assaying CoQ_{10} in Human Plasma

To determine the sensitivity and detection limit for assessing CoQ_{10} in human plasma, the assay is performed within a concentration range of 0–1.0 µg/mL. A representative plot of concentration vs PH and PR (**Fig. 2**) shows excellent detection of CoQ_{10}. With either the PH or PR method of CoQ_{10} detection, 0.025 µg/mL in 50 µL injection volume is detected, which is equivalent to 1.3 ng of CoQ_{10} for each assay determination.

3.6.4. CoQ_{10} Content in Different Rat Tissues

1. Ubiquinone components could be easily identified within the HPLC profile from a change in absorption ratio at 275 and 290nm or by coinjection of a standard (**Fig. 3**). Both approaches are taken initially to identify CoQ_9 and CoQ_{10} peaks. Using a Supercosil LC-18-DB column and separation conditions described in the Methods section, retention times for CoQ_9 and CoQ_{10} are 7.6 and 9.4. min, respectively (**Fig. 3**).

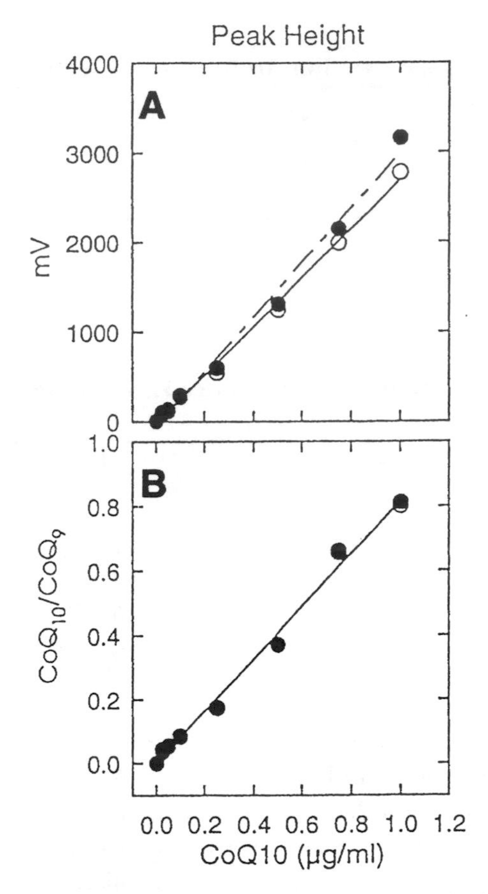

Fig. 2. Concentration-response curve of CoQ_{10} in lower alcohol extraction method for human serum, 0–1 μg/mL range. The peak height response of CoQ_9 is fixed at a concentration of 0.5 μg/mL whereas varying amounts of CoQ_{10} concentrations (circles) are presented as mV at 275 nm (absorbance) and plotted in panel **(A)**. The CoQ_{10} to CoQ_9 ratios are plotted in panel **(B)**. Open and closed symbols indicate the two different preparations of samples; the two least square fitted curves are presented as continuous dotted lines.

2. In rat tissues, CoQ_{10} represents only a minor component of the ubiquinone fraction. The prevalent ubiquinone form is CoQ_9 (14–21, **Fig. 4**). The ratio of CoQ_9 to CoQ_{10} varied for different tissues (**Fig. 4**), and according to our estimations it ranged from 2:1 in brain, 3:1 in PC12 cells, 6:1 in plasma to 10:1 in liver.
3. It is difficult to compare directly our results for different tissue concentrations of CoQ_{10} with those published in the literature, because they are obtained by different extractions methods, sampling of materials, and techniques for determining the concentration of CoQ_{10}. As presented in **Table 3**, CoQ_{10} concentration in rat

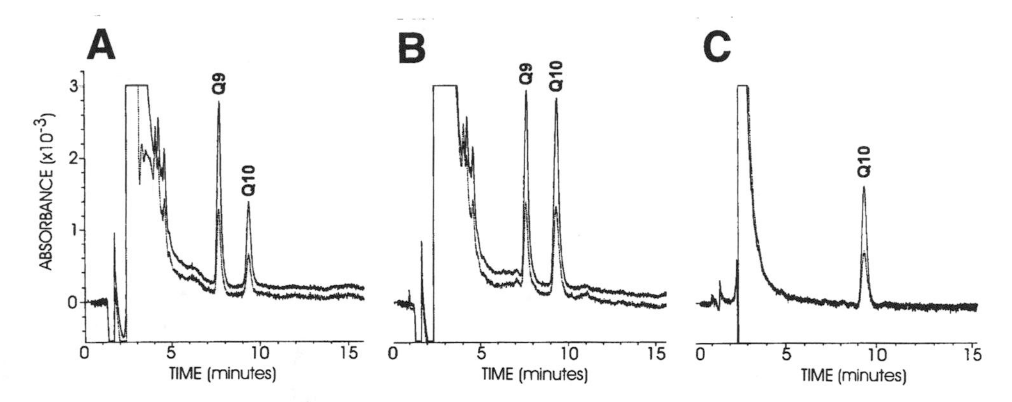

Fig. 3. Identification of CoQ_{10} peak in HPLC profile. **(A)** 10 µL of CoQ_{10} standard solution in ethanol (3.34 ng/µL) is injected onto the column under conditions described in the Methods Section. **(B)** 100 mg of brain tissue is processed, extracted with 1-propanol-n-hexane, and reconstituted in 60 µL of ethanol. 12 µL of the final ethanol fraction injected onto the column and analyzed as described in the Methods Section. **(C)** To the same brain sample (Panel B) is spiked 25 ng of standard CoQ_{10}.

brain and liver obtained here are within the range detected by others *(6,13,15)*. However, the procedure that we described here is much less complicated. A sample treatment with 1-propanol followed by n-hexane could be done in 1.5-mL Eppendorf tubes, eliminating the necessity of handling large volumes of solvents. A single-step extraction of CoQ_{10}, as opposed to multiple extractions with other solvent systems that, additionally, are often combined with saponification or TLC purification as an intermediate step before final HPLC analysis, is sufficient to reliably analyzed a diverse biological material, including fluids, tissues, and cultured cells.

3.6.5. Efficiency of Recovery

In order to establish CoQ_{10} recovery during the 1-propanol extraction procedure, a known amount of CoQ_{10} is added to the rat plasma samples and tissue homogenate before their 1-propanol/n-hexane treatments (**Table 4**). The samples are then processed and analyzed as described above (*see* **Note 1**). The obtained results indicated that the CoQ_{10} extraction is quantitative and its recovery from biological samples is close to 90%. Therefore, this extraction procedure can be applied without the use of internal standard and CoQ_{10} concentration read directly from a standardized concentration curve.

3.6.6. Reproducibility

1. Reproducibility of the 1-propanol procedure is established from analysis of CoQ_{10} in replicate extracts obtained from the same biological sample (**Table 5**). Typi-

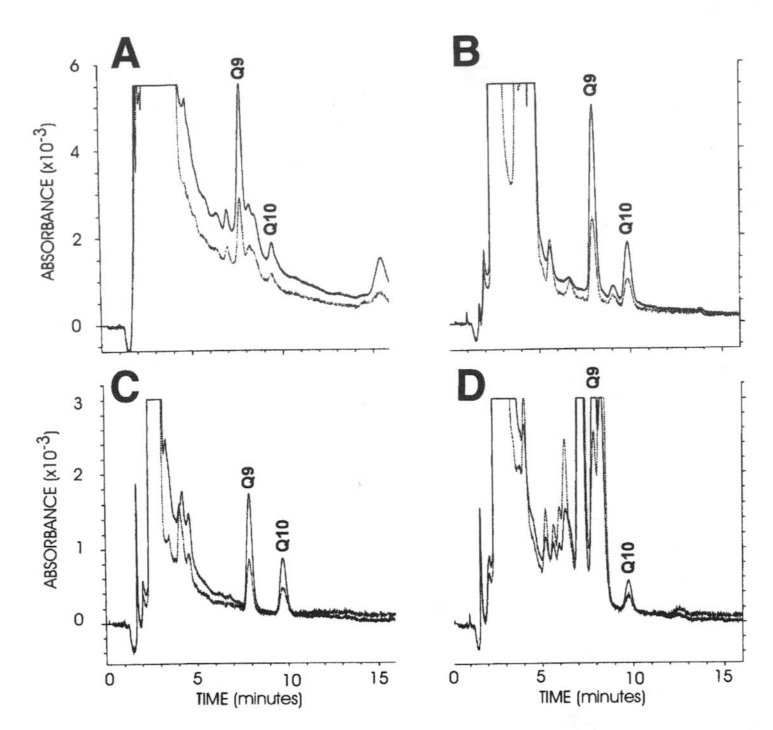

Fig. 4. Representative HPLC profiles of rat tissues. **(A)** 300 μL of plasma sample is processed, the final extract made in 60 μL of ethanol and 40 μL sample injected onto the column. **(B)** 105 mg of brain tissue is homogenized in 1 mL of double-distilled (DD) water. 100 μL of homogenate is further extracted and processed into 60 μL of ethanol, out of which 12 μL is injected onto the column. **(C)** 106 mg of liver tissue is homogenized in 1 mL of DD water, 100 μL of homogenate is further extracted, reconstituted in 60 μL and 12 μL injected onto the column. **(D)** 32 × 106 cells are lysed in 100 μL of DD water, processed and extracted as described in the Methods Section. The final extract is made in 60 μL of ethanol, out of which 12 μL is injected onto the column.

cally, a rat plasma aliquot or solid tissues (brain and liver) are divided into 3 samples which are extracted and analyzed separately.

2. An average relative standard deviation (RSD) of 5% of CoQ_{10} quantitation. We obtained similar values, indicative of excellent reproducibility of the method, for both tissue and plasma samples. It is important to note that rat plasma often contains a very low level of CoQ_{10}. For example, the content of CoQ_{10} in 300 μL of plasma III sample (**Table 5**) approaches the detection limit of this method (*see* **Note 2**).
3. For this sample, the RSD value indicative of the reproducibility is 15% higher, and this error is most likely, owing to interference from other lipids present in the sample. In such cases, the reproducibility is improved by processing a larger volume of plasma.

Table 3
CoQ_{10} Content in Rat Tissues

Sample	CoQ_{10} content[a]	n
Plasma	13.9–67.7	20
Brain	7.8–10.9[b]	7
Liver	4.1– 7.3[b]	7
PC12 cells	2.7– 8.4	5

[a]CoQ_{10} content is calculated as ng/mL of plasma, g/g of tissue and ng/106 cells

[b]Values compatible with those published in ***(6,13,15)***.

Table 4
Recovery Tests

CoQ_{10} content (ng)	Mean ± SD	RSD (%)	n	Recovery (%)
Plasma II				
CoQ_{10} added	62.9 ± 1.1		3	
CoQ_{10} present	17.2 ± 1.3	7.5	3	
CoQ_{10} found	72.2 ± 1.4	2.0	3	87.5
Brain II				
CoQ_{10} added	708.9			
CoQ_{10} present	219.7 ± 13.8	6.3	3	
CoQ_{10} found	855.4 ± 27.5	3.2	3	89.7

4. Notes

Using the alcohol/hexane extraction method, the actual detection limit of CoQ_{10} in human plasma depends on a number of variables, such as extraction efficiency, and assay interference by other exogenous substances, such as drugs, which may produce high background. In addition, preliminary studies showed that alcohol-extraction techniques are less efficient in extracting CoQ_{10} from biological tissues other than serum. Therefore, for CoQ_{10} extraction in other tissues, we defer to the 1-propanol method.

When using the 1-propanol extraction method, we found it practical, especially for processing minimal amounts of biological material, to apply a working range of detection of 3–50 ng of CoQ_{10} per injection. By using this concentration range, the relationship of absorption to concentration is linear. Samples containing 1–2 ng per injection can also be evaluated. However, sample reproducibility is lower as the detection limit for sensitivity of the applied HPLC system and its software is apparently reached. Measurements of low levels of CoQ_{10} in plasma samples with high lipid contents are more likely

Table 5
Reproducibility Tests Using 1-Propanol Extraction of Rat Tissues[a]

Sample	mean ± SD	RSD(%)	n
Plasma I	43.2 ± 2.5	5.8	4
Plasma II	57.3 ± 4.3	7.5	3
Plasma III	15.5 ± 2.3	14.8	3
Liver	4.93 ± 0.2	4.0	3
Brain I	8.50 ± 0.3	3.5	5
Brain II	9.80 ± 0.6	6.3	4

[a]Total Q_{10} content is calculated as ng/mL of plasma and mg/g of wet tissue.

to be error prone. Rat plasma samples, it is found, sometimes contain unknown impurities which followed the CoQ_{10} peak during HPLC analyses, obscuring accurate estimation (**Fig. 2**). Such problems, however, do not hamper analysis of samples such as human plasma, which contain significantly more CoQ_{10}.

References

1. Lenaz, G., Battino, M., Costeluccio, C., Fato, R., Cavazzoni, M., Lauchova, H., Bovina, C., Formiggini, G., and Prenteca-Castellig, G. (1990) Studies on the role of ubiquinone in the control of the mitochondrial respiratory chain. *Free Rad. Res. Commun.* **8(4–6),** 317–327.
2. Frei, B., Kim, M. C., and Ames, B. N. (1990) Ubiquinol-10 is an effective lipid-soluble antioxidant at physiological concentrations. *Proc. Natl. Acad. Sci. USA* **87(12),** 4879–4883.
3. Littaru, G. P., Ho, L., and Folkers, K. (1972) Deficiencies of Coenzyme Q_{10} in human heart disease II. *Int. J. Vitam. Nutr. Res.* **42,** 413–434.
4. Hansen, I. L. (1976) Bioenergetics in clinical medicine: ginival leucocyte deficiencies of coenzyme Q10 in patients with periodontal disease. *Res. Comm. Chem. Path. Pharm.* **14(4),** 792–738.
5. Folkers, K., Hanioka, T., Li-Jun, X., McRee, J. T., and Langsjoen, P. (1991) Coenzyme Q10 increases T4/T8 ratios of lymphocytes in ordinary subjects and relevance to patients having the AIDS related complex. *Biochem. Biophys. Res. Comm.* **176(2),** 786–791.
6. Abe, K., Ishibashi, K., Ohmae, M., Kawabe, K., and Katsui, G. (1978) Determination of ubiquinone in serum and liver by high-speed liquid chromatography. *J. Nutr. Sci. Vitaminol.* **24,** 555–567.
7. Abe, K., Katayama, K., Ikenoya, S., Takada, M., Yazuriha, T., Hanamura, K., Nakamura, T., Yusuda, K., Yoshida, S., Watson, B., and Kogure, K. (1981) Levels of coenzyme Q in tissue : recent analytical studies on coenzyme Q in biological materials, in *Biomedical and Clinical Aspects of Coenzyme Q,* (vol. 3),

(Folkers, K. and Yamamura, Y., eds.), Elsevier/North Holland Press, Amsterdam, pp. 53–66.
8. Takada, M., Ikenoya, S., Yazuriha, T., and Katayama, K. (1982) Studies on reduced and oxidized coenzyme Q(ubiquinones). II: the determination of oxidation-reduction levels of coenzyme Q in mitochondria, microsomes and plasma by high-performance liquid chromatography. *Biochim. Biophys. Acta* **679,** 308–314.
9. Vadhanavikit, S., Sakamoto, N., Ashada, N., Kishi, T., and Folkers, K. (1984) Quantitative determination of coenzyme Q_{10} in human blood for clinical studies. *Anal. Biochem.* **142,** 155–158.
10. Vadhanavikit, S., Morishita, M., Duff, G. A., and Folkers, K. (1984) Micro- analysis for coenzyme Q_{10} in endomyocardial biopsies of cardiac patients and data on bovine and canine hearts. *Biochem. Biophys. Res. Commun.* **123(3),** 1165–1169.
11. Okamoto, T., Fukui, K., Nakamoto, M., and Kishi, T. (1985) High-performance liquid chromatography of coenzyme Q-related compounds and its application to biological materials. *J. Chromatogr.* **342,** 35–46.
12. Burton, G. W., Webb, A., and Ingold, K. U. (1985) A mild, rapid, and efficient method of lipid extraction for use in determining vitamin E/lipid ratios. *Lipids* **20,** 29–39.
13. Lang, J. K., Kishorchandra, G., and Packer, L. (1986) Simultaneous determination of tocopherols, ubiquinols, and ubiquinones in blood, plasma, tissue homogenates, and subcellular fractions. *Anal. Biochem.* **157,** 106–116.
14. Battino, M., Ferri, E., Gorini, A., Villa, R. F., Huertas, J. F. R., Fiorella, P., Genova, M. L., Lenaz, G., and Marchetti, M. (1990) Natural distribution and occurrence of coenzyme Q homologues. *Membrane Bioch.* **9,** 179–190,
15. Berg, F., Appelkvist, E-L., Dallner, G., and Ernster, L. (1992) Distribution and redox state of ubiquinones in rat and human tissues. *Arch. Biochem. Biophys.* **295,** 230–234.
16. Ye, C.-Q., Folkers, K., Tamagawa, H., and Pfeiffer, C. (1988) A modified determination of coenzyme Q_{10} in human blood and CoQ_{10} blood levels in diverse patients with allergies. *BioFactors* **1,** 303–306.
17. Edlund, P. O. (1988) Determination of coenzyme Q_{10}, α-tocopherols and cholesterol in biological samples by coupled-column liquid chromatography with coulometric and ultraviolet detection. *J. Chromatogr.* **425,** 87–97.
18. Crane, F. L. and Barr, R. (1971) Determination of ubiquinones. *Methods Enzymol.* **XVIIIc,** 137–165.
19. Hatefi, Y. (1963) Coenzyme Q (ubiquinone). *Adv. Enzymol.* **25,** 275–328.
20. Lenaz, G. and Degli Eposti, M. (1985) Physical properties of ubiquinones in model systems and membranes, in *Coenzyme Q,* (G. Lenaz, ed.), Willey & Sons, Chichester, UK, pp. 83–105.
21. Kroger, A. (1985) Determination of content and redox state of ubiquinone and menaquinone. *Methods Enzymol.* **53,** 579–591.

III

Techniques Applicable to Both Oxidative Stress and Antioxidant Activity

35

A Simple Luminescence Method for Detecting Lipid Peroxidation and Antioxidant Activity In Vitro

Minoru Nakano, Takashi Ito, and Tadahisa Hiramitsu

1. Introduction

It has been reported that in a system containing microsomal phospholipids, nicotinomide adenine dinucleotide phosphate (NADPH), Fe^{3+}-ADP complex and cytochrome P-450 reductase (a reconstituted microsomal lipid peroxidation system), excited carbonyl species and singlet molecular oxygen, ($^1O_2$) may generate according to Russell's proposed mechanism (Reaction 1). In the present study, generation of excited carbonyls was verified by their energy being transferred to dyes containing heavy atoms in their molecules. A possible mechanism for the generation of excited species and the energy transfer from excited carbonyls to dye molecules leading to the formation of excited dye molecules is outlined on p. 370 where LH, L•, LOO•, L′ = O* and Dye* represent unsaturated lipids, lipid radicals, lipid peroxy radicals, excited carbonyls in triplet states, and excited dye in singlet state Reaction 4 is forbidden, but proceeds because of heavy atom perturbation *(1)*. Thus, the excited carbonyl species formed during lipid peroxidation may transfer their energy to dye molecules containing heavy atoms such as Rose Bengal to produce singlet dye molecule in excited state. The excited Rose Bengal then emits a photon to return to its ground state. The aim of this chapter is to describe the relationship between Rose Bengal-dependent luminescence and O_2 consumption in a methemoglobin (MetHb)-induced lipid peroxidation system and demonstrate how the reaction is quenched by α-tocopherol or probucol present in phospholipid liposomes.

2. Materials

2.1. Reagents

All chemicals used are of reagent grade. Buffer solution was path-through a Chelex 100 column to eliminate iron contaminate. Commercial chloroform was

From: *Methods in Molecular Biology, vol. 108: Free Radical and Antioxidant Protocols*
Edited by: D. Armstrong © Humana Press Inc., Totowa, NJ

$$LH \longrightarrow L\bullet \quad (1)$$

$$L\bullet + O_2 \longrightarrow LOO\bullet \quad (2)$$

$$LOO\bullet + LOO\bullet \longrightarrow \text{tetraoxide} \longrightarrow L'=O^* + {}^1O_2 + LOH \quad (3)$$

$$L'=O^* + Dye \longrightarrow L'=O + Dye^* \quad (4)$$

$$Dye^* \longrightarrow Dye + h\nu \quad (5)$$

Reactions

distilled and mixed with methanol to obtain chloroform–methanol (2:1, v/v) solution.

2.2. Instruments

A photon counter equipped with a photomultiplier R3550 (Hamamatsu Photonics) is used for the detection of Rose Bengal dependent luminescence. An oxygenometer with Clark type electrode is used for the measurement of oxygen consumption.

3. Methods

3.1. Preparation of Biochemical Samples

3.1.1. Microsomal Phospholipids and Liposomes

Microsomal phospholipids are extracted from rat liver microsomes by the method of Folch et al. *(2)*. These are freed from contaminants, i.e., free fatty acids, cholesterol, esters, and alpha-tocopherols (α-T) by silicic acid column chromatography using chloroform as washing agent and methanol as an eluting agent *(3)*. Commercial Yolk lecithin (Wako Chemicals) is used for the lipid sources instead of microsomal phospholipids after removing the contaminants by the above procedures.

After the removal of methanol in the eluate by a rotary evaporator, the residue is dissolved in chloroform/methanol (2:1 v/v) and stored at –20°C under N_2 atmosphere.

Just before the experiments, the phospholipids are then mixed with or without water-insoluble antioxidants in chloroform-methanol (2:1 v/v). The organic solvents are subsequently evaporated, and liposomes are prepared by sonicating the residues in 100 m*M* Tris-HCl buffer, pH 7.4 ***(4)***.

3.1.2. MetHb

Purified oxyhemoglobin (HbO_2) is prepared from peripheral human blood erythrocytes by ion-exchange chromatography on DEAE-Sephadex A-50 ***(5)***. HbO_2 is then converted to the ferric form by addition of slight excess of potassium ferricyanide ***(6)***. MetHb obtained herein is desalted with Sephadex G-25 and then purified with Dowex 1-8 ***(7)***. The concentration of HbO_2, MetHb, and hemichrome in the purified HbO_2 or MetHb are determined from the millimolar extinction coefficients of the tetrameric species at the following wavelengths in solution at pH 7.4 ***(8)***.

$$[HbO_2] = (-9.8\ A_{630} + 29.8\ A_{577} - 22.2\ A_{560}) \tag{1}$$

$$[MetHb] = (76.8\ A_{630} + 7.0\ A_{577} - 13.8\ A_{560}) \tag{2}$$

$$[Hemichrome] = (-36.0\ A_{630} - 33.2\ A_{577} + 58.2\ A_{560}) \tag{3}$$

3.2. Assay Conditions

The standard reaction system for the measurement of Rose Bengal luminescence contains 250 μ*M* microsomal phospholipid liposomes, 1 μ*M* MetHb, 1 μ*M* Rose Bengal, and 100 m*M* Tris-HCl pH 7.4 in a total volume of 2.0 mL. The reaction mixture in plastic cells is placed just in front of a photomultiplier in a photon counter. The reaction is initiated by the addition of 0.1 mL of MetHb and maintained without agitation for the time indicated. The same mixture without MetHb is used as control. A test mixture for the measurement of O_2 consumption contains all the aforementioned constituents except Rose Bengal.

3.3. Results

3.3.1. Comparison of the Rose Bengal Luminescence and O_2 Consumption

1. When MetHb is added to the phospholipid liposomes in Tris-HCl buffer, in the presence of Rose Bengal, there is an immediate burst of luminescence, which reaches a

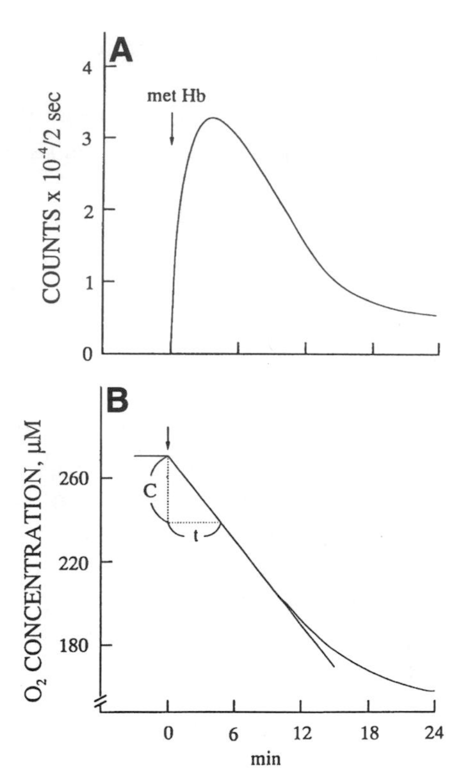

Fig. 1. Time courses of Rose Bengal luminescence (**A**) and oxygen consumption (**B**) in MetHb-induced lipid peroxidation. The reaction mixture contained 425 μ*M* phospholipid, 1 μ*M* MetHb, 1 μ*M* Rose Bengal in (A) or none in (B), and 0.1*M* Tris-HCl buffer. Arrow indicates the time at which MetHb was added. The maximum rate of O_2 consumption was calculated from C/t.

peak value and then begins to decline, as seen in **Fig. 1A**. Excited Rose Bengal emits light peaking at 575 nm (a fluorescence). On the other hand, there is a fall in O_2 consumption in the reaction mixture (without Rose Bengal), reflecting O_2 consumption according to the reaction schemes presented in the introduction section (**Fig. 1B**).

2. From the results shown in **Fig. 1A,B**, the following relationship can be derived: max. luminescence (counts $\times 10^{-4}/2$ s) = k × maximal rate of O_2 consumption (C/t). No marked luminescence or oxygen consumption is observed with a test mixture from which MetHb is excluded. Under these conditions, the maximum rate of O_2 consumption calculates from the slope of the O_2 consumption curve (C/t) and maximum luminescence intensity increases with the amount of phospholipid liposomes.

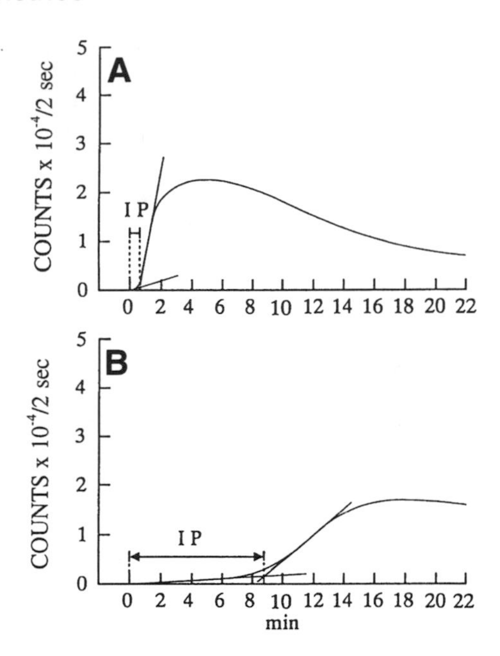

Fig. 2. Effect of α-T on Rose Bengal luminescence; prolongation of the induction period (IP) and suppression of maximum luminescence intensity. The incubation mixture contained 250 μ*M* phospholipid, 1 μ*M* Rose Bengal, 1 μ*M* MetHb, 0.75 μ*M* α-T (**B**) or none (**A**), in 0.1 *M* Tris-HCl buffer.

3. Some deviations from a paralleled rise between O_2 consumption and luminescence intensity is attributable to several minor side reactions of MetHb, in addition to driving the reactions shown in the **Introduction.**

Relatively good parallelism between these could be obtained using, less than 450 μ*M* phospholipid liposomes in the reaction system containing the MetHb + Rose Bengal system. When another triplet sensitizer, 9:10 dibromoanthracene sulfonate (Reaction 1*)* is used, no marked enhancement of the original luminescence is observed, which may suggest partial absorption of sensitized 9:10 dibromoanthracene sulfonate luminescence by MetHb (*see* **Note 1**).

3.3.2. Effects of Water Insoluble Antioxidant on the Rose Bengal Luminescence and Oxygen Consumption

When 250 μ*M* phospholipid liposomes containing 0.75 μ*M* α-tocopherol (α-T) is exposed to the MetHb + Rose Bengal system, the appearance of luminescence is prolonged and the peak luminescence intensity is dose-dependently suppressed by α-T. Typical results obtained in the absence or presence of a

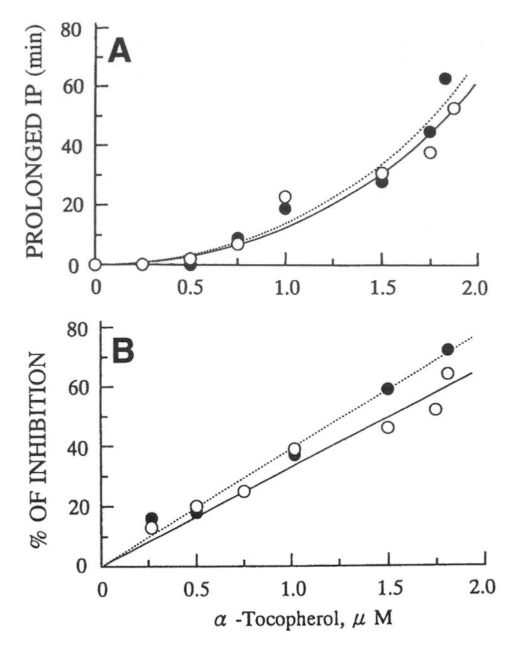

Fig. 3. Relationship between prolongation of induction period (**A**) or inhibition (**B**) and the concentration of α-T on lipid peroxidation monitored by Rose Bengal luminescence (○) or O_2 consumption (●). The reaction mixture was essentially the same as described in the legend to **Fig. 2**, except that Rose Bengal was excluded from the test mixture for the measurement of O_2 consumption and α-T at varying concentration. Prolongation of the induction period was calculated by subtracting the induction period in control system without α-T from that containing α-T.

fixed concentration of α-T are presented in **Fig. 2A,B** representing the procedures used to measure the induction period of luminescence. The data in the these figures are used for calculating percentage inhibition of luminescence using the formula:

$$\% \text{ of inhibition} = \frac{A - C}{A - B} \times 100$$

where A, B, and C are maximum luminescence intensities without MetHb and α-T and with α-T, respectively. A similar trend is also observed in O_2 consumption in the presence of α-T. Indeed, under the present conditions, the prolongation of the induction period, the % inhibition of luminescence intensity and O_2 consumption in the presence of α-T are well in line with each other as seen in **Fig. 3A,B**. From these figures, the amounts of α-T required for a 30 min prolongation of the induction period (induction index) and a 50% inhibition of lipid peroxidation (inhibition index) as determined from the Rose Bengal luminescence are, both calculated to be 1.5 μ*M*.

Under identical experimental conditions, probucol produced a more striking antioxidant effect than α-T, with 0.5 μ*M* for the induction index and 0.6 μ*M* for the inhibition index.

4. Notes

MetHb in the reaction mixture gives no fluorescence without the Rose Bengal. In contrast to a water-insoluble antioxidant emulsified with microsomal phospholipid, a water-soluble antioxidant added to the reaction mixture does not possess a prolonged induction period.

References

1. Sugioka, K. and Nakano, M. (1976) A possible mechanism of the generation of singlet molecular oxygen in NADPH-dependent microsomal lipid peroxidation. *Biochim. Biophys. Acta* **423,** 203–216.
2. Folch, J., Lees, M., and Sloane-Stanley, G. H. (1957) A simple method for the isolation and purification of total lipides from animal tissues. *J. Biol. Chem.* **226,** 497–509.
3. Hanahan, D. J., Dittmer, J. C., and Warashina, E. (1957) A column chromatographic separation of classes of phospholipides. *J. Biol. Chem.* **228,** 685–700.
4. Sugioka, K. and Nakano, M. (1982) Mechanism of phospholipid peroxidation induced by ferric ion-ADP-adriamycin-coordination: complex. *Biochim. Biophys. Acta* **713,** 333–343.
5. Williams, Jr., R. C. and Tsay, K. Y. (1973) A convenient chromatographic method for the preparation of human hemoglobin. *Anal. Biochem.* **54,** 137–145.
6. Tomoda, A., Matsukawa, S., Takeshita, M., and Yoneyama, Y. (1976) Effect of organic phosphates on methemoglobin reduction by ascorbic acid. *J. Biol. Chem.* **25,** 7494–7498.
7. Hegesh, E. and Avron, M. (1967) The enzymatic reduction of ferrihemoglobin. *Biochim. Biophys. Acta* **146,** 91–101.
8. Szebeni, J., Winterbourn, C. C., and Carrell, R. W. (1984) Oxidative interaction between haemoglobin and membrane lipid. *Biochem. J.* **220,** 685–692.

36

Analysis of Oxidation and Antioxidants Using Microtiter Plates

Germán Camejo, Boel Wallin, and Mervi Enojärvi

1. Introduction

Measurement of thiobarbituric acid reacting substances (TBARS), diene derivatives, and hydroperoxides are probably the most widely used procedures for evaluation of free radical-mediated oxidation of biological or model systems containing polyunsaturated fatty acids (PUFA). It is possible to write a sequence of rational reactions in which dienes formation comes first, followed by lipid hydroperoxides (LOOHs) production, and that terminates with fragmentation of the PUFA chains to carbonyl compounds that are TBARS. However, it appears that after the first few min, species of the three types co-exist even in simple suspensions of linoleic acid *(1,2)*. These compounds are very reactive, especially in biological systems, and follow different rates of formation and conversion. Therefore, it is more informative to follow their kinetics of appearance and disappearance than to measure single time-points *(3–5)*.

Extension of conventional photometric procedures for TBARS and LOOHs to a microtiter-plate format requires relatively simple alterations *(6)*. In most of the described uses, aliquots of samples in which oxidation is conducted are transferred to plates, where oxidation products are measured after colorimetric reactions. The microtiter adaptations offer several advantages over conventional spectrophotometry:

1. Possibility of following discontinuous or continuous oxidation kinetics,
2. Rapid preparation of samples,
3. Multi-sample format ideal for evaluation of pro-oxidants or anti-oxidants, and
4. Minimal reagent quantities. The presence in many laboratories of computer-controlled, thermostated microplate readers, with filter or tunable wavelength selection, facilitates the use of plate-adapted procedures for TBARS and LOOHs.

From: *Methods in Molecular Biology, vol. 108: Free Radical and Antioxidant Protocols*
Edited by: D. Armstrong © Humana Press Inc., Totowa, NJ

We further simplified the use of microtiter plates for measurement of oxidation susceptibility by conducting the oxidation step and the photometric analysis in the same plate. TBARS and LOOHs are endpoint procedures. Therefore, to follow kinetics, aliquots of the oxidizing samples are taken after stopping the process at different times with a free-radical scavenger, butylated hydroxytoluene (BHT), and allowing oxidation to proceed in others. Once the desired time points are reached, the colorimetric reactions for TBARS and LOOHs are carried out *(7,8)*. The methods can be used to evaluate Cu^{2+}- and hemin-mediated oxidation susceptibility of plasma lipoproteins and phospholipid vesicles. In addition, the procedures are convenient for measuring the effect of antioxidants in such systems ***(9,10)***.

In these models, evaluation of dienes in microtiter format can provide, on the other hand, continuous evaluation of oxidation without the need of an additional colorimetric step. Diene formation is followed in conventional spectrophotometers using the peak maxima at 234 nm. However, differential spectrophotometry indicates that the changes in absorption associated with free radical-mediated oxidation of PUFA involves the production of conjugated double-bond chromophores with a broad absorption spectra ***(11)***.

Microtiter spectrophotometers with tunable wavelength down to 250 nm are commercially available together with ultraviolet (UV)-transparent plates. This allows one to follow the increase in conjugated-diene absorption at wavelengths on the shoulder of the absorption maximum of "pure" dienes (250 nm). In addition, formation of carbonyl-containing conjugated products at 270 nm, and in hemin-mediated oxidation, the important 407 nm band can be followed in the same experiments. We believe that extension to the microtiter format of these useful methods can be used for following oxidation of PUFA and is described here for the first time. Details are given for UV evaluation of oxidation of low-density lipoproteins (LDL) and phospholipid vesicles. LDL oxidation is under study in many laboratories for its potential relation to atherogenesis *(12)*. Evaluation of hemin-mediated oxidation is included here because evidence suggests that this may be a more biologically significant oxidizing agent than Cu^{2+} *(13,14)*.

2. Materials

2.1. Instruments

1. Computer controlled, Vmax Kinetic microplate reader (Molecular Devices, Sunny Valley, CA).
2. Computer controlled Spectra Max 250 (Molecular Devices, Sunny Valley CA).
3. For determinations of TBARS and hydroperoxides formation, polystyrene microtiter plates with flat bottom holes are used.
4. For the measurement of dienes, UV-transparent plates (Molecular Devices) are used.

2.2. Reagents

1. Solvents, and salts of analytical grade are used (Sigma, St. Louis, MO or Merck, Rahway, NJ). Gel-exclusion columns to equilibrate the lipoproteins in the desired buffer (Pharmacia, Piscataway, NJ).
2. UV-transparent silicon oil, (Dupont Chemicals, Boston, MA) 100 cSt: $\rho = 0.97$ g/mL (KEBO Lab., Spånga, Sweden).
3. Hemin (Sigma).
4. CHOD-iodide reagent (Merk, Darmstad, Germany).
5. Hydroperoxyeicosatetraenoic, 15(S)-HPTE, (Cayman Chem. Co., Ann Arbor, MI).
6. Hyperoxyoctadecadienoic, 13(S)-HPODE, (Cayman Chem. Co.).
7. 1,1,3,3-tetraethoxypropane (Sigma).
8. Phosphate buffered saline (PBS), without Ca^{2+} and Mg^{2+} (Sigma).
9. PBS-EDTA.

3. Methods

3.1. Biological Materials

3.1.1. Plasma Lipoproteins

Human lipoproteins are prepared from fresh plasma containing 1 mg/mL Na_2-ethylenediamine tetracetate (Na-EDTA) by differential centrifugation with use of KBr to adjust densities at 1.006, 1.019, 1.063, and 1.210 g/mL for isolation of very low density lipoproteins (VLDL), LDL, and high density lipoproteins (HDL). Isolation of characterization of lipoproteins is not a trivial procedure, but detailed descriptions are available ***(9)***. To prevent oxidation, the lipoproteins can be stored with 1 mg/mL Na-EDTA under argon or nitrogen and used within 2 wk. If different lipoprotein preparations are to be compared, we strongly recommend equilibrating the lipoproteins using gel-exclusion, manufacturer-made columns, with PBS containing 1 μ/L Na.-EDTA (PBS-EDTA).

3.1.2. Phospholipid Vesicles

1. To run multiple analysis, such as when evaluating the pro-oxidant or anti-oxidant properties of several substances, it is convenient to use as substrate for oxidation, phospholipid vesicles, which are of more constant composition than plasma lipoproteins. Pure phospholipids, like phosphatidyl ethanolamine or phosphatidylcholine with linoleic acid as the main fatty acid, are good models for oxidation experiments and can be obtained commercially from different sources; however, they are expensive. Partially purified mixtures of phospholipids from soybean are good starting materials to prepare better than 90% pure phosphatidylcholine. We use the following modification method of a described procedure ***(8,15)***.
2. One g of Azolectin, crude soybean phospholipids (Sigma) is dissolved in 2 mL of chloroform, 100 μL of 1 *M* Na-EDTA, and 8 mL of acetone in screw-capped, Teflon-lined glass tubes.

3. After centrifugation, at 750*g*, the acetone phase is decanted and the sediment extracted once with 10 mL of chloroform/acetone (8:2 v/v).
4. After centrifugation at 750*g*, the sediment is extracted again with 8 mL of pure acetone. After a new centrifugation step, the remaining pellet is dried under N_2 for 1 h and dissolved in chloroform at approx 250 mg/mL.
5. From the chloroform mixture, 1 mL is transferred to a glass tube with Teflon-lined screw cap, dried completely under N_2, and dissolved in 80 mg of warm ethanol and 200 mg PBS. The solution in a capped tube is placed in a bath at 60°C for 10 min and cooled to 20°C. This pro-liposome mixture can be stored under N_2 for 48 h at 2°C.
6. The pro-liposome mixture is converted to a liposome suspension by dropwise addition of 10 mL of PBS-EDTA. This is best done with a buret or with a peristaltic pump.

3.1.3. Oxidation Step on Microtiter Plates with Copper or Iron

1. For oxidation of lipoproteins, add to each well 100 µL of lipoprotein solution containing 0.22 mg protein/mL, mix with 10 µL of 55 µmol/L of $CuSO_4$, and complete to 120 µL with PBS containing 1 mmol/L PBS-EDTA (*see* **Note 1**).
2. For oxidation of phospholipid vesicles, add 50 µL/well of 2 mg/mL of liposomes suspension, mix with 100 µL/PBS, 10 µL of 0.18 mmol/L $FeNH_4(SO_4)_2$. 12 H_2O and 10 µL of 3.5 mmol/L ascorbic acid.
3. The antioxidants when tested are added in 10 µL ethanol aliquots of different concentration.
4. After additions, the microtiter plates are covered with adhesive film and placed in a forced-air oven at 37°C for the desired period, usually for a maximum of 4 h.
5. To stop oxidation at earlier times, selected wells containing 10 µL of 1 mmol/L BHT in ethanol are added. We use 10- or 15-min intervals in order to follow the oxidation between 0 and 4 h.
6. Standards and blank wells with the same buffer are included in the incubation process. They should contain the same amount of BHT as the experimental samples and ethanol, if antioxidants are being tested. The blanks should contain the lipids or lipoproteins being tested, but no oxidizing agent.

3.1.4. Evaluation of TBARS in Microtiter Plates

The reagents and conditions are essentially those described by Yagi for conventional spectrophotometry ***(16)***. Straightforward spectrophotometric methods have been developed for evaluation of TBARS, dienes and LOOH. Each of these procedures has its limitations with regard to the sample to which they can be applied, to the selection of blanks and standards, and about results interpretation. The reader is urged to study the specific chapters in the present volume that discuss the use of the methods for measurements of dienes, LOOH and TBARS.

1. After BHT is added to all wells to stop the oxidation step, including the blanks and standards, volumes are adjusted to 190 µL with PBS-EDTA.
2. To each well are added 50 µL of 50% (w/v) trichloroacetic acid and 75 µL of 1.3% (w/v) thiobarbituric acid dissolved in 0.3% (w/v) NaOH.
3. The 1,1,3,3-tetraethoxypropane standard is dissolved to a final concentration of 0.5–10 nmol/mL in a total volume of 180 µL.
4. The plate is covered with tape and placed in a forced-air oven at 60°C for 60 min.
5. The plate is cooled over ice-water and the absorbance at 530 nm minus the absorbance at 650 nm is measured in the microplate reader. The reading at 650 nm serves to eliminate differences in turbidity.

3.1.5. Measurement of Hydroperoxides in Plates

This procedure is based on the iodometric analysis of LOOH described by El-Saadani et al. ***(17)***. In this reaction, LOOH oxidize iodide to iodine, then iodine reacts further to form triiodide, which absorbs strongly at 365 nm.

1. After oxidation and addition of BHT at the selected times to stop oxidation, 250 µL of the reagent is added to each plate well.
2. Plates are incubated for 60 min at 37°C and covered with adhesive film.
3. The absorbance at 365 nm minus that at 650 is read.
4. Standards are prepared by addition of 100 µL of PBS-EDTA and 20 µL of ethanolic solutions of 13(S)-HPODE containing from 5 to 250 nmol/120 µL.

BHT should be included in standards and blanks, as well as ethanol if solutions of antioxidants have been added to the solvent (*see* **Note 2**).

3.1.6. Hemin Oxidation of LDL and Phospholipid Vesicles on Microtiter Plates and Continuous Dienes Formation Measurement

1. Hemin: dissolve to 1 mmol/L in 0.02N NaOH by gently shaking.
2. Workings solution of 65 µmol/L hemin and 130 µmol/L H_2O_2 in PBS: Prepare and use the same day (*see* **Note 3**).
3. To each sample well, add 200 µL of LDL, 0.26 mg protein/mL. Immediately add 20 µL of hemin and 20 µL H_2O_2 working solutions. This gives final concentrations of 5 and 10 µmol/L, respectively.
4. If antioxidants are to be tested, they are added in 10 and 20 µL volumes as ethanolic solutions.
5. Blank samples in which no oxidation takes place contain LDL and hemin, but no H_2O_2.
6. If antioxidants are to be tested, the blanks should contain the respective ethanol aliquots.
7. All the well volumes are adjusted to 260 µL with PBS.
8. It is important to assure that solutions in the wells are well mixed.
9. Immediately after the additions, 30 µL of silicone oil are added to all wells, the plates are read at 250 and 405 nm, and maintained at 37°C.

A similar protocol is used for hemin oxidation of phospholipid vesicles, except that the final concentrations of hemin and H_2O_2 in each well are 10 and 20 µmol/L.

In our experiments we use the microtiter-plate reader in the kinetics mode and take measurements at 250 and 405 nm in each well, every 5 or 10 min for 10 h (*see* **Note 4**). These conditions should be adapted to each system because the oxidation kinetic could vary.

3.2. Results

3.2.1. TBARS and LOOH in Microtiter Plates

The results of these two methods for evaluation of oxidation of LDL and lipid vesicles have been published in detail *(7,8)*. There is good correlation between the conventional spectrophotometric procedure and the microtiter modification in terms of standard curves and evaluation of TBARS formed after Cu^{2+}-mediated oxidation of LDL and Fe^{2+}-oxidation of phospholipid vesicles. Furthermore, because several time points can be included, the susceptibility to oxidation of multiple samples of antioxidants on the lag phase, or the linear-propagation phase of TBARS formation and hydroperoxide formation can be readily evaluated. Also, the microtiter format allows rapid comparison of several antioxidants at several concentrations. **Table 1** presents the pIC_{50} of several antioxidants in the Cu^{2+}-mediated oxidation of LDL evaluated at the lag phase or at the propagation phase using the kinetics of TBARS and LOOH production. It can be observed that there is good agreement between TBARS and LOOH measurements of oxidation susceptibility and in evaluation of the efficacy of the antioxidants tested.

3.2.2. Conjugated Dienes in Microtiter Plates

Figure 1 shows the different spectra of human LDL after different times of oxidation with hemin, H_2O_2, and Cu^{2+} (*see* **Note 5**). In the reference cell is present LDL/hemin (no H_2O_2) in experiment A and in the reference cell of experiment B, LDL with no Cu^{2+}. It can be observed that the "pure" diene band (220–255) increases with time, accompanied by a lower band at 260–300 nm. This last band is probably caused by conjugated structures that include carbonyl compounds *(3,4,11)* The ratio of these chromophores at different times is not the same for hemin- and Cu^{2+}-mediated oxidation. In the conditions used, "pure" dienes reach a maximum and begin to diminish earlier with Cu^{2+} than with hemin. During hemin oxidation, a negative band at 407 nm follows closely diene formation. This is caused by the free radical-mediated degradation of hemin.

The kinetics of increase absorbance at 250 nm for LDL oxidation induced by hemin/H_2O_2 is shown in **Fig. 2** (panel 1). Readings taken every 10 min

Table 1
Effect of Antioxidants on the Cu(II)-Catalyzed Oxidation of LDL Measured in the Microtiter Version of Measurements of TBARS and LOOH

	pIC_{50} (-log M)		
Antioxidant	Lag ph.[a] (TBARS)	Prp. ph.[b] (TBARS)	Prp. ph.[c] (LOOH)
α-Tocopherol	5.6	5.7	interf.[d]
Probucol	6.1	6.1	5.8
BHT	5.2	6.4	6.2

[a]Lag phase measured by extrapolation of the linear propagation phase to the time axis of the oxidation kinetic curve obtained with different concentrations of the antioxidants.
[b]Propagation phase values for TBARS and LOOH measured after 90 min of oxidation.
[c]interf.: α-tocopherol markedly reduces the color yield in the LOOH determination.

allow precise evaluation of the lag phase, propagation phase, and termination phase *(4)*. Using the different lag-phase values (in min) obtained for the different concentration of BHT used in **Fig. 2**, accurate values of antioxidant efficacy, expressed as pIC_{50}, can be obtained. The oxidation-mediated degradation of hemin causes the absorbance decrease at 405 nm (**Fig. 1A**) and in the microtiter procedure this change closely follows diene formation (**Fig. 2**, panel 2). This gives useful information about the fate of hemin in the system. Similar curves for conjugated-diene formation are obtained during Cu^{2+}-catalyzed oxidation of LDL. The microtiter-plate method can be used also to follow hemin/ H_2O_2- or Cu^{2+}-mediated oxidation of lipid vesicles, as long as they are not turbid (*see* **Note 6**).

4. Notes

1. In our original description of the method for TBARS and LOOH, the buffer used to adjust volumes and to equilibrate the isolated lipoproteins was HEPES *(7,8)*. However, for evaluation of dienes, cell culture-grade PBS, without Ca^{2+} and Mg^{2+}, is more convenient for its transparency in the UV range. PBS can also be used for TBARS and LOOH.
2. In the evaluation of antioxidants or pro-oxidants with TBARS or LOOH measurements, it is important that the added substances do not interfere with the color of the colorimetric reaction. This is best investigated by comparing standard curves with or without the potential interfering substance. Such a preliminary test is especially important in the iodimetric measurements of LOOH because the colorimetric reaction involves a redox process.
3. Concerning the preparation of hemin stock solutions, it should be noted that this substance takes time to dissolve in 0.01 NaOH, and care should be taken that all hemin is dissolved before preparing the working dilution. Hemin solutions are

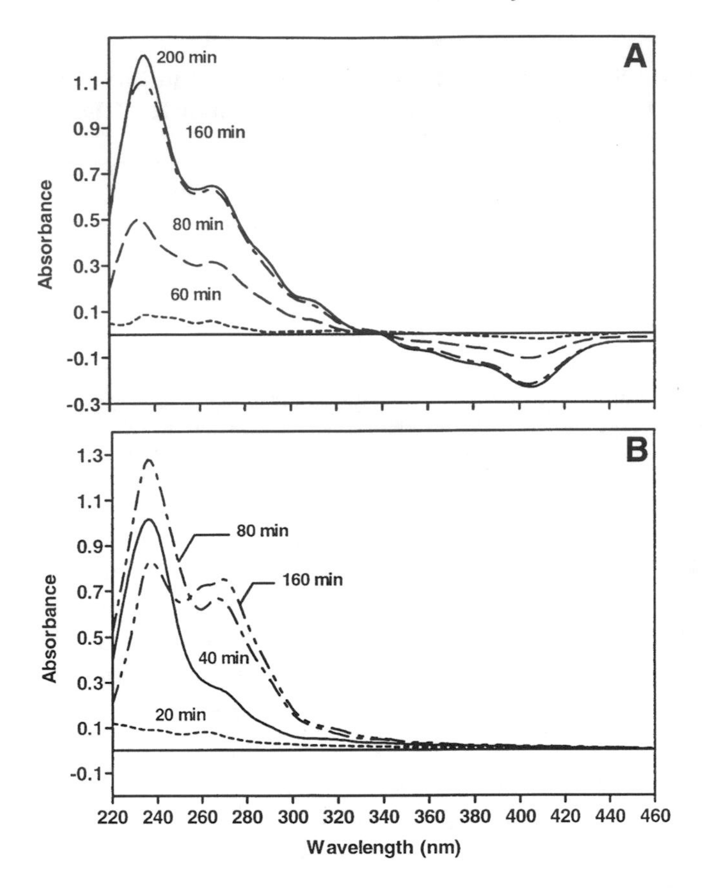

Fig. 1. Difference spectra of the hemin- and Cu^{2+}-mediated oxidation of human low density lipoproteins (LDL) taken at different times. (**A**) Hemin oxidation of LDL. The reference cell contained LDL and hemin but not H_2O_2. The sample cell contained the same LDL concentration (200 µg/mL protein), 5 µmol/L hemin and 10 µmol/L H_2O_2.(**B**) Cu^{2+} oxidation of LDL. The reference cell contained LDL and no $CuSO_4$. The sample cell contained: 200 µg/mL LDL protein and 10 µmol/L $CuSO_4$.

unstable and light-sensitive, and should be protected accordingly and used the same day. Some authors use dimethyl sulfoxide (DMSO) to dissolve hemin, however DMSO is an antioxidant and can interfere with oxidation reactions.

4. Measurements of diene formation in the microtiter reader at 37°C can take several h. Therefore, it is important to reduce evaporation of the small volumes in the plates. We found that silicone oil of low viscosity and very low vapor pressure, added as a thin upper layer, controls evaporation very well. It is also transparent in the UV range and improves the optical reproducibility within samples, probably by homogenizing the shape of the interfacial meniscus of the liquid.

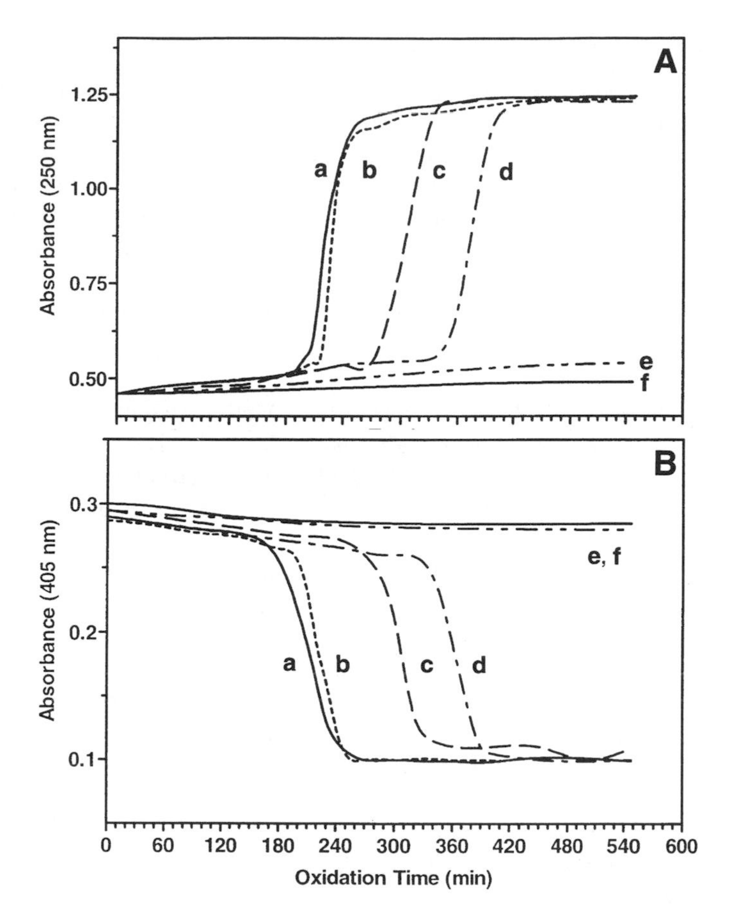

Fig. 2. Effect of BHT on oxidation of LDL followed by changes in conjugated dienes absorption (**A**) and of hemin disappearance (**B**) using a UV-tunable microtiter-plate reader. The letters indicate the concentration of BHT corresponding to each curve: (A), Curve BHT, a, 3.8 ¥ 10^{-8} *M*; b, 7.7 ¥ 10^{-8} *M*; c, 3.8 ¥ 10^{-7} *M*; d, 7.7 ¥ 10^{-7} *M*; e, 3.8 ¥ 10^{-6} *M*; f, 7.7 ¥ 10^{-6} *M*. (B), Curve BHT, a, 3.8 ¥ 10^{-8} *M*; b, 7.7 ¥ 10^{-8} *M*; c, 3.8 ¥ 10^{-7} *M*; d, 7.7 ¥ 10^{-7} *M*; e, 3.8 ¥ 10^{-6} *M*; f, 7.7 ¥ 10^{-6} *M*.

5. When comparing lipoproteins or vesicles, it is mandatory to equilibrate them in the same solutions. We recommend PBS with 1 μmol/L Na-EDTA and the use of gel exclusion, manufacturer-made columns pre-equilibrated in this buffer. We do not use dialysis because of the prolonged periods required and the possibility of transition metal contaminants binding to membranes.
6. We find that oxidation kinetics is affected by the geometry of the microtiter plate wells. We prefer flat-bottomed ones. A final note of caution: lack of proper mixing in microtiter-format reactions is the main source of intra- and inter-sample variability; special attention should be given to this practical aspect.

References

1. Halliwell, H. and Gutteridge, J. (1990) Role of free radicals and catalytic metal ions in human disease: an overview. *Methods Enzymol.* **186(B),** 1–88.
2. Song Kim, R. and LaBella, F. (1987) Comparison of analytical methods for monitoring autoxidation profiles of authentic lipids. *J. Lipid Res.* **28,** 1110–1117.
3. Esterbauer, H., Schaur, R. J., and Zollner, H. (1990) Chemistry and biochemistry of 4-hydroxynonenal, malondialdehyde and related aldehydes. *Free Rad. Biol. Med.* **11,** 81–128.
4. Esterbauer, H., Gebicki, J., Puhl, H., and Jürgens, G. (1992) The role of lipid peroxidation and antioxidants in oxidative modification of LDL. *Free Rad. Biol. Med.* **13,** 341–390.
5. Puhl, H., Waeg, G., and Esterbauer, H. (1994) Methods to determine oxidation of low-density lipoproteins. *Methods Enzymol.* **233,** 425–441.
6. Auerbach, B., Kiely, J., and Cornicelli, J. (1992) A spectrophotometric microtiter-based assay for the detection of hydroperoxy derivatives of linoleic acid. *Anal. Biochem.* **201,** 375–380.
7. Wallin, B. and Camejo, G. (1994) Lipoprotein oxidation and measurement of hydroperoxide formation in a single microtitre plate. *Scand. J. Clin. Lab. Invest.* **54,** 341–346.
8. Wallin, B., Rosengren, B., Shertzer, H., and Camejo, G. (1993) Lipoprotein oxidation and measurements of thiobarbituric acid reacting substances formation in a single microtiter plate: its use for evaluation of antioxidants. *Anal. Biochem.* **208,** 10–15.
9. Hallberg, C., Håden, M., Bergström, M., Hanson, G., Pettersson, K., Westerlund, C., Bondjers, G., Östlund-Lindqvist, A-M., and Camejo, G. (1994) Lipoprotein fractionation in deuterium oxide gradients: a procedure for evaluation of antioxidant binding and susceptibility to oxidation. *J. Lipid Res.* **35,** 1–9.
10. Carmena, R., Ascaso, J., Camejo, G., Varela, G., Hurt-Camejo, E., Ordovas, J., Martinez-Valls, M., Bergström, M., and Wallin, B. (1996) Effect of olive and sunflower oils on low density lipoprotein level, composition, size, oxidation and interaction with arterial proteoglycans. *Atherosclerosis* **125,** 243–256.
11. Watson, A. D., Berliner, J., Hama, S., La Du, B., Faull, K., Fogelman, A., and Navab, M. (1995) Effect of high density lipoprotein associated paraoxonase: inhibition of the biological activity of minimally oxidized low density lipoprotein. *J. Clin. Invest.* **96,** 2882–2891.
12. Navab, M., Berliner, J., Watson, A., Hama, S., Territo, M., Lusis, A., Shih, D., Van Lenten, B., Frank, J., Demer, L., Edwards, P., and Fogelman, A. (1996) The Yin and Yang of oxidation in the development of the fatty streak. *Arterioscler. Thromb. Vasc. Biol.* **16,** 831–842.
13. Balla, G., Jacob, H. S., Eaton, J., Belcher, J., and Vercellotti, G. (1991) Hemin: a possible physiological mediator of low density lipoprotein oxidation and endothelial injury. *Arterioscler. Thromb.* **11,** 1700–1711.
14. Tribble, D., Kraus, R., Lansberg, M., Thiel, P. M., and van den Berg, J. J. (1995) Oxidative susceptibility of low density lipoprotein subfractions is related to their ubiquinol-10 and α-tocopherol content. *Proc. Natl. Acad. Sci. USA* **91,** 1183–1187.

15. Perret, S., Golding, M., and Williams, P. (1990) A simple method for the preparation of liposomes for pharmaceutical application: characterization of the liposomes. *J. Pharm. Pharmacol.* **43,** 154–161.
16. Yagi, K. (1976) A simple fluorometric assay for lipoperoxide in blood plasma. *Biochem. Med.* **15,** 212–216.
17 El-Saadani, M., Esterbauer, H., El-Sayed, M., Goher, M., Nasser, A., and Jürgens, G. (1989) A spectrophotometric assay for lipid peroxides in serum lipoproteins using a commercially available reagent. *J. Lipid Res.* **30,** 627–630.

37

Trace Element Analysis of Biological Samples by Analytical Atomic Spectroscopy

Hiroki Haraguchi, Eiji Fujimori, and Kazumi Inagaki

1. Introduction

It is known that Hg and Cd causes Minamata and Itai-Itai (Auch-Auch) diseases because of the environmental pollution problems in some local areas of Japan. As a result, the high toxicity of compounds like Hg and Cd are now well-recognized. In addition, Pb, As, Cr(VI), Se, Be, T1, Sn, etc. have also caused environmental problems because of their high toxicity.

On the other hand, biological and bio-medical research have elucidated that some elements in animals and plants are essential, even at trace or ultratrace levels *(1)*, Such elements, for example, Fe, Zn, Mn, Cu, Se, I, Mo, Cr(III), and Co are essential for humans. Pb, As, and Se are sometimes toxic, although they are essential trace elements for humans and/or animals. These contradictory concepts are ascribed to the concentration levels of the elements in biological fluids and organs. Extremely large bio-accumulation of some toxic elements often occurs in animals and fish, which are foods. In general, high concentrations of elements cause high toxicity, whereas low concentrations of same elements cause deficiency. It is important, therefore, that the concentrations of elements in animals and their organs, as well as plants, should be determined. Thus, sensitive and accurate analytical methods are required for bio-medical research of trace and ultratrace elements.

In the last three decades, various analytical methods have been developed in the field of analytical atomic spectroscopy *(2)*, and they are now extensively used in bio-medical and environmental studies on trace elements *(1)*. They are mainly atomic absorption spectrometry (AAS; **Figs. 1** and **2**), inductively coupled plasma atomic emission spectrometry (ICP-AES; **Fig. 3**) and inductively coupled plasma mass spectrometry (ICP-MS; **Fig. 4**) *(2–4)*. Instruments

From: *Methods in Molecular Biology, vol. 108: Free Radical and Antioxidant Protocols*
Edited by: D. Armstrong © Humana Press Inc., Totowa, NJ

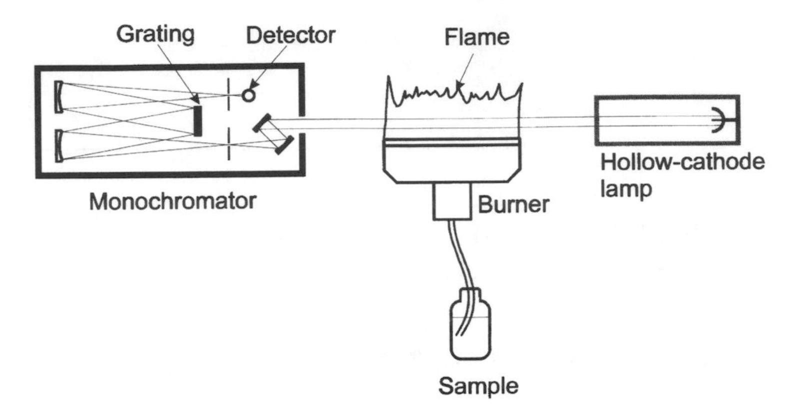

Fig. 1. Schematic diagram of a flame atomic absorption spectrometry (FAAS).

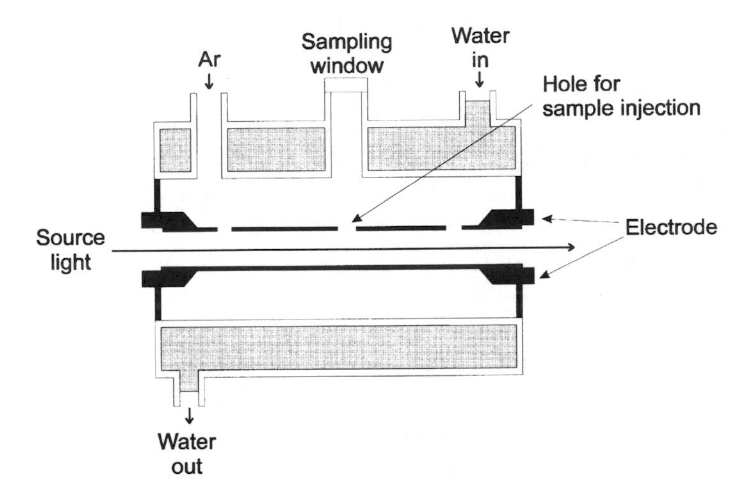

Fig. 2. Graphic furnace.

for these methods are now commercially available. In the present chapter, the analytical features, instrumentation, and some application work are described in relation to bio-medical and environmental research.

1.2. Trace Elements in Biology

It is known that animals and plants contain various elements *(5)*. Among them, the elements which have some biological functions are called "bio-elements." The concentration of elements are determined by various spectrochemical methods. Furthermore, many metallo-proteins and metallo-enzymes have been found and their biological functions are now elucidated on a molecular basis.

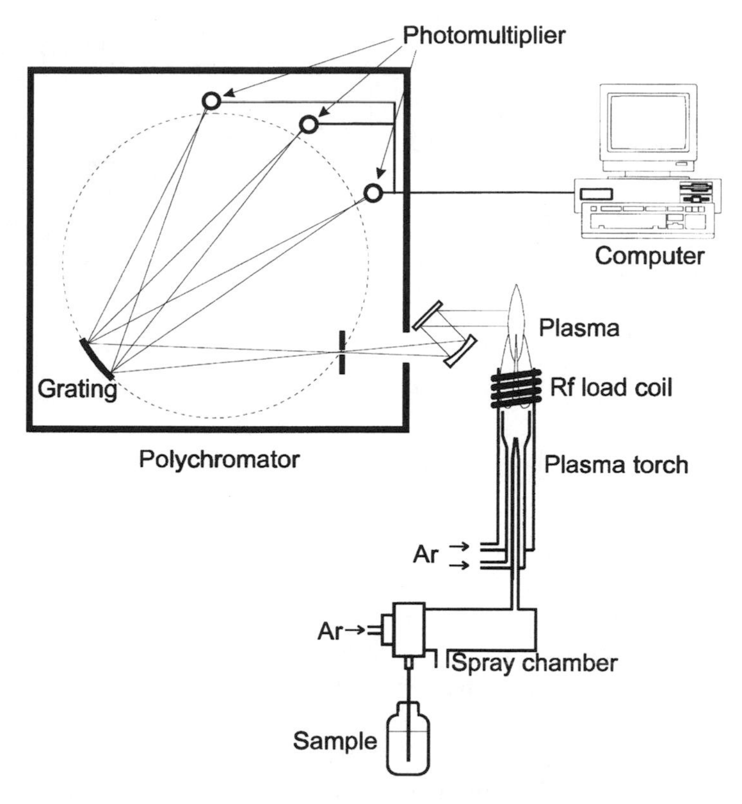

Fig. 3. Schematic diagram of a multichannel ICP-AES instrument.

In **Table 1**, elements that are contained in the human body are shown together with their amounts and their concentrations per gram of body weight *(6)*. Elements are classified according to concentration levels used in analytical chemistry into four groups, i.e., major elements, minor elements, trace elements, and ultratrace elements.

Major elements (constituents):	100–1%
Minor elements (constituents):	1–0.01%
Trace elements (constituents):	0.01–0.0001% (100 μg/g–1 μg/g)
Ultratrace elements (constituents):	>0.0001% (>1 μg/g)

As seen in **Table 1**, O, C, H, N, Ca, P, S, K, Na, Cl, and Mg are major and minor elements. All 11 elements are essential to the human body as well as most animals. Consequently, they are hereafter called essential major elements. The total percentage of essential major elements is almost 99.4%.

Among trace elements, Fe is at the highest concentration of 85.7 ppm (1 ppm = 1 μg/g). As shown in **Table 1**, Fe, F, Si, Zn, Sr, Pb, Mn, Cu, Sn, Se, I,

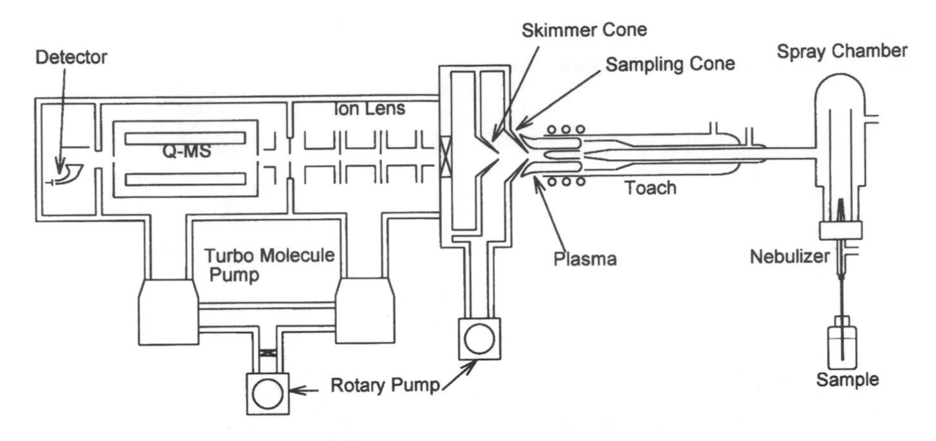

Fig. 4. Schematic diagram of an ICP-MS instrument.

Mo, Ni, Cr, As, Co, and V are essential to experimental animals and, Fe, Zn, Mn, Cu, Se, I, Mo, Cr, and Co are essential to humans.

The role and function of various trace elements have been elucidated in relation to health science, toxicology, epidemiology, nutrition science, biochemistry, molecular biology, and environmental science, although the biological functions of each trace element are not explained here. Thus, simultaneous multi-element determination of the elements in the concentration range from major to ultratrace level is another requirement (or trend) in modern trace-element research. Analytical atomic spectroscopy, especially ICP-AES and ICP-MS, may be the most promising method for such requirement ***(1,2)***, as will be described in the following sections.

1.3. Analytical Atomic Spectroscopy

1.3.1. General Features of Trace-Element Analysis

Fundamentally, samples are required to be in solution for analysis. Biological samples in the solid state must be digested by proper methods to prepare the analysis solution prior to determination of the elements of interest. Even in the cases of body-fluid samples such as blood and urine, some inorganic salts and organics (e.g., proteins) are contained at the high concentration. Such high contents of inorganic salts and proteins are prone to analytical interferences (causes of analytical errors). Thus, blood and urine samples are generally decomposed with mineral acids and diluted by 10–100 times with water or 0.1 *M* HNO_3 prior to analysis. As a result, the original concentrations of the elements become significantly lower by 10–100 times in the final analysis solution through digestion and dilution procedures. In other words, the concentrations of trace elements in biological samples become extremely low in the analysis solutions. These

Table 1
Elements in Human Body and Their Concentrations[a]

	Element	Weight percentage in body (%)	Amount in body per 70 kg weight	Concentration per 1 g body weight
Major element	O	65.0	45.5 kg	650 mg/g weight
	C	18.0	12.6	180
	H	10.0	7.0	100
	N	3.0	2.1	30
	Ca	1.5	1.05	15
	P	1.0 (98.5%)	0.70	10
Minor element	S	0.25	175 g	2.5
	K	0.20	140	2
	Na	0.15	105	1.5
	Cl	0.15	105	1.5
	Mg	0.15 (99.4%)	105	1.5
Trace element	Fe[b,c]		6 g	85.7 μg/g weight
	F[b]		3	42.8
	Si[b]		2	28.5
	Zn[b,c]		2	28.5
	Sr[b]		320 mg	4.57
	Rb		320	4.57
	Pb[b]		120	1.71
	Mn[b,c]		100	1.43
	Cu[b,c]		80	1.14
Ultratrace element	Al		60 mg	857 ng/g weight
	Cd		50	714
	Sn[b]		20	286
	Ba		17	243
	Hg		13	186
	Se[b,c]		12	171
	I[b,c]		11	157
	Mo[b,c]		10	143
	Ni[b]		10	143
	Cr[b,c]		2	28.5
	As[b]		2	28.5
	Co[b,c]		1.5	21.4
	V[b]		1.5	21.4

[a]Cited from **ref.** ***6***.
[b] Trace elements are known to be essential in experimental animals.
[c]Trace elements are known to be essential in man.

situations in sample pretreatment for biological samples make it difficult to determine trace elements even by various sensitive analytical methods.

X-ray fluorescence spectroscopy (XRF), electron microscopy, secondary ion mass spectroscopy (SIMS), and neutron activation analysis (NAA) can be used for direct analysis of solid samples, or after freeze-drying The direct analysis is a great advantage over analytical atomic spectroscopy. However, XRF, electron microscopy, and SIMS are not so sensitive for a variety of the elements. It is generally stated that XRF and electron microscopy are available for the determination of major/minor elements and some trace elements (**Table 1**), although they can be used for the two-dimensional analysis of biological organs in combination with electron microscopy. On other hand, NAA is generally a very sensitive method for various elements, almost comparable to ICP-AES and ICP-MS, and thus it has been extensively applied to the analysis of biological samples *(2)*. However, NAA has a disadvantage that it requires a nuclear reactor for irradiation of the samples by neutron fluxes. Thus, NAA is not a routine laboratory instrument. Although sample pretreatment is not necessary, long measurement times for some elements are required in NAA analysis.

In **Table 2**, the detection limits obtained by GFAAS (graphite furnace atomic absorption spectrometry), ICP-AES, and ICP-MS are summarized for general comparison. As for the detection limits and other analytical figures of merit, more detailed explanation will be given in **Subheadings 3.–4.**

1.3.2. ICP-AES

ICP-AES provides the following analytical feasibilities *(2)*:

1. High sensitivity for almost all metallic elements, including phosphorus and sulfur;
2. Wide dynamic range (4–5 orders) of the calibration curve;
3. Multielement detection capability; and
4. Less chemical and ionization interferences.

The detection limits obtained by ICP-AES are summarized in **Table 2.** Many elements are detectable at the ppb (ng/mL) or sub-ppb level and so ICP-AES is suitable for the accurate and precise determination of major and minor elements and is applicable to the determination of some trace elements whose concentrations are higher than 10 ng/mL.

A schematic diagram of an ICP-AES instrument is shown in **Fig. 2**, where an argon plasma is used as the atomization/ionization source as well as excitation source. The argon plasma is sustained on a plasma torch by applying the RF (radiofrequency) power (27.12 M Hz with 1–2 kW). The plasma torch is a co-axial three-cylinder tube made of quartz, into which argon gas flows in three different ways. The outer gas, intermediate gas, and inner gas are called "coolant gas (15–20 L/min; flow rate)," "auxilliary gas (0–1 L/min)," and "carrier gas (0.5–2 L/min),"

Table 2
Detection Limits in GFAAS, ICP-AES, and ICP-MS

[Remarks]

1. In each element box, the numbers at the upper, middle and lower positions indicate the detection limits obtained by GFAAS, ICP-AES and ICP-MS, respectively.
2. Unit: ppb (10^{-9} g/mL).
3. The sign – in the table indicates the impossible detection..
4. The elements without any numbers cannot be detected or are radioactive elements.

Element	GFAAS	ICP-AES	ICP-MS
H	–	–	–
He			
Li	0.3	1	0.027
Be	0.003	0.1	0.05
B	20	2	0.1
C	–	10	–
N			
O			
F			
Ne			
Na	0.02	10	0.03
Mg	0.1	0.1	0.018
Al	0.1	4	0.015
Si	10	5	5
P	0.3	30	5
S	–	20	–
Cl			
Ar			
K	0.1	40	–
Ca	0.04	0.1	0.5
Sc	6	0.4	0.015
Ti	40	0.8	0.011
V	0.3	1	0.008
Cr	0.2	2	0.04
Mn	0.02	0.3	0.006
Fe	1	0.62	0.58
Co	0.2	0.85	0.005
Ni	0.9	3	0.013
Cu	0.1	1	0.04
Zn	0.003	1	0.035
Ga	–	7	0.009
Ge	3	13	0.013
As	0.8	10	0.031
Se	0.9	15	0.37
Br	–	1.6	–
Kr			
Rb	1	–	0.005
Sr	0.1	0.1	0.003
Y	–	0.8	0.004
Zr	–	1.9	0.005
Nb	–	10	0.002
Mo	0.2	2	0.006
Tc			
Ru	–	7	0.05
Rh	0.8	8	0.002
Pd	0.4	10	0.009
Ag	0.01	1	0.005
Cd	0.008	1	0.012
In	0.04	20	0.002
Sn	0.1	10	0.01
Sb	0.5	10	0.012
Te	0.1	10	0.032
I	–	50	0.8
Xe			
Cs	0.04	4000	0.002
Ba	0.6	0.2	0.006
La*			
Hf	–	5	0.005
Ta	–	8	0.002
W	–	10	0.007
Re	–	2	0.005
Os	–	0.5	–
Ir	–	7	0.005
Pt	1	10	0.005
Au	0.1	3	0.005
Hg	0.2	5	0.018
Tl	1	10	0.003
Pb	2	20	0.01
Bi	0.4	5	0.004
Po			
At			
Rn			
Fr			
Ra			
Ac**			
* La	–	2	0.002
Ce	–	9	0.004
Pr	–	9	0.004
Nd	–	10	0.007
Pm			
Sm	–	8	0.013
Eu	0.5	0.45	0.007
Gd	–	3	0.009
Tb	–	5	0.002
Dy	–	2	0.007
Ho	–	1	0.002
Er	–	2	0.005
Tm	–	1.3	0.002
Yb	0.07	0.4	0.005
Lu	–	0.3	0.002
** Ac			
Th	–	14	0.000
Pa			
U	–	50	0.000
Np			
Pu			
Am			
Cm			
Bk			
Cf			
Es			
Fm			
Md			
No			
Lr			

respectively. The sample solutions are aspirated continuously into a spray chamber through a nebulizer and then introduced into the plasma through the inner tube together with the carrier argon gas. Only 1% of the sample solution nebulized into the spray chamber is usually introduced into the plasma in ICP-AES. The sample aspiration rate is almost 1–2 mL/min, and the optical measurement is performed for 1–2 min.

In the argon plasma, elements in the sample solutions are atomized and ionized spontaneously at 15–20 mm above the RF load coil. Thus, the plasma position at 15–20 mm is called "the observation height" in ICP-AES. The emission signals of atoms or ions in the plasma are observed by the optical detection system. A monochromator or a polychromator is used as the optical detection system. The former system allows sequential multi-element measurement and the latter system does simultaneous multi-element measurement. At present, the multi-element determination of up to 50 elements is possible by using 1–5 mL of the sample solution for 1–5 min by both sequential and simultaneous multi-element ICP-AES.

1.3.3. ICP-MS

In the last decade, ICP-MS has been newly developed as a sensitive method for trace analysis *(2,3)*. In ICP-MS, the same argon plasma as used in ICP-AES is utilized as an ionization source with a mass spectrometer (MS) for ion detection. Usually, ion detection is superior to optical detection employed in ICP-AES. Thus, ICP-MS provides much lower detection limits by 3–4 orders of magnitude, compared to ICP-AES, as seen in **Table 2**. The analytical feasibilities, such as wide dynamic ranges and multi-element detection capability, are also obtained in ICP-MS, in a manner similar to ICP-AES.

In ICP-MS, matrix effects, in which the signal intensities of the analytes are suppressed by major elements in the samples, are a serious problem *(3)*. Polyatomic ion interferences are also caused by the co-existing major elements. The matrix effects (mass discrimination) are usually corrected by internal standards. On the contrary, the polyatomic-ion interferences are difficult to correct and the use of the ICP-MS instrument using a high-resolution MS is the only way to avoid such interferences. Another method to avoid the matrix effects and the polyatomic-ion interferences is to separate the analyte elements chemically from major and minor elements, which are prone to these interferences.

A schematic diagram of an ICP-MS instrument is shown in **Fig. 2**. As mentioned earlier, the sample introduction system (nebulizer) and the argon plasma are the same as those in ICP-AES. The analyte ions produced in the plasma are introduced into a quadrupole-type mass spectrometer (Q-MS) through the sampling interface (sampling cone and skimmer cone) and ion lens. The ions are finally detected by an electron multiplier and provide the signals at the corresponding m/z positions, where m and z are the mass and electric charge of the ions, respectively.

Very recently, a high-resolution double-focusing MS using electric field and magnetic field has been developed for ICP-MS. This high-resolution ICP-MS (HR-ICP-MS) is very efficient to resolve signals with close m/z values *(see* **Fig. 5**) *(11)*. Thus, the HR-ICP-MS is useful to avoid polyatomic-ion interferences. The MS and the interface part after the sampling cone are usually kept at 10^{-5} Torr and 10^{-1} Torr, respectively.

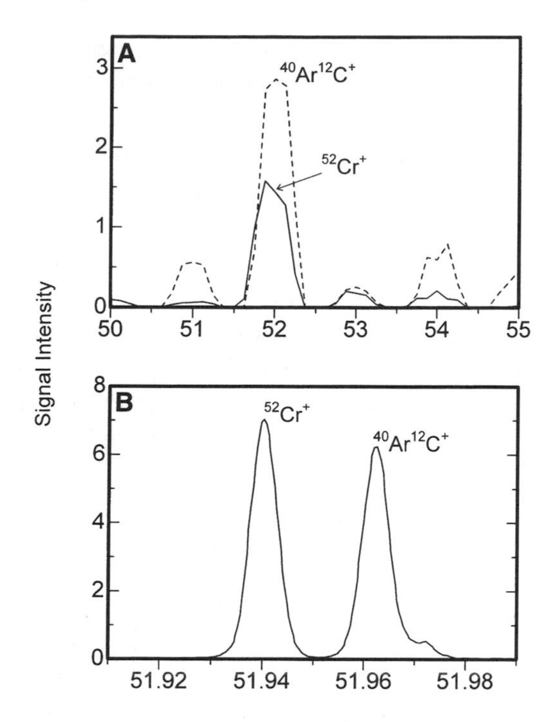

Fig. 5. Mass spectrum obtained by ICP-MS. (**A**) Mass spectrum of chromium 10 ng/mL (solid line) and 5% glycine (dashed line) obtained by quadrupole-type ICP-MS. (**B**) Mass spectrum of *m/z* 51.91 to *m/z* 51.99 obtained by HR-ICP-MS. The sample was SRM 2670 diluted fivefold.

In ICP-MS, the solution samples are introduced into the argon plasma through a nebulizer. Usually, the sample introduction rate is 1 mL/min, for 1% of the nebulized sample introduced into the plasma. For quantitative analysis, 1–2 mL of the sample solution is required in the continuous pneumatic nebulization. As a result, 10–30 elements are simultaneously determined by ICP-MS.

1.3.4. AAS

AAS is extensively applied to the determination of trace bio-elements. In AAS, the optical absorption of the neutral atoms produced in the high-temperature flames, electrothermal graphite furnace, or metal filaments are generally observed *(2,4)*.

AAS provides good sensitivity for a variety of the elements. The detection limits obtained by GFAAS are summarized in **Table 2**. As seen, most elements can be detected at the ppb or sub-ppb level. However, AAS has some problems or disadvantages *(4)*:

1. Chemical interferences owing to co-existing major elements,
2. Large background absorptions owing to matrix elements, and
3. Narrow dynamic ranges of the calibration curves.

Furthermore, the sample pretreatment procedures such as separation and preconcentration of the analyses are often required to obtain accurate analytical values.

The AAS consists of light source, atomization sources, monochromator, detection device, electronic devices, and display (or recorder). In **Fig. 1**, a schematic diagram of an AAS instrument is shown for a single-beam optical system. The AAS instrument of a double-beam optical system is often used because of better instrumental stability. AAS is basically a single-element detection system, because the optical beam from the light source to the monochromator is arranged on a straight line.

In AAS, the hollow cathode lamp is used as a light source, which consists of a cathode with metallic hollow cup and an anode. Electric voltage (100–500 V) is applied between the cathode and the anode, and 5–20 mA of electric current flows between the electrodes. The emission lines of each element emitted from the hollow cathode lamp are characteristically very narrow, within 0.001 nm as the half linewidth. In AAS, a second light source is used for background correction and is usually a deuterium or hydrogen lamp (continuum light source). This background correction system is required in electrothermal GFAAS because the molecular band absorptions, owing to major constituents in the sample, such as NaCl, KCl, and SO_2 are significantly large in the graphite furnace.

As atomization sources in AAS, the chemical flames and electrothermal graphite furnace are generally used. Air-acetylene flame (ca. 2500 K) and dinitrogen monoxide (N_2O)-acetylene flame (ca. 3000 K) are employed, depending on the kind of elements. In the flame AAS instrument, 2–5 mL of the sample solution is required for each analysis.

A schematic structure of the graphite furnace is shown in **Fig. 2**. In GFAAS, 5–20 μL of the sample solution is injected on the center of the furnace through the sampling window. Then, the furnace is heated through three stages, i.e., at 80–150°C (drying stage) for 1–2 min, 400–1000°C (ashing stage) for 1–3 min, and 2000–2800°C (atomization stage) for 5–20 s. At the atomization stage, because the analyte elements in the samples are rapidly atomized, the atomic absorptions are observed as the transient peak signals.

In the determination of Hg, a cold vapor generation method is employed for atomization, where Hg(II) ions in the sample solution are chemically reduced to Hg atoms with $NaBH_4$ at room temperature. Furthermore, in the case of As, Se, Sb, Sn, Te, Ge, Bi, and Pb, a hydride generation method is used to release volatile elements (e g., AsH_3, SeH_2) by reduction with $NaBH_4$ from the sample solutions. Then, the hydrides are introduced into the hydrogen-argon flame or a heated quartz tube, where atomization by heating occurs and the atomic absorptions are observed. In the cold-vapor method for Hg and the hydride-generation method for As, Se, Ge etc., the detection limits of these elements are markedly improved at the ppb or sub-ppb level, compared to the flame AAS method.

1.3.5. Analytical Figures of Merit

In analytical chemistry, the terminology of "sensitivity" is used, and, we say, for example, AAS is a sensitive analytical method. In these cases, "sensitivity" means the slope of the calibration curve. If the slope is large, the sensitivity is high. On the other hand, if the slope is small, the sensitivity is poor.

In **Table 2**, the detection limits obtained by GFAAS, ICP-AES, and ICP-MS are summarized as the indices of the sensitivity of each method. In instrumental analysis, the detection limits are generally used instead of "sensitivity." When the detection limit of a method is low, it is said that the method is sensitive. This concept of the sensitivity is commonly employed in a relative analytical method (instrumental analysis).

In AAS, the detection limits are defined as the concentration of the analyte, which provides a signal intensity corresponding to S/N = 3 (S, signal intensity of the analyte; N, background noise) or 3σ (σ: standard deviation [SD]) of the background intensity, when the blank solution is measured. In addition, the determination limit is defined as the analyte concentration corresponding to S/N = 10 or 10σ. In general, when the analyte element is determined near the detection limit, the relative standard deviation (RSD) of the repeated measurements (n = 10) is 50–100%, whereas RSD is at 10% or less, when determined above the determination limit. Thus, it is desirable that the determination of trace elements is performed at the concentration level over the determination limit, when good analytical precision is required.

As seen in **Table 2**, the detection limits obtained by GFAAS, ICP-AES, and ICP-MS are generally superior in the following order:

ICP-MS > GFAAS > ICP-AES

The following process are recommended, when all these methods are available.

1. ICP-AES is suitable for the determination of the analyses in the concentration range from 10 ppb to 1000 ppm with good precision. Thus, the major and minor elements in the samples should be first determined by ICP-AES, and we can find the matrix concentrations.
2. ICP-MS is suitable for the determination of trace elements in the concentration range from 1 ppt (1 pg/mL) to 1 ppm. Thus, as the second step, all other trace

elements should be surveyed by ICP MS. In the ICP-MS measurement, matrix interferences (so-called "matrix effects") owing to major elements are often serious, and so it is desirable that total concentration of matrix elements is less than 100 ppm.

3. GFAAS is suitable for the determination of volatile elements such as Zn, Cd, Pb, As, Se, Ge, and Sb. Because GFAAS requires the sample amount (volume) of only 5–50 μL for one analysis, the determination of the elements previously described is performed by GFAAS, especially when the sample amount available for analysis is limited. However, only one element can be determined at one measurement by AAS.

2. Materials

2.1. Reagents

All reagents are of ACS grade or better and were purchased from Kanto Chemical Co. Inc. (Tokyo, Japan). Metal standards for stock solution were purchased from Wako Pure Chemical Co. (Osaka, Japan). Pure water used throughout the present experiment was prepared by a Milli Q purification system (Nihon Millipore Kogyo Co., Tokyo, Japan).

2.2. Standards

1. National Institute for Environmental Studies (NIES; Tsukuka, Japan) No. 4 (serum).
2. National Institute of Standards and Technology (NIST; Gaithersberg, MD) SRM 909b, (serum).
3. Second-generation human serum reference sample (Ghent University).
4. International Atomic Energy Agency (IAEA; Vienna, Austria) A-13 (whole bovine blood).
5. NIST SRM 2670 (urine).
6. NIST SRM 1577 (bovine liver).
7. NIES No. 13 (human hair).

2.3. Instrmuments

1. ICP-AES Model Plasma AtomComp Mk II (Jarrell-Ash, Franklin, MA).
2. ICP-MS Model SPQ 8000A (Seiko Instruments Inc., Tokyo, Japan).

3. Methods

3.1. Pretreatment of the Samples

1. Serum samples:
   a. NIES No. 4: For analysis, a serum reference sample (0.8 g) from 2 sample vials, which corresponds to 10 mL of original blood serum, is placed in a Teflon beaker (100 mL in volume), and 6 mL of conc. HNO_3 is added. After standing overnight, the solution is heated at 100°C for 6 h on a hot plate. Then, 3 mL of conc. HNO_3 is added and heated at 150°C for 5 h on a hot plate. This procedure is repeated two times. The sample solution is dried at 250°C,

and the residue is dissolved with 0.76 mL of conc. HNO_3 and diluted to 100 mL with pure water. This final solution, which is hereafter referred to as "the analysis solution," is used for ICP-AES measurements. The solution is further diluted twofold and internal standard elements (Ge, Rh, Re, and Tl; 10 ppb each are added for measurements by ICP-MS (*see* **Note 1**).

b. Second-generation human serum reference sample and NIST SRM 909b: The freeze-dried samples (correspond to 8 g of the original serum samples) are reconstituted by diluting with water and put into Teflon beakers. One mL of conc. HNO_3 is added to the sample and heated to near dryness. Then, 2 mL of conc. HNO_3 is added to the solution and heated at 150°C for ca. 2 h on a hot plate. Then, 2 mL of conc. HNO_3 and conc. 1 mL of 60% $HClO_4$ are added and heated on a hot plate at 150°C for 4 h to near dryness (*see* **Note 2**). Again, 1 mL of conc. HNO_3 and 1 mL of 60% $HClO_4$ are added and heated for ca. 4 h on a hot plate, until a white fume appears. The residue is dissolved with 0.76 mL of conc. HNO_3, and diluted to 100 mL with pure water. The solution is further diluted twofold and internal standard elements (Ge, Rh, Re, and Tl) are added for the measurements by ICP-AES and ICP-MS.

2. Whole blood sample (IAEA A-13): The whole-blood sample (0.5 g) is placed in a Teflon beaker, and 5 mL of conc. HNO_3 added. After standing overnight, 2 mL of conc. HNO_3 and 2 mL of 60% $HClO_4$ are added to the sample and heated on a hot plate at 200°C for 8 h. Then, 1 mL of conc. HNO_3 is added and heated at 200°C for 4 h. Again, 1 mL of conc. HNO_3 is added to the solution and heated at 200°C for 1 h to near dryness (*see* **Note 2**). Finally, the residue is dissolved by adding 1.52 mL of conc. HNO_3 and pure water with heating at 100°C. The dissolved solution is diluted to 20 mL, which provides the analysis solution for measurement by ICP-AES. The solution is further diluted by fivefold and internal standard elements (In, Re, and Tl; 10 ppb each) are added for measurements by ICP-MS.
3. Urine sample (NIST SRM 2670): The freeze-dried human reference sample is reconstituted by diluting with water and diluted 25-fold with 0.1 *M* HNO_3 and the internal standards (8 ng/mL of Ge, In, and Tl) are added for the internal standard correction in the ICP-MS measurements.
4. Bovine liver sample (NIST SRM 1577): The sample (0.5 g) is placed in a Teflon beaker, and 7 mL of conc. HNO_3 added. After standing overnight, the sample is heated at 100°C for 6 h, and then 3 mL of conc. HNO_3 is added and heated at 150°C for 4 h. Then, 2 mL of conc. HNO_3 and 2 mL of 60% $HClO_4$ are added into the residual solution and heated at 150°C for 4 h almost to dryness (*see* **Note 2**). This procedure is repeated two times to decompose bovine liver as much as possible. Finally, the residue is dissolved by adding 1.52 mL of conc. HNO_3 and pure water with heating at 110°C. The dissolved solution is diluted to 100 mL, which provides the analysis solution. The solution is further diluted by twofold and internal standard elements (Ge, Rh, Re, and Tl; 10 ppb each) are added for the measurements by ICP-MS.
5. Human-hair sample (NIES No. 13): The sample (0.2 g) is placed in a Teflon beaker, and 5 mL of conc. HNO_3 is added. After standing overnight, 5 mL of

conc. HNO_3 is added and heated at 180°C for 10 h. Then, 5 mL of conc. HNO_3 and 1 mL of (1+1) HF are added in the residual solution, heated at 180°C for 2 h almost to dryness, and 5 mL of conc. HNO_3 and 1 mL of 60% $HClO_4$ added and heated on a hot plate, until a white fume appears. Finally, the residue is dissolved by adding 1.52 mL of conc. HNO_3 and pure water with heating at 110°C. The dissolved solution is diluted to 20 mL, which provides the analysis solution for the measurements by ICP-AES. The solution is further diluted by twofold and internal standard elements (Ge, In, Re, and Tl; 10 ppb each) are added for the measurements by ICP-MS.

3.2. Results

3.2.1. Application to Trace-Element Analysis

In the previous sections, analytical features, characteristics, and analytical figures of merit in analytical atomic spectroscopy have been described from the viewpoints of bio-trace element analysis. It is stressed again that the methods in analytical atomic spectrometry are available as high-sensitive methods with multi-element detection capability. However, analytical atomic spectrometry often suffers from matrix interferences owing to major and minor elements (constituents) co-existing in the samples. Thus, sample pretreatment such as separation of major elements and preconcentration of trace elements is necessary ***(1,2,4)***. In the following description, the applications of analytical atomic spectroscopy to the multi-element determination of major-to-ultratrace elements in some biological samples will be introduced (*see* **Note 3**).

3.2.2. Blood-Serum Analysis

The blood serum samples are analyzed daily in clinical laboratories for medical diagnosis. Trace metals such as Fe, Zu, and Cu, and major metals such as Na, K, Ca, and Mg are the subject of clinical analysis. In addition, Pb, Cd, Hg, Cr (VI), and As in blood serum are often determined to check health conditions caused by environmental pollutions. So far, the determination of elements previously listed have been usually carried out by flame AAS and GFAAS.

1. In **Table 3**, the analytical results for a blood-serum reference sample (NIES No. 4), is summarized together with the acceptable values reported by NIES. The blood-serum reference samples in the experiment in **Table 3** are the freeze-dried samples.
   As seen in **Table 3**, Na, K, P, Ca, Mg, Fe, Cu, Zn, and A1 are determined by ICP-AES using the analysis solution without further dilution, whereas other elements (Ba ~ Pr) are determined by ICP-MS, where the analysis solution is further diluted by a factor of 2. In **Table 3**, the concentrations of A1, Ba, Rb, Sr, Au, Mo, Pb, Ag, and Cs are still the information values, and there is no acceptable or information value for Zr, Ce, Bi, Pt, La, Nd, Th, U, and Pr. It is noted that the analytical

Table 3
Analytical Results for Blood-Serum Reference Sample (NIES No. 4) Obtained by ICP-AES and ICP-MS

Element	Wavelength[a] or m/z	Observed value[b]	Acceptable value[c]
Na	589.0 nm I	2750 ± 30 μg/mL	3000 μg/mL
K	766.4 nm I	169 ± 1	173
P	213.6 nm II	108 ± 0	107
Ca	317.9 nm II	77.7 ± 0.8	78
Mg	279.0 nm II	17.0 ± 0.1	19.5
Fe	259.9 nm II	0.98 ± 0.01	1.07
Cu	324.7 nm I	0.97 ± 0.00	1.04
Zn	213.8 nm I	0.93 ± 0.02	0.91
Al	308.2 nm I	0.37 ± 0.04	(0.3–1.5)
Ba	138	848 ± 103 ng/mL	(510–620) ng/mL
Rb	85	178 ± 0	(200–320)
Sr	88	67.9 ± 1.4	(50.0)
Au	197	5.3 ± 0.2	(6)
Sb	121	2.4 ± 0.3	(0.8)
Mo	98	2.21 ± 0.10	(3.59)
Zr	90	2.11 ± 0.86	
Pb	208	2.0 ± 0.9	(101)
W	184	1.70 ± 0.00	
Ag	107	1.23 ± 0.02	(1.32)
Cs	133	0.59 ± 0.00	(0.8–1.1)
Ce	140	0.22 ± 0.01	
Bi	209	0.137 ± 0.007	
Pt	195	0.13 ± 0.01	
La	139	0.10 ± 0.01	
Nd	146	0.07 ± 0.01	
Th	232	0.07 ± 0.00	
U	238	0.06 ± 0.00	
Pr	141	0.019 ± 0.001	

[a]I and II indicate atomic and ionic lines, respectively.
[b]Mean ± SD (n = 3).
[c]Values in the parentheses are the information values.

values obtained in **Table 3** agree with the acceptable or information values except for the value of Ba.

2. The second-generation human serum sample *(7)* is considered to be the most reliable blood-serum reference sample for trace elements. The analytical results for the second-generation human serum, which are analyzed by ICP-AES and ICP-MS in the authors' laboratory, are summarized in **Table 4**, together with the

Table 4
Analytical Results for Second-Generation Human Serum Obtained by ICP-AES and ICP-MS

Element	Wavelength[a] or *m/z*	Observed value[b]	Reference value
Na	589.0 nm I	3134 ± 28 µg/g	
P	213.6 nm I	119 ± 5	
K	766.4 nm I	148 ± 14	
Ca	396.8 nm II	103 ± 5	
Mg	279.5 nm II	16.8 ± 0.4	17.1 ± 0.8[d]
Fe	259.9 nm II	2.3 ± 0.1	2.35[c]
Cu	324.7 nm I	0.93 ± 0.03	1.01[c]
Zn	213.8 nm I	0.83 ± 0.04	0.87[c]
Rb	87	179 ± 1 ng/g	168[c]
Sr	88	22.5 ± 0.4	(22.7)[e]
Li	7	1.6 ± 0.2	(1.59)[e]
Pb	208	1.20 ± 0.10	4.21 ± 0.42[d]
Mo	98	0.95 ± 0.24	0.68[c]
Cs	133	0.95 ± 0.01	0.91[c]
Ba	137	0.48 ± 0.43	1.01 ± 0.28[d]
Bi	209	0.060 ± 0.004	0.063 ± 0.01[d]

[a]I and II indicate atomic and ionic lines, respectively.
[b]Mean ± SD (n = 4).
[c]Cited from **ref. *8***.
[d]Cited from **ref. *9***.
[e]Information values.

reference values ***(8,9)***. The sample pretreatment carried out in **Table 4** is the same as that in **Table 3**. These analytical values agree well with the reference values except for Pb, Mo, and Ba.

3. The analytical results for blood-serum reference sample (NIST SRM 909b) obtained by ICP-AES and ICP-MS are shown in **Table 5 *(10)***. The blood-serum sample used is usually for clinical analysis standardization of Na, K, Mg, and Ca, and so the certified values are provided only for them. The analytical values in **Table 5** are obtained in the authors' laboratory, where the sample pretreatment and the measurements of ICP-AES and ICP-MS are performed in a similar manner to those in **Table 3**. The data for Li and Y in **Table 5** are newly added to the elements examined in **Table 3**. It is noted that major and minor elements are almost at the same level (within a factor of 2) in both **Table 3** and **Table 5**, but some of the elements such as Li, Ba, and Mo are markedly large in **Table 5**, compared to the values in **Tables 3** and **4**. These elements in **Table 5** are perhaps contaminated in the sample preparation process.

Table 5
Analytical Results for Blood-Serum Reference Sample (NIST SRM 909b) Obtained by ICP-AES and ICP-MS-MS

Element	Wavelength[a] or *m/z*	Observed value[b]	Certified value
Na	589.0 nm I	2940 ± 40 μg/mL	2780 μg/mL
K	766.4 nm I	130 ± 3	134
P	213.6 nm II	90.4 ± 0.7	
Ca	317.9 nm II	81.3 ± 1.0	88.9
Mg	279.0 nm II	17.4 ± 0.3	18.6
Fe	259.9 nm II	1.27 ± 0.02	
Zn	324.7 nm I	1.11 ± 0.05	
Cu	213.8 nm I	0.96 ± 0.01	
Al	308.2 nm I	0.15 ± 0.10	
Li	7	3030 ± 60 ng/mL	
Ba	137	153 ± 4	
Sr	88	112 ± 1	
Mo	98	13.6 ± 0.2	
Rb	87	6.59 ± 0.47	
Sb	121	2.25 ± 0.08	
Zr	90	1.89 ± 0.09	
Ce	140	1.31 ± 0.06	
Pb	208	1.09 ± 0.24	
La	139	1.05 ± 0.03	
Cs	133	0.94 ± 0.04	
Y	89	0.730 ± 0.010	
Ag	107	0.64 ± 0.02	
Nd	146	0.566 ± 0.004	
Th	232	0.495 ± 0.017	
W	184	0.344 ± 0.033	
U	238	0.308 ± 0.005	
Bi	209	0.250 ± 0.007	
Pr	141	0.166 ± 0.001	

[a]I and II indicate atomic and ionic lines, respectively.
[b]Mean ± SD (n = 3).

The analytical results for all three kinds of human blood-serum reference samples, issued in Japan, Belgium, and USA, are determined using the same experimental procedures obtained in the same laboratory. The concentration levels of trace elements are not consistent with each other in the results for the three samples. These results suggest that the blood-serum samples are easily contaminated in the stages of sampling and the sample pretreatments. Thus, the determination of trace elements in blood samples should be very carefully performed in sampling stage and pretreatment procedures, as well as in the analysis by ICP-AES and ICP-MS.

3.2.3. Whole-Blood Analysis

Whole blood consists of erythrocytes, leukocytes, blood plasma, and other components in addition to blood serum. Thus, the whole blood samples are digested with acids before analysis by ICP-AES and ICP-MS.

The analytical results are summarized in **Table 6**, along with the certified values. As seen in **Table 6**, the analytical values agree well with the certified values, although the data in the present experiment are slightly lower than the certified values. In general, the multi-element analysis of the whole-blood sample is not as easy because the elemental concentrations of whole blood are generally lower than those of blood serum.

3.3.4. Urine Sample Analysis

The urine sample is another subject of clinical analysis. Analytical results for the human-urine reference sample (NIST SRM 2670; elevated) are summarized in **Table 7** ***(11)***, together with the certified and information values. The certified values for Na, Ca, Mg, Cu, Pb, Cd, and Cr are provided from NIST, and the present analytical values for these elements agree with the certified values. In addition, the analytical values for K, Mn, Ni, and Au are almost consistent with the information values. **Table 7** shows that the data for P, Si, Ba, Rb, Zn, Sr, Co, Bi, Mo, and Cs are newly obtained by ICP-MS in the present experiment.

In **Fig. 5A**, an example of the polyatomic-ion interference in the ICP-MS measurement is shown, where the mass spectrum in the *m/z* range between 50 and 55 is illustrated in terms of the solution containing 10 ng/mL of Cr and 5% glycine. The ICP-MS spectrum in **Fig. 5A** is measured by the Q-MS (resolution ca. 300) ***(11)***. It is seen that the signal of $^{52}Cr^+$ overlaps with the apparent signal (dashed line), which corresponds to $^{40}Ar^{12}C^+$. That is, in the determination of Cr in the biological sample by ICP-MS, $^{40}Ar^{12}C^+$ interferes with $^{52}Cr^+$, when low-resolution ICP-MS is used. In **Fig. 5B,** the mass spectrum in the *m/z* range between 51.91 and 51.99 measured by HR-ICP-MS (resolution ca. 5000) is shown ***(11)***, in which the SRM 2670 urine sample is diluted 5-fold. As is clearly seen in **Fig. 5B**, the signals of $^{52}Cr^+$ (amu 51.9405) and $^{40}Ar^{12}C^+$ (amu 51.9624) are baseline-separated. The analytical value for Cr in **Table 7** was obtained under the mass resolution of 5000.

3.3.5. Bovine Liver Analysis

As an example of organ analysis, the determination of major-to-ultratrace elements in bovine-liver sample by ICP-AES and ICP-MS is described here.

Table 6
Results for Whole Blood Reference Sample (IAEA A-13) Obtained by ICP-AES and ICP-MS

Element	Wavelength[a] or *m/z*	Observed value[b], µg/g dry	Certified value[c], µg/g dry
Na	589.0 nm I	12600 ± 200	12600
S	180.7 nm I	6230 ± 50	6500
K	766.4 nm I	2310 ± 40	2500
Fe	259.9 nm II	2360 ± 30	2400
P	213.6 nm I	820 ± 10	(940)
Ca	317.9 nm II	269 ± 3	286
Mg	279.0 nm II	87 ± 2	(99)
Zn	213.8 nm I	10.6 ± 0.3	13
Cu	324.7 nm I	3.53 ± 0.04	4.3
Rb	85	2.31 ± 0.02	
Pb	208	0.177 ± 0.023	(0.18)
Sr	88	0.245 ± 0.004	
Mo	98	0.0151 ± 0.0032	
Cs	133	0.0040 ± 0.0003	

[a]I and II indicate atomic and ionic lines, respectively.
[b]Mean ± SD (n = 3).
[c]Values in the parentheses are information values.

The digested solution of the bovine-liver reference sample is also analyzed by ICP-AES and ICP-MS. The analytical values are summarized in **Table 8** ***(10)***, together with the certified and reference values ***(12,13)***. In general, the analytical values are consistent with the certified values. However, the elements for which the reference values are available provided somewhat inconsistent values between the observed and reference values. These results indicate that further investigation is required for elements such as Al, Ba, Ni, W, Cs, Th, Y, Ce, Pr, and Nd. Most of the reference values are obtained by ICP-AES, whereas the observed values in **Table 8** are done by ICP-MS. Thus, the present observed values may be more reliable than the reference values.

3.3.6. Human-Hair Analysis

It is well known that various heavy metals accumulate in human hair, and are sometimes used for diagnosis of health conditions. As an example of such hair analysis, the results for multi-element analysis of the human-hair reference sample (NIES No. 13) by ICP-AES and ICP-MS are summarized in **Table 9**, together with the certified, information, and reference values ***(14)***. The

Table 7
Analytical Results for Human Urine Reference Sample (NIST SRM 2670 Elevated) Obtained by ICP-AES and ICP-MS

Element	Wavelength[a] or *m/z*	Observed value[b]	Certified value[c]
Na	589.0 nm I	2.61 ± 0.05 mg/mL	2.62 ± 0.14 mg/mL
K	766.4 nm I	1.38 ± 0.05	(1.5)
P	213.6 nm I	0.614 ± 0.002	
Ca	393.4 nm II	0.107 ± 0.001	0.105 ± 0.005
Mg	279.5 nm II	0.0601 ± 0.0001	0.063 ± 0.003
Si	251.6 nm I	0.0102 ± 0.0001	
Ba	138	1.08 ± 0.06 μg/mL	
Rb	85	1.02 ± 0.01	
Zn	66	0.925 ± 0.016	
Cu	63	0.387 ± 0.004	0.37 ± 0.03 μg/mL
Mn	55	0.350 ± 0.004	(0.33)
Ni	60	0.253 ± 0.002	(0.30)
Sr	88	0.193 ± 0.001	
Au	197	0.174 ± 0.001	(0.24)
Pb	208	0.118 ± 0.001	0.109 ± 0.004
Co	59	0.113 ± 0.002	
Cd	114	0.0909 ± 0.0006	0.088 ± 0.003
Cr	52	0.0826 ± 0.0015	0.085 ± 0.006
Bi	209	0.0563 ± 0.001	
Mo	98	0.0402 ± 0.0039	
Cs	133	0.0215 ± 0.0001	

[a]I and II indicate atomic and ionic lines, respectively.
[b]Mean ± SD (n = 3).
[c]Values in the parentheses are information values.

reference values in **Table 9** are obtained by neutron activation analysis. A large number of the elements are contained in human-hair reference samples, as seen in **Table 9**. However, the certified values only for nine elements (Zn, K, Cu, Sr, Pb, Ni, Cd, and Sb) have been provided from NIES, and the values for all other elements are information or reference values. These results indicate that the analysis of the human-hair samples are not so easy, because of quite difficult sample pretreatment (digestion) owing to the complicated composition of human hair. In the present experiment, in **Table 9**, 33 elements, including all rare earth elements, are determined by ICP-AES and ICP-MS as the observed values. In general, the present observed values are quite consistent with certified, information, and reference values except for K, Sr, Ni, Sb, U, and La. The analytical values for Bi, U, Pr, Nd, Gd, Dy, Ho, Er, Tm, Yb, and Lu, which are

Table 8
Results for Bovine-Liver Reference Sample (NIST SRM 1577) Obtained by ICP-AES and ICP-MS

Element	Wavelength[a] or *m/z*	Observed value[b]	Certified value
P	213.6 nm II	11600 ± 200 μg/g	10500[c] μg/g
K	766.4 nm II	9420 ± 440	9700
Na	589.0 nm I	2500 ± 100	2430
Mg	279.0 nm II	571 ± 9	605
Fe	259.9 nm II	261 ± 1	270
Cu	324.7 nm I	183 ± 4	193
Zn	213.8 nm I	133 ± 1	130
Sr	407.7 nm II	128 ± 6	140
Ca	396.8 nm II	120 ± 2	123
Mn	257.6 nm II	9.67 ± 0.05	10.3
Al	308.2 nm I	2.41 ± 1.95	10–50[c]
Ba	493.4 nm II	0.057 ± 0.009	0.1–123[c]
Rb	87	16700 ± 200 ng/g	18300 ng/g
Pb	208	324 ± 7	340
Ag	107	62.9 ± 1.6	60
Ni	60	48.8 ± 11.7	400 ± 400[c]
W	184	6.70 ± 0.36	14
Cs	133	11.1 ± 0.3	17 ± 7[c]
U	238	0.67 ± 0.07	0.8
Th	232	0.50 ± 0.06	<1000[c]
Y	89	1.75 ± 0.09	<1000[c]
La	139	16.4 ± 0.2	28[c], 17[d]
Ce	140	23.0 ± 0.2	46[c], 48[d]
Pr	141	3.00 ± 0.07	4.6[c], 4.9[d]
Nd	146	9.85 ± 0.78	17.0[c], 15.9[d]

[a]I and II indicate atomic and ionic lines, respectively.
[b]Mean ± SD (n = 3).
[c]Cited from **ref. *12***.
[d]Cited from **ref. *13***.

obtained by ICP-MS, were for the first time obtained in the present experiment. These results indicate that the analytical sensitivity of ICP-MS is excellent, compared to other methods.

4. Notes

1. The technology for sample pretreatment must be improved and developed to obtain good precision control in trace analysis of various biological samples, even if state-of-the-art analytical instruments are available.

Table 9
Analytical Results for Human-Hair Reference Sample (NIES No. 13) Obtained by ICP-AES and ICP-MS

Element	Wavelength[a] or *m/z*	Observed value[b]	Certified value[c]
Ca	317.9 nm II	812 ± 6 µg/g	(820) µg/g
P	213.6 nm II	179 ± 2	
Zn	213.8 nm I	152 ± 1	172 ± 11
Fe	259.9 nm II	143 ± 4	(140)
Mg	279.0 nm II	143 ± 1	(160)
Al	308.2 nm I	119 ± 3	(120)
Na	589.0 nm I	67.4 ± 0.8	(61)
K	766.4 nm I	30.9 ± 2.5	70 ± 8[d]
Cu	324.7 nm I	14.8 ± 0.2	15.3 ± 1.3
Sr	407.7 nm II	2.49 ± 0.02	8.7 ± 0.8[d]
Ba	455.4 nm II	1.83 ± 0.07	(2.0)
Pb	208	4400 ± 80 ng/g	4600 ± 400 ng/g
Ni	60	1140 ± 80	1950 ± 13[d]
Cd	114	213 ± 4	230 ± 30
Sb	121	94.8 ± 2.6	42 ± 8
Mo	98	57.3 ± 0.3	<36[d]
Bi	209	14.4 ± 0.4	
Cs	133	7.51 ± 0.53	8.9 ± 0.5[d]
U	238	5.10 ± 0.36	<4.1[d]
La	139	81.6 ± 4.1	128 ± 30[d]
Ce	140	139 ± 9	184 ± 32[d]
Pr	141	17.2 ± 1.0	<130[d]
Nd	143	49.1 ± 2.5	<64[d]
Sm	147	6.2 ± 0.5	7.0 ± 1.1[d]
Eu	151	1.49 ± 0.07	2.0 ± 0.2[d]
Gd	157	8.5 ± 0.4	<120[d]
Tb	159	1.06 ± 0.04	1.32 ± 0.26[d]
Dy	163	5.73 ± 0.27	<6.9[d]
Ho	165	1.16 ± 0.05	<97[d]
Er	166	3.03 ± 0.17	
Tm	169	0.48 ± 0.03	<4.2[d]
Yb	174	2.99 ± 0.22	<3.5[d]
Lu	175	0.439 ± 0.028	<0.82[d]

[a]I and II indicate atomic and ionic lines, respectively.
[b]Mean ± SD (n = 3).
[c]Values in the parentheses are the information values.
[d]Cited from **ref. *14***.

2. To avoid the risk of explosion, the perchloric solution should not be taken to complete dryness.
3. The reference samples (blood serum, whole blood, urine, bovine liver, and human hair) are used to provide a guideline for establishing values for individual laboratories.

Acknowledgments

E. Fujimori expresses his thanks to the Japan Society for Promotion of Science for a Research Fellowship for Young Scientists (1995–1996).

References

1. Alfassi, Z. B. (ed.) (1994) *Determination of Trace Elements.* VCH Publishers, NY.
2. Vandecastele, C. and Block, C. B. (1993) *Modern Methods for Trace Element Determination.* John Wiley & Sons, Chichester, UK.
3. Montaser, A. and Golighthly, D. W. (eds.) (1992) *Inductively Coupled Plasmas for Analytical Atomic Spectrometry.* VCH Publishers, NY.
4. Haswell, S. J. (ed.) (1991) *Atomic Absorption Spectrometry: Theory, Design and Applications.* Elsevier, Amsterdam.
5. Iyenger, G. V. (1989) *Elemental Analysis of Biological Systems.* (vol. 1), CRC Press, Boca Raton, FL.
6. Sakurai, H. and Tanaka, H. (eds.) (1994) *Bio-trace Elements* (in Japanese). Hirokawa Shoten, Tokyo.
7. Versieck, J., Hoste, J., Vanballenberghe, L., De Kesel, A., and Van Renterghem, D. (1987) Collection and preparation of a second-generation biological reference material for trace element analysis. *J. Anal. Nucl. Chem.* **113,** 299–304.
8. Versieck, J. et al. (1988) Certification of a second-generation biological reference material (freeze-dried human serum) for trace elements determinations. *Anal. Chim. Acta* **204,** 63–75.
9. Vandecasteele, C. et al. (1993) Inductively coupled plasma mass spectrometry of biological samples. (Invited lecture.) *J. Anal. Atom. Spectrom.* **8,** 781–786.
10. Inagaki, K. (1997) Master Thesis, Department of Applied Chemistry, Nagoya University.
11. Fujimori, E., Sawatari, H., Chiba, K., and Haraguchi, H. (1996) Determination of minor and trace elements in urine reference sample by a combined system of inductively coupled plasma mass, spectrometry and inductively coupled plasma atomic emission spectrometry. *Anal. Sci.* **12,** 465–470.
12. Gladney, E. S. (1980) Elemental concentrations in NBS biological and environmental standard reference materials. *Anal. Chim. Acta* **118,** 385–396.
13. Meloni, S., Genova, N., and Oddone, M. (1987) Rare-earth elements in the NBS standard reference materials spinach, orchard leaves, pine needles and bovine liver. *Sci. Total Environ.* **64,** 13–20.
14. Suzuki, S. et al. (1994) Multielement determination of human hair reference material by instrumental NAA. *Bunseki Kagaku* **43,** 845–849.

38

Energy Dispersive X-Ray Microanalysis

Don A. Samuelson

1. Introduction

In the past, the principal techniques of choice for analyzing tissue for various inorganic atomic species have been colometric and via atomic absorption spectrophotometry *(1)*. Although both methods can be accurate and precise when performed in the same laboratory, there are disadvantages, which include interlaboratory variability in results owing to sample preparation and instrumentation, and the need for relatively large amounts of tissue to be analyzed. As a result, one is unaware of exactly what structure or cell type in the sample is the source of ions. An alternative technique is the use of energy dispersive X-ray microanalysis (EDX), which allows multi-element analysis at the subcellular level.

The principal advantage of EDX, which is also referred to as energy dispersive spectroscopy (EDS), is that each structure is able to be examined morphologically in an electron microscope and undergo rapid and highly sensitive simultaneous multi-element analysis (with a Z number of 11 or above). Consequently, specimens as small as a few square micrometers can be prepared and studied. Moreover, serial sections can be evaluated by other methods, e.g., immunocytochemistry or histochemistry. It has been estimated that the minimum concentration of an ion that can be detected is in the range of 100–500 ppm, and because the total mass of a thin section (200–400 nm thickness) is only about 10^{-15} gm, it is possible that an elemental mass of 10^{-19} could be detected *(2)*.

EDX is possible to perform in all electron microscopes (scanning electron microscope [SEM], transmission electron microscope [TEM], scanning transmission electron microscope [STEM]) in that as an electron beam passes into a sample, X-rays are produced. With the assistance of appropriate detection equipment, EDX works by detecting characteristic X-rays emitted by atoms

From: *Methods in Molecular Biology, vol. 108: Free Radical and Antioxidant Protocols*
Edited by: D. Armstrong © Humana Press Inc., Totowa, NJ

that are bombarded with an electron beam *(3)*. When an electron is ejected from an inner shell (K or L), it is replaced by one from an outer shell with the emission of an X-ray that has a characteristic amount of energy. The resulting X-rays are measured as energy spectra (counts vs KeV) and peaks are displayed and categorized via a multichannel analyzer. Although the technique has been used extensively in materials science research, EDX allows elemental mapping on either bulk biological specimens or on thin-sectioned tissue on a SEM or STEM, respectively.

This chapter will describe several EDX procedures for analyzing ions important in studies involving free radicals and antioxidants. Methods include examining specimens of different sizes and that have been prepared in different ways.

2. Materials

2.1. Electron Microscope

Hitachi H7000 transmission electron microscope with Kevex Super 8000 X-Ray Microanalysis unit. Hitachi S450 scanning electron microscope with Kevex 7000 X-Ray Spectrometer.

2.2. EM Preparative Equipment

1. Denton CPD-1 critical point drying apparatus.
2. Lyophilizer/Denton Vacuum Evaporator.
3. Reichert Ultracuts ultramicrotome.
4. Lipshaw Crytostat cryotome.
5. Life Cell ultrarapid-freezing device.
6. LKB glass knifemaker.
7. Dewar flasks with aluminum stage for cryofracture and copper cup (1.5–2.0 cm in diameter for liquid propane cryofixation.

2.3. Plastic Embedding

1. Araldite 501 epoxy resin (JJ Ladd, Inc).
2. Low-viscosity plastic resin/LX-112 (Polysciences, Inc).
3. LR White methacrylate resin (Ted Pella, Inc).

2.4. Fixatives

1. Formaldehyde/10% buffered formalin.
2. Glutaraldehyde in phosphate buffer. The strength of the fixative can vary according to the level of tonicity that is desired.
3. Acrolein.

2.5. Freezing Compounds

Liquid nitrogen. Propane (to be chilled by liquid N_2).

3. Methods

3.1. Specimen Preparation

When examining ions, especially metal ions as they relate to free radicals and antioxidants, specimens are prepared several different ways, depending on whether all ions are to be observed or those that are bounded or a combination of all vs bounded. The size of the specimen is also a determining factor of how the tissue is to be prepared.

3.1.1. Preparation for the Analysis of All Ions

3.1.1.1. Large Specimens

Specimens that are greater than 20 µm in thickness and 1 mm in length or longer (up to 1 cm) can be processed in the following manner. Freshly collected specimens are rapidly frozen by immersing them in liquid propane for at least 30 s and then placing them in a prechilled cryotome where they are then mounted onto a specimen holder and sectioned at a thickness of 6–10 µm (*see* **Note 1**). Sections are placed on carbon planchettes attached to aluminum stubs and viewed by SEM with EDX capability. The planchettes with sections are placed in a lyophilizer and freeze-dried overnight (**Fig. 1**). Serial sections are placed on albumin-coated glass slides and stained for histological evaluation. By making a small notch at one corner of the specimen before sectioning, it is possible to go to various regions in the specimen when viewing by SEM in the absence of morphological features.

3.1.1.2. Small Specimens

Specimens that are 20 µm in thickness or thinner are processed by a procedure that is sometimes referred to as "flash freezing." This procedure is performed so that the leaching of min quantities of free elements out of cells and tissues is prevented or greatly reduced. Using a copper anvil, a specimen is placed on a molybdenum strip and immersed in liquid nitrogen for a minimum of one min and then stored in liquid nitrogen until lyophilization. Flash-freezing stabilizes biological structures up to three orders of magnitude faster than chemical fixation *(4)* and thus minimizes ice-crystal formation and eliminates changes in the cellular fine structure caused by chemical fixatives. Flash-freezing is performed by an ultrarapid-freezing device, such as made by Life Cell, Inc., or by a simple copper-plated tongs. After the specimens are dried, they are placed in a paper pouch, which is similar to a tea bag, and suspended in a capped scintillation vial that contains a small amount of crystallized acrolein. Acrolein vapors tend to stabilize the membranes of cells and organelles. This further stabilization is important if the specimen is to be additionally analyzed immunocytochemically. Specimens are immersed in plastic under vacuum. The

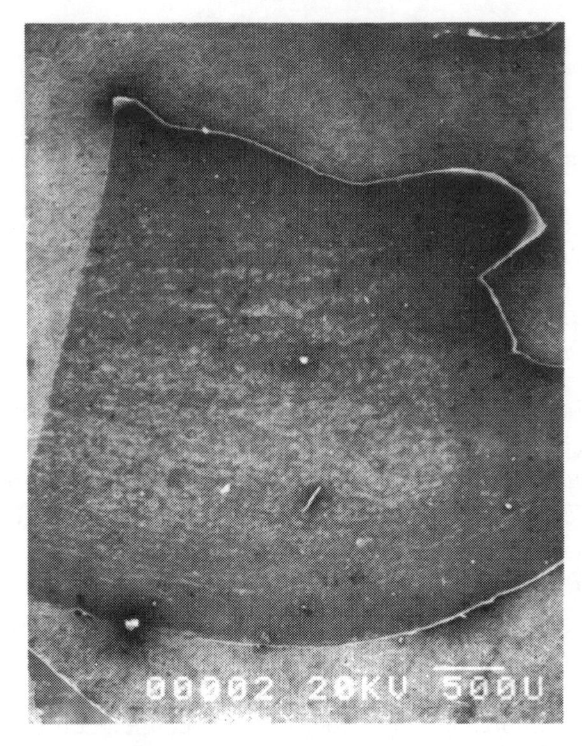

Fig. 1. Frozen section of a specimen (equine cornea) that has been placed on a carbon planchette, lyophilized, and coated with carbon.

plastic of choice is a low-viscosity resin such as LX-112 and Araldite (*see* **Note 2**). The epoxy resin is polymerized over a 48-h period in a vacuum oven at 50°C. To enhance penetrance of the plastic, it is recommended that the temperature is turned up gradually from room temperature over this period of time. If soluble electrolytes, such as sodium, potassium, and chlorine, need to be analyzed, cryosectioning may be required, which necessitates the use of a modified ultramicrotome for such purposes ***(5,6)***. The cryosections are lyophilyzed or transferred and examined in the frozen-hydrated state with an accessory sample holder for the STEM.

Plastic-embedded specimens are sectioned 1 μm in thickness and stained (1% toluidine blue) in order to evaluate morphologically the appearance of the specimen and subsequently select one or more areas to be analyzed by EDX. Specimens of adequate size (typically less than 0.5 mm in height) are sectioned at a chosen "semi-thin" thickness, between 200 and 400 nm, and placed on either nylon grids or formvar-coated copper-slot grids. Nylon grids have the advantage of not contributing to the energy spectrum except for an artifactual titantium peak. Specimens on copper slot coated-grids are unimpeded by supporting mesh and consequently allow the observer to examine the entire area, which is best for X-ray mapping (*see* **Note 3**).

3.1.2. Preparation for Analysis of Bounded Ions

If the ions that are to be examined by EDX are firmly bound intracellularly or extracellularly, the specimens are preserved chemically rather than frozen. Chemical fixation offers the advantages of being able to collect tissues easily and observe their morphology under optimal conditions. If one is unsure of the bounded stability of the ions to be examined, it would be wise to prepare the samples by cryofixation. If one wishes to examine both total and bounded elements, then adjacent specimens can be preserved by cryo- and chemical fixation (*see* **Note 4**).

3.1.2.1. Large Specimens

Specimens of the size described in **Subheading 3.1.1.1.** are preserved in any number of fixatives that are used for light microscopic observations, such as 10% buffered formalin. However, some fixatives contain metals such as mercuric chloride and should not be used. Specimens are either frozen and sectioned with a cryotome (as in **Subheading 3.1.1.1.**), cryofractured *(7)*, or dehydrated, embedded in paraffin, and sectioned. Cryofracture is a process where a previously fixed specimen is dehydrated to 100% ethanol and immersed in liquid N_2. After freezing the specimen for at least one min (no bubbles should come from the specimen), a prechilled, new single-edged razor blade is tapped through the center of the specimen along the preferred plane of orientation *(7)*. The cyrofracture technique offers the advantage of being able to recognize easily, morphological detail by SEM (**Fig. 2**). As a result, individual cells and their organelles can be analyzed by EDX *(7)*. As before, the sections and cryofractured specimens are mounted on carbon planchettes, which, in turn, are cemented to aluminum stubs.

3.1.2.2. Small Specimens

Specimens of the size described in **Subheading 3.1.1.2.** are preserved by the traditional methods for TEM, such as glutaraldehyde or a mixture of paraformaldehyde and glutaraldehyde *(8,9)*. Bear in mind that buffers used with these fixatives may contain metals, e.g., arsenic in sodium cacodylate, which may or may not have an effect on the analysis. Specimens prepared in the manner that is traditional for TEM have the advantage of being gradually dehydrated and embedded in plastic, which results in the thorough penetration of the plastic medium.

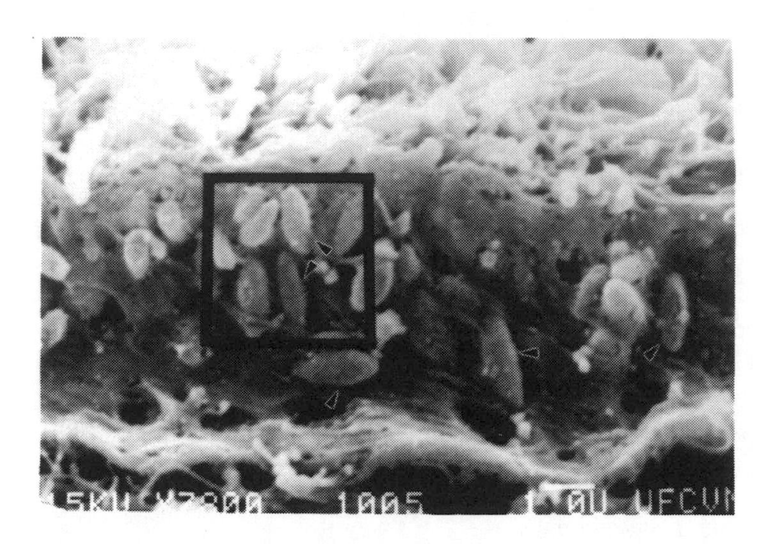

Fig. 2. Cryo-fractured sample of posterior sheep eye reveals the presence of numerous melanosomes (arrows) within the retinal pigment epithelium. The square indicates the size of the area being analyzed with the SEM.

3.2. EDX Operation

3.2.1. SEM

Sections and cryofractured specimens on carbon planchettes that are attached to aluminum stubs need to be coated with a conducting element in order to eliminate charging. It is best to coat each specimen with carbon, which allows excess electrical current (charging) to be conducted to ground potential ***(10)***. However, if a typical coating element such as gold is used and it does not interfere with the measurement of another (for example, silver interferes with chloride analysis), then its use is acceptable. When performing EDX on the scanning electron microscope, it is important to maintain the same working parameters as much as possible for each specimen. The same working distance should be maintained as well as accelerating voltage (e.g., 1–30 kV, typically 20 kV), take-off angle (45°), and analysis time or total counts. The accelerating voltage influences both the efficiency with which X-rays are excited and how far the electrons penetrate the sample and analytical spatial resolution. Because most quantitative-analysis programs assume a single source of X-rays beneath the sample surface, the take-off angle markedly influences the absorption path. The path length is minimized by using high take-off angles. When performing EDX in a study that requires more than 1 d to perform analyses, one sample is re-analyzed as a point of reference (*see* **Note 5**).

3.2.2. STEM

EDX of small specimens is best performed by STEM. Semi-thin sections (200–400 nm in thickness) that are on nylon or single-slot formvar-coated copper grids are placed in a carbon grid holder before being inserted into the electron microscope. With the STEM system turned on and the objective aperture removed from the microscope's column, the specimen is visible on the cathode ray tube (CRT) monitor. It is important to have first viewed the tissue by light microscopy, using a stained, 1 μm-thick section so that the investigator is familiar with all structures that will be encountered by STEM. The semi-thin sections that are viewed by STEM are unstained and consequently often difficult to interpret. As specific cellular and noncellular structures become identified, one can proceed to perform EDX. The limit of size resolution is a concern when performing EDX on the STEM. This limit determines how small a biological structure can be measured without contamination from neighboring area. The contamination is caused by the spreading of the electron beam when it hits the specimen and the scattering of X-rays, both from the beam and those ejected by the beam, within the specimen. As sections become thicker, their mass increases and resultant background radiation (Bremsstrahlung) becomes higher. By reducing section thickness and operating the scope at higher accelerating voltages (75 or 100 kV), the spread of the electron beam is reduced and the electron scattering is lowered. However, as the sections become thinner (less than 200 nm in thickness), they become less stable under the electron beam, causing the sections to drift, melt, or split. Thicker sections (200–400 nm) are more stable under the beam and improve contrast in the section and as a result justify a loss in spatial resolution (**Fig. 3**) (*see* **Notes 7** and **8**). Recent advances in the physics of EDX systems, such as bringing the detector closer to the specimen, have improved resolution. Structures as small as 30 nm in diameter can be analyzed when the STEM raster is restricted to a "spot" mode. A resolution of at least 30 nm is more than acceptable in most studies that wish to analyze the multi-elemental content of subcellular structures.

As in SEM, when performing EDX on more than one specimen it is important to keep the working parameters the same including accelerating voltage, dead time, input range, lifetime or total counts, and spot size. Dead time is the time taken by the EDX system to digitize the input pulses and is the time when the analyzer will not record detected X-rays. Substantial differences in dead time could indicate differences in section thickness, specimen composition, and so forth (*see* **Note 9**). Input range is the rate of X-ray acquisition, i.e., counts per second, which should be kept the same throughout a study, such as at 2500 or 3000 counts/s. Lifetime is the time that the analyzer is recording detected X-rays. Typically, a specific lifetime is preset, such as 100 s, for each

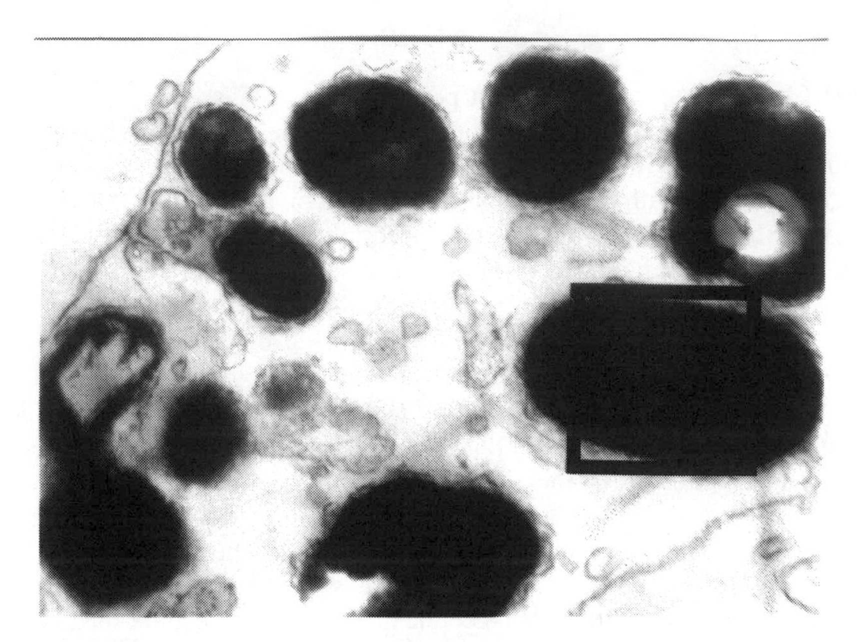

Fig. 3. Semi-thin section of a cryo-preserved pigmented cell that is examined by STEM at a magnification of ×20,000. The square indicates the size of the area being analyzed with the STEM.

analysis. Instead of keeping the lifetime the same, the total count is kept identical for all analyses in a study.

3.3. Data Analyses

3.3.1. Qualitative

Much of the analyses performed on biological specimens are of a qualitative nature. This is especially true when specimens, large or small, are being examined in order to assess the presence or absence of one or more specific elements in a particular cell type, tissue, or region of an organ. The major concern of analyses of this type is to reduce the potential for detecting artifacts that could be introduced (solvents, embedding media, grids, supporting films, etc.) (*see* **Note 4**). Preserving the specimen in a way that immobilizes elements as rapidly as possible, i.e., cryofixation, optimizes qualitative analyses. The conventional chemical preparation procedures as previously described are used best in instances where elements are insoluble or firmly bound as in bone, teeth, abnormal mineral deposits in various tissues (e.g., kidney stones), and foreign objects. When the acquisition of a spectrum is begun, each element of specimen is identified by the location of its energy peak(s) along the horizontal axis (**Fig. 4**). The energy peaks are usually displayed over an energy range of 0–10.23 keV.

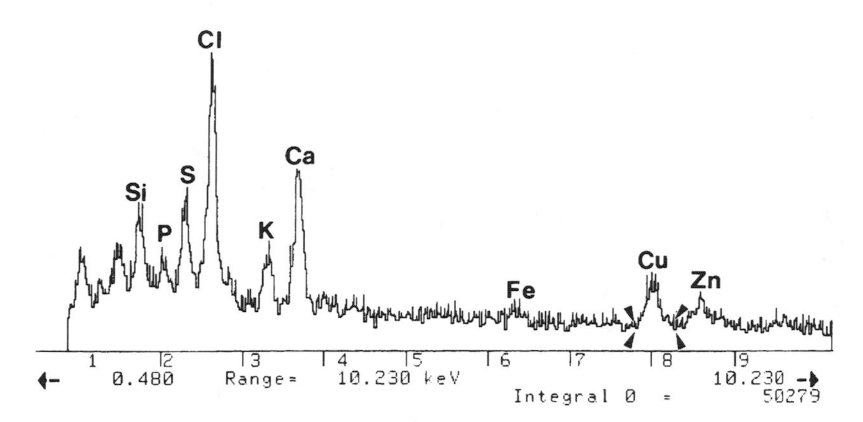

Fig. 4. Graph of elemental spectral microanalysis of a cryo-preserved specimen exhibits the presence of numerous elements, including those that are easily soluble such as potassium (K). The presence of silicon (Si), which is owing to the X-ray detector having a silicon-lithium crystal, demonstrates the need to analyze nonspecimen areas in order to subtract extraneous data. Arrows point to background measurements for copper (Cu), for example, after extraneous contributions from the support film, system background, etc., have been made.

3.3.2. Quantitative

In instances where it is desirable to determine the amount of one or more elements within a specimen, quantitative analysis is necessary. Bulk specimens that have an uneven surface and variable thicknesses are not amenable to measuring the amounts of any specific element. The topography of a sample will produce a variety of artifacts that arise from absorption and blockage of X-rays. Specimens to be quantitatively analyzed by SEM should be flat and even. The computers and software associated with EDX systems enable the user to determine peak-intensity measurements by removing background counts via background filtering or background modeling. If the area under the peaks is to be measured, it may be necessary to separate overlapping peaks of two or more elements by a process called deconvolution. In a popular method called super multiple linear least squares, standards for the elements that are being analyzed (and deconvoluted) are required ***(11)***.

The semi-thin sections for EDX on the STEM are more suitable for measuring elements in a sample. In these specimens, the electron beam is able to pass through with little loss of energy and the X-rays are able to escape with little absorption. A common method for quantitating semi-thin specimens employs the measurement of peak-to-background ratios and is sometimes called the "Hall method" *(11)*. In theory, the amount of X-rays from an element being measured is proportional to the atoms of that element in the region that is being analyzed. As a result, the quantity of X-rays is proportional to the thickness of the section. Also, the quantity of Bremsstrahlung or X-ray background is proportional to the total number of all atoms in the region being analyzed. The peak-to-background ratio for an element that is being analyzed is a measure of concentration because it directly compares the number of atoms being counted to the total atoms present. Because peak-to-background measurements are relative measurements, variations in section thickness and beam current are compensated for. Background measurements should be made in the part of the spectrum that lacks characteristic peaks and as close to the selected elemental peak as possible (*see* **Fig. 4**). Analyses should also be done away from the specimen where only the influences of the embedding media, support film, grid, and system background are assessed and subtracted from the measured background. Standards to be used for quantitation can consist of salts that are dissolved in the embedding medium. For frozen sections the embedding medium could be gelatin, whereas for plastic sections the embedding medium is the same resin used for specimen preparation.

4. Notes

1. For the measurement of total elements, tissues are preserved by cryopreservation immediately in order to minimize the effects of autolysis and the potential leaching of soluble ions.
2. If further immunocytochemical analysis is to be performed, a methacrylate-based plastic such as LR White is recommended.
3. If copper is being mapped, it will be necessary to place sections on beryllium grids, which are very expensive and should be used only for this purpose.
4. When a chemical fixative is selected to preserve a specimen, the potential for introducing additional amounts of elements exists and should be considered. In order to assess the effect of different fixations on elemental binding within a sample, spectral analyses is made from two specimens that are collected from the same tissue but prepared differently, such as one being chemically fixed, with the other being cryopreserved. Specimens should then be prepared identically (embedded in the same plastic, sectioned at the same thickness, etc.). In analyses of that type, the peaks for most of the observed elements of the cryofixed specimen can be 20–200% greater than those from the chemically fixed specimen (**Fig. 4**). Moreover, easily solubized elements such as potassium are usually absent in chemically fixed samples.

5. Measurements made from the same specimen on the same microscope on two different d will vary as much as 20% and should be taken into account. The routine use of standards (specimens with known concentrations of specific metals) is recommended in order to keep track of instrument variation and to calibrate if necessary, so that accurate elemental identification is always maintained.
6. In cryopreserved specimens, pigment compounds, such as melanin and lipofuscin, have the capacity to act as ion-exchange polymers, even when embedded in plastic ***(12)***. Consequently, plastic sections that are used for EDX should be collected on highly purified water (high-pressure liquid chromatography [HPLC] grade) as quickly as possible (within 30 s), so that native elements can be accurately measured ***(13)***.
7. When performing EDX by STEM, the possibility for the section to drift constantly or split under the electron beam can be minimized by:
   a. Thorough embedding;
   b. Making the sections relatively thick (400–500 nm);
   c. Using formvar or some other support film, which, in turn, can be coated with carbon; and
   d. Keeping the cold trap on the STEM filled with liquid N_2.
8. The solutions to make support films, such as formvar, should be made fresh. Old solutions become impure and will have holes formed in the film, which makes viewing and analysis on the STEM considerably more difficult.
9. During quantitative analyses on multiple specimens over a period of w, it is important that any substantial fluctuation (greater than 20%) in the spot size, input rate, and dead time measurements are recorded. It is equally important to select one specimen to be analyzed each time EDX is being performed and record four to five spectra of the same structure. The average of the peak-to-background measurements of selected elements of this specimen will allow one to compare present data to that previously gathered.

References

1. Potts, A.M. and Au, P.C. (1976) The affinity of melanin for inorganic ions. *Exp. Eye Res.* **22,** 487–491.
2. Russ, J. C. (1973) Microanalysis of thin sections in the TEM and STEM using energy-dispersive X-ray analysis, in *Electron Microscopy and Cytochemistry* Wisse, E., Daemes, W. T., Molenaar, I., and van Duijn, P., eds.), North-Holland Publishing Co., Amsterdam, pp. 223–228.
3. Chandler, J. A. (1973) The use of wavelength dispersive X-ray microanalysis in cytochemistry, in *Electron Microscopy and Cytochemistry* (Wisse, E., Daemes, W. T., Molenaar, I., and van Duijn, P., eds.), North-Holland Publishing Co., Amsterdam, pp. 203–222.
4. Gilkey, J. C. and Staehelin, L. A. (1986) Advances in ultrarapid freezing for the preservation of cellular ultrastructure. *J. Elec. Microsc. Tech.* **3,** 177–210.
5. Hutchinson, T. E. (1977) Energy dispersive X-ray microanalysis, in *Analytical and Quantitative Methods in Microscopy* (Meek, G. A. and Elder, H. Y., eds.), Cambridge University Press, Cambridge, UK, pp. 213–226.

6. Appleton, T. C. (1977) The use of ultrathin frozen sections for X-ray microanalysis of diffusible elements, in *Analytical and Quantitative Methods in Microscopy* (Meek, G. A. and Elder, H. Y., eds.), Cambridge University Press, Cambridge, UK, pp. 247–268.
7. Samuelson, D. A., Armstrong, D., and Jolly, R. (1990) X-ray microprobe analysis of the retina and RPE in sheep with ovine ceroid-lipofuscinosis. *Neurobiol. Aging* **11,** 663–667.
8. Hyatt, M. A. (1970) *Principles and Techniques of Electron Microscopy. (vol. 1), Biological Applications.* Van Nostran Reinhold Co., NY, pp. 13–104.
9. Samuelson, D. A., Smith, P., Ulshafer, R. L., Hendricks, D. G., Whitley, R. D., Hendricks, H., and Leone, N. C. (1993) X-ray microanalysis of ocular melanin in pigs maintained on normal and low zinc diets. *Exp. Eye Res.* **56,** 63–70.
10. Dykstra, M. J. (1992) *Biological Electron Microscopy.* Plenum Press, NY.
11. Flegler, S. L., Heckman, Jr., J. W., and Klomparens K. L. (1993) *Scanning and Transmission Electron Microscopy: An Introduction.* W. H. Freeman and Co., NY.
12. Samuelson, D. A., Smith, P. J., Lewis, P., Chisholm, M., and Orr, E. (1995) The effects of varying Zn nutrition, light damage and aging on cationic binding in RPE melanin. *Invest. Ophthalmol. Vis. Sci.* **36(Suppl.),** 5769.
13. Samuelson, D. A., Konkal, S., Lewis, P., and Chisholm, M. (1996) Changes of metal binding in human RPE melanin and lipofuscin with aging and drusen. *Invest. Ophthalmol. Vis. Sci.* **37(Suppl.),** 535.

39

Pro-Oxidant and Antioxidant Effects of Estrogens

Joachim G. Liehr and Deodutta Roy

1. Introduction

1.1. Natural and Synthetic Estrogens

Many compounds of great structural diversity possess varying degrees of estrogenic activity (**Fig. 1**). Estrogenic substances may contain a steroidal ring system as do, for instance, the ovarian hormones estradiol[3] (E2), estrone (E1), or the synthetic estrogens 17α-ethinylestradiol or mestranol. Nonsteroidal phenols with estrogenic activity include the potent synthetic estrogen diethylstilbestrol (DES) and environmental substances produced by plants (equol, coumestrol), or microbes (enterolactone). Some industrial chemicals have estrogenic activity, such as bisphenol A and nonylphenol. In addition to hormonal activity, both natural and synthetic estrogenic compounds containing a phenolic structure possess pro-oxidant and/or antioxidant activities.

1.2. Antioxidant Activity

The antioxidant activity of estrogens and their metabolites has been demonstrated in vivo and in vitro by determining the inhibition of cholesterol oxidation or peroxidation of polyunsaturated lipids (diene conjugation or malonaldehyde formation) either in low density lipoprotein (LDL), in microsomes, or in other biological preparations *(1–3)*. This antioxidant effect is associated with the phenol structure of E2 and its metabolites, likely resulting in the intermediate formation of phenoxy radicals, which may be reduced again to phenol by other cellular antioxidants. Hindered phenols such as t-butyl hydroxyanisol or t-butyl hydroxytoluene, are commonly used anti-oxidants and react, for instance, with hydroxy radicals by forming the more stable phenoxy radicals. Phenolic estrogens, their catechol or methyl ether metabolites, may also be regarded as substituted, hindered phenols with strong anti-oxidant activity.

From: *Methods in Molecular Biology, vol. 108: Free Radical and Antioxidant Protocols*
Edited by: D. Armstrong © Humana Press Inc., Totowa, NJ

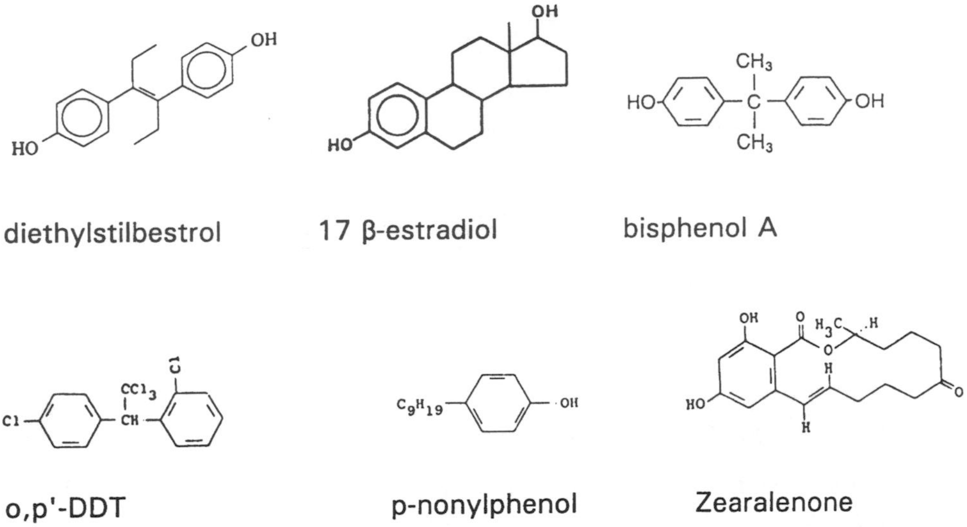

Fig. 1. Structures of selected natural and environmental estrogens.

1.3. Pro-Oxidant Activity

In the presence of redox-active metal ions, catechol estrogens are established oxidants. Evidence of this pro-oxidant activity of estrogens is primarily based on various types of oxygen free radical-mediated toxicity, such as single-strand breaks of DNA *(4,5)*, lipid peroxidation *(6)*, 8-hydroxylation of guanine bases of DNA *(7,8)*, or chromosomal abnormalities induced in kidneys of hamsters treated with DES or E2 *(9)*. Oxygen-free radicals are thought to arise by cytochrome P450 and cytochrome P450 reductase-catalyzed redox cycling between hydroxylated catecholestrogen metabolites and their corresponding quinones as illustrated in **Fig. 2** *(10)*. In this process, semiquinone intermediates may react with molecular oxygen to form superoxide radicals, which may be converted to hydrogen peroxide (H_2O_2) and reduced further by metal ions to hydroxy radicals *(10–12)*. For instance, 4-hydroxyestradiol (4-OHE2), a metabolite of E2, increases the concentration of 8-hydroxyguanine bases of DNA over control values by hydroxy radical addition to guanine in hamsters in vivo, or after microsomal activation in vitro as shown in **Fig. 2** *(7,13)*. Hydroxy radicals are generated by reduction of H_2O_2 by Fe^{2+}, or any other redox-active metal ion. Consistent with this notion, the addition of the iron chelator deferoxamine to estrogen quinone substrate incubated with a microsomal activation system restored control levels of 8-hydroxyguanine (8-OHdG) bases of DNA *(13)*. In addition, Cu(II) can oxidize 2- and 4-OHE2 by a nonenzymatic process resulting in the production of single- and double-strand breaks of supercoiled plasmid DNA *(14)*.

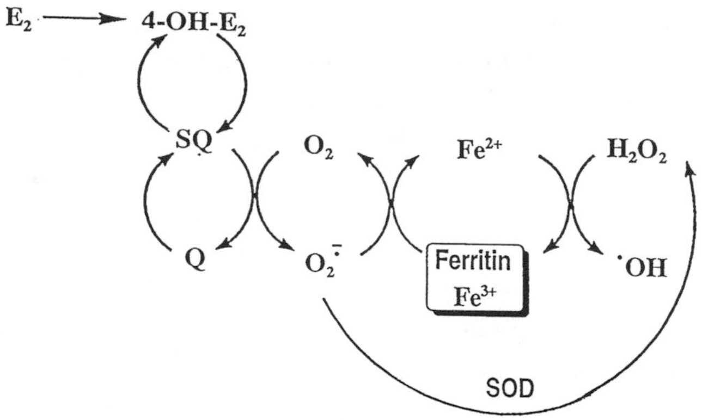

Fig. 2. Proposed formation of hydroxy radicals by steroid estrogens. Steroid estrogens are converted to catechol metabolites by cytochrome P450-mediated oxidation. 4-Hydroxylated estrogens are oxidized by organic hydroperoxide-dependent microsomal enzymes to semiquinones (SQ) and quinones (Q) at higher rates than 2-hydroxyestrogens. Quinone metabolites (Q) may be reduced by NADPH-dependent cytochrome P450 reductase ***(10,11)***. This metabolic redox-cycling may generate superoxide radicals ***(10,11)***, which reduce Fe^{3+} bound to ferritin to Fe^{2+} and thus mobilize it. Hydrogen peroxide, formed from superoxide by superoxide dismutase (SOD), may be reduced by Fe^{2+} to hydroxy radicals in a Fenton reaction. Reproduced from Han and Liehr ***(13)*** by permission of Oxford University Press.

1.4. Pro- and Antioxidant Activity

Based on these reported data of pro-oxidant and anti-oxidant activity of estrogens, it is likely that estrogens and/or their hydroxylated metabolites have both pro-oxidant and antioxidant properties, depending on the availability of metal ions or on their concentrations, as has recently been shown for the catecholamines L-DOPA and dopamine ***(15)***. Therefore, the pro-oxidant and anti-oxidant activities of estrogens and their metabolites need to be examined at all concentration ranges and under various physiological and pharmacological experimental conditions. This chapter describes several experimental conditions that are useful for the examination of pro-oxidant and anti-oxidant activities of estrogen substrates. As experimental parameter of free-radical action, we used the enhanced formation of 8-OHdG bases of DNA by hydroxy radicals generated during metabolic redox cycling of estrogens or its inhibition in the presence of estrogens.

1.5. Experimental Procedures

Oxygen radicals may be generated by microsome-mediated redox cycling between estrogen quinone and semiquinone intermediates as shown in **Fig. 2**. Conversely, estrogens may function as radical scavengers and reduce free radical-induced damage. Procedures for each of these possibilities are outlined. An indicator of free-radical damage, the hydroxylation of guanine bases of DNA by hydroxy radicals was chosen, because this type of free-radical modification of DNA may produce mutations with possibly serious biological consequences. Two different methods of analysis of 8-OHdG bases of DNA have been used and are described. One is a high pressure liquid chromatography (HPLC) procedure with electrochemical detection as introduced by Floyd and Wong ***(16)***. The other is a ^{32}P-postlabeling procedure of Gupta and Randerath ***(17)*** as modified by Devanaboyina and Gupta ***(18)***. The advantages and disadvantages of each procedure are discussed.

2. Materials

2.1. Chemicals

1. Deoxyguanosine 3′-monophosphate (dGMP), E1, 2-hydroxyestrone (2-OHE1), E2, 2-hydroxyestradiol (2-OHE2), 4-OHE2, nicotinamide adenine dinucleotide phosphate (NADPH), desferrioxamine mesylate, cumene hydroperoxide, dicumarol, nuclease P1, and *Escherichia Coli* alkaline phosphatase (Sigma Chemical Co., St. Louis, MO)
2. 4-OHE1 (Steraloids, Wilton, NH).
3. Equilenin and Fremy's salt (potassium nitrosodisulfonate) (Aldrich Chemical Co., Milwaukee, WI).
4. 8-OHdG (kindly provided by Dr. Robert A. Floyd, Oklahoma Medical Research Foundation, Oklahoma City, OK).
5. Carrier-free [^{32}P]-phosphate, (specific activity 285 Ci/mg at 100% isotopic enrichment or 9120 Ci/mmol) in water (ICN Radiochemicals, Irvine, CA).
6. Materials and chemicals needed for the ^{32}P-postlabeling assays are obtained from the sources described previously ***(17,18)***. Only the application of the ^{32}P-postlabeling procedure for the detection of 8-OHdG is described here, because the general procedure has been described in detail previously ***(17,18)*** and is beyond the scope of this paper.
7. All other reagents and chemicals used were of analytical grade or of highest grade available (*see* **Notes 1** and **2**).

2.2. Equipment

A Waters (Bedford, MA) solvent delivery system, a UV detector (Waters, model 440), and a Coulochem detector (ESA, Bedford, MA; model 5100A) are used (potentials set at +0.12 and +0.35 V for electrodes 1 and 2, respectively).

3. Methods

3.1. Synthesis of Estrogen Quinones

1. DES quinone is synthesized from DES by oxidation with silver oxide, as described previously ***(19)***. DES (0.03 mmol, 8.0 mg) and 80 mg of silver oxide in

1.2 mL of chloroform (anhydrous) is stirred at room temperature for 20 min. The mixture is then filtered through celite and this solution used as a stock solution.
2. Ultraviolet: (H_2O) λmax 310, 345 nm; (CH_3OH) λmax 296, 354 nm. Infrared: ($CHCl_3$) 1640 cm^{-1} (unsaturated carbonyl).
3. Equilenin-3,4-quinone is synthesized as described previously ***(20)***. Briefly, 0.2 g equilenin, 40 mL acetone, 60 mL 10% acetic acid, and 0.4 g Fremy's salt are mixed and agitated for 15 min. A chloroform (40 mL) extract of this mixture is washed with 30 mL 1.0*N* HCl. The chloroform is evaporated and the concentrated residue separated on a silica thin-layer plate using a hexane:ethyl acetate mobile phase (1:1, v:v). The band containing equilenin-3,4-quinone is scraped off the plate.
4. An acetone solution containing equilenin-3,4-quinone is concentrated using nitrogen gas, resulting in the formation of dark red crystals of equilenin-3,4-quinone (*see* **Note 3**) λmax for CH_3OH are 265 and 382 nm as reported previously ***(20)***.

3.2. DNA Preparation

DNA is isolated from hamster- or rat-liver tissues by phenol/chloroform/isoamyl alcohol extraction by the method of Gupta ***(21)***. The concentration of DNA or dGMP is estimated based on its absorption at 260 nm. The samples are stored at –70°C unless used immediately for 8-OHdG adduct analysis (*see* **Note 4**).

3.3. Generation of Free-Radical Damage to DNA by Microsome-Mediated Redox Cycling of Estrogen

1. Hamster-liver microsomes are prepared by the procedure of Dignam and Strobel ***(22)***. Purified DNA is sheared by three passages through a 25-gauge needle and then denatured by heat treatment at 100°C for 15 min prior to reaction with the substrate.
2. The incubation mixture consists of 1 mg DNA, 0.5 mg microsomal protein, 1 m*M* NADPH in 0.01 *M* phosphate buffered saline (PBS), pH 7.5, in a final volume of 1 mL. The incubations are started by addition of 200 μ*M* estrogen quinone substrate and carried out at 37°C for 60 min.
3. After incubation, the DNA is re-purified by extraction with diethyl ether (three times) and chloroform/isoamyl alcohol (twice). The aqueous phase is precipitated with ice-cold 100% ethanol. The DNA pellet is reconstituted in 20 m*M* sodium acetate buffer, pH 4.8.

3.4. Inhibition of 8-OHdG Formation in DNA by Estrogens and Their Metabolites

dGMP or DNA (1 μg/μL) is incubated with H_2O_2 (10 m*M*) and L-ascorbic acid (6 m*M*) in the dark for 1 h at 37°C in the presence of 50 μ*M* E2 or catecholestrogen. After incubation, the estrogen or catecholestrogen is extracted with ethyl acetate (three times). The aqueous phase is precipitated with ice-cold 100% ethanol and the DNA pellet reconstituted in 20 m*M* sodium acetate buffer, pH 4.8.

3.5. DNA Digestion

DNA is digested to a nucleotide 3′-monophosphate mixture with micrococcal nuclease and spleen phosphodiesterase (enzyme:substrate, 1:5 w/w, 37°C, 5 h). This digest is used for ^{32}P-postlabeling. For electrochemical detection, this DNA digest is further treated with nuclease P1 (enzyme substrate ratio 1:20) at 37°C for 40 min. After adjustment of the pH to 7.5 by addition of 0.1 *M* Tris-HCl, the DNA digest is incubated with alkaline phosphatase (1 unit/4 µg DNA) at 37°C for 60 min. This DNA digest is used for analysis of 8-OHdG by electrochemical detection.

3.6. Electrochemical Detection of 8-OHdG

8-OHdG concentrations in DNA hydrolysate are determined by HPLC with ultraviolet (UV) and electrochemical detectors linked in tandem according to the method of Floyd and Wong ***(16)***. Nucleotides are separated on a Microsorb C18 4.6 mm × 25 cm column (Rainin, Woburn, MA) under isocratic conditions with a mobile phase containing 12.5 m*M* citric acid, 25 m*M* sodium acetate, 30 m*M* sodium hydroxide, 10 m*M* acetic acid, and 10% methanol, pH 5.1, at a flow rate of 1 mL/min. Concentrations of 8-OHdG are expressed relative to the concentrations of dG detected by UV absorbance at 254 nm (*see* **Note 5**).

3.7. ^{32}P-Postlabeling of DNA

1. Analysis of 8-OHdG formation in DNA is carried out by the ^{32}P-postlabeling technique developed by Devanaboyina and Gupta ***(18)***. DNA digest (50 ng in 5 µL) is mixed with 5 µL of "hot mix," prepared by mixing 1 µL of 10X kinase buffer: 300 m*M* Tris-HCl, pH 9.5, 100 m*M* $MgCl_2$, 100 m*M* dithiothreitol, 10 m*M* spermidine, 0.5 µL [γ-^{32}P]-ATP, 0.2 µL of T4 polynucleotide kinase (10 U/mL), and 2.8 µL water.
2. The mixture is incubated at room temperature for 45 min. Labeled digest is further treated with nuclease P1 (2.5 µL, 1 mg/mL) containing sodium acetate (0.5 *M*, 2.5 µL), pH 4.5, zinc chloride (1 m*M*, 2.5 µL), and water (7.5 µL). The final pH of the reaction mixture is 5.0–5.5. The mixture is incubated again at room temperature for 1 h.
3. The resolution of ^{32}P-postlabeled adduct spots is carried out with unmodified nucleotides on polyetheleneimine-cellulose thin-layer chromatography (TLC) plates using the following solvents: dimension 1, 1.5 *M* formic acid; dimension 2, 0.6 *M* ammonium formate, pH 6.0.
4. The ^{32}P-labeled adduct spots are detected by autoradiography using Kodak X-Omat films. The spots containing ^{32}P-postlabeled adducts are excised from TLC plates, and levels of radioactivity determined by Cerenkov counting (*see* **Note 6**). The amount of 8-OHdG is determined as relative adduct labeling by measurement of the radioactivity in labeled 8-OHdG and normal nucleotide spots as described by Devanaboyina and Gupta ***(18)***.

Table 1
Microsome-Mediated 8-hydroxylation of Guanine Bases of DNA by Estrogens[a]

Estrogen	8-OHdG/10^5 dG
Experiment I	
Control	1.76 ± 0.99
E1	1.73 ± 1.14
Experiment II	
Control	6.30 ± 0.45
Equilenin	5.33 ± 2.22
Equilenin-3,4-quinone	9.68 ± 1.47[b]
Equilenin-3,4-quinone+deferoxamine	5.16 ± 0.74
Experiment III	
Control	0.68 ± 0.25
2-OHE1	1.00 ± 0.53
4-OHE1	1.62 ± 0.28[b]
2-OHE2	0.71+±+0.28
4-OHE2	1.27+±+0.31[b]

[a]DNA (1 mg) and 200 μ*M* substrate were incubated with 0.5 mg liver microsomal protein and 1 mM NADPH at 37°C for 60 min. DNA was extracted, hydrolyzed and its 8-OHdG content determined as described in Materials and Methods Sections. Values are expressed as means ± SD (n = 4). Data taken from Han and Liehr *(13)*.

[b]Significantly different from the control value ($p < 0.05$).

3.8. Results

In the presence of microsomes and NADPH, equilenin-3,4-quinone or DES quinone, substrates for redox cycling, raised 8-OHdG levels by more than 50% over controls *(8,13)*. A representative chromatogram of the measurement of 8-OHdG is shown in **Fig. 3**. The increase in 8-hydroxylation of guanine bases of DNA by equilenin quinone is inhibited by the addition of deferoxamine, a chelator of iron ions *(13)*. In the presence of a microsomal activation system, 4-hydroxylated catecholestrogens, the precursors of quinones in the metabolic-redox cycling process, also produced an increase in the level of 8-OHdG (**Table 1**). The 8-hydroxylation of guanine bases of DNA is specifically increased by the 4-hydroxylated estrogens and not affected by the corresponding 2-hydroxylated estrogens *(13)*.

The antioxidant activity of estrogens and their metabolites is assayed by inhibition of 8-OHdG formation using the ^{32}P-postlabeling assay. When rat-liver DNA is incubated with H_2O_2 and ascorbic acid and analyzed by ^{32}P-postlabeling assay, a major adduct is observed. A representative autoradiogram

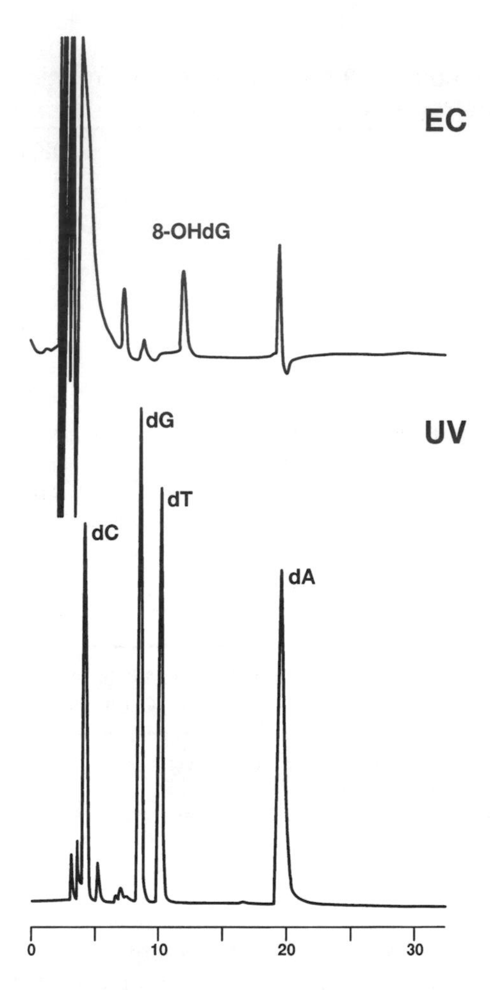

Fig. 3. Representative HPLC chromatogram of a hydrolysate of hamster-liver DNA incubated with microsomal protein, NADPH, and equilenin-3,4-quinone as described in Material and Methods Sections. The upper tracing is the electrochemical detector (EC) response, the lower the UV detector response set at 253 nm. Peaks identified are those of the major nucleosides, deoxycytidine (dC), deoxyguanosine (dG) (retention time: 8.5 min), thymidine (dT), deoxyadenosine (dA), and 8-OHdG (retention time: 12.1 min).

is shown in **Fig. 4**. dGMP is modified in an incubation with H_2O_2 and ascorbic acid under similar conditions and produced a similar adduct pattern. The addition of 50 μ*M* E2 or of hydroxylated catechol metabolite of E2 (2-OHE2 or 4-OHE2) inhibited the formation of this adduct by more than 50% (representative autoradiograms are shown in **Fig. 4**).

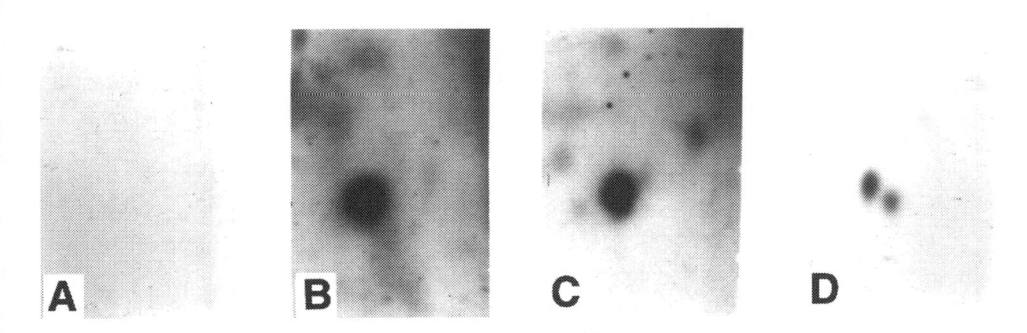

Fig. 4. Representative autoradiograms of 5′-^{32}P-labeled 8-OHdG 3′-monophosphate. (**A**), Labeled DNA digest (50 ng) from untreated rat-liver DNA; (**B**), DNA treated with H_2O_2 and ascorbic acid; (**C**), DNA treated with H_2O_2 and ascorbic acid plus 17β-estradiol; or (**D**) 4-hydroxyestradiol were developed in 1.5 *M* formic acid in the first dimension and in 0.6 *M* ammonium formate, pH 6.0, in the second dimension. Dried chromatograms were autoradiographed at –40°C for 15 h.

4. Notes

1. Use either ultrapure molecular biology grade or freshly distilled phenol, because impurities in phenol and autoxidation of phenol can also interfere in the measurement of 8-OHdG *(23)*.
2. H_2O_2, ascorbic acid, and catecholestrogen solutions must be freshly prepared because they are unstable and undergo auto-oxidation.
3. Unlike equilenin quinone, DES quinone is highly unstable and spontaneously rearranges to dienestrol in protic solvent systems. Therefore, it must be prepared fresh. When incubated with microsomes, phosphate buffer (10–20 m*M*) is used for incubations.
4. The purity of DNA is very important for 8-OHdG formation. The presence of RNA and protein interferes in the evaluation of 8-OHdG formation. In addition to absorption analysis of DNA at 230, 260, and 280 nm, we also check the purity of DNA by 1% agarose-gel electrophoresis and visualization of the band by ethidium bromide. The DNA must remain in solution and should not be lyophilized, because that will increase 8-OHdG artifacts *(23)*.
5. After turning on the electrochemical detector, 8-OHdG values in the first 2–3 chromatograms are usually too high and not reproducible. In subsequent chromatograms, the sensitivity of the detector stabilized and accurate 8-OHdG readings are obtained. The reproducibility of electrochemical-detector readings may also be affected by changing solvent batches. It is recommended to keep such changes at a minimum.
6. The postlabeling method requires <1 μg DNA, whereas for the HPLC-electrochemical detection procedure, 40–100 μg of DNA are needed. Thus, the sensitivity of detection of 8-OHdG by ^{32}P-postlabeling is comparable to that of GC/MS methods.

Acknowledgments

The authors are grateful to Dr. Xueliang Han for selecting a suitable chromatogram for preparation for **Fig. 3** and Kathie Obeck for preparing the manuscript.

References

1. Lacort, M., Leal, A. M., Liza, M., Martin, C., Martinez, R., and Ruiz-Larrea, M. B. (1995) Protective effects of estrogens and catecholestrogens against peroxidative membrane damage in vitro. *Lipids* **30,** 141–146.
2. Taniguchi, S., Yanase, T., Kobayashi, K., Takayanagi, R., Haji, M., Umeda, F., and Nawata, H. (1994) Catechol estrogens are more potent antioxidants than estrogens for the Cu(2+)-catalyzed oxidation of low or high density lipoprotein: antioxidant effects of steroids on lipoproteins. *Endocrine J.* **41,** 605–611.
3. Sack, M. N., Rader, D. J., and Cannon, III, R. O. (1994) Oestrogen and inhibition of oxidation of low-density lipoproteins in postmenopausal women. *Lancet* **343,** 269–270.
4. Han, X. and Liehr, J. G. (1994) DNA single strand breaks in kidneys of Syrian hamsters treated with steroidal estrogens: hormone-induced free radical damage preceding renal malignancy. *Carcinogenesis* **15,** 997–1000.
5. Nutter, L. M., Ngo, E. O., and Abul-Hajj, Y. J. (1991) Characterization of DNA damage induced by 3,4-estrone-o-quinone in human cells. *J. Biol. Chem.* **266,** 16,380–16,386.
6. Wang, M-Y. and Liehr, J. G. (1994) Identification of fatty acid hydroperoxide cofactors in the cytochrome P450-mediated oxidation of estrogens to quinone metabolites: role and balance of lipid peroxides during estrogen-induced carcinogenesis. *J. Biol. Chem.* **269,** 284–291.
7. Han, X. and Liehr, J. G. (1994) 8-Hydroxylation of guanine bases in kidney and liver DNA of hamsters treated with estradiol: role of free radicals in estrogen-induced carcinogenesis. *Cancer Res.* **54,** 5515–5517.
8. Roy, D., Floyd, R. A., and Liehr, J. G. (1991) Elevated 8-hydroxydeoxyguanosine levels in DNA of diethylstilbestrol-treated Syrian hamsters: covalent DNA damage by free radicals generated by redox cycling of diethylstilbestrol. *Cancer Res.* **51,** 3882–3885.
9. Banerjee, S. K., Banerjee, S., Li, S. A., and Li, J. J. (1994) Induction of chromosome aberrations in Syrian hamster renal cortical cells by various estrogens. *Mutat. Res.* **311,** 191–197.
10. Liehr, J. G. and Roy, D. (1990) Free radical generation by redox cycling of estrogens. *Free Rad. Biol. Med.* **8,** 415–423.
11. Roy, D. and Liehr, J. G. (1988) Temporary decrease in renal quinone reductase activity induced by chronic administration of estradiol to male Syrian hamsters. *J. Biol. Chem.* **263,** 3646–3651.
12. Nutter, L. M., Wu, Y-Y, Ngo, E. O. Sierra, E. E., Gutierrez, P. L., and Abul-Hajj, Y. J. (1994) An o-quinone form of estrogen produces free radicals in human breast cancer cells: correlation with DNA damage. *Chem. Res. Toxicol.* **7,** 23–28.

13. Han, X. and Liehr, J. G. (1995) Microsome-mediated 8-hydroxylation of guanine bases of DNA by steroid estrogens: correlation of DNA damage by free radicals with metabolic activation to quinones. *Carcinogenesis* **16,** 2571–2574.
14. Li, Y., Trush, M. A., and Yager, J. D. (1995) DNA damage caused by reactive oxygen species originating from a copper-dependent oxidation of the 2-hydroxy catechol of estradiol. *Carcinogenesis* **15,** 1421–1427.
15. Spencer, J., P. E., Jenner, A., Butler, J., Aruoma, O. I., Dexter, D. T., Jenner P., and Halliwell, B. (1996) Evaluation of the pro-oxidant and antioxidant actions of L-DOPA and dopamine in vitro: implications for Parkinson's disease. *Free Rad. Res.* **24,** 95–105.
16. Floyd, R. A. and Wong, P. K. (1988) Electrochemical detection of hydroxyl free radical adducts to deoxyguanosine, in *DNA Repair: A Laboratory Manual of Research Procedures* (Friedberg, E. C, and Hanawalt, P. C., eds.), (vol. 3), Marcel Dekker, Inc., NY, pp. 419–426,
17. Gupta, R. C. and Randerath, K. (1988) Analysis of DNA adducts by ^{32}P-labeling and thin layer chromatography, in *DNA Repair: A Laboratory Manual of Research Procedures* (Friedberg, E. C. and Hanawalt, P. C., eds.), (vol. 3), Marcel Dekker, Inc., NY, pp. 399–418.
18. Devanaboyina, U. S. and Gupta, R. (1996) Sensitive detection of 8-hydroxy-2-deoxyguanosine in DNA by ^{32}P-postlabeling assay and the basal levels in rat tissues. *Carcinogenesis* **17,** 917–924.
19. Liehr, J. G., DaGue, B. B., Ballatore, A. M., and Henkin, J. (1983) Diethylstilbestrol (DES) quinone: a reactive intermediate in DES metabolism. *Biochem. Pharmacol.* **32,** 3711–3718.
20. Teuber, H.-J. (1953) Reaktionen mit Nitrosodisulfonat, III. Mitteil.: Equilenin-chinon. *Chem. Berichte* **86,** 1495–1499.
21. Gupta, R. C. (1984) Nonrandom binding of the carcinogenic N-hydroxy-2-acetylaminofluorene to repetitive sequences of rat liver DNA *in vivo. Proc. Natl. Acad. Sci. USA*, **81,** 6943–6947.
22. Dignam, J. D. and Strobel, H. W. (1977) NADPH-cytochrome P-450 reductase from rat liver: purification by affinity chromatography and characterization. *Biochemistry* **16,** 1116–1122.
23. Claycamp, H. G. (1992) Phenol sensitization of DNA to subsequent oxidative damage in 8-hydroxyguanine assays. *Carcinogenesis* **13,** 1289–1292.

40

Artificial Radical Generating and Scavenging Systems

Photo-Fenton Reagent and Caged Compounds

Seiichi Matsugo and Ken Fujimori

1. Introduction

1.1. Caged Compounds

A photo-labile compound that is bioinactive but upon irradiation with light yields bioactive species, is called as "caged compound"*(1)* (*see* **Scheme 1**). Photolysis of caged compounds generating bioactive species, i.e., "uncaging," has become a general method to produce a desired amount of bioactive species in a specific time interval at desired place or area of the target biological systems *(1,2)*. This chapter deals with two caged free radicals: a caged hydroxyl radical, i.e., photo-Fenton reagent, and caged nitric oxides (NOs). Uncaging of these caged free radicals yield hydroxyl radical and NO, respectively, which are believed to be associated with free-radical tissue damage.

The requirements for the caged compounds for biological and medicinal applications are as follows:

1. They must be bio-inactive and stable under the biological (or physiological) conditions without light.
2. The caged compounds are requested to dissolve specifically in a desired compartment of the biological system, i.e., water phase or lipid phase.
3. The longer wavelength absorption band the better, because longer wavelength light is less toxic than the shorter one.
4. They require a large molar extinction coefficient (ε).
5. Uncaging proceeds in a high chemical yield and a high quantum yield (Φ). Thus, the quality of caged compound is expressed by $\varepsilon\Phi$ value.

From: *Methods in Molecular Biology, vol. 108: Free Radical and Antioxidant Protocols*
Edited by: D. Armstrong © Humana Press Inc., Totowa, NJ

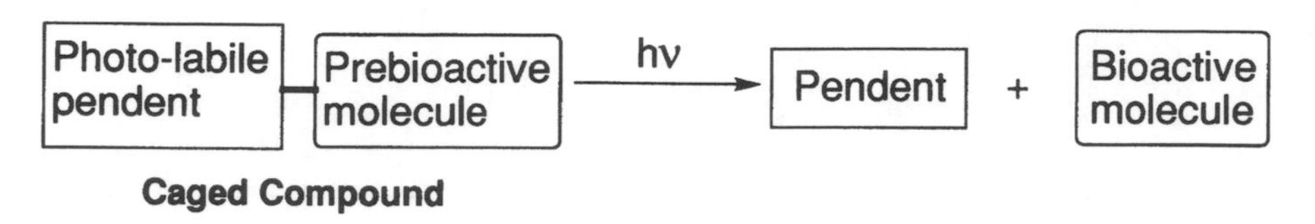

Scheme 1. Uncaging of caged compound.

This chapter describes protocols for the oxidation of biological materials by photo-Fenton reactions and evaluation of the scavenging activity of various biological antioxidants typically shown here, using α-lipoic acid and dihydrolipoic acid, and photochemical NO generation from caged NOs in a biological system.

1.2. Photo-Fenton Reagent

Hydroxyl radical is so reactive that it abstracts hydrogen atom or electron from organic molecules and adds to unsaturated bonds of organic compounds with almost diffusion-controlled rates in the vicinity of the generation. However, because hydroxyl radical is electrophilic, it reacts more readily with electron-rich compounds than with electron-deficient compounds, although the selectivity is low. There are several hydroxyl-radical generating methods:

1. Radiolysis of water,
2. Redox reactions of hydrogen peroxide (H_2O_2),
3. Photolysis of organic hydroperoxides or H_2O_2.

The second method involves the Harber-Weiss reaction and the Fenton reaction. The third approach is interesting because the hydroxyl-radical generation can be controlled externally by optical devices. Matsugo et al. designed and synthesized caged hydroxyl radical, i.e., "Photo-Fenton Reagent" NP-III (**Scheme 2**). NP-III has a strong absorption maximum at 377 nm (ε = 28.2 cm^{-1} mM^{-1}), and yields hydroxyl radicals upon ultraviolet (UV) light irradiation via intramolecular hydrogen abstraction in photo-excited NP-III followed by O-O bond homolysis (Φ_{OH} = 0.18 with 366 nm light in acetonitrile) *(3)*.

1.3. Caged NO

In contrast to hydroxyl radical, NO is a rather stable free radical. NO plays important roles in mammals, i.e., blood-pressure control, platelet aggregation inhibition, neurotransmission, immune regulation, and penile erection. NO is believed to damage electron-transfer systems and to yield highly toxic peroxynitrite by the reaction with superoxide anion radical during phagocytosis with immune cells *(4)*. To explore these physiological activities of NO,

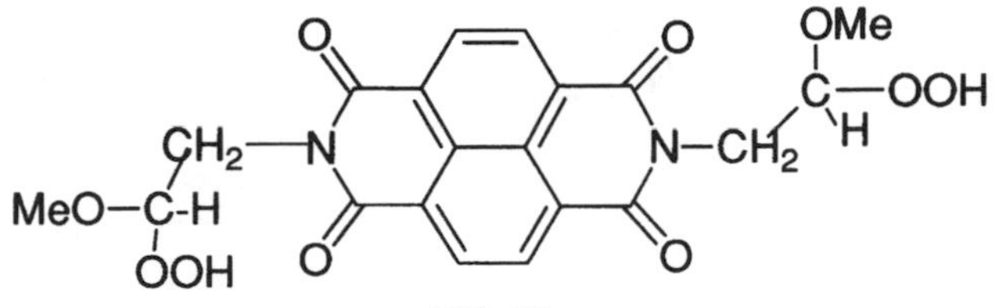

NP-III

Scheme 2. Photo-Fenton reagent.

Namiki and Fujimori synthesized caged NOs, i.e., N,N´-dinitroso-N,N´-dimethylphenylenediamine (BNN3), as a lipid soluble caged NO; N,N´-dinitrosophenylenediamine-N,N´-diacetic acid sodium salt (BNN5Na) as a water-soluble caged NO; and a dimethyl ester of BNN5 (BNN5M) as a lipid-membrane-permeable caged NO (**Scheme 3**) *(5)*.

These caged NOs have a strong UV-absorption band (λmax = 300 nm, ε = 13 cm^{-1} mM^{-1}). Photochemical NO extrusion proceeds stepwise (**Scheme 1**). Photo-excited BNN molecules yield one NO molecule in a time interval less than a few ns and provided the second NO molecule through the first-order reaction of the intermediary N-radical of the half-lives of 23 µs in water and 3.2 µs in benzene (**Scheme 3**). When concentrations of BNN reagents are lower than 10 µ*M*, chemical yield of NO in BNN5Na uncaging in water are quantitative. When good NO scavengers are used, the quantum yield for BNNNa uncaging (Φ_{NO}) in water is found to be 1.87 and that for BNN3 in organic solvents to be 2.00 *(5)*.

Caged NOs introduced here are quite new reagents and photo-vasorelaxation assay is the only known example of those biological application *(6)*. Therefore, assay conditions for photo-vasorelaxation are described. The other biological application of BNN reagents may be conducted in the same line described here.

2. Materials

2.1. Instruments

2.1.1. Instruments for Photo-Fenton Experiments

UV spectra are measured using Shimadzu UV 160 U spectrophotometer (Kyoto, Japan). Irradiation is carried out using an Oriel Corporation 68820 instrument (Stratford, CT) with UVB and UVC cut-off filters.

2.1.2. Instruments for Caged NO Experiments

JASCO high-power monochrome light source (Japan Spectroscopic Co. Hachiouji, Japan) is used as an uncaging light source. It is equipped with a 500 W xenon lamp, gratings, a light-path shutter, and an 0.8-mm quartz optical

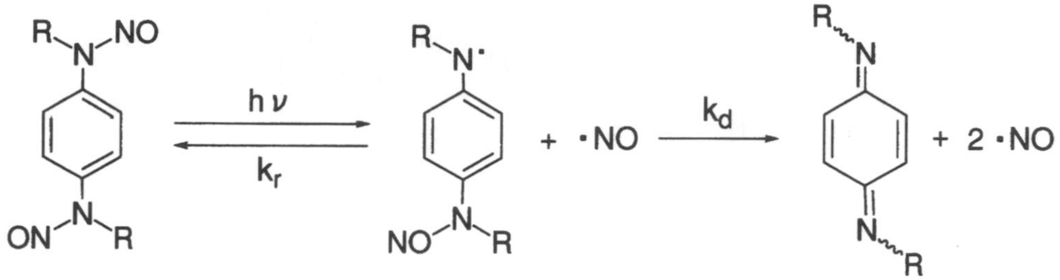

BNN5 ($R=CH_2COOH$)
BNN5Na ($R=CH_2COONa$)
BNN5Me ($R=CH_2COOCH_3$)
BNN3 ($R=CH_3$)

BNN5Na in water at 23°C: $k_d = 2.96 \times 10^4\ s^{-1}$; $k_r = 1.38 \times 10^8\ M^{-1}s^{-1}$
BNN3 in benzene at 23°C: $k_d = 2.15 \times 10^5\ s^{-1}$

Scheme 3. Photochemical generation of NO from caged NOs.

fiber. The light source has two slits to adjust light strength and width of light wavelength window. A force-displacement transducer (Nihon Kohden TB-611T, Tokyo, Japan) and thermal pen recorder (Nihon Kohden WT-647G, Tokyo, Japan) are used to read contractile responses.

2.2. Reagents

2.2.1. Photo-Fenton Reaction Experiments

1. All reagents used are of reagent grade and used without purification. except where indicated.
2. To a solution of N, N´-(bis-2,2-dimethoxyethyl)-1,4,5,8-naphthalimide (1.11 g, 2.5 mmol) (prepared from the reaction of naphthalene-1,4,5,8-tetracarboxylic anhydride and 2,2-dimethoxyethylamine) in dry dichloromethane (50 mL), is added excess H_2O_2 in dry ether, prepared by extracting 30% aqueous H_2O_2 with ether (*see* **Note 1**). Triflic acid (0.24 mL, 2.5 mmol) is added to the resulting solution at 0°C with a syringe (*see* **Notes 2** and **3**). The solution is stirred for 1 h at 0°C. After the reaction, the mixture was poured into ice water. The dichloromethane layer is separated, and washed twice with water. Evaporation of the dichloromethane layer at 0°C yields a yellow solid, which is washed with cold ethyl ether to give analytically pure NP-III as a yellow solid (*see* **Note 4**) *(3)*.
3. To prepare low-density lipoprotein (LDL) particles, the density of whole plasma is adjusted to 1.24 gm/mL by adding KBr (0.3816 g/mL plasma, *see* **Note 5**). To dissolve the KBr, the mixture is stirred gently with a magnetic flea at 4°C. Eight centrifuge tubes are charged with 27 mL of deaereated argon or nitrogen mock

buffer (density 1.006, 34.32g NaCl, 0.300g ethelynediamine tetraacetic acid [EDTA] dissolved in 3L distilled water). Twelve mL of plasma per tube is underlayed using a long Lucr fitting needle. The tubes are centrifuged in a Vti 50 vertical rotor at 50,000 rpm at 10°C for 2 h. Crude LDL, which appears as a yellow band in the center of the tube, is removed by aspirating through the side of the tube. From 12 mL of plasma, 3 mL of LDL is usually obtained. The crude LDL (25 mL) is re-centrifuged at 10°C, at 51,000 rpm for 24 h to obtain pure LDL as a yellow band at the top of the tube *(7)*. EDTA contaminating the LDL sample solution is removed by dialysis for 24 h against phosphate buffered saline (PBS) buffer, pH 7.4.

2.2.2. Caged NO Experiments

1. Methoxamine and Cremophor-EL are purchased from Sigma Chem Co. (St. Louis, MO). Acetylcholine and other reagents of special reagent grade are obtained from Wako Pure Chemical Industries Co. (Osaka, Japan).
2. Caged NOs are synthesized by the methods described in the literature ***(5,6)***.
3. BNN3 or BNN5M solution (5 m*M*) is prepared by adding a solution of 1.46 mg of BNN3 or 2.33 mg of BNN5M in 50 µL of dimethyl sulfoxide(DMSO) into 1.45 mL of 10% aqueous Chremophor-EL solution, followed by gentle ultrasound sonication for 20 s. The BNN5Na solution (5 m*M*) is prepared by dissolving 2.46 mg of BNN5Na in 1.5 mL of distilled water (*see* **Notes 6** and **7**). These BNN solutions are stored for a few mo in a dark freezer at a temperature lower than –20°C.
4. Krebs-Ringer solution: 113 m*M* NaCl, 4.8 m*M* KCl, 25 m*M* $NaHCO_3$, 1.4 m*M* KH_2PO_4, 2.2 m*M* $CaCl_2$, 1.4 m*M* $MgSO_4$, and 5.5 m*M* glucose.
5. Rat aortic strips are obtained from 12-w-old 400–450 g male Wister rats. After washing inside with Krebs-Ringer solution, the aorta segments are cut into 3 × 13 mm helical strips. The endothelium is denuded by gentle rubbing with small swabs.

3. Methods

3.1. Oxidation of Protein and Apo-B Protein of Human LDL by NP-III (8)

1. For bovine serum albumin (BSA) oxidation by NP-III, the concentration of BSA is adjusted to 2 mg/mL in 0.1 *M* sodium phosphate buffer, pH 7.2. NP-III solutions are prepared in acetonitrile to 10-fold their final concentration. Solutions containing 900 µL of BSA and 100 µL of NP-III to final concentrations of 0, 1, 5, 10, and 25 µ*M* are directly irradiated for 15 min. A duplicate set of samples covered with foil serve as dark controls. After irradiation, the sample solution is reacted with dinitrophenylhydrazine for 1 h at room temperature, then 10% trichloroacetic acid (final concentration) is added to precipitate protein. The protein pellet is washed three times with ethanol-ethyl acetate (v/v, 1:1) and then dissolved in 6 *M* guanidine-HCl solution. Readings are taken at 360 nm using 22 $mM^{-1}cm^{-1}$ of DNPH to calculate protein-carbonyl content ***(9)*** (*see* **Fig. 1**).

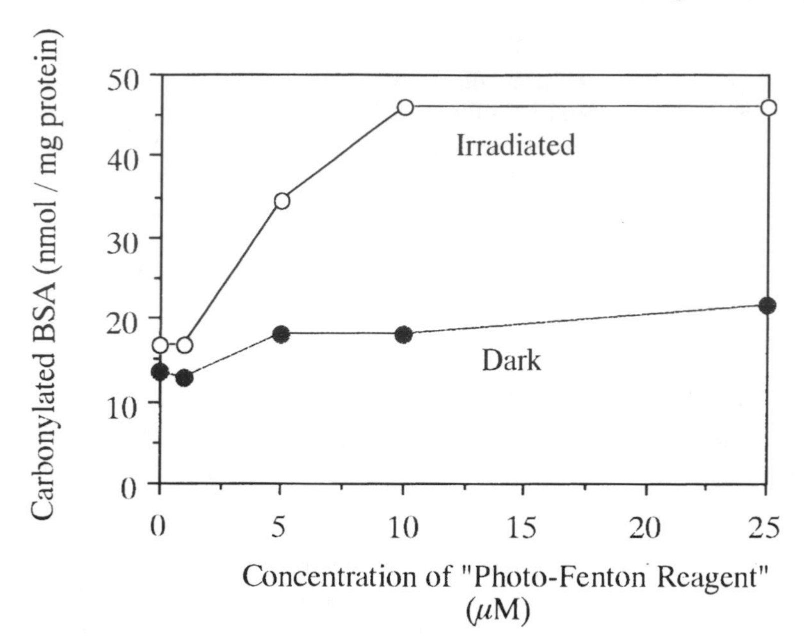

Fig. 1. Oxidation of bovine serum albumin by NP-III. BSA (2 mg/mL) was incubated with various concentrations of NP-III in 0.1 *M* phosphate buffer solution, pH 7.0. The carbonyl content of BSA was measured as described in **Subheading 3.1., step 1**. The points indicated in the figure are the mean of duplicate experiments.

2. LDL oxidation by NP-III at defined concentrations is carried out under almost the same reaction conditions as those described for the BSA reaction. 900 µL of LDL solution is added 100 µL of NP-III to final concentrations of 1, 2, 5, and 10 µ*M*. The protein concentration of LDL is 1 mg/mL, which is confirmed by the Lowry method ***(10)***. The irradiation is carried out in the same manner as with BSA for 15 min. After the photo-reaction, 1 mL of the sample solution is reacted with 200 µL of dinitrophenylhydrazine (10 m*M*) for 1 h, to which is added 0.6 mL of denaturing buffer (150 m*M* sodium phosphate buffer containing 3% sodium dodecylsulfate [SDS], pH 6.8). The samples are vigorously vortexed followed by the sequential addition of 3.6 mL of ethanol:heptane (1:1 v/v). The protein pellet is then three times washed with 3 mL ethanol:ethyl acetate (1:1 v/v) and dissolved in 1 mL of denaturing buffer. Each sample is scanned from 320 nm to 410 nm against HCl corresponding sample and the peak absorbance used to calculate protein carbonyls (*see* **Fig. 2**).
3. For evaluation of hydroxyl-radical scavenging activity of α-lipoic acid by using NP-III ***(11)***, the concentration of BSA is adjusted to 2 mg/mL in 0.1*M* sodium phosphate buffer (pH 7.2) ***(10)***. The NP-III solution of 100 µ*M* is prepared by diluting a 1 m*M* NP-III solution in acetonitrile in re-distilled water. Solutions containing 800 µL of BSA, 100 µ*M* of 100 µL of NP-III, and 100 µL of α-lipoic acid at final concentrations of 5, 2.5, 1, 0.5, 0.25, 0.1, and 0.05 m*M* are directly irradi-

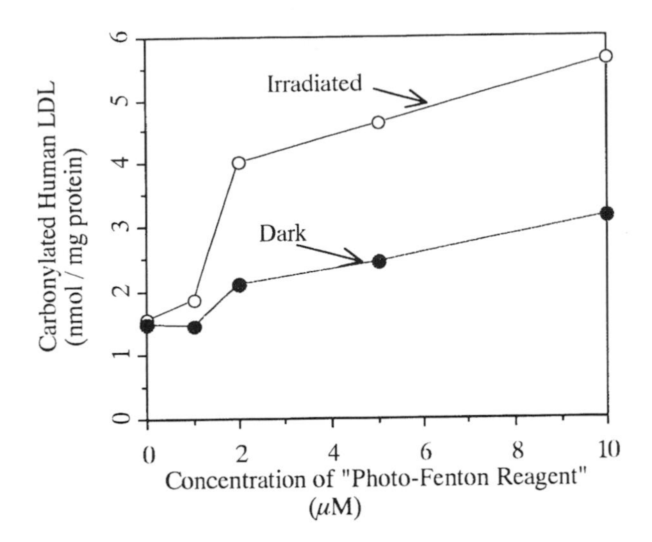

Fig. 2. Human LDL oxidation by NP-III. Human LDL (1.5 mg protein/mL) was incubated with various concentrations of NP-III in saline solution, pH 7.4. The carbonyl content of LDL was measured as described in **Subheading 3.1., step 2**. The points indicated in the figure are the mean of duplicate experiments.

ated. After the irradiation, protein carbonyl content in the sample solution is determined in the same manner as described in **Subheading 3.1., step 1** (*see* **Fig. 3**).

4. Human LDL oxidation by NP-III at 10 μ*M* concentration in the presence of a defined concentration of α-lipoic acid is carried out under the same reaction conditions as those for BSA reaction ***(11)***. To 800 μL of LDL solution is added to 100 μL of 100 μ*M* of NP-III and 100 μL of α-lipoic acid at 50, 25, 10, 5, 2.5, and 1 m*M* concentrations. The protein concentration of LDL is adjusted to 1 mg/mL, which is confirmed by the Lowry method ***(10)***. Irradiation is carried out in the same manner as BSA. After the photoreaction, protein-carbonyl content in the sample solution is determined by the same method mentioned in **Subheading 3.1., step 2** (*see* **Fig. 4**).
5. For the oxidation of BSA by NP-III in the presence of dihidrolipoic acid (DHLA) ***(12)***, the concentration of BSA is adjusted to 2 mg/mL in 0.1 *M* sodium phosphate buffer (pH 7.2) ***(10)***. NP-III solutions (100 μ*M*) are prepared by diluting a 1 m*M* NP-III solution in acetonitrile re-distilled water. Solutions containing 800 μL of BSA, 100 μL of 100 μ*M* NP-III, and 100 μL of DHLA at final concentrations of 50, 25, 10, 5, 2.5, 1, 0.5, 0.25, and 0.1 m*M* are directly irradiated with an Oriel Corporation 68820 apparatus, as previously described. Duplicate set of samples covered with foil serve as dark controls. After the irradiation, protein-carbonyl content in the sample solution is determined in the same manner as described in **Subheading 3.1., step 1** (*see* **Fig. 5**).

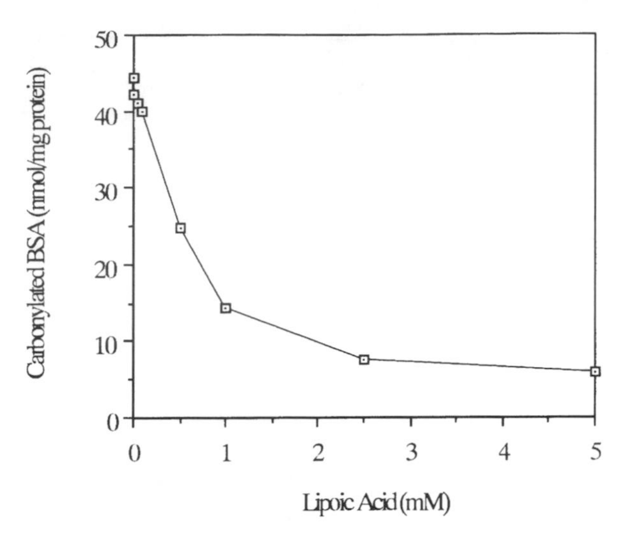

Fig. 3. Oxidation of bovine serum albumin by NP-III in the presence of α-lipoic acid. BSA (2 mg/mL) was incubated with 10 μ*M* of NP-III in the presence of various concentrations of α-lipoic acid in 0.1 *M* phosphate buffer solution, pH 7.0. The carbonyl content of LDL was measured as described in **Subheading 3.1., step 3**. The points indicated in the figure are the mean of duplicate experiments.

6. Oxidation of human LDL oxidation is carried out by 10 μ*M* NP-III in the presence of a defined concentration of DHLA using the same reaction conditions as those for the BSA reaction ***(12)***. 800 μL of LDL solution is added to 100 μL of 100 μ*M* NP-III and 100 μL of DHLA at 50, 25, 10, 5, 2.5 1, 0.5, 0.25, and 0.1 m*M* concentrations. The protein concentration of LDL is 1 mg/mL, as confirmed by the Lowry method ***(10)***. Irradiation is carried out in the same manner as BSA. After the photoreaction, protein-carbonyl content in the sample solution is determined by the same method mentioned in **Subheading 3.1., step 2** (*see* **Fig. 6**).

3.2. Caged NOs

3.2.1. Tension Readings

Arterial strips are placed by means of thread in a jacketed drop-away organ bath containing 20 mL Krebs-Ringer solution maintained at 37°C and perfused with 95% O_2-5% CO_2. In the beginning, the removal of endothelium is confirmed by the loss of a relaxant response to acetylcholine. Arterial stripes are equilibrated at a passive tension of 0.4 g until the contractile response caused by 50 m*M* K^+ attains a steady state. The upper thread of each strip is connected to the force-displacement transducer, whereas the lower thread is anchored at the bottom. Contractile responses are recorded by the thermal pen recorder.

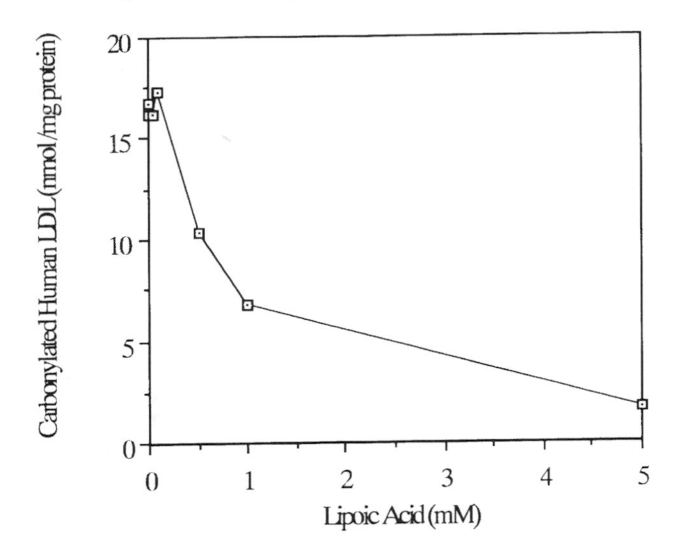

Fig. 4. Human LDL oxidation by NP-III in the presence of α-lipoic acid. Human LDL (1.0 mg protein/mL) was incubated with 10 μ*M* of NP-III in the presence of various concentrations of α-lipoic acid in saline solution, pH 7.4. The carbonyl content of LDL was measured as described in **Subheading 3.1.**, **step 4**. The points indicated in the figure are the mean of duplicate experiments.

3.3. Photo-Vasorelaxation Assay

All procedures are done in a dark room under red light (*see* **Notes 6** and **7**). The BNN5M solution (1 m*M*) is diluted with 10% aqueous Cremophor-EL solution, so that when 40 μL of the solution is added to 20-mL organ bath containing aorta strips filled with Krebs-Ringer solution, the final BNN5M concentration becomes the desired value. Loading of BNN is performed by adding 40 mL of the BNN solution in the organ bath and equilibrated for predetermined time intervals. After washing twice with normal Krebs-Ringer solution, 10 m*M* methoxamine is applied to the strips, which develop and reaches tension at plateau state within 20 min. In Entry 3 only, strips are not rinsed with BNN-free Krebs-Ringer solution after BNN5Na loading, i.e., irradiation is carried out in a solution that contains BNN5Na. The tissue is irradiated with a 300- or 360-nm light introduced in the solution through an 0.8-mm quartz optical fiber from a JASCO high-power monochrome light source equipped with a 500 W xenon lamp and a light-path shutter (*see* **Note 8**). The top of the optical fiber is continuously moved by hand so that the isostrength light reaches all tissue. The BNN3 and BNN5Na assay is conducted in the same manner as applied to BNN5M, whereas a BNN5Na solution is prepared by diluting 1 m*M* BNN5Na solution with distilled water.

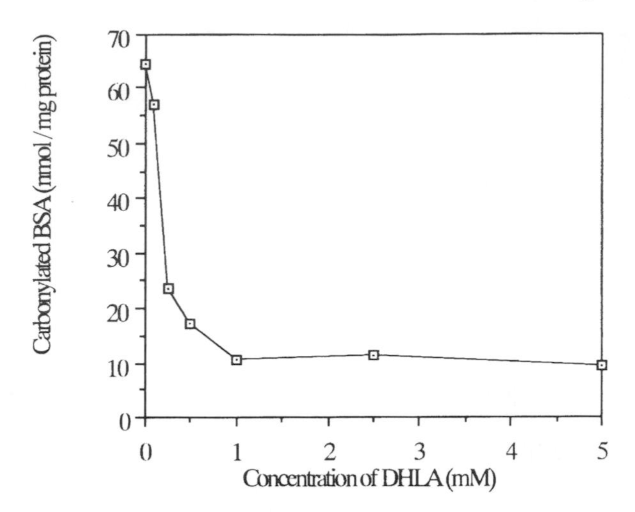

Fig. 5. Oxidation of bovine serum albumin by NP-III in the presence of DHLA. BSA (2 mg/mL) was incubated with 10 μ*M* of NP-III in the presence of various concentrations of DHLA in 0.1 *M* phosphate buffer solution, pH 7.0. The carbonyl content of BSA was measured as described in **Subheading 3.1., step 5**. The points indicated in the figure are the mean of duplicate.

Because irradiation with UV light is known to induce rat-aorta relaxation, BNN-induced photorelaxation is obtained by subtracting the photorelaxation extent of carefully controlled experiments from that of BNN-loaded strips. Some typical examples are listed in **Table 1**. Substantial relaxation effects of BNNs appear even when aortic tissues are treated with only 1 μ*M* concentration of BNN solution. BNN5Na, does not relax aorta stripe when the strips are rinsed with BNN5Na-free Krebs-Ringer solution (Entry 2), whereas in the presence of 10 m*M* BNN5Na relaxation of strips takes place (Entry 3). BNN3 (Entry 1) and BNN5M (Entries 4 and 6) relax strips after rinsing with BNN-free Krebs-Ringer solution.

Three types of BNNs may be characterized as follows:

1. Lipid-soluble BNN3 dissolves into the lipid membrane of the tissues.
2. Water-soluble BNN5Na remains in the aqueous solution and does not permeate the tissue.
3. BNN5M is permeable through a tissue-cell membrane and is hydrolyzed with esterases to yield BNN5Na, which is insoluble in the cell membrane and therefore remains in the cell.

Researchers can generate NO in a desired compartment of tissue by choosing one of these BNNs.

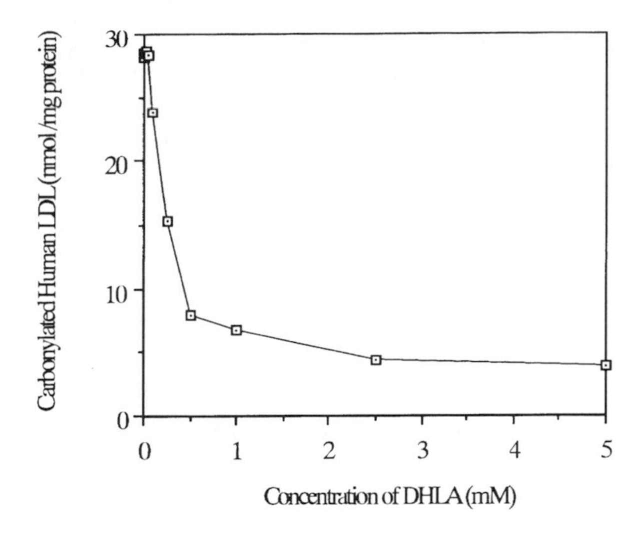

Fig. 6. Human LDL oxidation by NP-III in the presence of DHLA. Human LDL (1.0 mg protein/mL) was incubated with 10 μ*M* of NP-III in the presence of various concentrations of DHLA in saline solution, pH 7.4. The carbonyl content of LDL was measured as described in **Subheading 3.1., step 6**. The points indicated in the figure are the mean of duplicate.

Table 1
Photorelaxation of Caged NO-Loaded Rat Aorta Strips at 25°C[a]

Entry	Caged NO	Wavelength (nm)[b]	% Relaxation[c]
1	BNN3	300	30.1 ± 9.0
2	BNN5Na	300	10.4 ± 12.3
3	BNN5Na	300	45.7 ± 17.4[d]
4	BNN5M	300	58.1 ± 9.2
5	BNN5M	360	25.4 ± 13.0

[a]Concentration of loading solution = 1 μ*M*. After 30 min loading, strips were washed with Krebs-Ringer solution.

[b]Wavelength of light used for uncaging (slit-width of light path: 4 mm).

[c]n = 4.

[d]Photolysis was conducted in Krebs-Ringer solution containing 1 μ*M* of BNN5Na.

4. Notes

1. Use caution in the extraction of H_2O_2 from aqueous solution of 30% H_2O_2 because it is quite dangerous, needs careful treatment, and must be prepared at low temperature. The etheral solution of H_2O_2 is dried by sodium sulfate.
2. The NP-III preparation should be carried out in an oxygen-free vessel in a draft chamber under red light with the aid of experienced synthetic chemist.

3. The use of a strong acid (triflic acid) is the key step for the preparation of NP-III. This acid is very strong and gloves must be worn to handle this reagent.
4. NP-III is quite stable if kept it in a solid state at –70°C. In aqueous solution, NP-III decomposes within 24 h at room temperature. Handling of NP-III should be conducted under red light.
5. Because plasma is a biological sample, it is a potential biohazard. The isolation of LDL should be carried out using universal precautions. At the very least, one should use gloves, mask, and goggles to avoid infection from the plasma.
6. All experiments should be conducted under the light of wavelengths longer than 475 nm.
7. Handling of caged NOs must be done with gloves.
8. Because caged NOs' absorption bands extend to 400 nm, 300–360 nm can be applicable for uncaging of BNN reagents. The light for uncaging may be irradiated to a tissue or a cell on a microscope stage through quartz optical lens.

References

1. Adams, S. R. and Tsien, R. Y. (1993) Controlling cell chemistry with caged compounds. *Ann. Rev. Physiol.* **55,** 755–784.
2. Corrie, E. T. and Trentham, D. R. (1993) Caged nucleotides and neurotransmitters. *Bioorga. Photochem.* **2,** 243–305.
3. Matsugo, S., Yamamoto, Y., Kawanishi, S., Sugiyama, H., Matsuura, T., and Saito, I. (1991) Bis(hydroperoxy)naphtahlimide as a "Photo-Fenton Reagent": sequence-specific photocleavage of DNA. *Angew. Chem. Int. Ed. Engl.* **30,** 1351–1353.
4. Beckman, J. F., Beckman, T. W., Chen, J., Marshall, P. A., and Freeman, B. A. (1990) Apparent hydroxyl radical production by peroxynitrite: implications for endotherial injury from nitric oxide and superoxide. *Proc. Nat. Acad. Sci. USA* **87,** 1620–1624.
5. Namiki, S., Arai, T., and Fujimori, K. (1997) High performance caged nitric oxide. new molecular design, synthesis, and photochemical reaction, *J. Am. Chem. Soc.* **119,** 3840–3841.
6. Namiki, S., Asada, S., Hama, H., Kasuya, Y., Goto, K., and Fujimori, K. (1998) Assay of new high-performance caged nitric oxides to rat aorta relaxation, in press.
7. Havel, R. J., Eder, H. A., and Bragdon, J. H. (1955) The distribution and chemical composition of ultracentrifugally separated in human serum. *J. Clin. Invest.* **34,** 1345–1353.
8. Matsugo, S., Yan, L.-J., Han, D., and Packer, L. (1995) Induction of apo-B protein oxidation in human low density lipoprotein by the photosensitive organic hydroperoxide, N,N′-bis(2-hydroperoxy-2-methoxyethyl)-1,4,5,8-naphthalene-tetra-carboxylic diimide. *Biochem. Biophys. Res. Commun.* **206,** 138–145.
9. Reznick, A. Z. and Packer, L. (1994) Oxidative damage to proteins: spectrophotometric method for carbonyl assay. *Methods Enzymol.* **233,** 357–363.
10. Lowry, O. H., Rosenbrough, H. J., Farr, A. L., and Randall, R. J. (1951) Protein measurement with the folin phenol reagent. *J. Biol. Chem.* **193,** 265–275.

11. Matsugo, S., Yan, L.-J., Han, D., Tritschler, H. J., and Packer, L. (1995) Elucidation of antioxidant activity of α-lipoic acid toward hydroxyl radical. *Biochem. Biophys. Res. Commun.* **208,** 161–167.
12. Matsugo, S., Yan, L.-J., Han, D., Tritschler, H. J., and Packer, L. (1995) Elucidation of antioxidant activity of dihydrolipoic acid toward hydroxyl radical generator NP III. *Biochem. Mol. Biol. Int.* **37,** 375–383.

Index

U

V

X